BASIC

GEOMETRY

Deborah A. Hale
Shelby L. Hawthorne

Thomas Nelson Community College
Developmental Studies / Natural Science and Math Department
Hampton, Virginia 23670

ISBN 0-931647-00-2

10

TABLE OF CONTENTS

APPENDICES

PREFACE

Basic Geometry is designed for use in a one-semester course for students enrolled in either a high school or college class. This text is intended for students who have no previous experience with geometry or who need a thorough review.

Although it is assumed that the student has a working knowledge of algebra, a brief review of the rules for simplifying irrational numbers and solving quadratic equations is included in Unit 7. Also, a review of solving linear equations and operating on the set of rational numbers appears in the appendix.

There are thirteen units in the text. Each unit is thoroughly developed using simple, easily understood language. Postulates, theorems, and definitions are clearly identified. Detailed explanations, numerous step-by-step examples, and guiding questions accompany each new topic. Arithmetic and algebra are integrated into the presentations of the material to help a student in sharpening these skills. Applications of geometric principles in real life situations are incorporated throughout.

The units begin with learning objectives for the instructor's and student's convenience. Each topic is explained fully. Worked out examples are followed by practice problems for students to work. At the end of major topics, exercises are provided with answers available in the appendix. Extensive reviews with answers are provided at the end of each unit. The emphasis of the book is on applying geometric concepts as opposed to developing a theoretical course.

The book is adaptable to a variety of modes of instruction: self paced, lecture-discussion, lab courses, and independent contract.

The authors wish to thank the Thomas Nelson Community College Developmental Studies Mathematics Department's faculty and students for classroom testing the manuscript and for their valuable suggestions for improvement.

Our special thanks goes to our spouses, Vincent and William, for their encouragement and faith in our efforts and to our children, Vincie, Bryan, Alice, and Mary who spent many hours without their mothers.

TO THE STUDENT

> James is a student in a geometry class. He doesn't know what to expect, but he has heard all sorts of rumors about the course from other math students. He isn't sure how much of a challenge this class is going to be, but he suspects that it will be rigorous.
>
> Like most students, he opened the textbook to the table of contents and skimmed his finger over the material in each unit. After analyzing the amount of material expected to be covered, James has just about decided to let this course set sail with one less passenger . . . himself.

Perhaps you have felt the same way as James when you first entered a geometry class. You may have even entertained some of the same thoughts as James or experienced the same anxieties.

We are aware that students do now always have a positive attitude toward mathematics. Those students who do have a positive attitude usually have had some measure of success. Those students who have a somewhat negative attitude toward math have experienced frustrations and not the success they desired. This latter group of students may have gone so far as to even doubt their ability to deal with geometry or any form of mathematics.

Our primary concern is you, the student. We want you to have a positive experience in this study of geometry. It is important that you have a clear notion of each concept in order to help you build a strong, meaningful foundation. Upon that foundation, we intend to develop the manipulative skills or practical tools used in deriving logical conclusions.

After all, geometry is essentially a practical subject. Practical in the sense that it is primarily concerned with the basic observation and construction of things in our everyday world. Surely the construction of walls of a house or building clearly illustrate some elements of geometry using planes, straight lines, and area. The beaten path through school campus lawns shows that both man and animal instinctively seek out the shortest distance between two points . . . a straight line versus the awkward triangular turn.

So why do you need to study it? Of what value is this course to you? These are questions nearly every prospective student asks. Possibly the best answer is this. Geometry will give you the practice and the training necessary to think in a logical, ordered sequence.

Learning to study geometry can be compared to operating a business. You are going to have to learn to think. If you intend to be successful, necessarily you will have to use your head. Undefined policies, confusing data, and random guessing will only produce disastrous results.

This text is designed to teach you geometry from a non-rigorous approach. The concepts are presented in easy-to-understand language with numerous examples, illustrations, and exercises.

The material in this book has been written in a self-study format that will enable you to study and progress at your own rate if you desire.

We have found that the following steps provide the most effective approach toward gaining success.

STEP 1 KEEP AN OPEN MIND
You must want to learn. We can teach you how to do math only if you want to learn it. If you don't see the value in developing math skills or you don't know how these skills could possibly be of benefit to you, then you are defeated before you start.

STEP 2 RELAX
Basic math is not a raging beast. Don't go all to pieces at the mere mention of the word "math." Give yourself time to understand each concept. Review each concept before going on. If you have questions, ask someone to help make them clear to you. Building your understanding of mathematics is like constructing a solid foundation for your home; each brick must be mortared exactly right before you construct.

STEP 3 LEARN RULES
Mathematics is a logical system. It has rules and regulations that must be followed. People who don't like math usually don't understand the rules. No one appreciates watching an exciting game of football if he doesn't know how to play. You must study each concept and master each rule to be successful.

STEP 4 REPETITION
Practice does make perfect. Once you understand a concept, then drill it into your head. Repetition is truly the best teacher. Some individuals get the "hang" of a concept and quickly move on to a new one. By the end of the unit, they usually have forgotten what they learned at the beginning. Math rules will not stick with you unless you practice them thoroughly.

Be sure to read each concept carefully, study each example thoroughly, and cover the answers that are provided with a separate sheet of paper. Commit yourself to an answer by writing your answer down before you check it with the answer given in the right margin. If you fail to write down your answer first before checking, you are not properly exercising you ability to think on your own. You are, in fact, merely shifting from question to answer without channeling information through your brain. You are exercising only your eyes. If you continue using the wrong method, at the conclusion of this course, you may develop 20/20 vision, but you may not have learned geometry.

Unit 1

Angles, Lines and Points

Learning Objectives:

1. The student will list the four elements of a deductive reasoning system.

2. The student will differentiate between deductive and inductive reasoning.

3. The student will differentiate between axioms or postulates, and theorems.

4. The student will demonstrate his mastery of the following definitions and theorems by writing them, or by applying them to solutions of selected problems.

   Definitions: line segment, compass, midpoint, ray, angle, vertex, vertical angles, degrees, perpendicular lines, protractor, right angle, acute angle, obtuse angle, straight angle, complementary angles, supplementary angles, adjacent angles, angle bisector, bisector of a line segment, and perpendicular bisector.

   Theorems:
   (a) Supplements of the same or equal angles are themselves equal.
   (b) Complements of the same or equal angles are themselves equal.
   (c) Pairs of vertical angles are equal.

The following constructions are required:

(a) Construct the perpendicular bisector of a line segment.

(b) Construct the bisector of a given angle.

(c) Construct a copy of an angle in another position.

ANGLES, LINES AND POINTS

In geometry, the thread that ties one idea with another is called deductive reasoning. Are there other types of reasoning processes? Yes, but these other processes are not used in geometry. One example is called inductive reasoning. For a moment, let us discuss the second type. What exactly is inductive reasoning? How does it differ from deductive reasoning?

Inductive Reasoning

Example: Study the circles below.

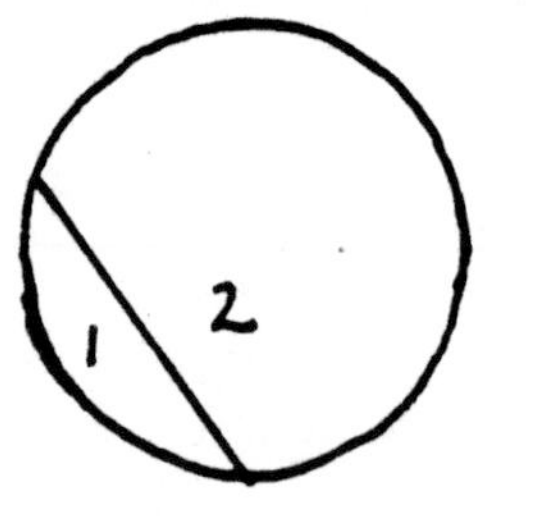

Figure A

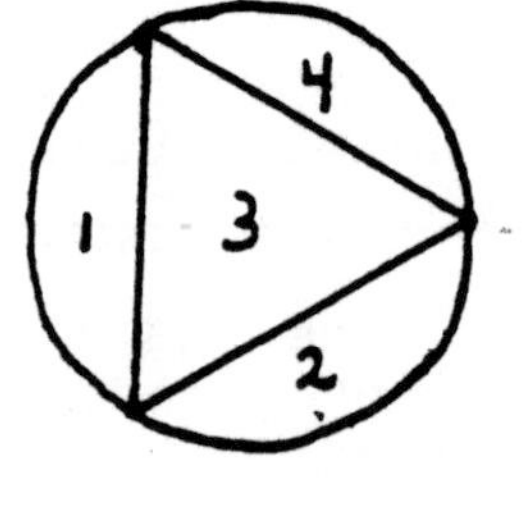

Figure B

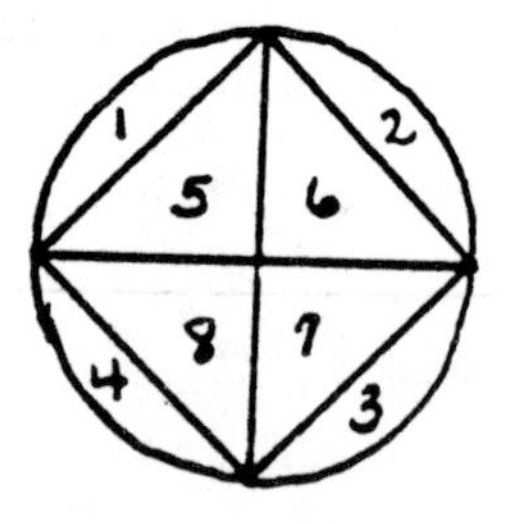

Figure C

1. In Figure A, two points have been chosen and one straight line drawn between them. The circle has been divided into _______ regions.
2. In Figure B, three points determine three straight lines. The circle has been divided into _________regions.
3. In Figure C, four points determine six straight lines. The circle has been divided into __________regions.

Fill in the table below.

Number of points on each circle	2	3	4	
Number of regions resulting				

1. 2

2. 4

3. 8

4. What is the pattern that your see?

2 points ______________ 2 regions
3 points ______________ 4 region
4 points ______________ 8 regions

Each time the number of regions seems to ______________.

5. Inductive reasoning involves observing patterns and drawing conclusions based upon this pattern. The patterns in this problem indicate this conclusion . . . "by adding one more point to the circle, the number of regions seems to double."

5 points would indicate ________regions
6 points would indicate ________regions

6. Now demonstrate that your answers are correct. The circle below has five dots. Connect each set of dots with straight lines in every direction. Count the number of regions resulting.

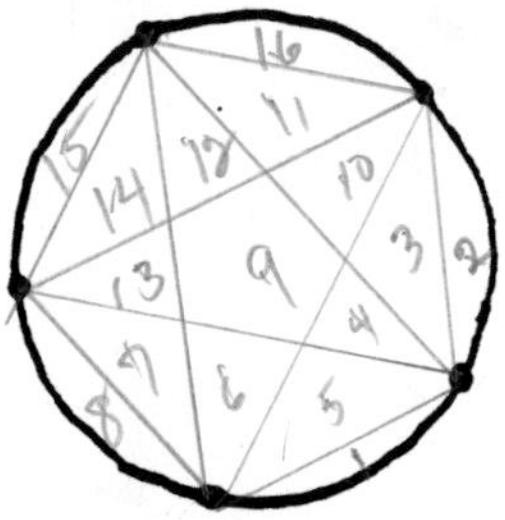

Number of regions _____

Was your guess correct?

____

7. Draw a large circle having six dots. Connect each pair of dots and count the number of regions resulting.

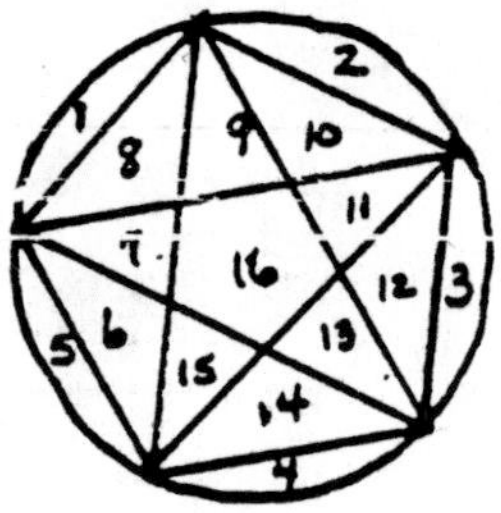

Number of regions _________
Was your guess correct? ________

4. double

5. 16; 32

6.

16; yes

7. 30, if dots are equally spaced; 31, if not.

No, your guess in Exercise 5 was 32.

As you demonstrated, using inductive reasoning can lead to false conclusions. Inductive reasoning by itself cannot serve as a mathematical proof for anything.

What are the elements of inductive reasoning? You move from the specific to the general by a three-step process.

1) Collect data
2) Observe patterns
3) Draw conclusions based upon these patterns

Here is another example of inductive reasoning. Follow all the instructions listed below.

Choose a number ________
Add four ________
Double the result ________
Subtract six ________
Divide by 2 ________
Subtract the original number ________

Your result is ... ________

Try the above process two more times and record your result. Have a friend or relative try a number and record the result.

8. If your addition, subtraction, division, and multiplication have been performed correctly, your result will always be ________________.

8. 1 (one)

9. Again you have demonstrated inductive reasoning by (1) starting with specific numbers (collecting data), (2) observing a pattern, and (3) drawing a conclusion based upon this pattern.

What was the conclusion drawn in this problem?

____

____

9. See exercise 8.

Deductive Reasoning

The inductive approach does not always give consistent results as answers. Therefore, it cannot be used as a geometric proof for anything.

For a mathematical proof, we need to use the deductive method for reasoning.

Deductive reasoning begins with the general and goes to the specific. Deductive reasoning involves mathematical rules and logic to arrive at its specific conclusions. We will be studying this type of reasoning throughout the course.

One example of deductive reasoning can be demonstrated by using the process found on page 4.

Remember to start with the general. What is the most general expression for numbers that you have used in algebra? _______________ Yes, use "x"!

General ↓ Specific			
General	Choose a number	x	x
	Add four	______	x + 4
	Double the result	______	2x + 8
	Subtract six	______	2x + 2
	Divide by 2	______	x + 1
	Subtract the original number	______	x + 1 - x
Specific	Your result is	______	1

You have just used a general expression to prove that your answer will always be 1 no matter what the original number.

We had to try this process with several specific examples in inductive reasoning before we arrived at this same conclusion.

In order for general statements to have meaning, each individual word or term that it contains must be understood. To get started at some point in our communication, we admit that some terms must be left undefined. However, we will define as many terms as possible.

General statements that must be accepted without proof will be called postulates. General statements that have been proven or will be proved will be called theorems. However, all theorems are not proved in this text.

Therefore, deductive reasoning involves four elements in moving from the general to the specific.

(1) Undefined terms
(2) Defined terms
(3) Postulates
(4) Theorems

Question	Answer
10. What process of reasoning	10.
(a) Moves from the specific to the general? ____________	(a) Inductive
(b) Moves from the general to the specific? ____________	(b) Deductive
(c) Involves undefined terms, defined terms, postulates, and theorems? ____________	(c) Deductive
(d) Involves collecting data? ________	(d) Inductive
11. What process of reasoning <u>can</u> be used as a mathematical proof? ____________	11. Deductive
12. What process will be used in studying geometry? ________________	12. Deductive
13. Why can't inductive reasoning be used in geometric proof? ____________ ____________	13. The conclusion will not always be correct.

We don't always think about the words that we use when we communicate with others in our everyday activities. We <u>assume</u> that the phrases we speak are generally understood and their meanings, we believe, are widely accepted.

Take the word "duck". Just the mere mention of the word "duck" does give rise to a variety of interpretations. Depending upon the perception of the listener, we can expect him (the listener) to interpret the word "duck" to mean one of the following:

1. A bird with webbed feet and a flat bill

2. A person with an unusual disposition (slang)

3. A sail cloth

4. The act of bending suddenly

Misunderstanding of phrases or terms occurs frequently in the course of a discussion. It is, therefore, very important for the people communicating to <u>agree</u> upon the terms or words involved in their exchange of ideas.

Geometry has its own language and it is important that you know the meanings of the terms involved in order to clearly understand what it is that we may be discussing.

The terms that will be defined in this unit are listed on the next page. As you read through this first unit, jot down definitions beside the respective term. You should refer to this list constantly until you have mastered each and every definition in the unit.

	Term	Definition
1.	line segment	is that set of all points on a line btw 2 pts and including those points
2.	midpoint	of a line segement is a point on the segment that devids it into 2 = parts
3.	ray	is the set of all points on a line beginning with the endpoint + cont. indefinitely thru a second point
4.	angle	is the union of 2 rays with a common endpoint
5.	vertex	is the common end point of a angle - 2 or more rays come together
6.	right angle	measures exactly 90°
7.	acute angle	less than 90° greater than ZERO
8.	obtuse angle	less than 180° greater than 90
9.	straight angle	an ∠ that measures = 180°
10.	adjacent angles	share common vertex and have a common side between them
11.	complementary angles	2 angles that add to 90°
12.	supplementary angles	2 angles that add to 180°
13.	vertical angles	a pair of nonadjacent angles formed by 2 intersecting lines
14.	perpendicular lines	2 lines are perpendicular if they form Rt. angles
15.	bisector of a line segment	a line that devides a line into 2 = parts
16.	perpendicular bisector	PER. LINE THAT DIVIDE A LINE INTO 2 EQUAL PARTS
17.	angle bisector	A LIN THAT DIVIDES A ANGLE INTO 2 EQUAL ANGLES

These definitions will be your key to interpreting each new geometric concept. In turn, it will be these concepts that will gracefully unlock the door to your understanding each new unit.

Point

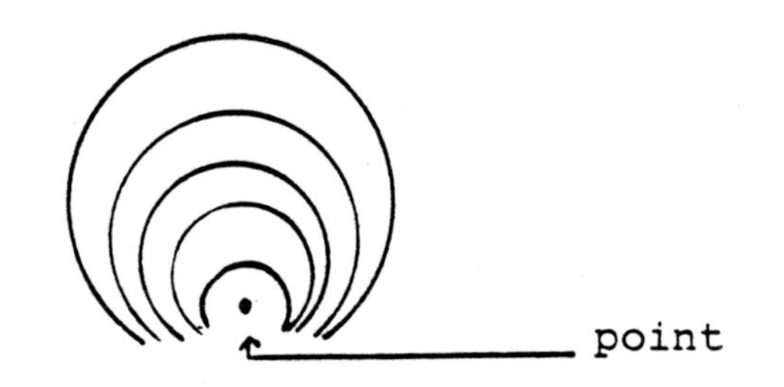

The diagram on the bottom of the previous page is a pictorial description of a point. If this figure continues to reduce in size, you will not be able to see it. The dimensions will shrink to no measure at all. It is this small dot that is used to represent a point.

Point is the first term we will not attempt to define because we already have some intuitive knowledge of what it means.

Other undefined terms are listed below:

line, between, closed, figure, intersection, measure, side,

interior, exterior, length, height, width, plane, endpoint

Line Segment

Below is a straight line with points A, B, and C on it. This can be denoted as $\overleftrightarrow{AB}$.

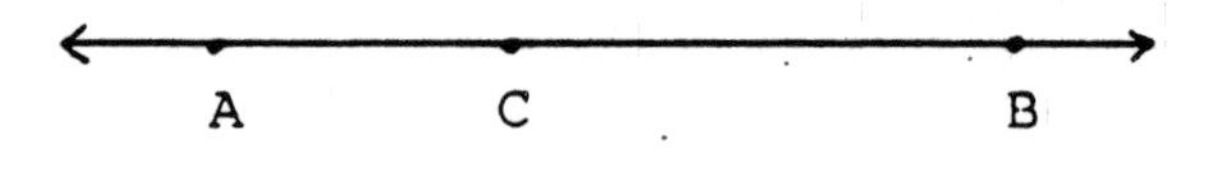

Figure 1.1

The line segment where A and B are the endpoints is denoted $\overline{AB}$ or $\overline{BA}$. The line segment is the set of all points on the line between A and B including both endpoints. See Figure 1.1 above.

Point C is a point on $\overline{AB}$ such that it divides $\overline{AB}$ into two smaller sections. These sections, $\overline{AC}$ and $\overline{CB}$, are also line segments.

The sum of these two segments can be expressed as:

$$\overline{AC} + \overline{CB} = \overline{AB}$$

In general, the sum of two (or more) segments of a line equals the measure of the entire segment.

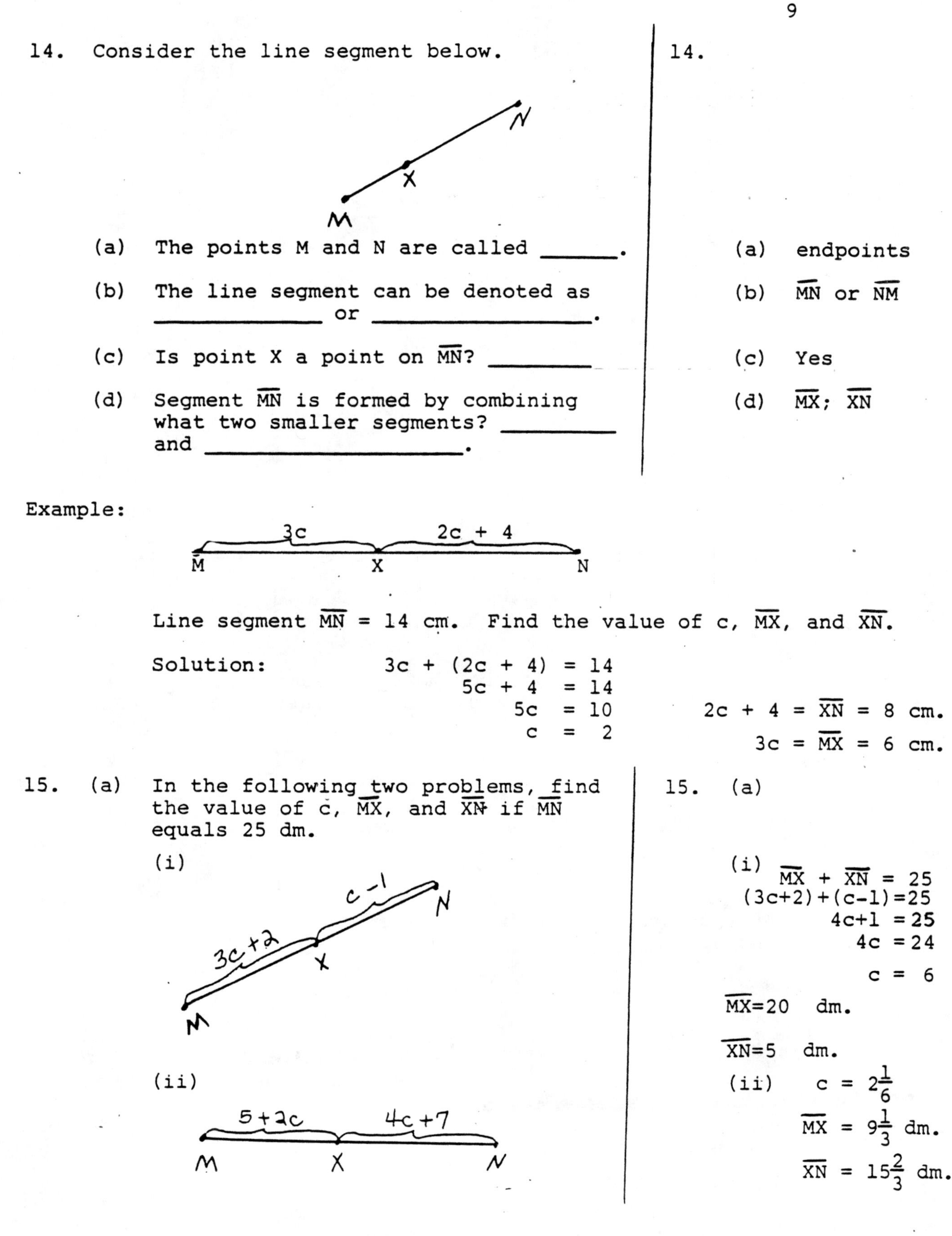

14. Consider the line segment below.

M X N

(a) The points M and N are called ______.

(b) The line segment can be denoted as ______________ or ______________.

(c) Is point X a point on $\overline{MN}$? ________

(d) Segment $\overline{MN}$ is formed by combining what two smaller segments? ________ and ________________.

14.

(a) endpoints

(b) $\overline{MN}$ or $\overline{NM}$

(c) Yes

(d) $\overline{MX}$; $\overline{XN}$

Example:

M —3c— X —2c + 4— N

Line segment $\overline{MN}$ = 14 cm. Find the value of c, $\overline{MX}$, and $\overline{XN}$.

Solution:

$$
\begin{aligned}
3c + (2c + 4) &= 14 \\
5c + 4 &= 14 \\
5c &= 10 \\
c &= 2
\end{aligned}
$$

$2c + 4 = \overline{XN} = 8$ cm.

$3c = \overline{MX} = 6$ cm.

15. (a) In the following two problems, find the value of c, $\overline{MX}$, and $\overline{XN}$ if $\overline{MN}$ equals 25 dm.

(i)

M —3c+2— X —c−1— N

(ii)

M —5+2c— X —4c+7— N

15. (a)

(i)

$$
\begin{aligned}
\overline{MX} + \overline{XN} &= 25 \\
(3c+2)+(c-1) &= 25 \\
4c+1 &= 25 \\
4c &= 24 \\
c &= 6
\end{aligned}
$$

$\overline{MX}=20$ dm.

$\overline{XN}=5$ dm.

(ii) $c = 2\frac{1}{6}$

$\overline{MX} = 9\frac{1}{3}$ dm.

$\overline{XN} = 15\frac{2}{3}$ dm.

(b). If $\overline{XN}$ = 30 ft., find the value of c, $\overline{XY}$, $\overline{YZ}$, and $\overline{ZN}$.

2c + 1 3c 2c + 1

X Y Z N

(b). $2(2c+1)+3c=30$
$7c+2=30$
$7c=28$
$c=4$

$\overline{XY}$ = 9 ft.

$\overline{YZ}$ = 12 ft.

$\overline{ZN}$ = 9 ft.

(c). If $\overline{AC}$ = 10 inches, find the value of x, $\overline{AB}$, and $\overline{BC}$.

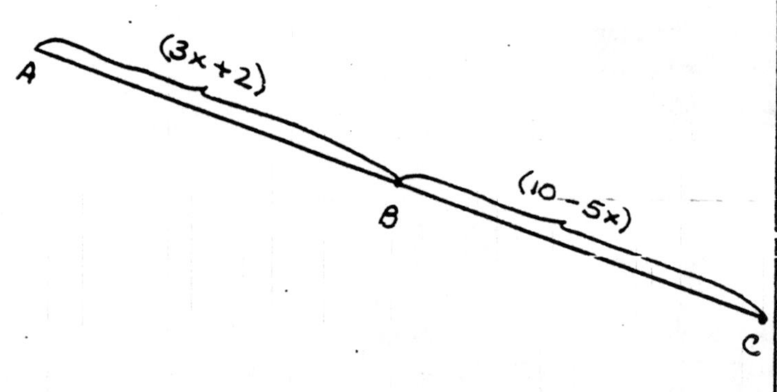

(c). $\overline{AB} + \overline{BC} = 10$

$(3x+2)+(10-5x)=10$

$12 - 2x = 10$

$-2x = -2$

$x = 1$

$\overline{AB}$ = 5 in.

$\overline{BC}$ = 5 in.

Two line segments are <u>equal</u> if they have the same measure. In the figure below $\overline{AT}$ has the same measure as $\overline{TB}$. Therefore, $\overline{AT} \doteq \overline{TB}$.

16.

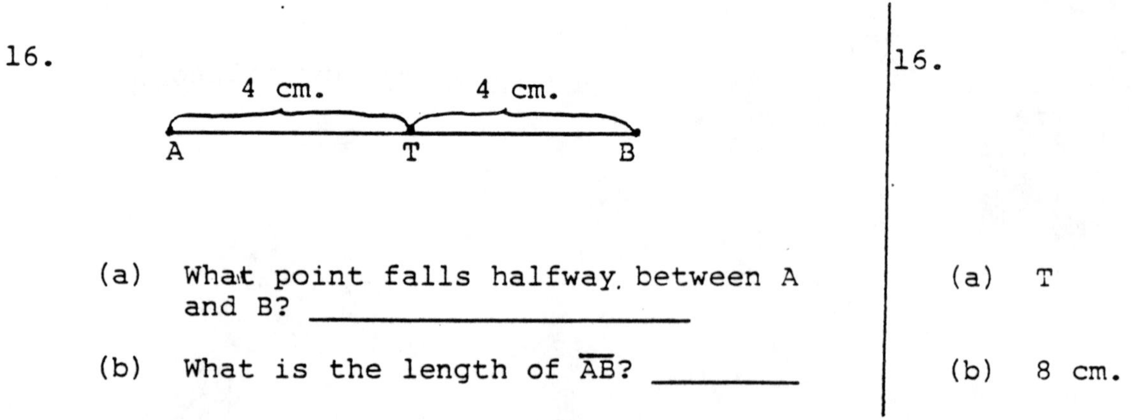

(a) What point falls halfway between A and B? ____________________

(b) What is the length of $\overline{AB}$? ________

16.

(a) T

(b) 8 cm.

Point T is called the midpoint of $\overline{AB}$. The <u>midpoint</u> lies exactly in the middle of the segment. It is the point that divides a line segment into two equal parts.

Exercise 1

1. Given line segment $\overline{AB}$.

2 cm. 2 cm.

A C B

a. What are the endpoints of $\overline{AB}$?

b. Is $\overline{AB} = \overline{BA}$?

c. Find the measure of $\overline{AB}$. Is C the midpoint?

2. In the figure below, if $\overline{XW}$ = 12 cm., find the length of $\overline{XY}$, $\overline{YZ}$, and $\overline{ZW}$ using algebra. (First find the value of c.)

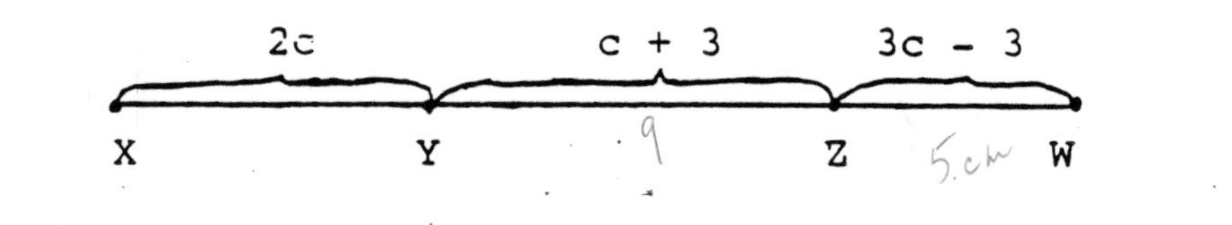

Rays

A figure T———S———→ is denoted as $\overrightarrow{TS}$ and called a ray.

Ray $\overrightarrow{TS}$ is the set of all points beginning at T on the same side as S and continuing indefinitely through S. (Notation $\overrightarrow{TS}$ or $\overleftarrow{ST}$.) Point T is called the endpoint of $\overrightarrow{TS}$.

17. The symbol $\overrightarrow{AB}$ indicates A———B———→.
The endpoint of the ray is ____________.

18. The symbol $\overleftarrow{QR}$ indicates that the endpoint of $\overleftarrow{QR}$ is ________ and the ray extends ____________.

17. Point A

18. Point R; indefinitely beyond Q.

19. In the figure (points M, N, O on a ray)

(a) Is $\overrightarrow{MN}$ the same ray as $\overrightarrow{MO}$? _____

(b) Is $\overrightarrow{MN}$ the same ray as $\overrightarrow{NO}$? _____

(c) Is $\overrightarrow{MO}$ the same ray as $\overleftarrow{MO}$? _____

20. Distinguish between a ray and a line segment in regards to endpoints.

21. Answer yes or no to the following statements.

(points C, D, E on a ray)

(a) $\overrightarrow{CD} = \overrightarrow{DE}$ _____

(b) $\overrightarrow{CD} = \overline{CD}$ _____

(c) $\overrightarrow{DE} = \overrightarrow{ED}$ _____

(d) $\overrightarrow{DE} = \overrightarrow{EC}$ _____

(e) $\overrightarrow{CD} = \overrightarrow{CE}$ _____

19. (a) yes, same endpoint and same direction.

(b) no, not the same end-points.

(c) no, not the same end-points.

20. A ray has only one endpoint and a line segment has two endpoints.

21.

(a) no

(b) no

(c) no

(d) no

(e) yes

Angle

Two rays may have a common endpoint as illustrated here.

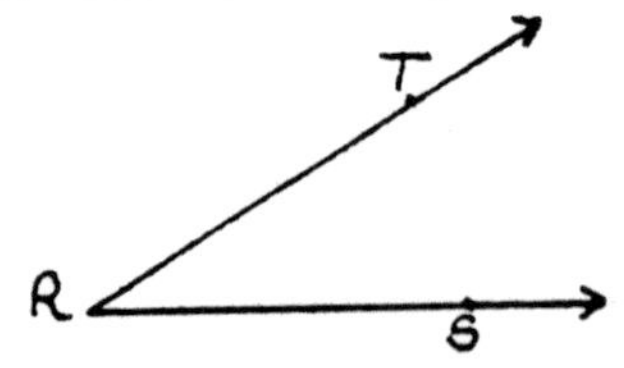

Rays $\overrightarrow{RT}$ and $\overrightarrow{RS}$ share a common endpoint R. The rays form a figure called an angle. The common endpoint R is called the vertex of the angle. The symbol for an angle is $\angle$. The above figure can be denoted as $\angle$TRS and read as "angle TRS".

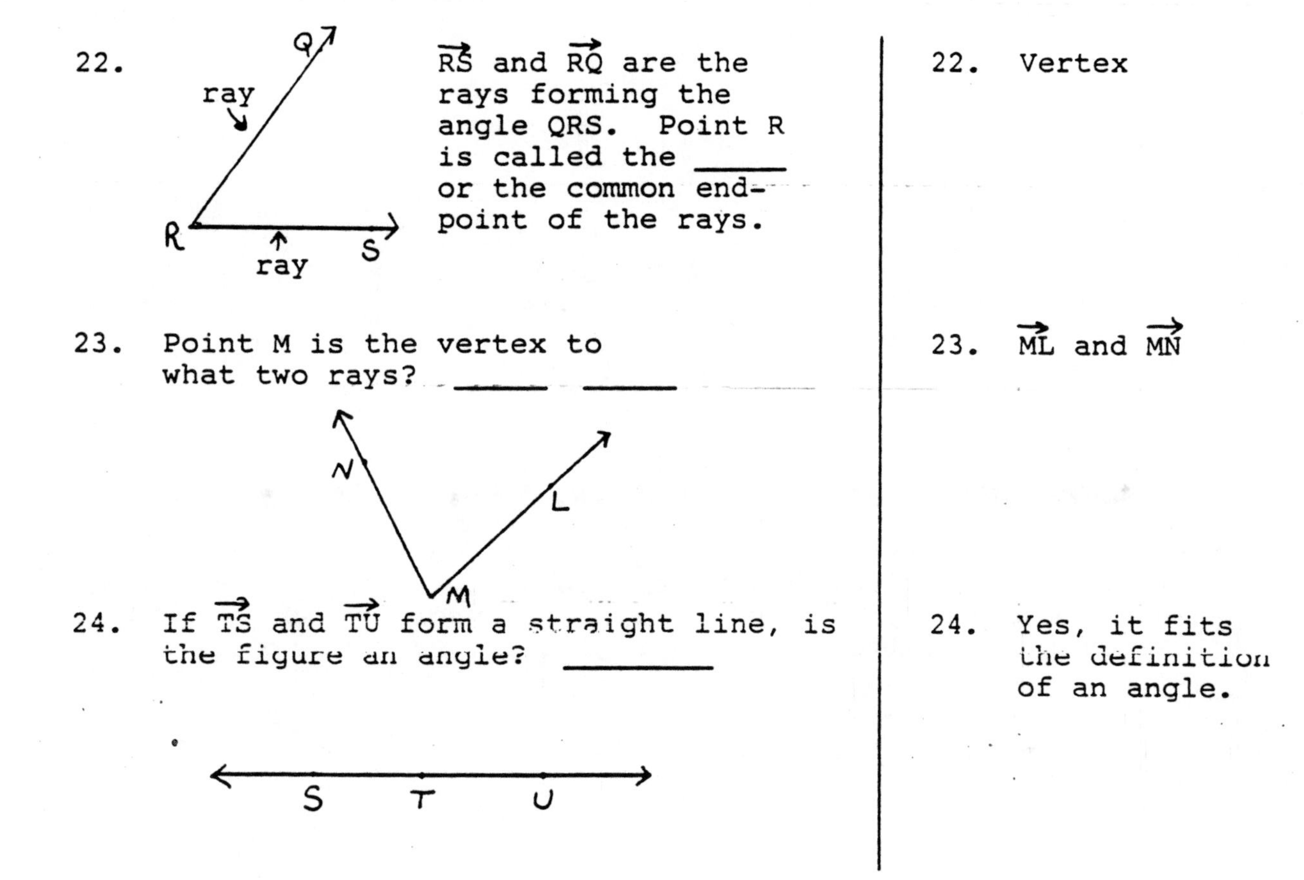

22. $\overrightarrow{RS}$ and $\overrightarrow{RQ}$ are the rays forming the angle QRS. Point R is called the _____ or the common end-point of the rays.

23. Point M is the vertex to what two rays? _____ _____

24. If $\overrightarrow{TS}$ and $\overrightarrow{TU}$ form a straight line, is the figure an angle? ________

22. Vertex

23. $\overrightarrow{ML}$ and $\overrightarrow{MN}$

24. Yes, it fits the definition of an angle.

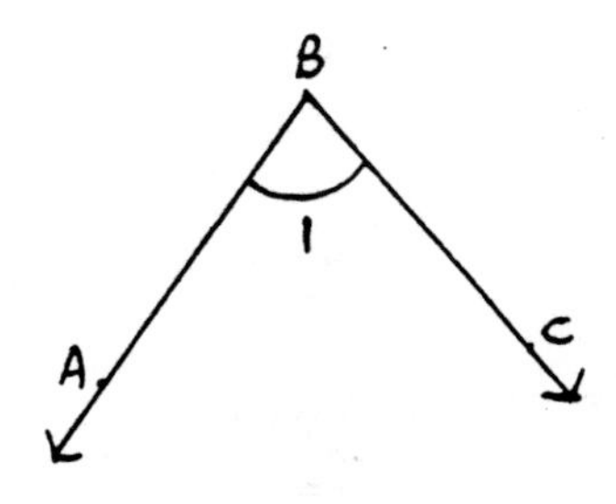

Figure 1.2

The middle letter of the notation ∠ABC is the vertex. The angle may be written as either ∠ABC, ∠CBA or ∠B.

Numbers may also be used to name an angle. Therefore, Figure 1.2 may be named in four different ways: $\angle 1$, $\angle ABC$, $\angle CBA$, and $\angle B$.

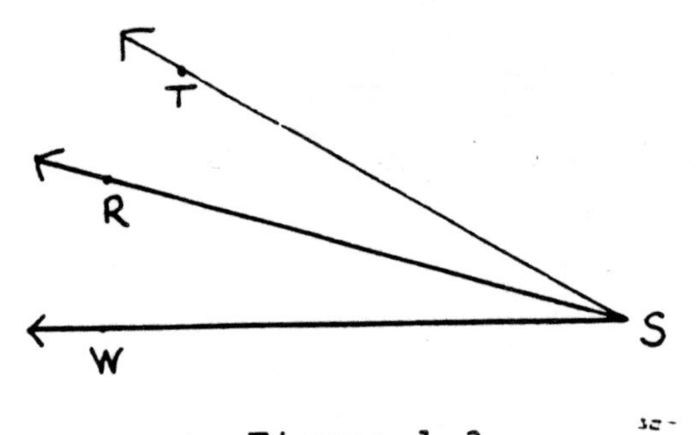

Figure 1.3

If the verticies of more than one angle meet at a point, then three letters or a number <u>must</u> be used to name the angle. In Figure 1.3, $\angle S$ refers to which angle? To avoid confusion, be very specific in problems where more than one vertex is involved. Instead of using only one letter, $\angle S$, be specific. Name $\angle TSW$, $\angle TSR$, or $\angle RSW$ if those are the angles you want to indicate.

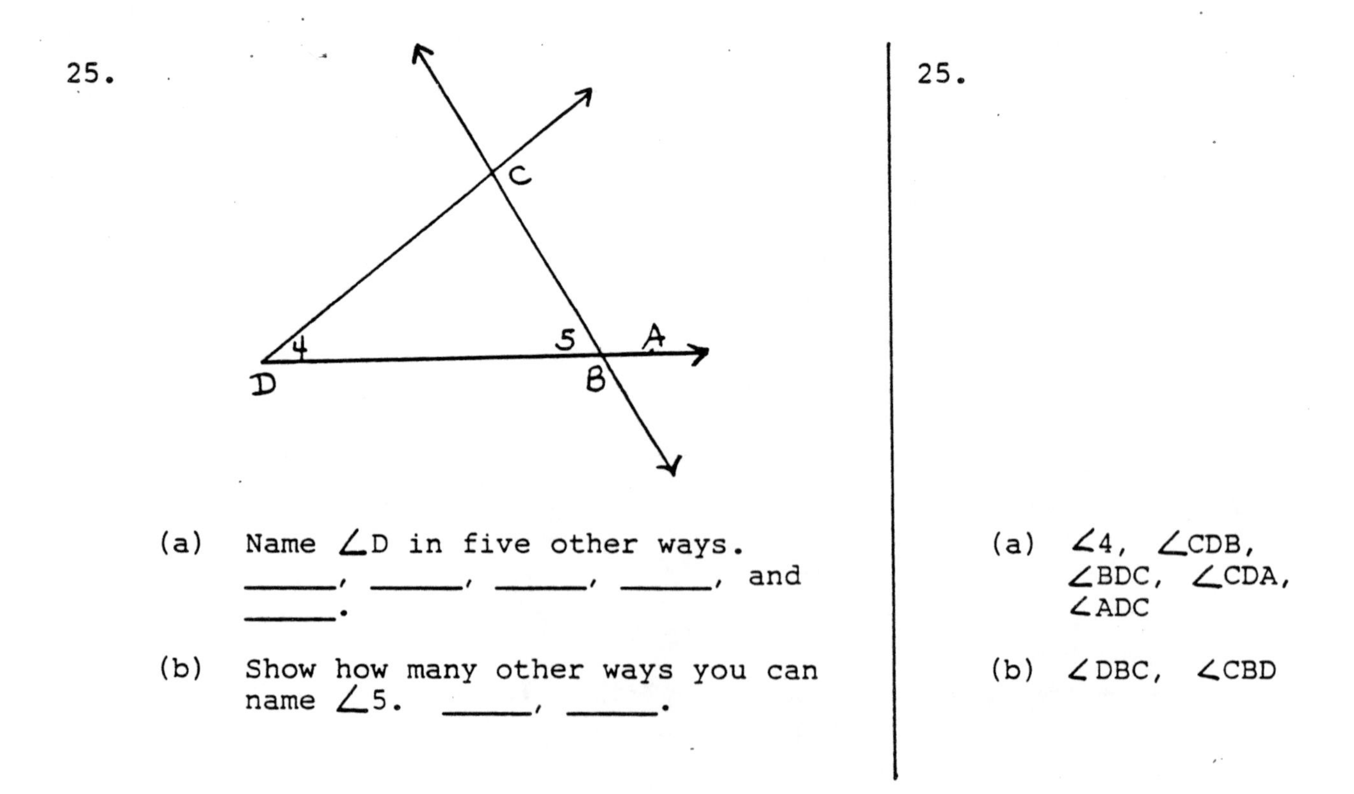

25.

(a) Name $\angle D$ in five other ways. ______, ______, ______, ______, and ______.

(b) Show how many other ways you can name $\angle 5$. ______, ______.

25.

(a) $\angle 4$, $\angle CDB$, $\angle BDC$, $\angle CDA$, $\angle ADC$

(b) $\angle DBC$, $\angle CBD$

Angular Measure

Angles are measured with an instrument called a protractor. The number of degrees in an angle is called its measure. Your instructor will demonstrate the method used in measuring angles with a protractor. In Appendix E, a detailed explanation of this procedure can be found.

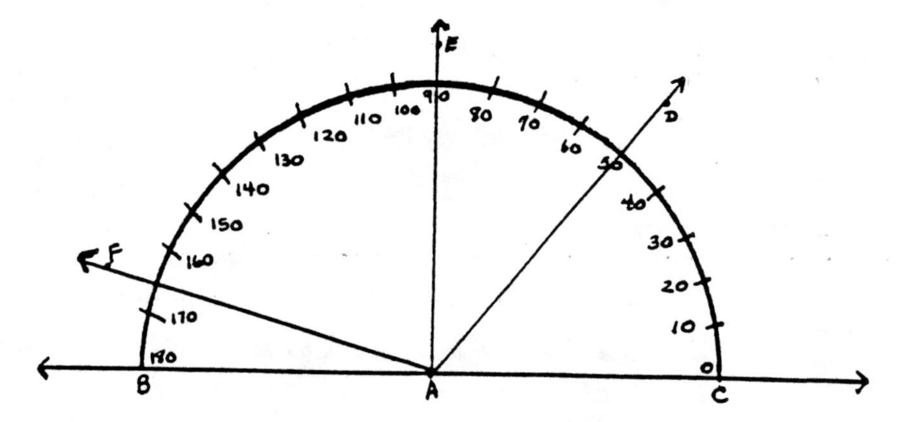

The measure of ∠CAD is equal to 50° (degrees).

∠CAE = 90°   ∠CAF = 165°   ∠CAB = 180°

26. Using a protractor, measure the given angles. (You may extend the sides of the angle if necessary.)

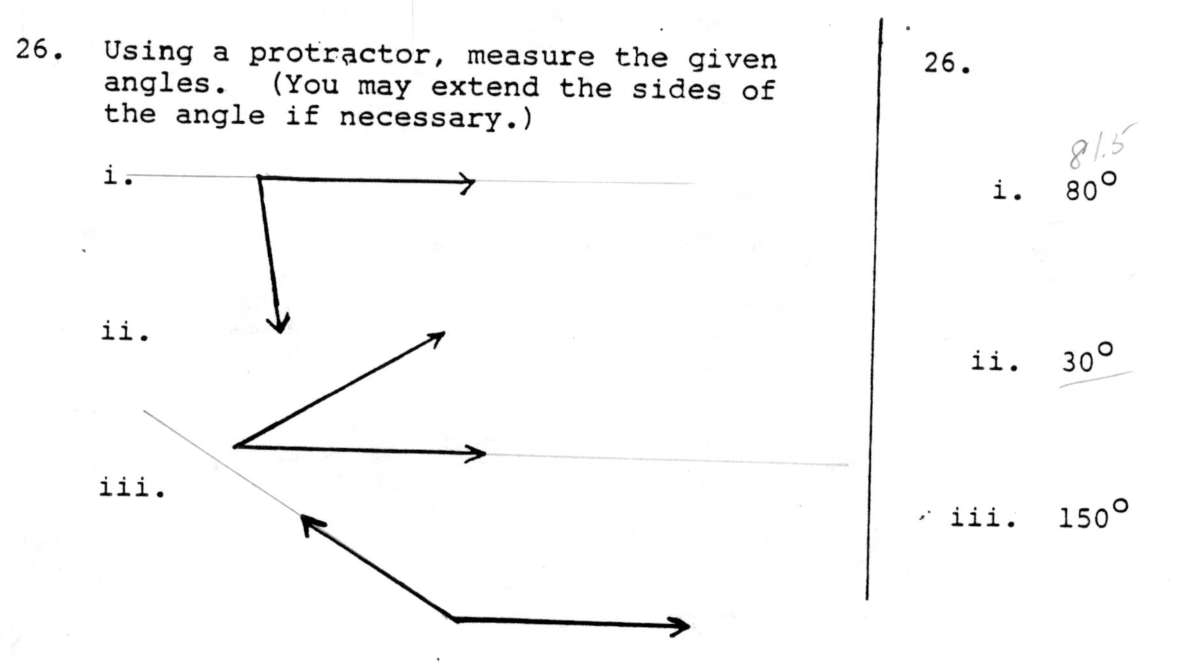

26.

i. 80° (81.5)

ii. 30°

iii. 150°

The measure of an angle in this text may vary anywhere from any degree greater than 0° through 180°. Angles are placed in categories depending upon their degree measure.

Definitions

(A) A right angle is an angle that measures exactly 90°.

You may compare this measure of 90° to the 3 o'clock position on your watch. Note the symbol, ⊾, denotes a right angle.

(B) An acute angle is an angle which measures more than 0° but less than 90°

(C) An obtuse angle is an angle which measures greater than 90° but less than 180°.

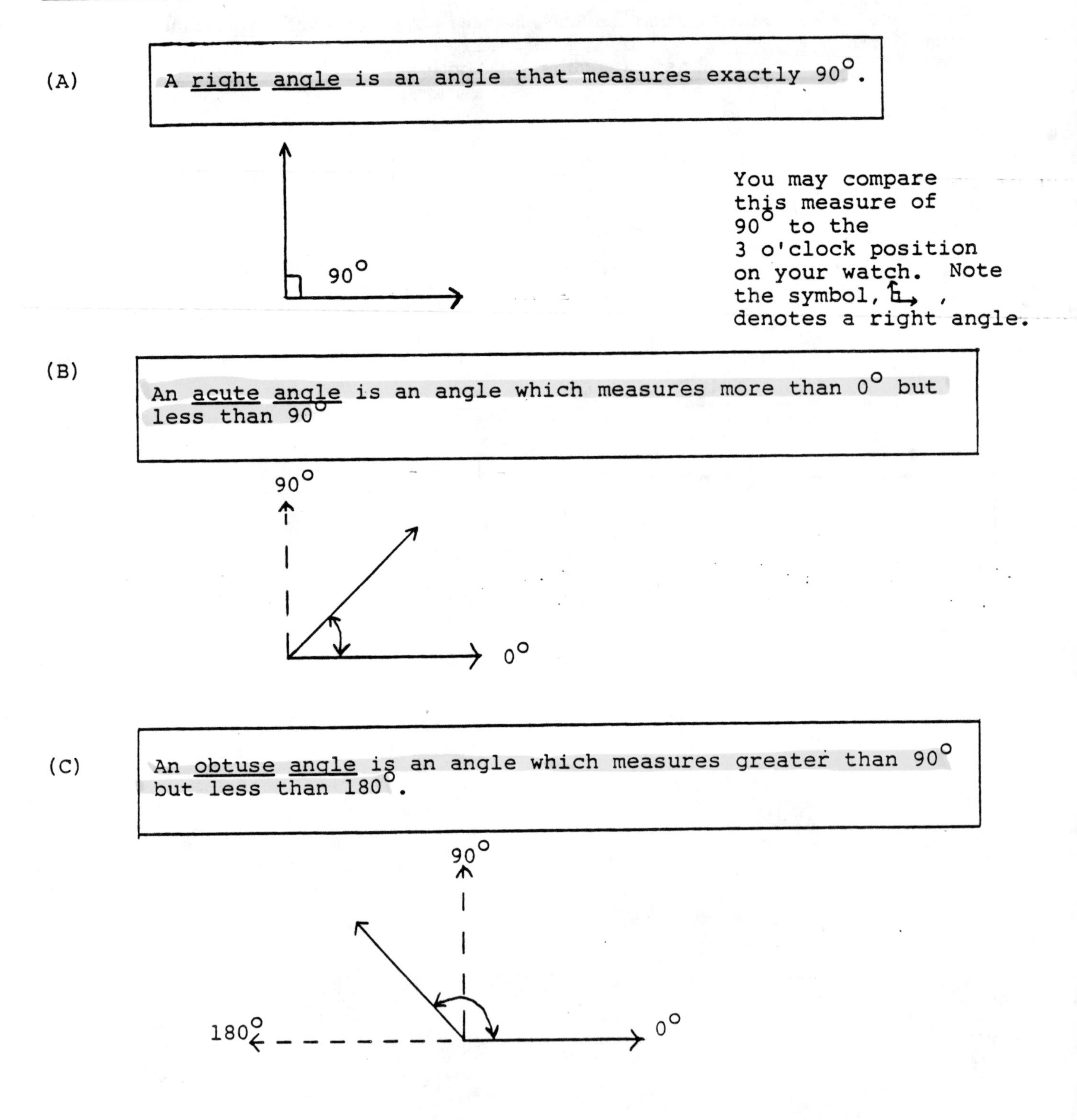

(D) A **straight angle** is an angle which measures exactly 180°

27. Using the preceding definitions, classify each of the following angles by the measures given:

(a) 160°
(b) 180°
(c) 3°
(d) 90°
(e) 110°
(f) 89°
(g) 91°

27. (a) obtuse
(b) straight
(c) acute
(d) right
(e) obtuse
(f) acute
(g) obtuse

28. Use a protractor to measure the angles given below:

______(a)

______(b)

28. (a) 120°
(b) 90°

(E) **Adjacent angles** are two angles that share a common vertex and a common side lying **between** them.

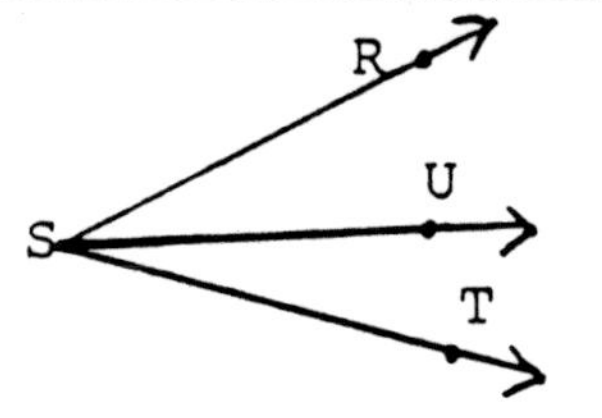

$\angle$ RSU and $\angle$ UST are adjacent angles. Common vertex is S. Common side between $\angle$ RSU and $\angle$ UST is $\overrightarrow{SU}$.

29.

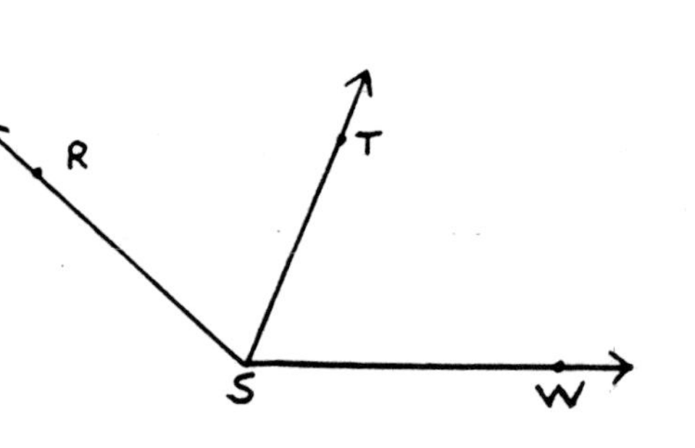

(a) ∠RST and ∠TSW share what common vertex? ____________

(b) ∠RST and ∠TSW share what common side? ____________

(c) ∠RST and ∠TSW are called adjacent angles because ____________.

(d) Are ∠RSW and ∠TSW called adjacent angles? ____________.

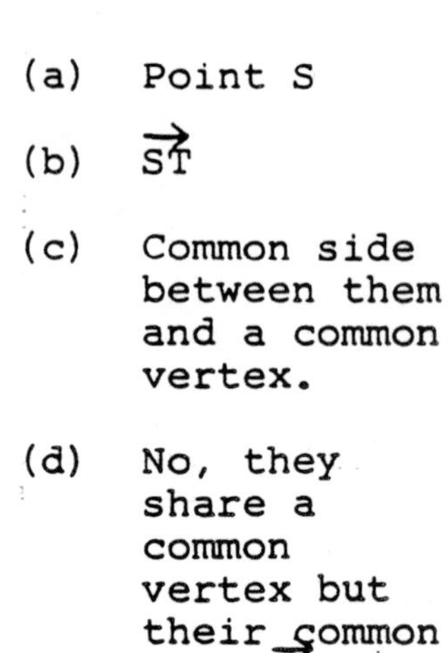

29. (a) Point S

(b) $\overrightarrow{ST}$

(c) Common side between them and a common vertex.

(d) No, they share a common vertex but their common side $\overrightarrow{SW}$ doesn't lie between the two angles.

Example:

∠XYZ is a straight angle. Find the measures of each angle.

9+c

2c

c+1

X Y Z

Solution: $(2c)+(9+c)+(c+1) = 180^{\circ}$

$4c+10 = 180$

$4c = 170$

$c = 42\frac{1}{2}^{\circ}$

$2c = 85^{\circ}$, $(9+c) = 51\frac{1}{2}^{\circ}$, $(c+1) = 43\frac{1}{2}^{\circ}$

30. Find the measure of each angle if $\overleftrightarrow{OP}$ is a straight line. ∠1, ∠2, and ∠3 are given the respective values in each part.

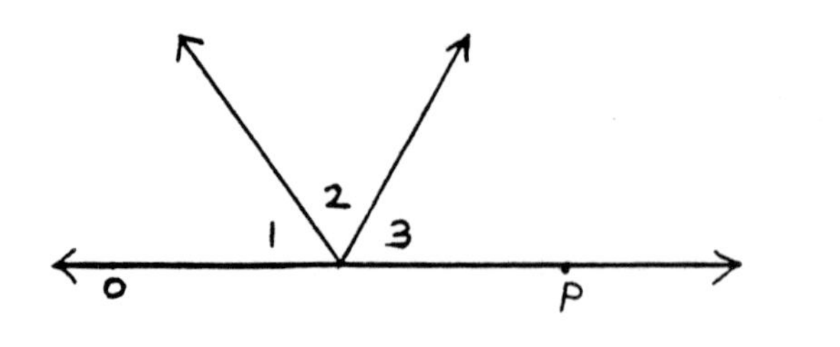

30.

(a) c; 4c; 7c

(b) 9c + 2; 8c - 15; 7c

30. (continued)

(a) $c+4c+7c=180^\circ$

$12c=180^\circ$

$c=15^\circ$

$4c=60^\circ$

$7c=105^\circ$

(b)

$(9c+2)+(8c-15)+7c=180^\circ$

$24c-13=180^\circ$

$24c=193^\circ$

$c=8\frac{1}{24}^\circ$

$9c+2=74\frac{3}{8}^\circ$

$8c-15=49\frac{1}{3}^\circ$

$7c=56\frac{7}{24}^\circ$

Supplementary and Complementary Angles

The measure of two angles may be added together to illustrate two relationships between angles. These angles may be adjacent or they may be on different planes but the relationship will continue.

The first relationship is as follows:

(F) Two angles are said to be complementary if their sum is 90°. That is, if $\angle A + \angle B = 90^\circ$, then $\angle A$ and $\angle B$ are complementary.

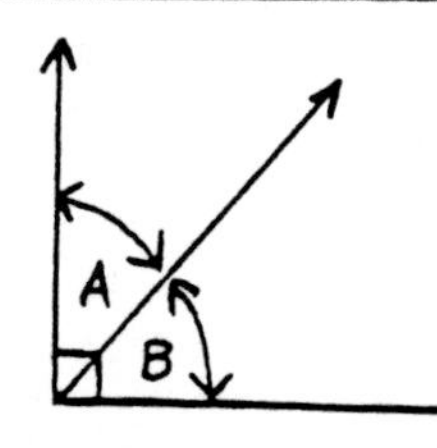

In the figure to the left $\angle A$ and $\angle B$ are complementary.

31. Two angles are said to be __________ if the sum of their measures equals 90°.

32. Angles A and B are called complements of one another. If $\angle A = 50^\circ$, then the complement, $\angle B$, must equal __________.

31. complementary

32. $\angle B = 40^\circ$, since $50^\circ + 40^\circ = 90^\circ$

33. If $\angle A = 70^\circ$ and $\angle B = 70^\circ$ are they complementary? ____________

34. If $\angle B = 35\frac{1}{7}^\circ$, how do you find its complement? ____________

The complement, $\angle A$, must equal what measure? ____________

33. No, $70^\circ + 70^\circ \neq 90^\circ$

34. Subtract $35\frac{1}{7}^\circ$ from 90°

$54\frac{6}{7}^\circ$

Example: $\angle ABC$ and $\angle CBD$ are complementary angles. Find the measure of each.

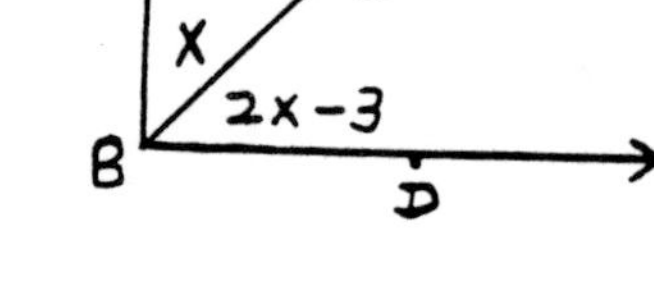

Solution:

$$(x) + (2x - 3) = 90^\circ$$
$$3x - 3 = 90$$
$$3x = 93$$
$$x = 31^\circ \text{ equals the measure of } \angle ABC$$
$$2x - 3 = 59^\circ \text{ equals the measure of } \angle CBD$$

35. Find the measure of each angle if these rays, $\overleftarrow{MN}$ and $\overrightarrow{NO}$, form a right angle at their intersection. $\angle 1$ and $\angle 2$ are the respective values in each part.

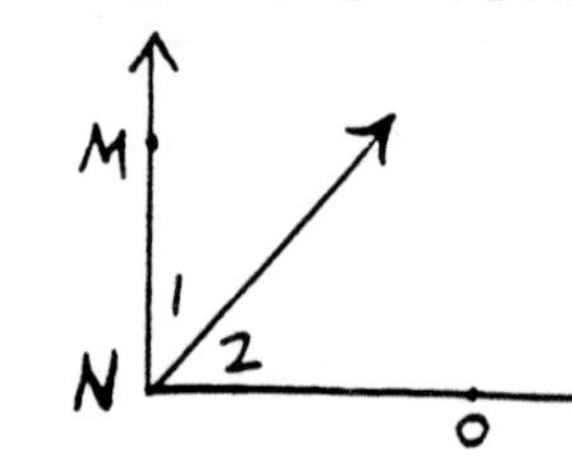

(a) 3x and 4x

(b) (3x + 2) and (3 + 2x)

35. (a) $3x+4x=90^\circ$

$7x=90^\circ$

$x=12\frac{6}{7}^\circ$

$4x=51\frac{3}{7}^\circ$

$3x=38\frac{4}{7}^\circ$

(b)

$(3x+2)+(3+2x)=90^\circ$

$5x+5 = 90^\circ$

$x=17^\circ$

$3x+2 = 53^\circ$

$3+2x = 37^\circ$

The second relationship is as follows:

(G)	Two angles are <u>supplementary</u> if their sum is 180°. That is, if $\angle A + \angle B = 180^\circ$, then $\angle A$ and $\angle B$ are <u>supplementary</u>.

In the figure to the left, $\angle A$ and $\angle B$ are supplementary.

36. Two angles are supplementary if the sum of their measures is ________________.

36. 180°

37. Angles A and B are called supplements of each other. If $\angle B = 50^\circ$, then the supplement, $\angle A$, must equal __________.

37. 130°

38. If $\angle B = 100^\circ$ and $\angle A = 80^\circ$ are they complementary angles? ______________

38. No, they are supplementary.

39. (a) If $\angle A$ equals 110°, how do you find its supplement? ______________ $\angle B$ equals ________________.

39. (a) Subtract 110° from 180°. 70°

(b) How do you find the complement of $\angle A$? ____________________________

(b) 110° has no complement since $90^\circ - 110^\circ$ can not be the measurement of an angle.

Supplementary and complementary angles can be expressed in verbal problems. Study the example shown on page 22.

Example: Two angles are complementary and one is 60° smaller than the other. Find the measures of the two angles.

Solution: Let x = first angle (unknown value)

$x - 60$ = second angle

$(x) + (x - 60) = 90$ since they are complementary angles.

$2x - 60 = 90$ simplifying algebraic expressions.

$2x = 150$ adding 60 to both members and simplifying.

$x = 75$ solving for "x", the first angle's measure.

$(x - 60) = 15$ evaluating $x - 60$ to find the second angle's measure.

Answers: $\underline{75^\circ}$, $\underline{15^\circ}$

40. Two angles are complementary. One angle is forty more than four times another. Find the measures of the two angles.

Solution:

x = one angle

____ = second angle (a)______

The equation $x + (4x+40) =$ __ (b)______

Simplifying ____________ $= 90^\circ$ (c)______

$5x =$ ___ (d)______

$x =$ ___ (e)______

The first angle is __________. (f)______

The second angle is

$4(10) + 40 =$ __________ (g)______

40.

(a) $4x + 40$

(b) 90°

(c) $5x + 40$

(d) 50°

(e) 10°

(f) 10°

(g) 80°

41. One angle is thirty-nine more than twice its supplement. Find the measure of both angles.

_____ = the supplement (a) _____
_____ = first angle (b) _____

$x + (2x + 39) = 180^\circ$
_____ = 180 (c) _____
3x = _____ (d) _____
x = _____ (e) _____
the supplement = _____ (f) _____
first angle = _____ (g) _____

41.

(a) x
(b) 2x + 39

(c) 3x+39=180
(d) 3x=141
(e) x=47
(f) 47°
(g) 133°

Exercise 2

1.

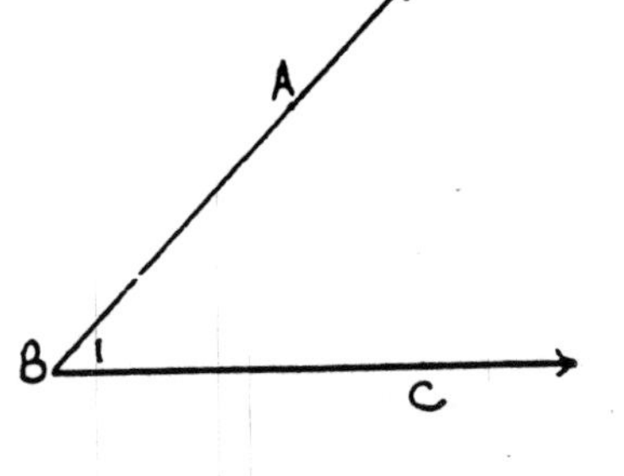

Indicate three possible labels for this angle

_____

_____

_____

2.

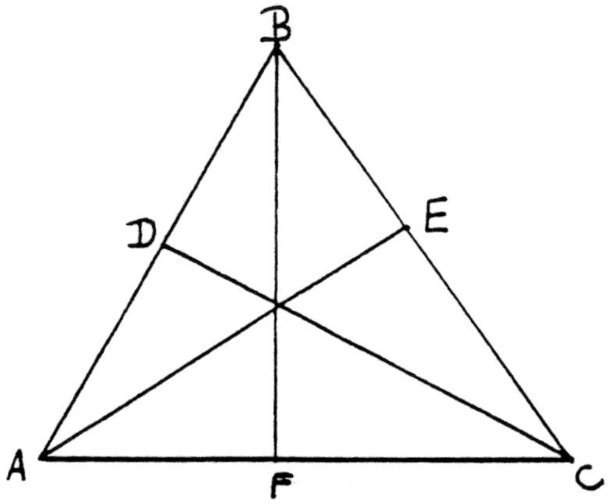

The angles listed below are numbered 1 through 8. Using the figure above, locate each angle and write its number on the figure in the order given.

1) ∠BAE
2) ∠ADC
3) ∠BDC
4) ∠ABC
5) ∠BEA
6) ∠EAC
7) ∠BFC
8) ∠DCF

3. Using a protractor, measure the given angles. Label each angle as straight, right, acute, or obtuse. Use a straightedge to extend the sides of the angles if necessary.

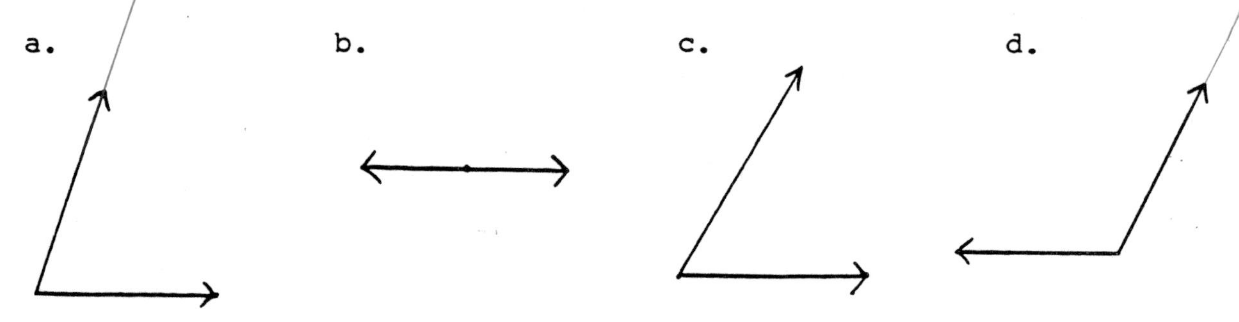

4. Find the supplement and complement of the given angle.

		Supplement	Complement
a.	30°	________	________
b.	120°	________	________
c.	x°	________	________
d.	45°	________	________

5. ∠XYZ is a straight line. Find the measures of each angle and classify each as straight, acute, obtuse, or right angle.

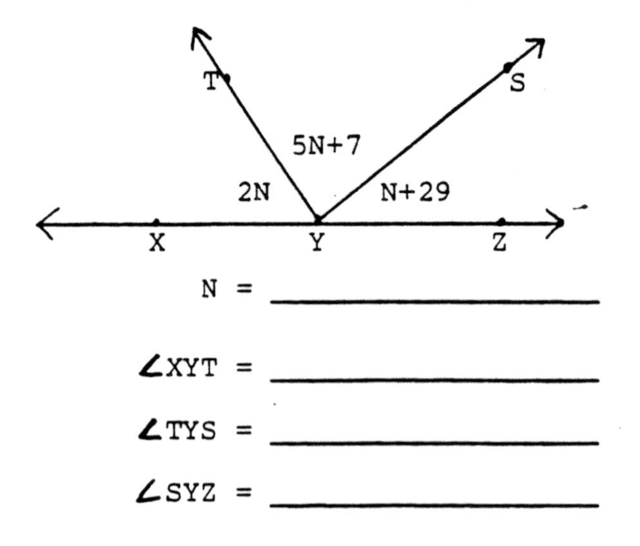

N = ________

∠XYT = ________

∠TYS = ________

∠SYZ = ________

6. Two angles are complementary and one is 30° less than the other. How many degrees are there in each angle?

7. Two angles are complementary. One angle is twelve more than five times the other. Find the measures of the two angles.

8. Two supplementary angles are equal in measure. Find their measure.

9. Two angles are supplementary. One angle is ten more than two thirds the other. Find the measures of both angles.

10. Find the measure of each angle.

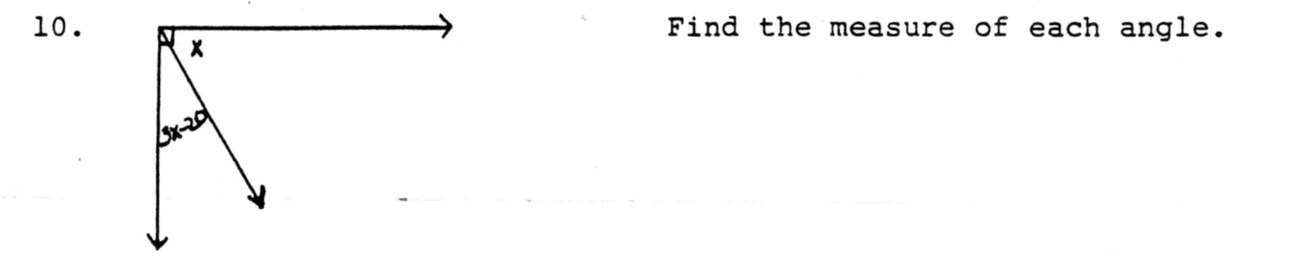

> Theorem 1: Supplements of the <u>same</u> (or <u>equal</u>) angles are themselves equal.

Each condition below is illustrated with an example.

> Condition 1: Supplements of the <u>same</u> angle are equal.

Condition 1 compares the measures of different supplements to the same angle. Let $\angle A$ and $\angle C$ be supplements of $\angle B$. Let $\angle B = 150^\circ$. If $\angle B = 150^\circ$, then $\angle A = 30^\circ$.

$\angle B + \angle A = 180^\circ$ by the definition of supplementary angles.

$$\angle B + \angle A = 180^\circ$$

$$150^\circ + 30^\circ = 180^\circ$$

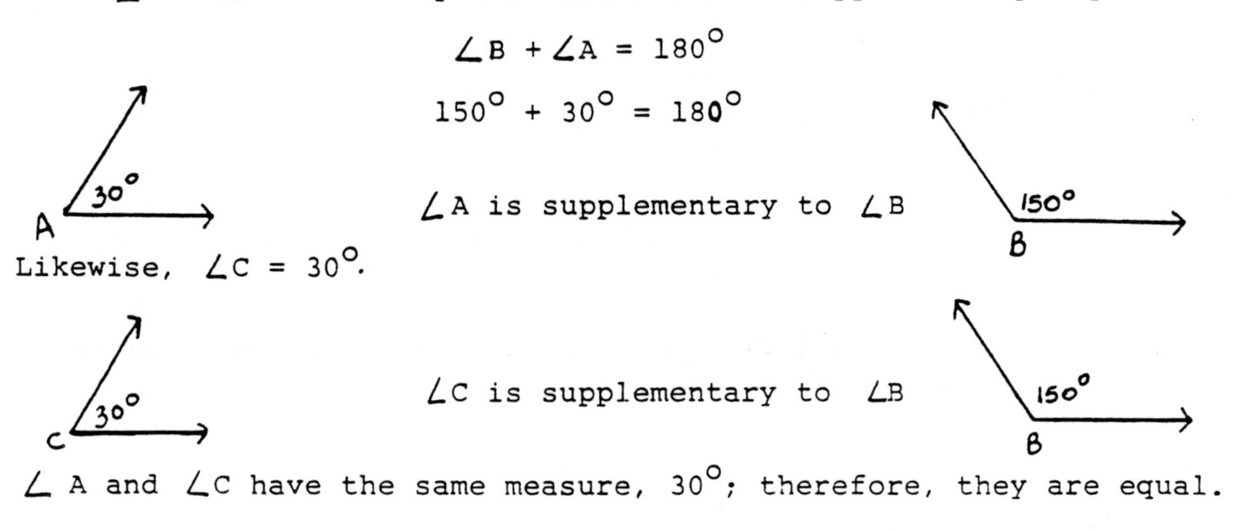

$\angle A$ is supplementary to $\angle B$

Likewise, $\angle C = 30^\circ$.

$\angle C$ is supplementary to $\angle B$

$\angle A$ and $\angle C$ have the same measure, 30°; therefore, they are equal.

Condition 2: Supplements of equal angles are themselves equal.

In this condition we discuss angles that are equal and compare the measures of the supplements to these angles.

Let $\angle S = \angle U$, $\angle S = 25^\circ$, and $\angle U = 25^\circ$. If $\angle S$ is supplementary to an angle, say $\angle T$, then $\angle S + \angle T = 180^\circ$ by the definition of supplementary angles.

$$25^\circ + \angle T = 180^\circ$$

$$\angle T = 155^\circ$$

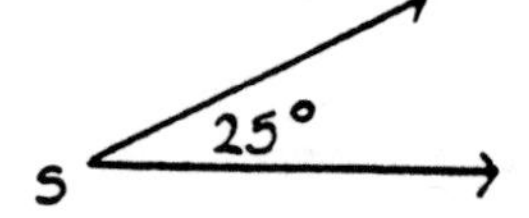

$\angle S$ is supplementary to $\angle T$

155°
T

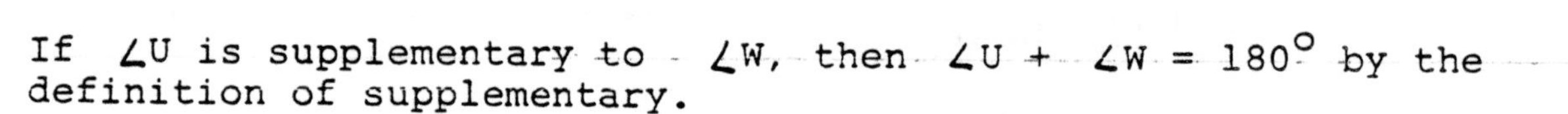

If $\angle U$ is supplementary to $\angle W$, then $\angle U + \angle W = 180^\circ$ by the definition of supplementary.

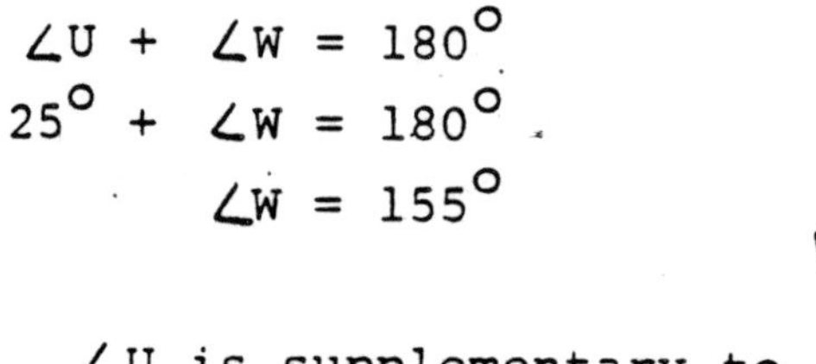

$$\angle U + \angle W = 180^\circ$$
$$25^\circ + \angle W = 180^\circ$$
$$\angle W = 155^\circ$$

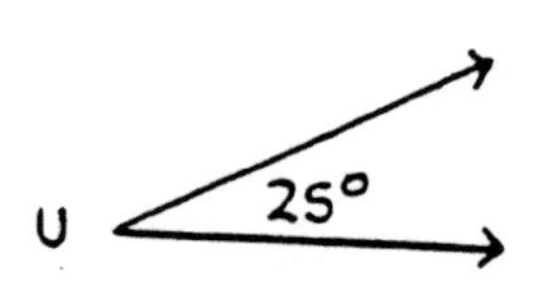

$\angle U$ is supplementary to $\angle W$

155°
W

$\angle T = 155^\circ$, $\angle W = 155^\circ$, so $\angle T = \angle W$. This demonstrates that supplements of equal angles are equal.

Theorem 2: Complements of the same (or equal) angles are themselves equal.

Each condition below will be illustrated with an example.

Condition 1: Complements of the same angles are equal.

Condition 1 compares the measures of different complements of a single angle. Let $\angle W$ and $\angle Y$ be complements of $\angle X$. Let $\angle X = 70^\circ$. If $\angle X = 70^\circ$, then $\angle W = 20^\circ$. $\angle X + \angle W = 90^\circ$ by the definition of complementary angles.

$$\angle X + \angle W = 90^\circ$$
$$70^\circ + \angle W = 90^\circ$$
$$\angle W = 20^\circ$$

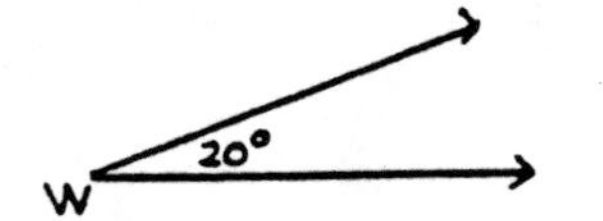

$\angle W$ is complementary to $\angle X$

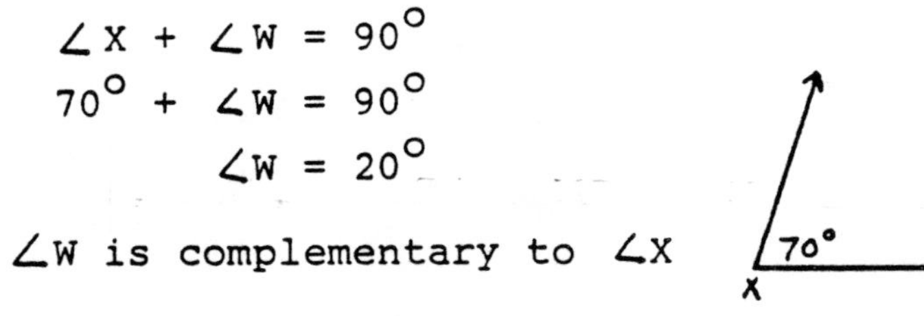

Likewise, $\angle Y = 20^\circ$

$\angle Y$ is complementary to $\angle X$

$\angle W$ and $\angle Y$ have the same measure, 20°; therefore, they are equal.

Condition 2: Complements of <u>equal angles</u> are themselves equal.

In this condition we discuss angles that are equal and compare the measures of the complements to these angles. Let $\angle R = \angle Q$, $\angle R = 40^\circ$ and $\angle Q = 40^\circ$. If $\angle R$ is complementary to $\angle P$, then $\angle R + \angle P = 90^\circ$ by the definition of complementary angles.

$$\angle R + \angle P = 90^\circ$$
$$40^\circ + \angle P = 90^\circ$$
$$\angle P = 50^\circ$$

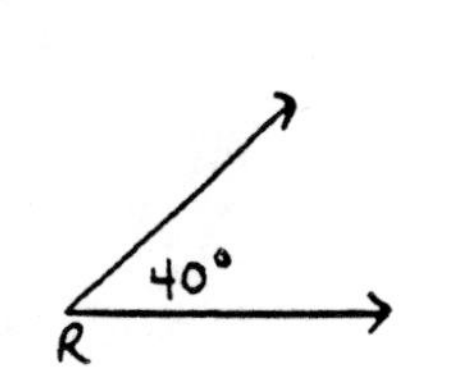

$\angle R$ is complementary to $\angle P$

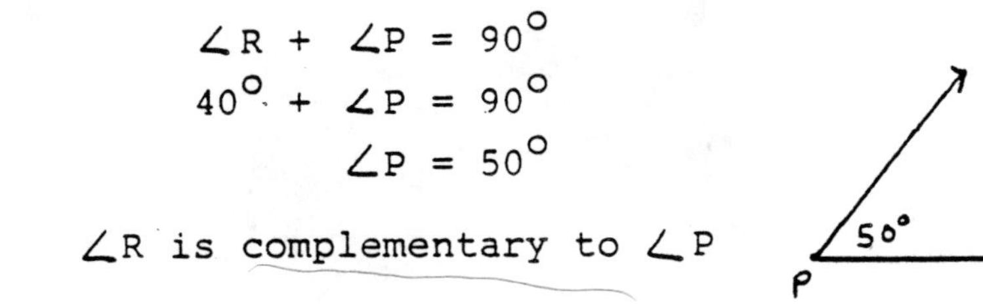

If $\angle Q$ is complementary to $\angle T$, then $\angle Q + \angle T = 90^\circ$ by the definition of complementary.

$$\angle Q + \angle T = 90^\circ$$
$$40^\circ + \angle T = 90^\circ$$
$$\angle T = 50^\circ$$

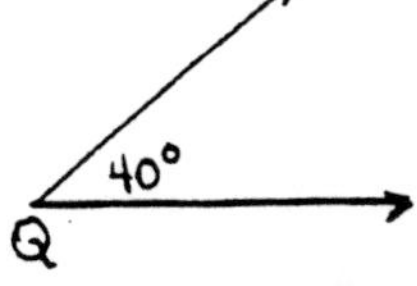

$\angle Q$ is complementary to $\angle T$

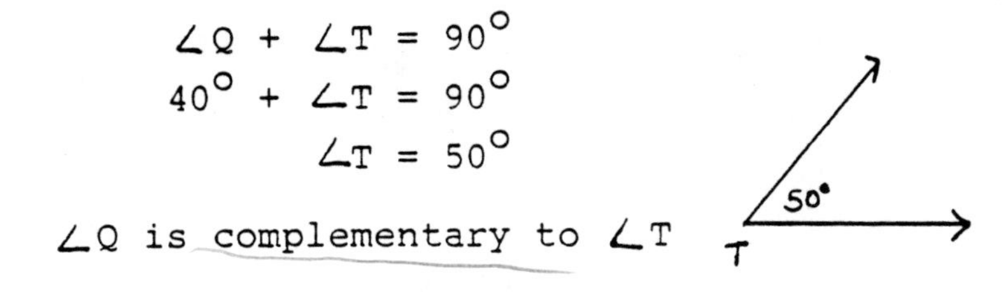

$\angle P = 50^\circ$ and $\angle T = 50^\circ$. Therefore, $\angle P = \angle T$. This demonstrates that complements of equal angles are equal.

Example: $\angle ABC$ and $\angle EBD$ are right angles. If $\angle 2 = 25^\circ$, show that $\angle 1$ equals $\angle 3$.

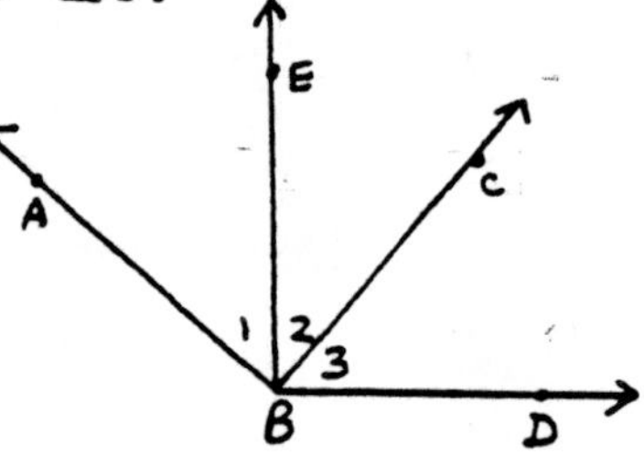

Solution:

1. $\angle 1$ and $\angle 2$ are complementary angles; $\angle 1 = 65^\circ$

2. $\angle 3$ and $\angle 2$ are complementary angles; $\angle 3 = 65^\circ$

3. $\angle 1 = \angle 3$ because complements of the same angle are themselves equal.

42. In the figure below, $\angle XYN$ and $\angle NYZ$ are right angles. Also, $\angle 2 = 44^\circ$ and $\angle 3 = 44^\circ$. Show that $\angle 1 = \angle 4$. For statements 2-4, give the reason why each statement is true.

1. $\angle 2 = \angle 3$, given

2. $\angle 1$ and $\angle 2$ are ________________.

3. $\angle 3$ and $\angle 4$ are ________________.

4. $\angle 1 = \angle 4$ because ________________.

42.

2. complementary angles; $\angle 1 = 46^\circ$

3. complementary angles; $\angle 4 = 46^\circ$

4. Complements of equal angles are themselves equal.

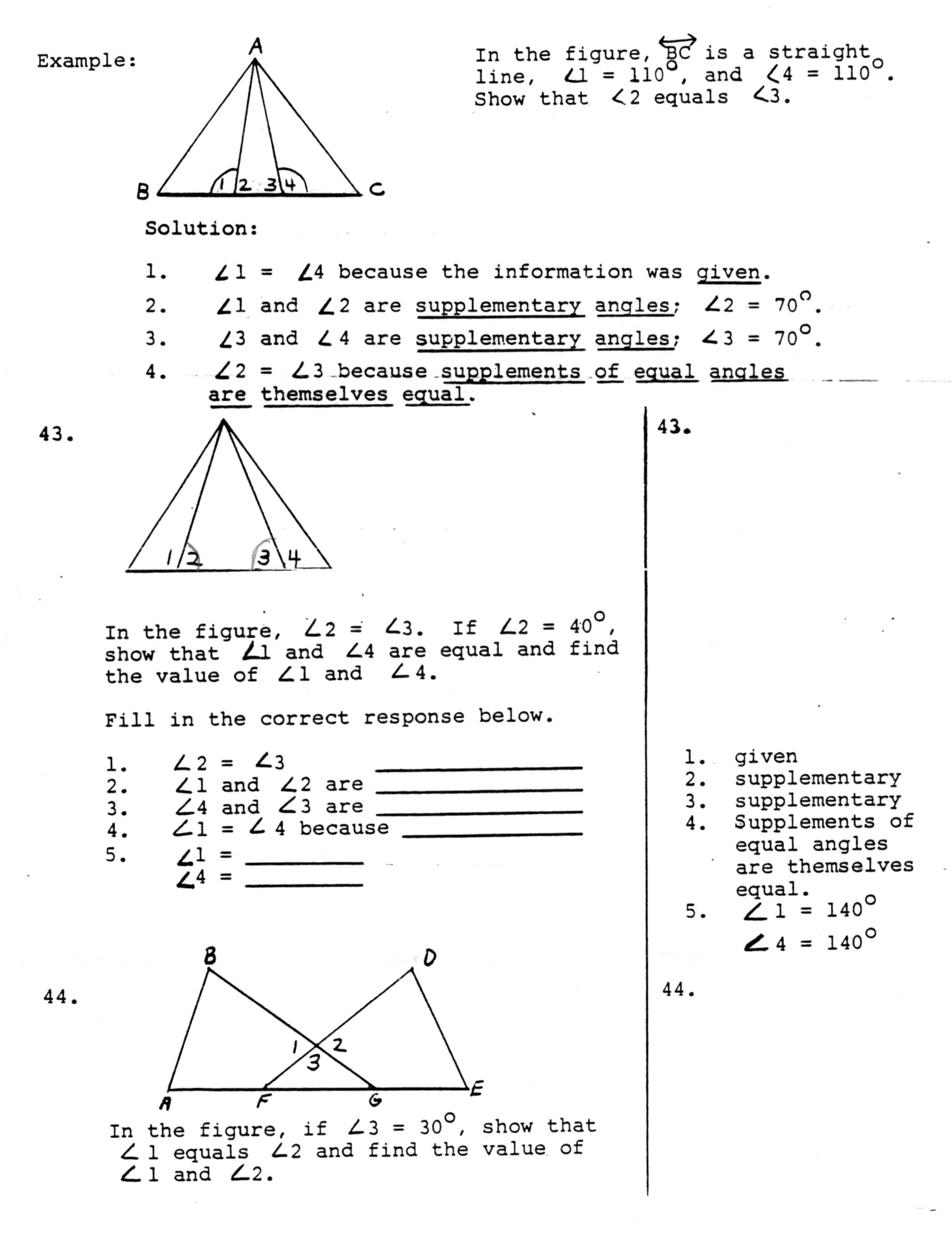

Example:

In the figure, $\overleftrightarrow{BC}$ is a straight line, $\angle 1 = 110^\circ$, and $\angle 4 = 110^\circ$. Show that $\angle 2$ equals $\angle 3$.

Solution:

1. $\angle 1 = \angle 4$ because the information was given.
2. $\angle 1$ and $\angle 2$ are supplementary angles; $\angle 2 = 70^\circ$.
3. $\angle 3$ and $\angle 4$ are supplementary angles; $\angle 3 = 70^\circ$.
4. $\angle 2 = \angle 3$ because supplements of equal angles are themselves equal.

43.

In the figure, $\angle 2 = \angle 3$. If $\angle 2 = 40^\circ$, show that $\angle 1$ and $\angle 4$ are equal and find the value of $\angle 1$ and $\angle 4$.

Fill in the correct response below.

1. $\angle 2 = \angle 3$ __________
2. $\angle 1$ and $\angle 2$ are __________
3. $\angle 4$ and $\angle 3$ are __________
4. $\angle 1 = \angle 4$ because __________
5. $\angle 1 =$ ________
   $\angle 4 =$ ________

43.

1. given
2. supplementary
3. supplementary
4. Supplements of equal angles are themselves equal.
5. $\angle 1 = 140^\circ$
   $\angle 4 = 140^\circ$

44.

In the figure, if $\angle 3 = 30^\circ$, show that $\angle 1$ equals $\angle 2$ and find the value of $\angle 1$ and $\angle 2$.

44.

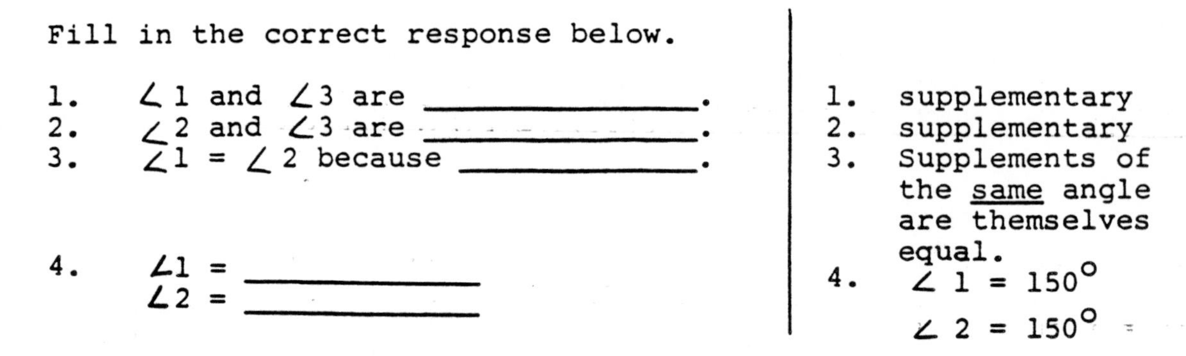

Fill in the correct response below.

1. ∠1 and ∠3 are ________________.
2. ∠2 and ∠3 are ________________.
3. ∠1 = ∠2 because ________________.

4. ∠1 = ________________
   ∠2 = ________________

1. supplementary
2. supplementary
3. Supplements of the <u>same</u> angle are themselves equal.
4. ∠1 = 150°
   ∠2 = 150°

Problems similar to #43-44 can be applied to <u>complementary angles</u>.

H. A pair of nonadjacent angles formed by two intersecting lines is called a pair of <u>vertical angles</u>.

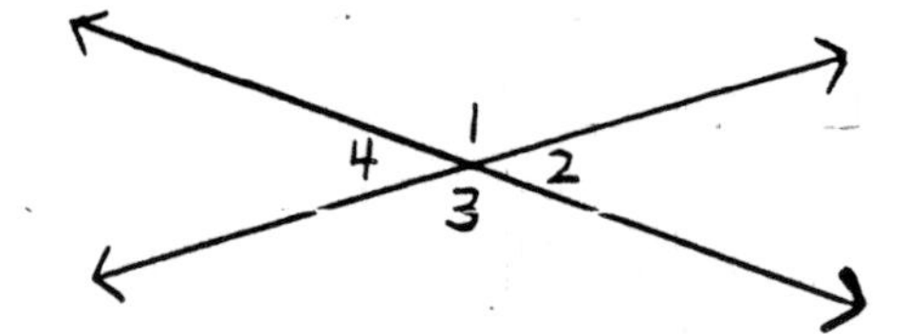

∠1 and ∠3 are vertical angles.

∠2 and ∠4 are vertical angles.

Theorem 3: Pairs of vertical angles are equal.

To show that pairs of vertical angles are equal, see problem #44, pages 29 and 30.

Definition: Two lines are <u>perpendicular</u> if they form right angles at their intersection.

The symbol for perpendicular is ⊥ . Each angle formed at the intersection of perpendicular lines is 90°. (Diagram follows on page 31.)

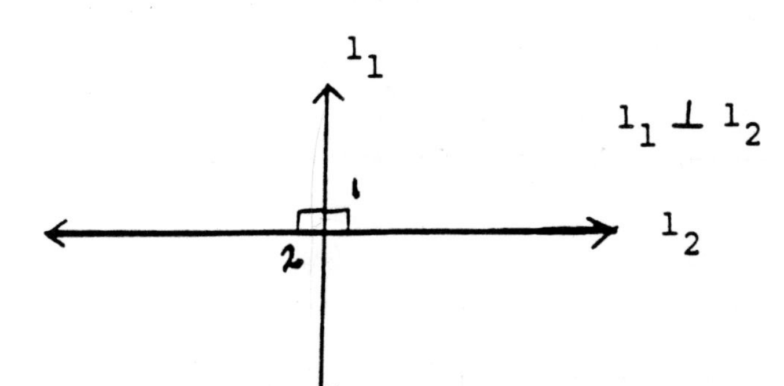

45. In the figure above, show that $\angle 1$ and $\angle 2$ are right angles.

1. Lines $l_1 \perp l_2$ ____________
2. $\angle 1$ is a right angle ____________
3. $\angle 1$ and $\angle 2$ are ____________
4. $\angle 1 = \angle 2$ because ____________

45.

1. Given
2. Perpendicular lines form right angles.
3. vertical angles by definition.
4. Vertical angles are equal.

46. If $\overleftrightarrow{EF}$ and $\overleftrightarrow{XY}$ are perpendicular and $\angle 3 = \angle 5$, find the measures of $\angle 1$, $\angle 2$, $\angle 3$, $\angle 4$, $\angle 5$, and $\angle 6$.

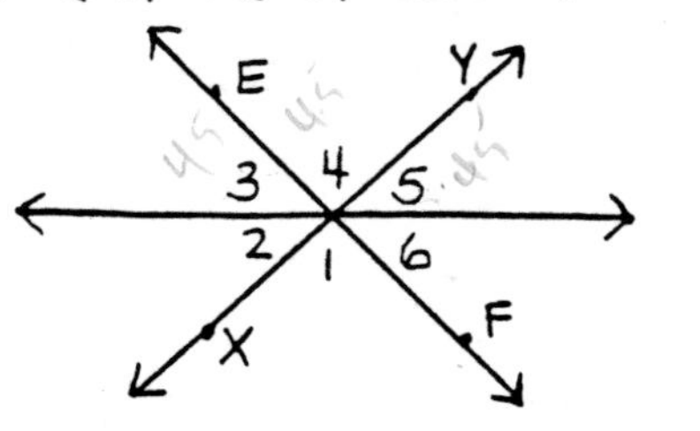

1. $\overleftrightarrow{EF} \perp \overleftrightarrow{XY}$ ____________
2. $\angle 4$ is 90° because ____________
3. $\angle 4$ and $\angle 1$ are ____________

   and $\angle 1 = 90^\circ$ because vertical angles are equal.
4. $\angle 3 = \angle 5$ ____________

46.

1. Given
2. perpendicular lines form right angles.
3. vertical angles by definition
4. Given

5. $\angle 3 + \angle 4 + \angle 5 = 180^\circ$ and

$x + 90^\circ + x = 180^\circ$ by substitution

$2x + 90 = 180$

$2x = 90$

$x =$ ______

6. $\angle 3$ and $\angle 6$ are ____________

and $\angle 6 =$ __________ because

vertical angles are equal.

7. $\angle 5$ and $\angle 2$ are ____________

and $\angle 2 =$ ________ because

____________________.

5. The angles all together form a straight line.

45°

6. vertical angles by definition. $\angle 6 = 45^\circ$

7. vertical angles by definition $\angle 2 = 45^\circ$

vertical angles are equal.

Definition:	A line that divides a line segment into two equal parts is called the <u>bisector</u> of a line segment.

Segment $\overline{EF}$ is bisected by line L in the figure below. Hence, $\overline{EM} = \overline{MF}$.

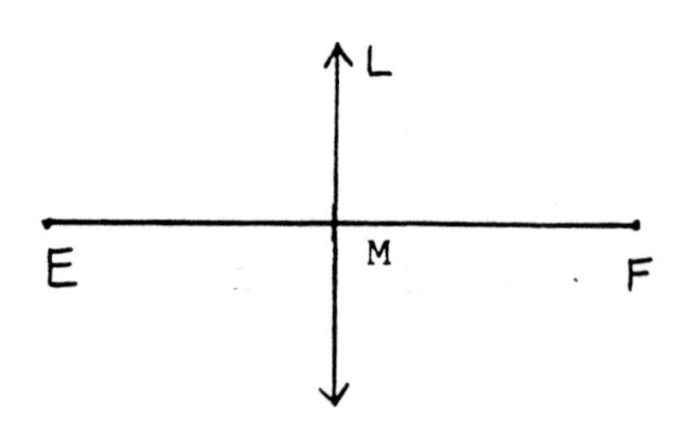

Constructions

With every geometric concept, there is a physical application of this concept. Constructions are only one method of demonstrating such applications. Since we are to make <u>no</u> measurements in a construction, we are limited to the use of two instruments... an unmarked ruler (<u>straight edge</u>) and <u>compass</u>. The straight edge will be used for drawing a line through two points and the compass will be used to draw circles and to make arcs.

> Definition: The <u>perpendicular</u> <u>bisector</u> of a line segment is a line which is perpendicular to that segment and divides that segment into two equal parts.

Construction #1: <u>To Find the Perpendicular Bisector of a Line Segment</u>

Procedure:

Step 1: Draw a line segment $\overline{MN}$.
Estimate the radius of your compass to be more than ½ the length of $\overline{MN}$.

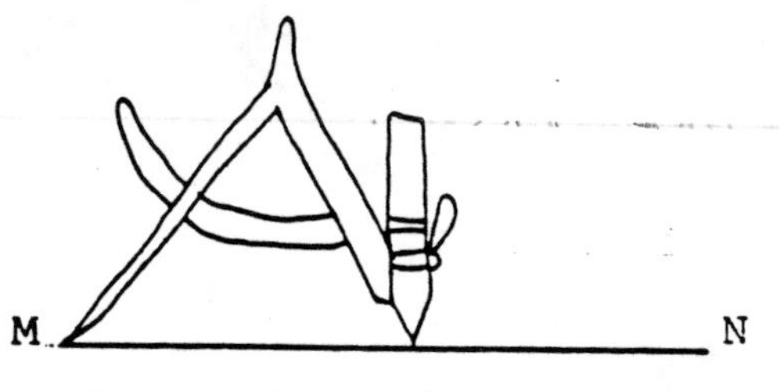

Step 2: With M as the center on your compass, draw a partial circle (arc) on the right side of the letter M.

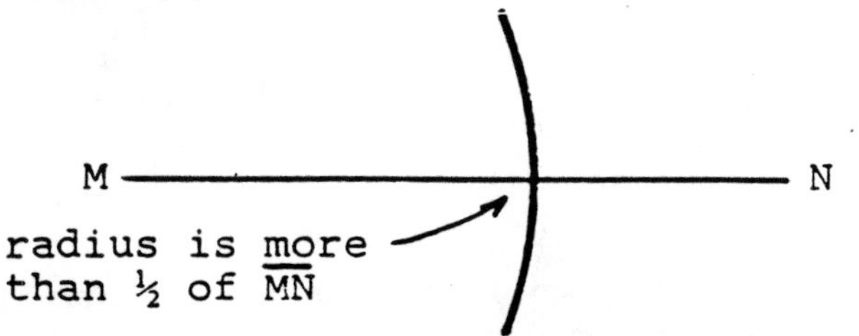

Step 3: With N as the center and using the same radius, draw a partial circle on the left side of letter N. (Note: If the arcs do not intersect the radius selected is less than half $\overline{MN}$ and needs to be increased.)

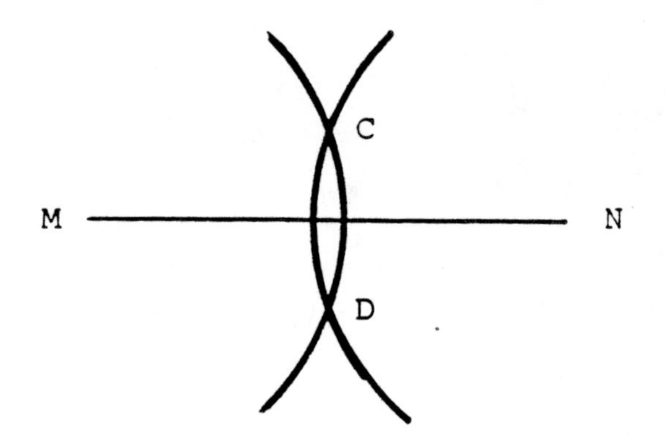

Step 4: With your straight edge, draw a line from point C to point D.

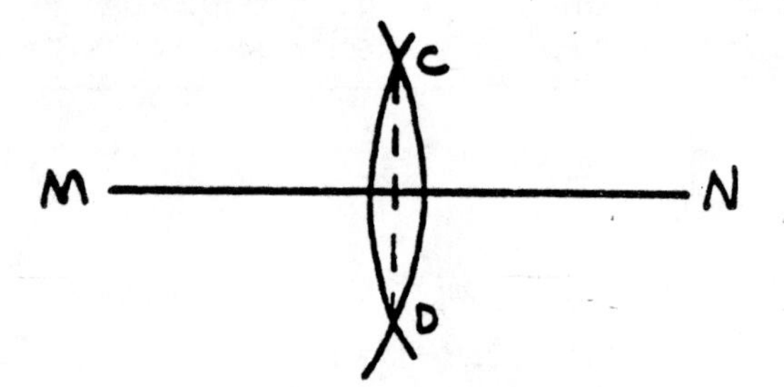

Step 5: $\overleftrightarrow{CD}$ intersects $\overline{MN}$ at a point we will call E. Point E is the midpoint of $\overline{MN}$ and line $\overleftrightarrow{CD}$ is the bisector of $\overline{MN}$.

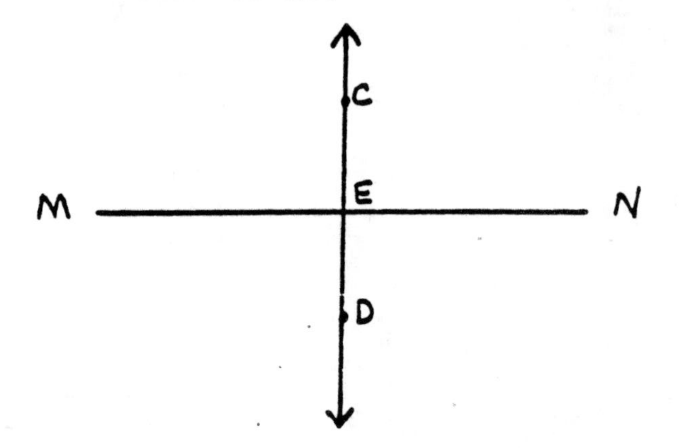

$\overleftrightarrow{CD}$ bisects $\overline{MN}$ and is perpendicular to $\overline{MN}$. $\overleftrightarrow{CD}$ is called the <u>perpendicular</u> <u>bisector</u> of the line segment.

47. Follow Steps 1-5 of Construction # 1. Find the midpoint and the perpendicular bisector of the segment $\overline{XY}$.

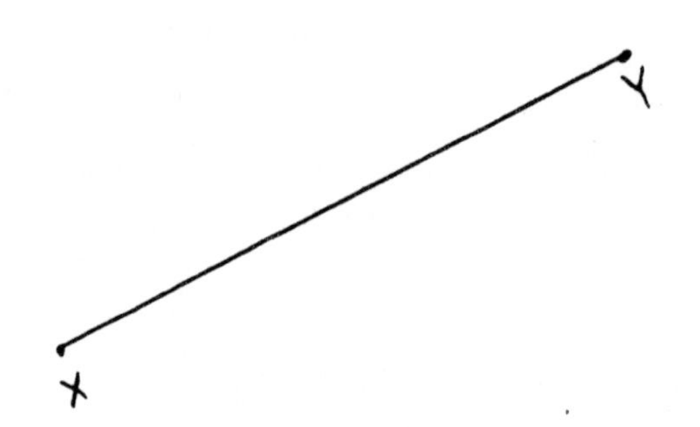

Definition:	The angle bisector is a ray that divides an angle into two smaller angles having equal measure.

Construction # 2: To Find the Angle Bisector

Procedure:

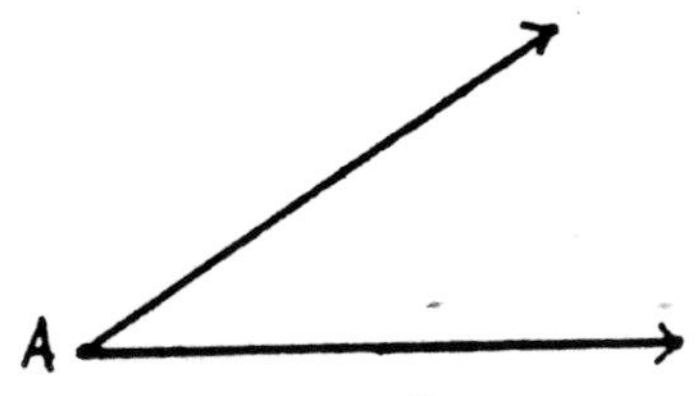

Step 1 With A as the center, draw a partial circle (arc) that crosses both arms of the angle and plot points B and C at the intersection.

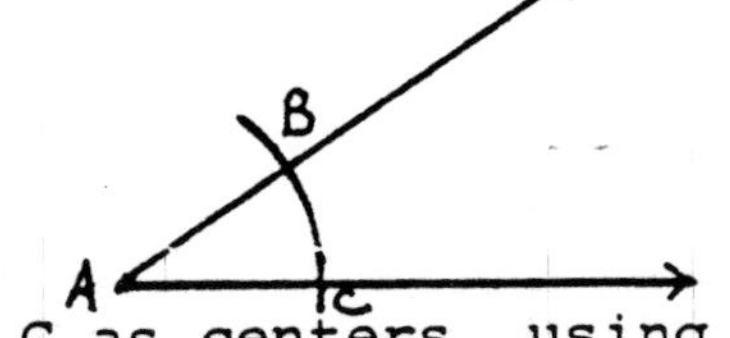

Step 2 With B and C as centers, using a radius greater than half the distance between B and C draw two arcs in the interior of the angle. Plot point D at the intersection.

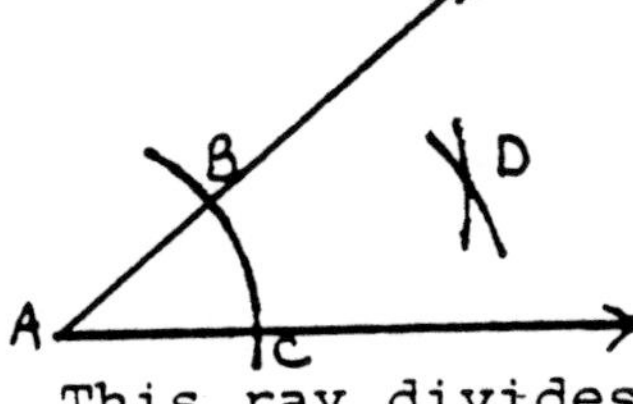

Step 3 Draw ray $\overrightarrow{AD}$. This ray divides $\angle A$ into two equal parts.

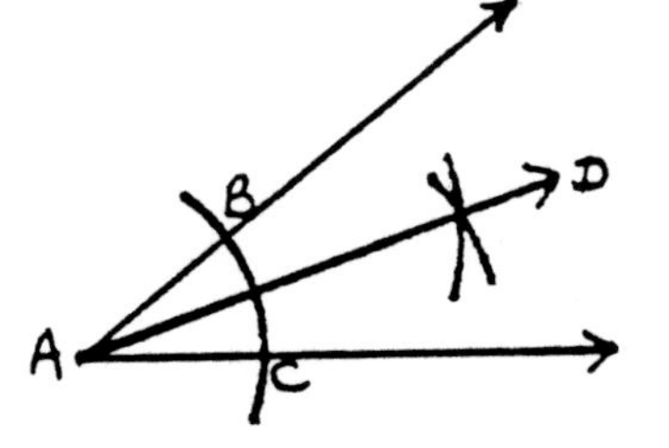

Using the above procedure, find the angle bisector of $\angle X$ below.

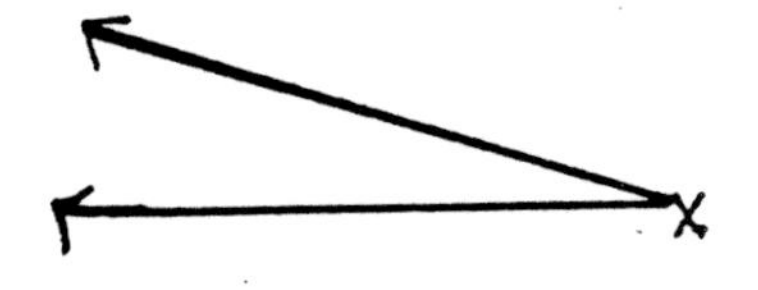

Construction # 3: <u>To Copy an Angle</u>

Step 1

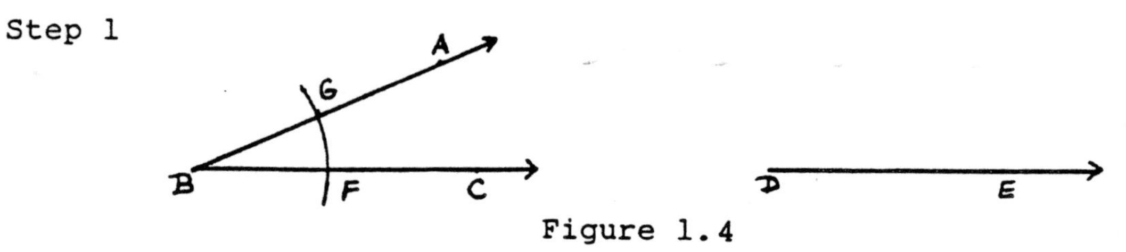

Figure 1.4

With B as center, draw an arc which intersects $\overrightarrow{BA}$ and $\overrightarrow{BC}$ at points G and F, respectively.

Step 2 With the same radius and with D as center, draw an arc which intersects $\overrightarrow{DE}$ (at point H) and extends above $\overrightarrow{DE}$ as shown below

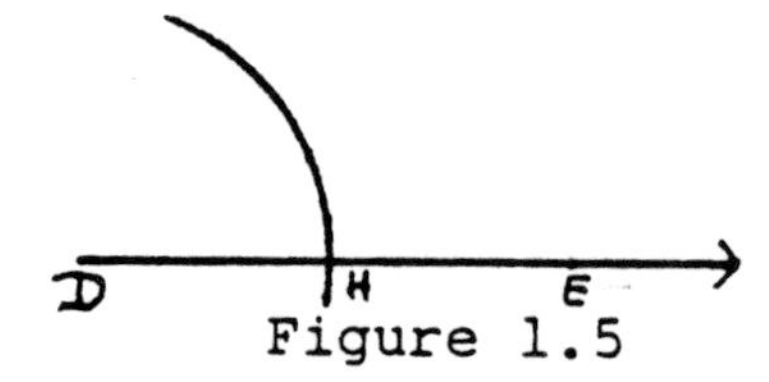

Figure 1.5

Step 3 In Figure 1.4, place one point of the compass at F and adjust the radius of the compass so the other point is at G. With this radius and using H as center in Figure 1.5, draw a second arc to intersect the previous arc to form point I as shown below.

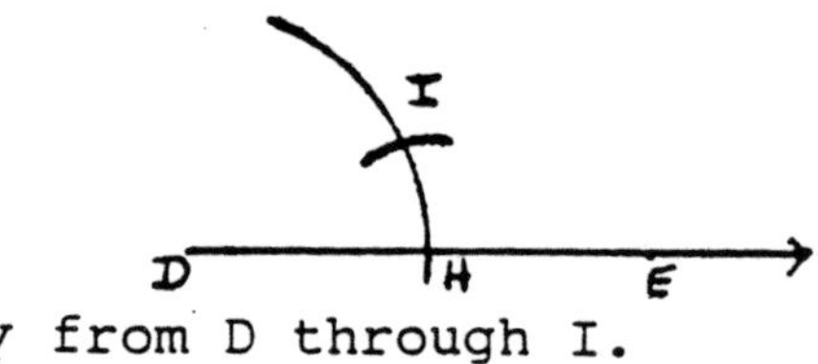

Step 4 Draw a ray from D through I.

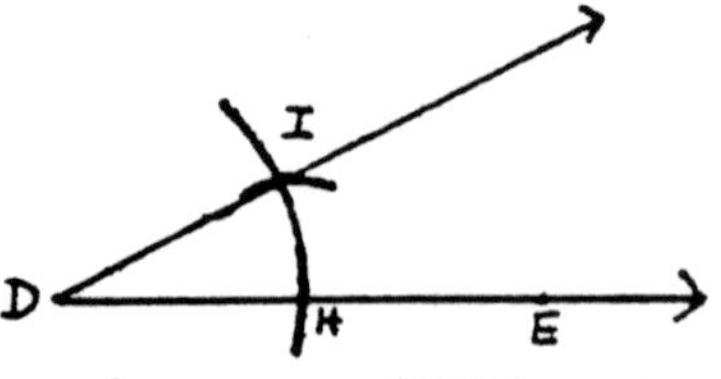

48. Follow Steps 1-4 and copy $\angle XYZ$ onto the segment $\overrightarrow{NO}$ by construction.

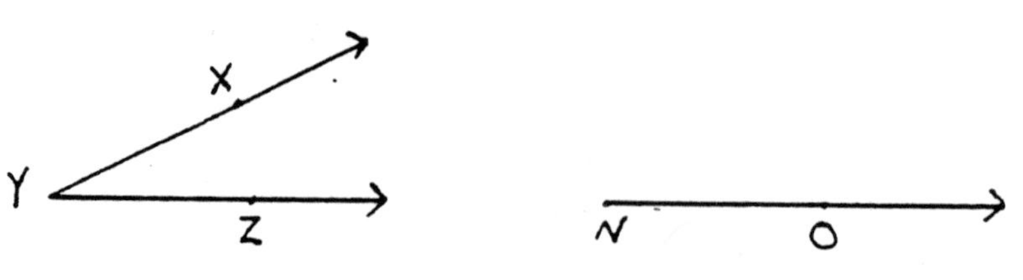

Exercise 3

1. Draw an acute angle and copy it onto segment $\overline{TR}$.

2. Find the perpendicular bisector of segments $\overline{MN}$ and $\overline{NO}$ by using only a straight edge and a compass.

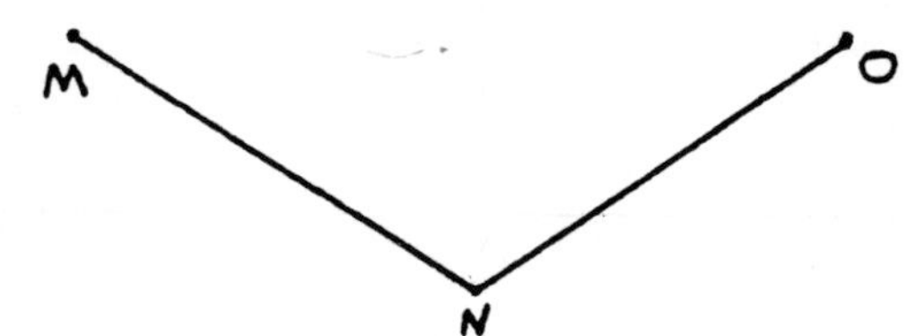

3. Using only a straight edge and compass, bisect $\angle$ABC.

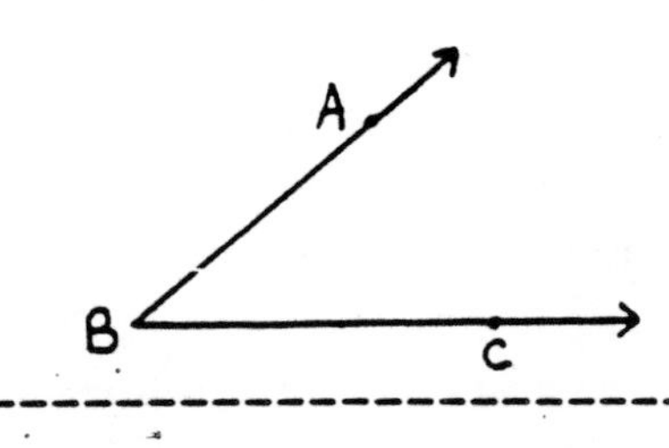

Unit 1 Review

1. a. Define the following:

postulate	deductive reasoning	straight angle
theorem	adjacent angles	midpoint
degrees	complementary angles	line segment
right angle	vertical angles	perpendicular bisector
acute angle	supplementary angles	bisector of a line segment
obtuse angle	angle bisector	vertex
angle	inductive reasoning	
ray	perpendicular lines	

   b. Describe the use of the following:

compass	straight edge	protractor

2. Fill in the blanks with the appropriate word or phrase to complete the definition.

   a. An __________ is formed by a pair of rays with a common endpoint called a ________________. The symbol for this figure is ________________.

b. If $\angle A + \angle B = 90^\circ$, the angles are said to be __________.

c. If $\angle A + \angle B = 180^\circ$, the angles are said to be __________.

3. The four elements of deductive reasoning are:

__________, __________, __________, __________

4. a. What is the degree of an angle whose measure is $\frac{2}{3}$ of its complement?

b. Two angles are supplementary. One angle is five times the measure of the other angle. Find the two angles.

c. Two angles are complementary. One angle is 20 more than four times as much as the other. Find the two angles.

d. Two angles are supplementary. One angle is twice the other. Find the angles.

5. Use only a straightedge and compass in the next two constructions.

a. Find the midpoint of $\overline{AB}$ by construction.

A B

b. Construct the bisector of the given angle.

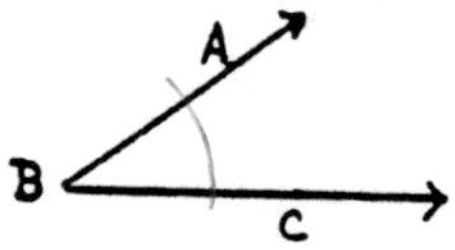

A B C

c. Copy the given angle.

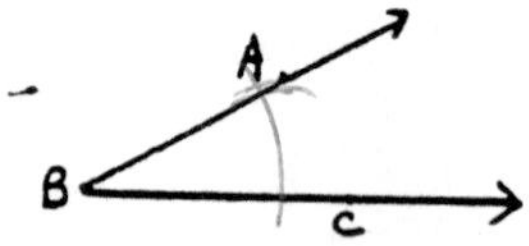

A B C

6. Using a protractor, find the measures of the given angles. Label each as straight, acute, obtuse, or right and write each of their degrees.

(i) (ii)

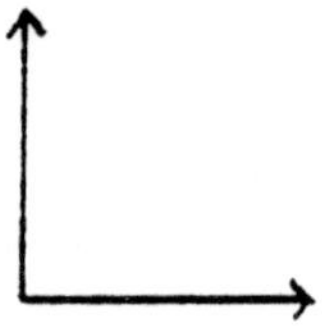

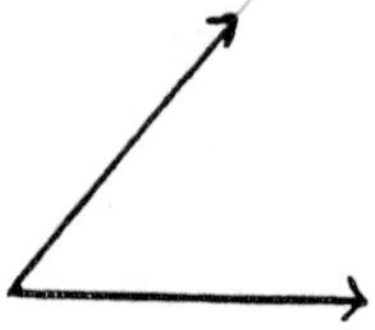

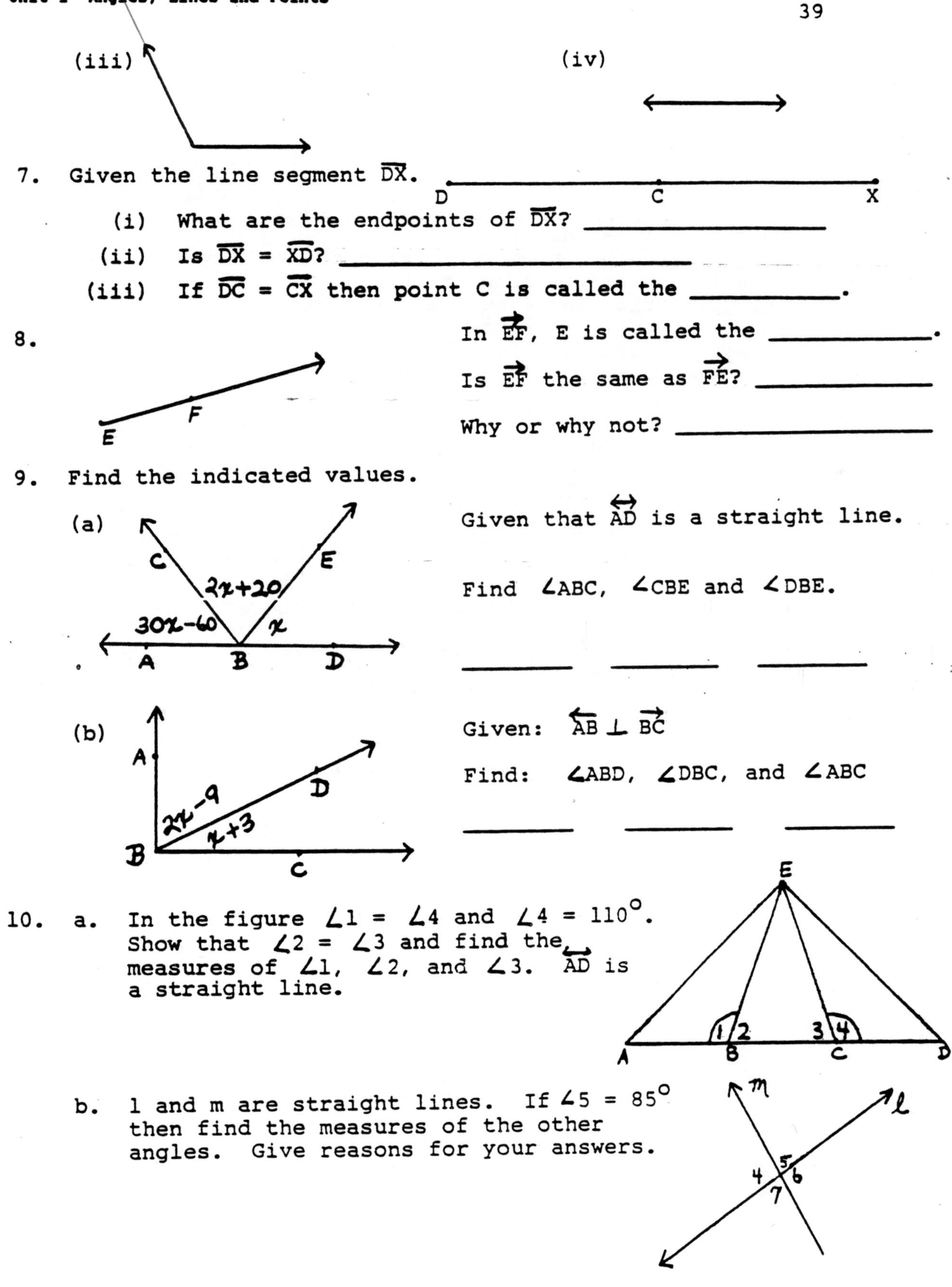

(iii)

(iv)

7. Given the line segment $\overline{DX}$.

(i) What are the endpoints of $\overline{DX}$? ____________

(ii) Is $\overline{DX} = \overline{XD}$? ____________

(iii) If $\overline{DC} = \overline{CX}$ then point C is called the ____________.

8. In $\overrightarrow{EF}$, E is called the ____________.

Is $\overrightarrow{EF}$ the same as $\overrightarrow{FE}$? ____________

Why or why not? ____________

9. Find the indicated values.

(a) Given that $\overleftrightarrow{AD}$ is a straight line.

Find $\angle ABC$, $\angle CBE$ and $\angle DBE$.

________ ________ ________

(b) Given: $\overleftarrow{AB} \perp \overrightarrow{BC}$

Find: $\angle ABD$, $\angle DBC$, and $\angle ABC$

________ ________ ________

10. a. In the figure $\angle 1 = \angle 4$ and $\angle 4 = 110^\circ$. Show that $\angle 2 = \angle 3$ and find the measures of $\angle 1$, $\angle 2$, and $\angle 3$. $\overleftrightarrow{AD}$ is a straight line.

b. l and m are straight lines. If $\angle 5 = 85^\circ$ then find the measures of the other angles. Give reasons for your answers.

Unit 2

Triangles and Congruence

Learning Objectives:

1. The student will demonstrate his mastery of the following definitions and postulates by writing them and by applying them to solutions of selected problems:
   <u>Definitions</u>: triangle, verticies, perimeter, isosceles triangle, equilateral triangle, scalene triangle, right triangle, obtuse triangle, acute triangle, equiangular triangle, hypotenuse, altitude, median, congruent triangles, and "≅".

   <u>Postulates</u>:
   (a) If three sides of one triangle are equal to three corresponding sides of a second triangle, the triangles are said to be congruent. (SSS = SSS)
   (b) If two sides and an included angle of one triangle are equal to two sides and an included angle of a second triangle, the triangles are congruent. (SAS = SAS)
   (c) If two angles and an included side of one triangle are equal to two angles and an included side of a second triangle, the triangles are congruent. (ASA = ASA)

2. The following constructions are required:

   (a) Construct the three medians of any triangle.

   (b) Construct the three altitudes of any triangle.

TRIANGLES AND CONGRUENCE

> **Definition:** A triangle is the union of three line segments determined by three points that are all not on the same line (noncollinear points). The symbol for a triangle is Δ .

A triangle is a figure that is shaped like one of the following. It is a closed three-sided figure.

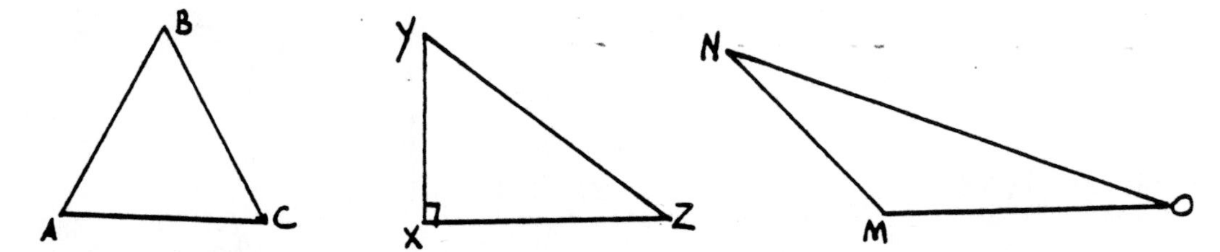

We name the first triangle by writing, Δ ABC. It may also be denoted as ΔBCA or ΔCAB. The order of the verticies does not matter. The three sides of ΔABC are $\overline{AB}$, $\overline{BC}$, and $\overline{AC}$. A triangle also has three angles, $\angle$A, $\angle$B, and $\angle$C.

1. What are the three sides of the second triangle, ΔXYZ? _____, _____, _____

2. What are the three angles in Δ XYZ? _____, _____, _____

In triangle, Δ ABC, the points A, B, and C are called verticies of the triangle. (Singular is called a vertex.)

F
7 cm.
4cm.
D
E
8cm.

3. a. Name the verticies of Δ DEF.
   b. Name the sides of Δ FED (same as Δ DEF).
   c. Name the angles of Δ EFD (same as Δ DEF).

---

1. $\overline{XY}$, $\overline{YZ}$, $\overline{XZ}$

2. $\angle$X, $\angle$Y, $\angle$Z

3. a. points D, E, F
   b. $\overline{DE}$, $\overline{EF}$, $\overline{DF}$
   c. $\angle$D, $\angle$E, $\angle$F

4.

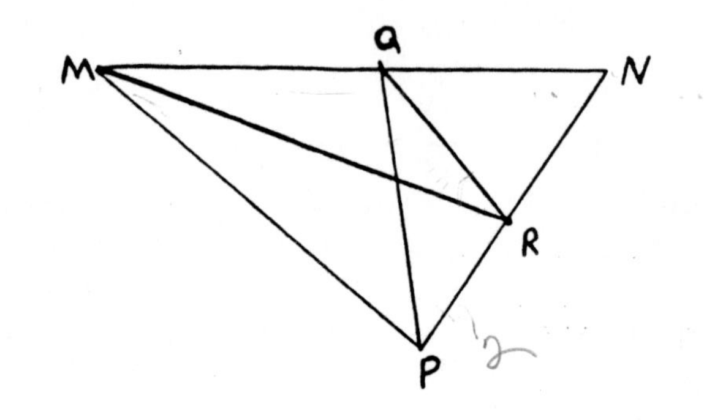

a. With your pencil, shade in ΔNQP.

b. Find angle NQP and number it with a 1.

c. With your pencil shade in ΔMPR.

d. Find angle MPR and number it with a 2.

e. Find angle MRQ and number it with a 3.

f. With your pencil shade in ΔMRQ.

4.

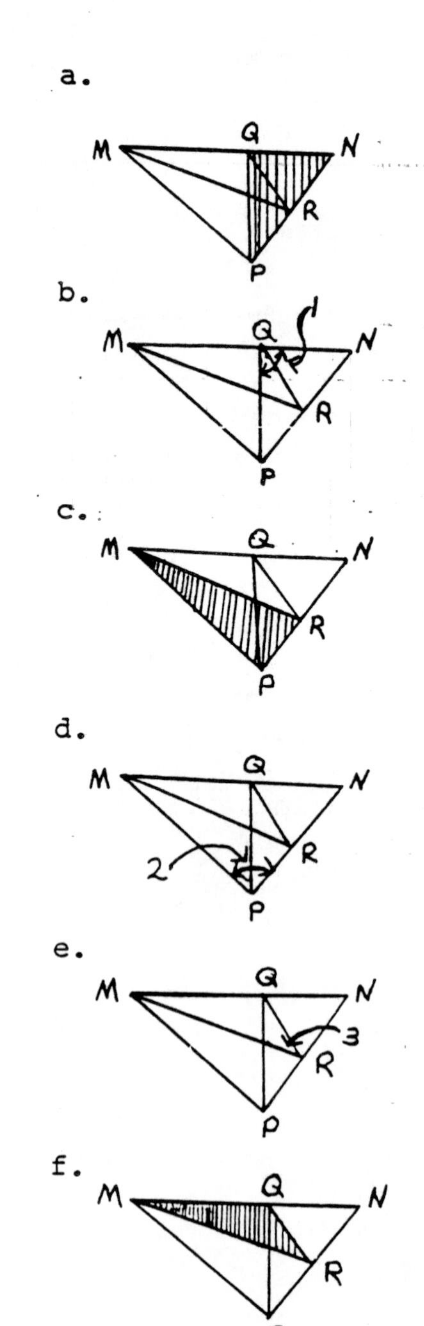

> Definition: The perimeter of a triangle is the sum of the measures of the sides of the triangle.

5. a. What is the perimeter of $\triangle$DEF on page 41 ?
   b. Find the perimeter of a triangle with sides $7\frac{1}{2}$ in., 3 in., and $2\frac{2}{3}$ in.

5. a. 19 cm.
   b. $13\frac{1}{6}$ in.

Exercise 1

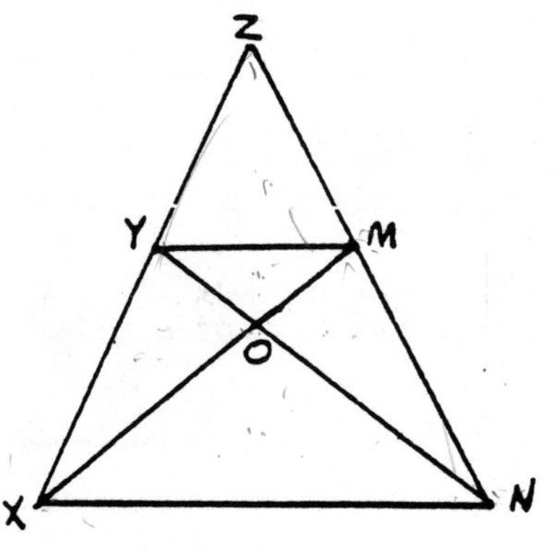

1. a. Name the verticies of Δ XZN.

   _____, _____, _____

   b. Name the twelve triangles that are in this figure. (triangles can overlap one another.)

   _____, _____, _____, _____, _____, _____,
   _____, _____, _____, _____, _____, _____.

2. Find the perimeter of this triangle.

   ____

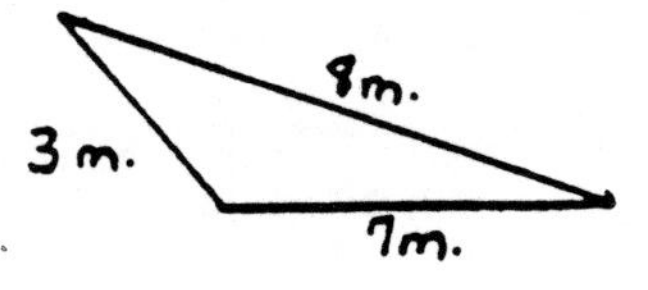

3. Is this figure a triangle?________

   Why or why not?________________

A D
B C

Exercise 3

CLASSIFICATION OF TRIANGLES

Triangles may be classified by two methods (I) sides or (II) angles.

I. The following definitions (a,b, and c) give the classification of triangles by lengths of the sides.

a.

An isosceles triangle is a triangle that has at least two sides equal.

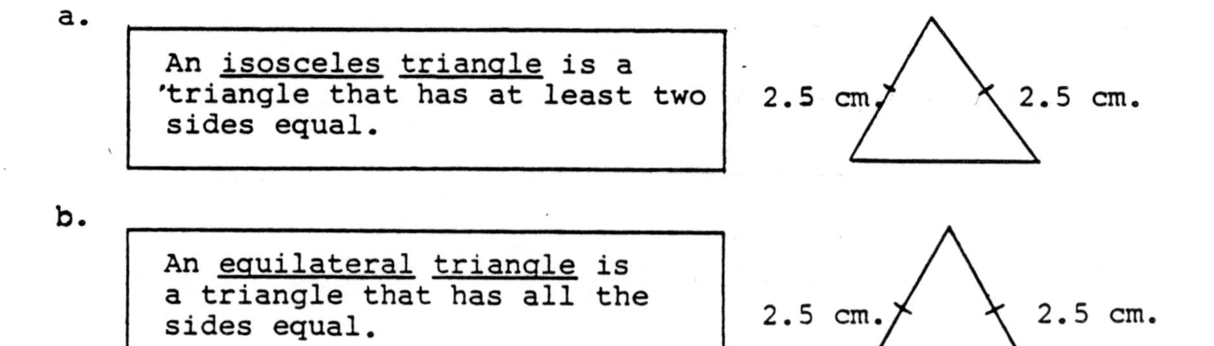

b.

An equilateral triangle is a triangle that has all the sides equal.

Can an equilateral triangle also be isosceles? Yes. An equilateral triangle has at least two sides equal.

Can an isosceles triangle also be equilateral? Yes. However not all isosceles triangles have all the sides of equal measure.

c.

A scalene triangle is a triangle that has no two sides of equal measure.

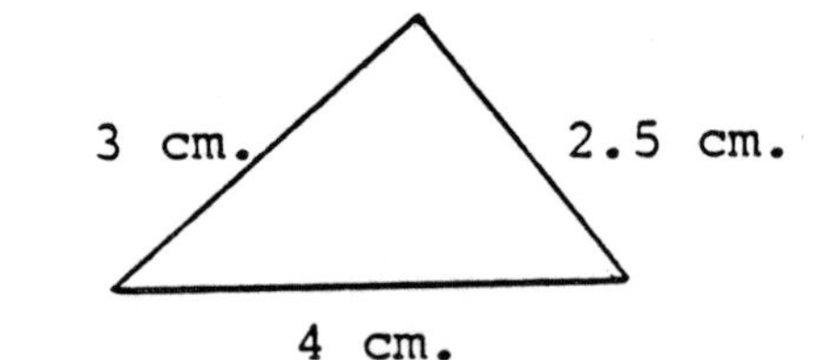

Using the above definitions, consider the following three triangles and classify each by its sides.

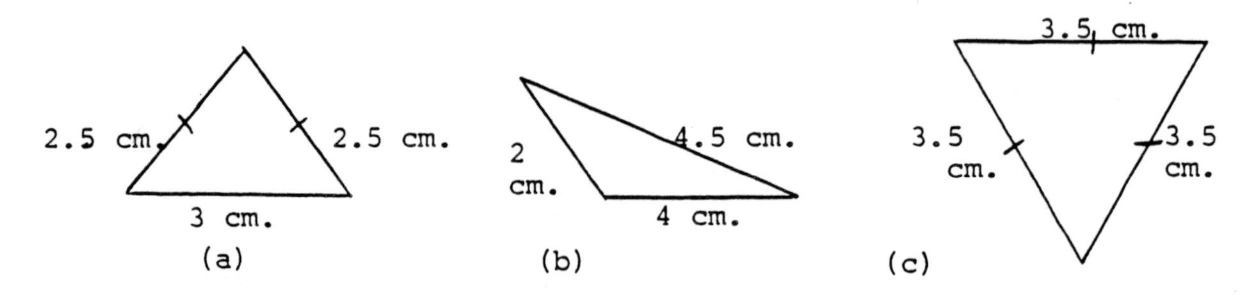

Figure (a) has exactly two sides the same length. This triangle is called isosceles. The equal sides are often called the legs of the triangle and the third side is called the base.

Figure (b) has no two sides equal. When all three sides of the triangle have different lengths, the triangle is called <u>scalene</u>.

Figure (c) has each side of length 3.5 cm. An <u>equilateral</u> triangle fits this description with all three sides the same measure.

Classify the triangles below according to the lengths of their sides.

6.

2 cm. 3.5 cm. 3.5 cm.

6. isosceles

7.

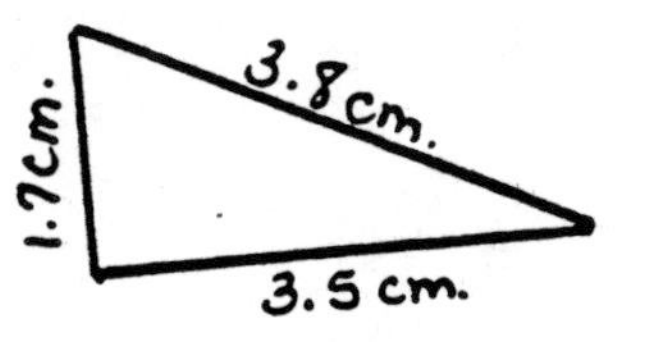

7. scalene

8.

8. equilateral

9. Classify each triangle by the length of its sides.

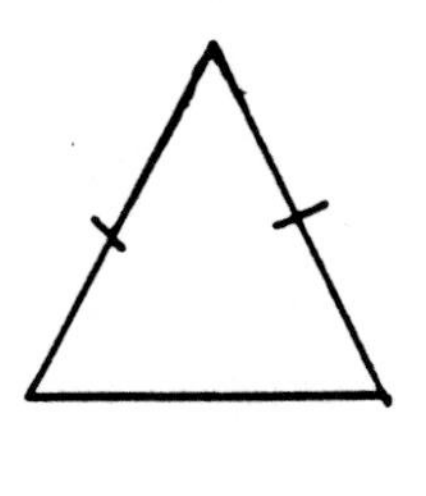

(a)

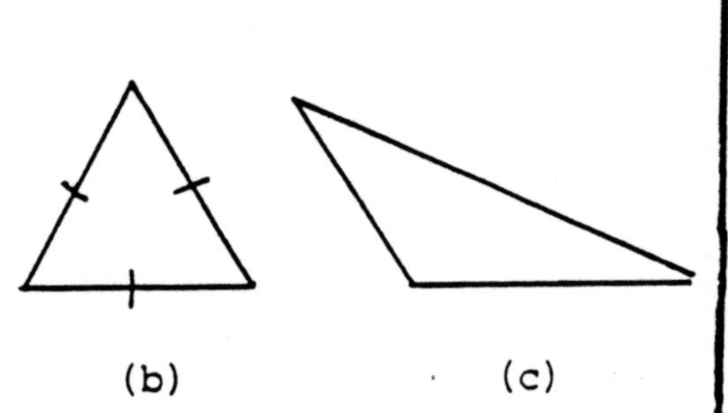

(b) (c)

9. (a) isosceles

(b) equilateral

(c) scalene

10. Can an equilateral triangle also be an isosceles triangle? ______________
Why? ______________

10. Yes, because an isosceles triangle has at least two equal sides.

11. Can a scalene triangle also be isosceles?
____
Why? ______________

11. No, because an isosceles triangle has at least two sides equal.

Classifying triangles by the measure of their angles is the second method of identifying types of triangles.

II. The following definitions give the classification of triangles by the measure of the angles.

(a)

A right triangle is a triangle that has one right angle.

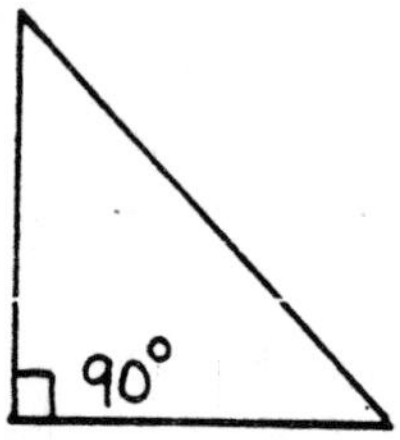

90°

(b)

An obtuse triangle is a triangle with one obtuse angle.

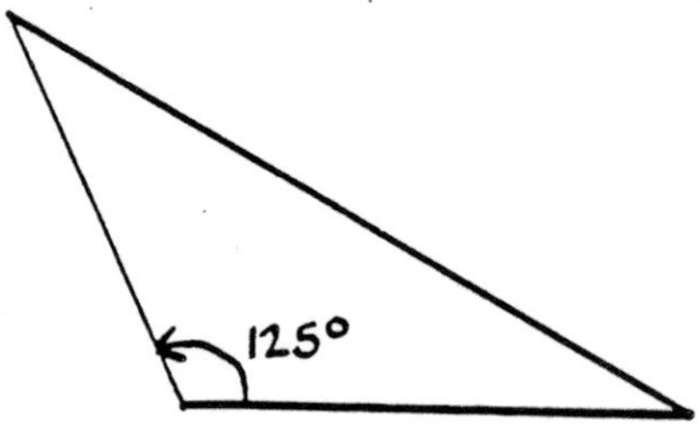

125°

(c)

An acute triangle is a triangle with all acute angles.

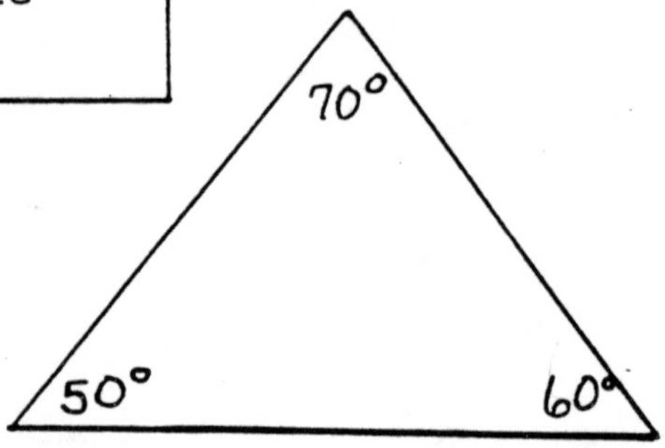

70°

50°

60°

Using the above definitions, consider the following three triangles and classify each by its angles.

Figure 2.1

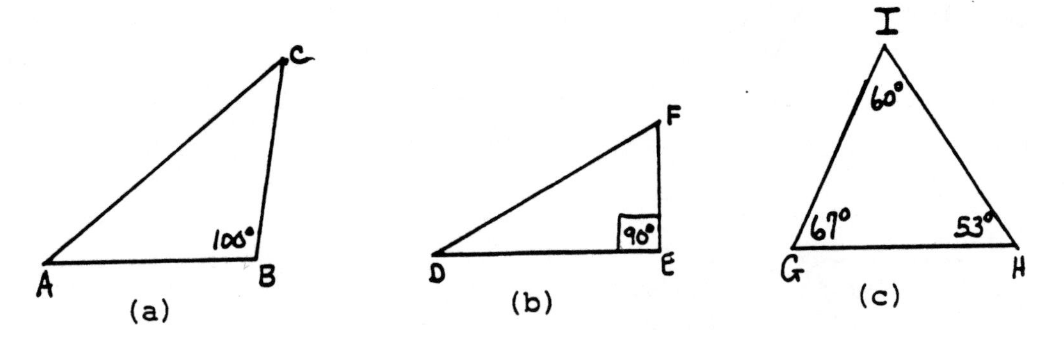

Δ ABC has one obtuse angle. This triangle is called an obtuse triangle.

Δ DEF has one right angle. Δ DEF is called a right triangle.

In Δ GHI, all three angles are acute angles. This triangle is called an acute triangle.

If an acute triangle has all the angles of equal measure, the triangle will be called an equiangular triangle.

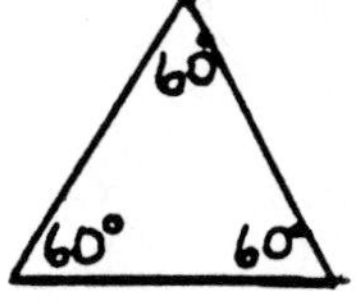

Classify the triangles below according to their angles or their sides.

12.

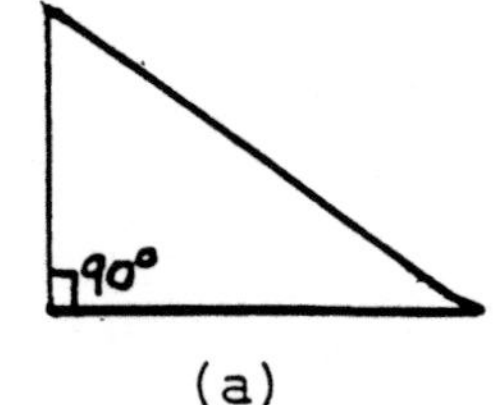

(a)

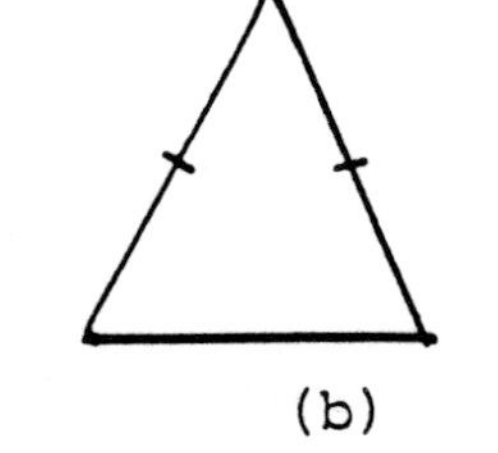

(b)

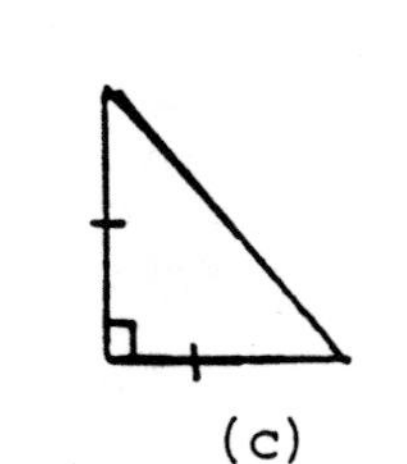

(c)

13.

(a)

60° 60° 60°

(b)

60° 60° 60°

(c)

12. (a) right triangle
(b) isosceles triangle
(c) isosceles right triangle

13. (a) equilateral triangle
(b) equiangular triangle
(c) equilateral and equiangular triangle

Acute

14.

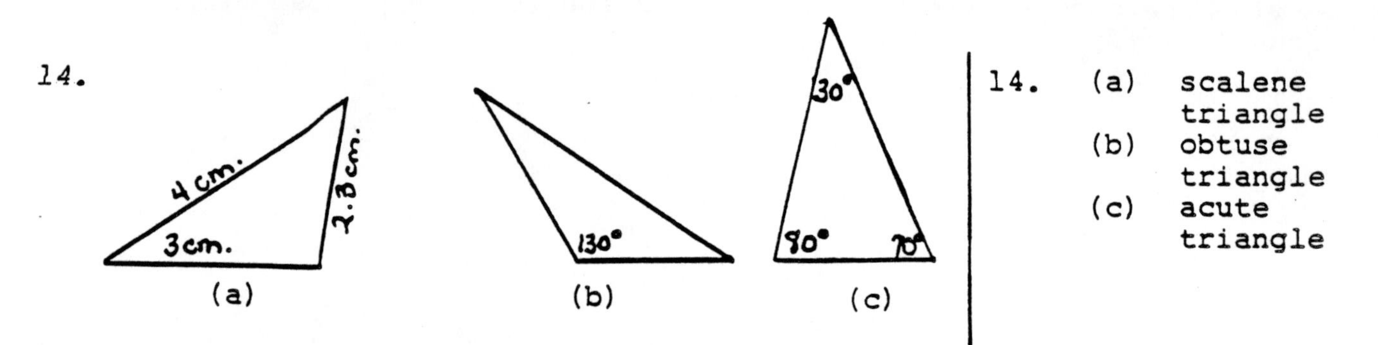

14. (a) scalene triangle
(b) obtuse triangle
(c) acute triangle

In a right triangle, sides $\overline{AB}$ and $\overline{BC}$ are called the legs of the triangle. However, the side opposite the right angle is the <u>hypotenuse</u> ($\overline{AC}$). The longest side of a right triangle is the hypotenuse.

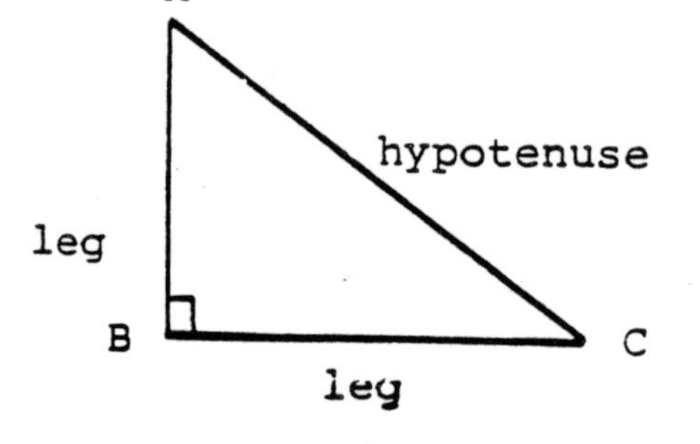

15. Find the hypotenuse in each of the following right triangles.

(a)

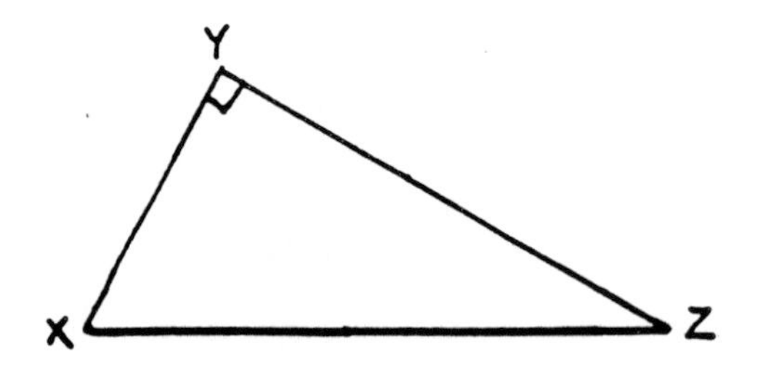

(b)

5
3
4

15. (a) $\overline{XZ}$ is opposite the right angle so XZ is the hypotenuse.

(b) The side of length 5 is the hypotenuse.

(c)

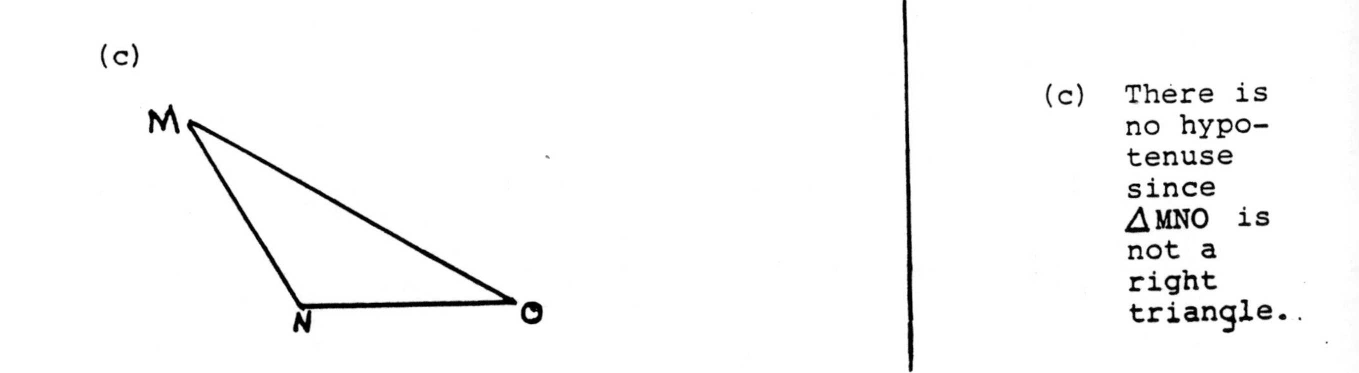

(c) There is no hypotenuse since ΔMNO is not a right triangle.

> Definition: An <u>altitude</u> of a triangle is a line segment that is drawn from the vertex perpendicular to the opposite side (or extension of that side).

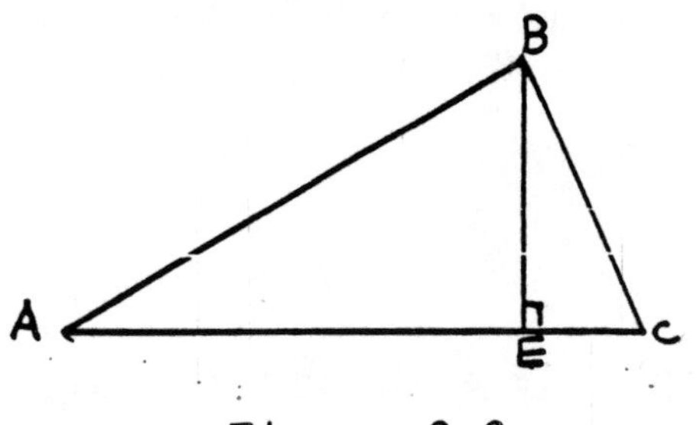

Figure 2.2

In Figure 2.2, $\overline{BE}$ is one altitude of ΔABC. It originates at vertex B and is perpendicular to side $\overline{AC}$. Please note that Δ ABC has three verticies. It also has three altitudes, one from each vertex.

In Figure 2.2,

16. (a) If you were to draw an <u>altitude</u> from vertex C, this altitude would be perpendicular to what side?_________

    (b) The altitude drawn from point A would be perpendicular to what side of the triangle? _________________________

16. (a) $\overline{AB}$

    (b) $\overline{BC}$

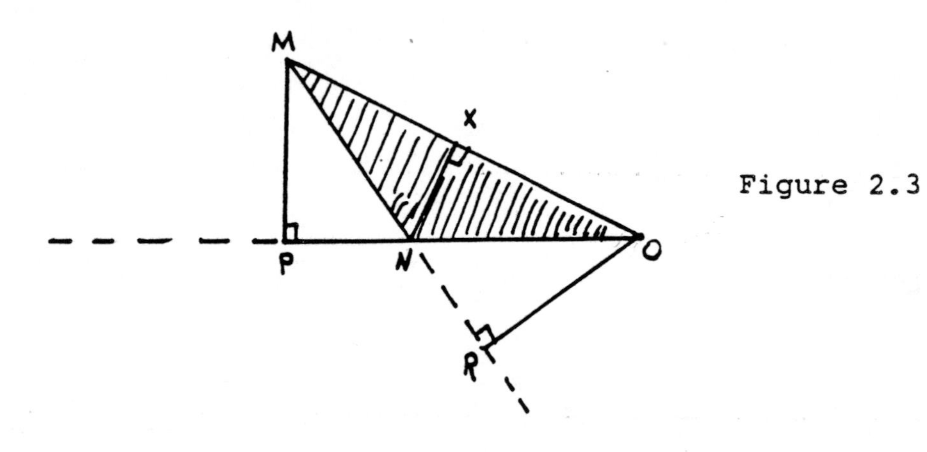

Figure 2.3

17. In Figure 2.3, the altitude dropped from point M is perpendicular to the extension of side $\overline{NO}$. What are the other altitudes found in $\triangle$ MNO?

17. $\overline{NX}$ from vertex N

$\overline{OR}$ from vertex O

Construction #1

To construct an altitude of a triangle follow the steps illustrated below:

Step 1: To construct an altitude from point M, extend side $\overline{NO}$.

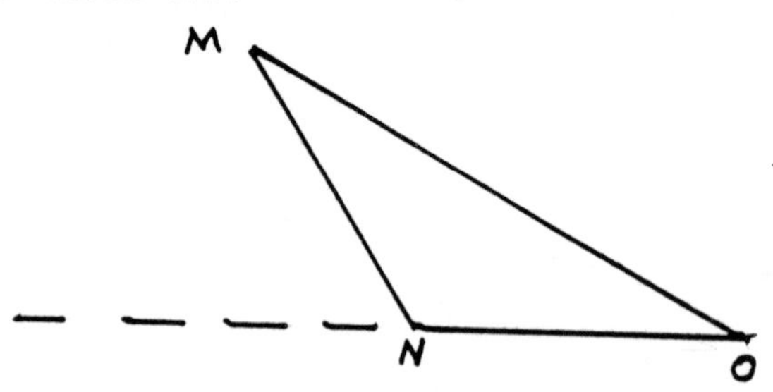

Step 2: Using point M as the center select a radius wide enough to intersect $\overline{NO}$ at two points, A and B.

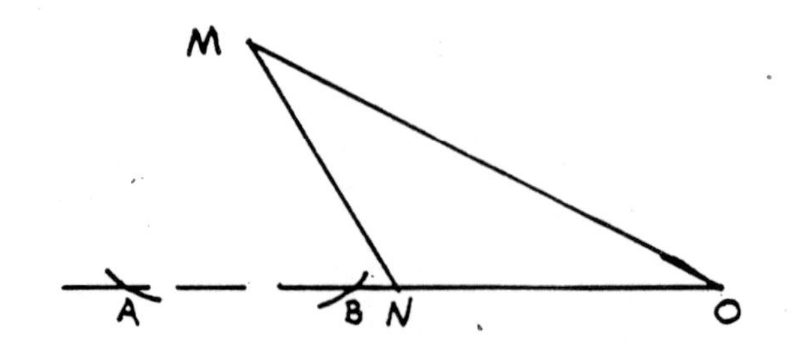

Step 3: Using points A and B as centers, construct point C below $\overline{NO}$ and draw a line from point M to C. $\overline{MP}$ is called the altitude.

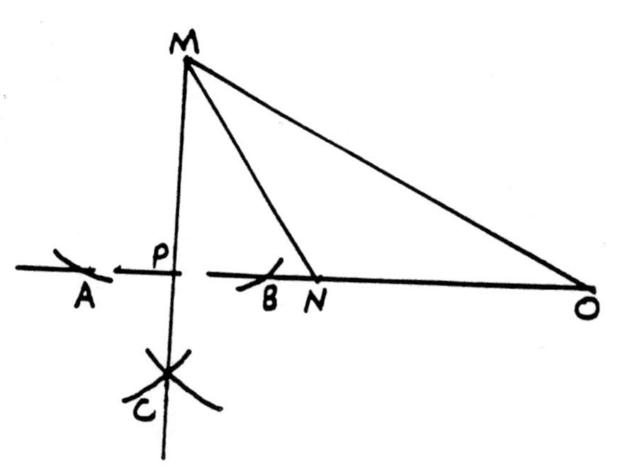

18. Now turn triangle $\triangle$MNO so that $\overline{MO}$ is the base. Using steps 1, 2, and 3, construct the altitude from point N to $\overline{MO}$. (You will not have to extend the base line for this vertex.) $\overline{NQ}$ is the altitude.

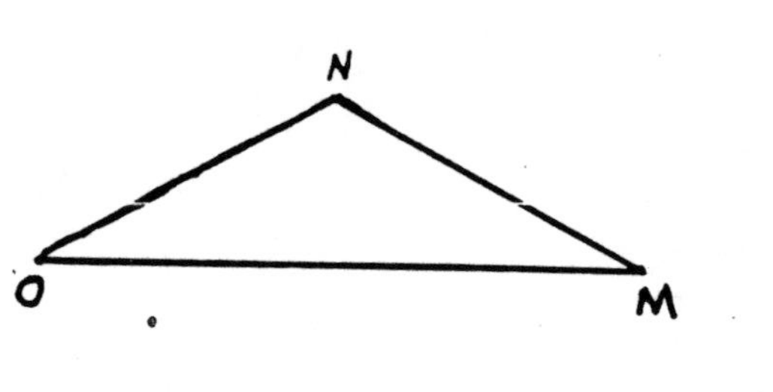

18.

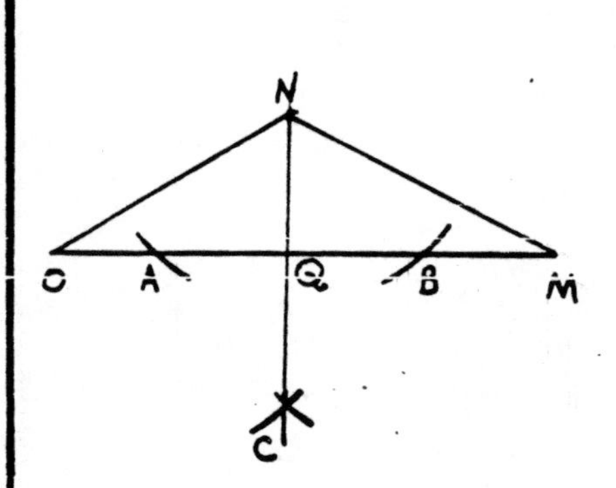

19. Finally, turn the $\triangle$ MNO again with side $\overline{MN}$ as the base. Using steps 1, 2, and 3, construct the altitude from point O to $\overline{MN}$. $\overline{OR}$ is the altitude.

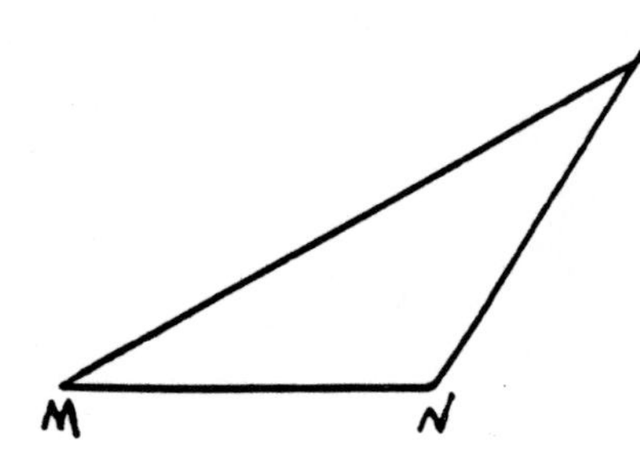

19.

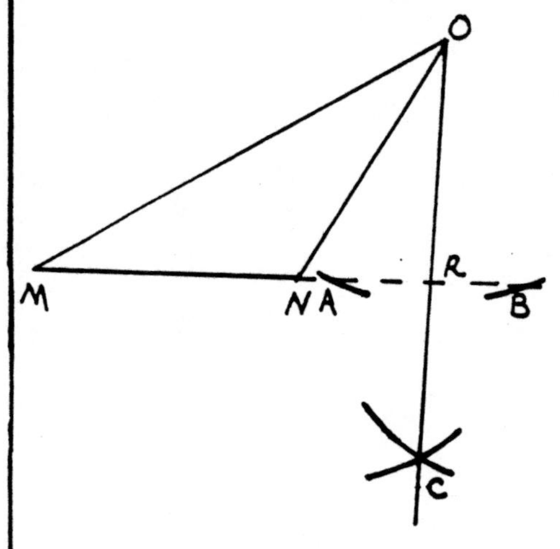

> Definition: A median of a triangle is a line segment that is drawn from the vertex to the midpoint of the opposite side.

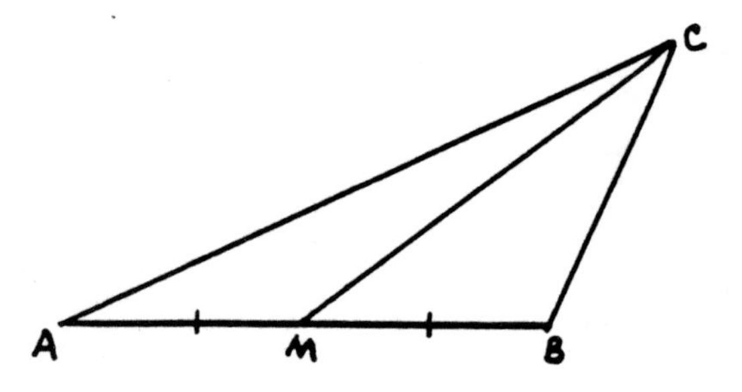

M is the midpoint of $\overline{AB}$ and $\overline{CM}$ is called a median of $\triangle ABC$.

A triangle has three altitudes and it has three medians.

20. (a) An altitude drawn from point C in $\triangle$ ABC is ______________ to side $\overline{AB}$.
    (b) A median drawn from point C in $\triangle$ ABC is drawn to the __________ of side $\overline{AB}$.

20. (a) perpendicular
    (b) midpoint

Construction # 2

To construct a median from point A to segment $\overline{BC}$ follow steps 1 and 2 as illustrated.

Step 1: Construct the perpendicular bisector of $\overline{CB}$ to find the midpoint of $\overline{CB}$. (See Unit 1, pages 33 and 34 .)

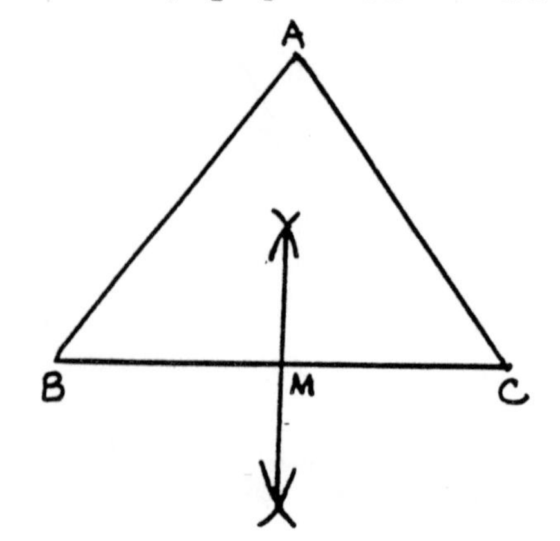

Step 2: Draw the line segment from the midpoint M to point A. $\overline{AM}$ is called the median of ΔABC.

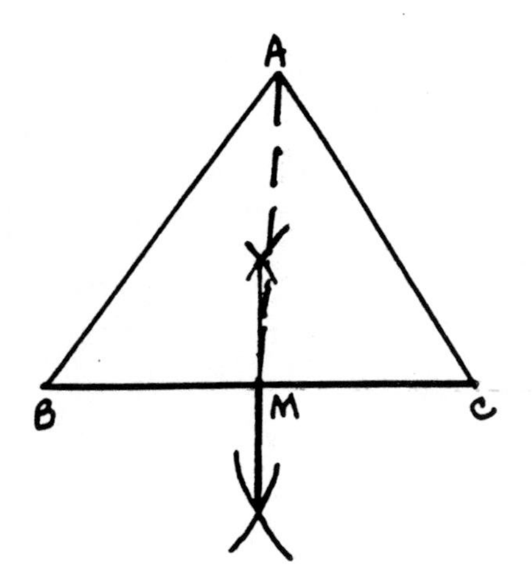

21. Using the same triangle as in figure 2.3 on page 50, construct the three medians of that triangle and check your answer on pages 53-54. Copies of triangle MNO are provided for your convenience.

(a) Construct the median from vertex M.

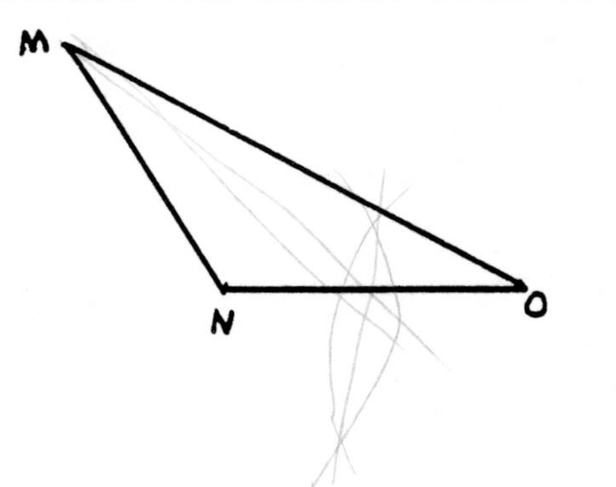

21.

(a)

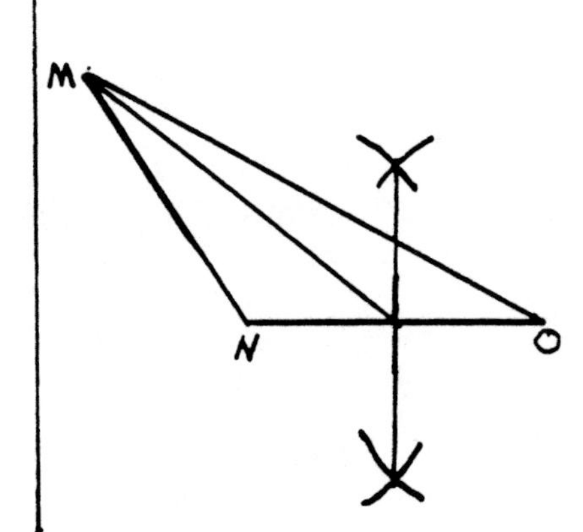

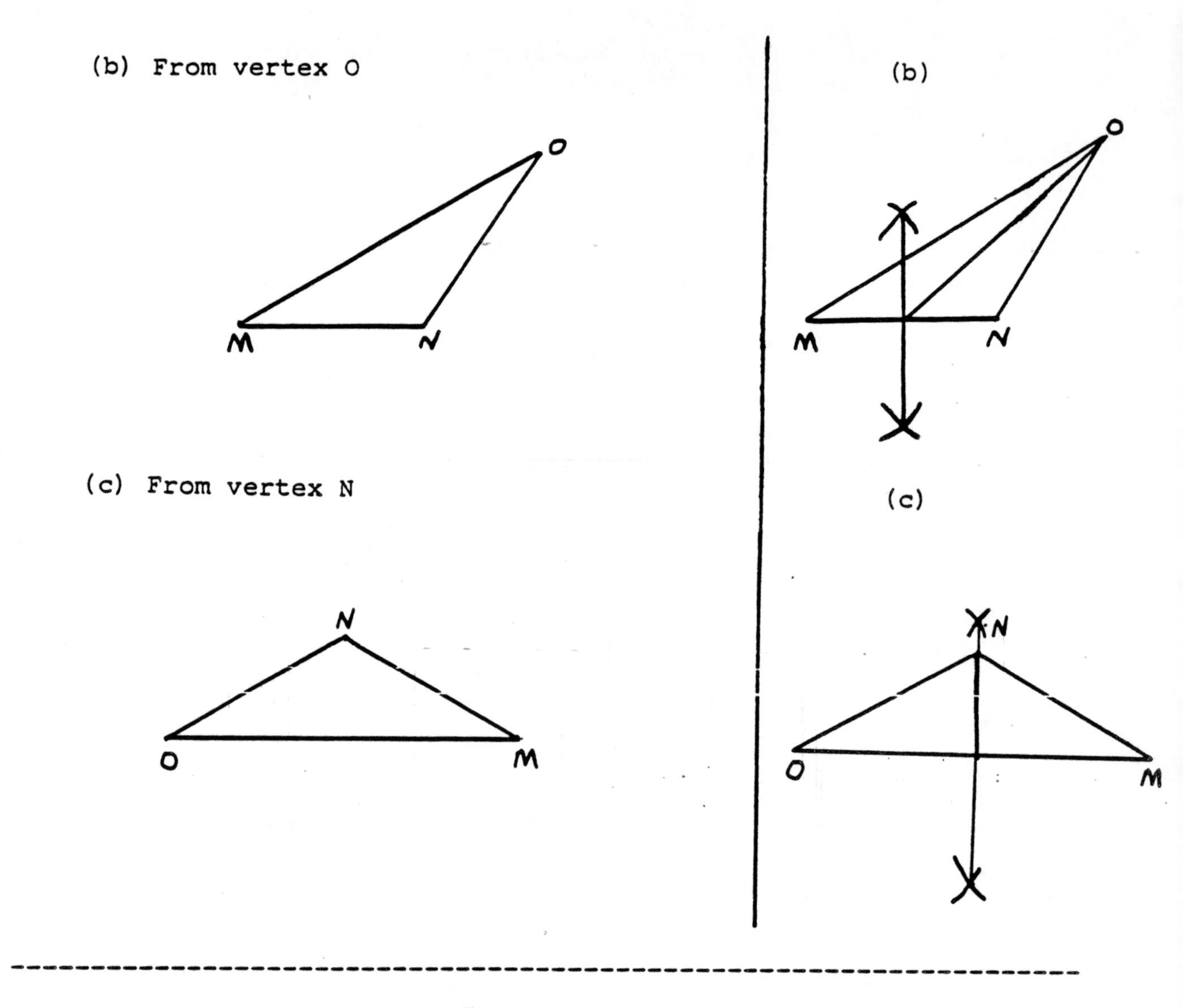

<u>Exercise 2</u>

1. Draw a triangle and construct the medians from each of its verticies.

2. Draw an obtuse triangle and construct the altitudes from each of its verticies.

3. Find the perimeter of an equilateral triangle with a side of 15 inches.

4. The perimeter of an isosceles triangle is 32 inches and its base is 8 inches. Find the length of one of its equal sides.

5. The base of an isosceles triangle is half the length of one of the equal sides. The perimeter is $41\frac{1}{2}$ inches. Find the length of the base.

6. a. Can a triangle have more than one 90° angle?______________
   b. More than one obtuse angle?______________
   c. More than one acute angle?______________

<u>Overlapping Triangles</u>

Geometric figures may overlap. It is important that you are able to see more than one figure in a drawing that overlaps such as the picture below. There are two ladies in this picture. Which do you see, the old lady or the young woman? Concentrate on trying to see each (one at a time). In order to recognize both, you must perceive a portion of the drawing in different relationships. Each figure is illustrated separately on the next page.

The young lady turned looking back over her right shoulder

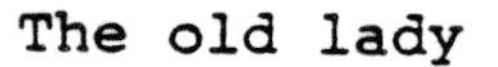

The old lady

The ability to see these relationships is important to geometry. Without this ability you will be unable to see the relationship of figures or the relationship of how parts in deductive reasoning fit together in a geometric proof.

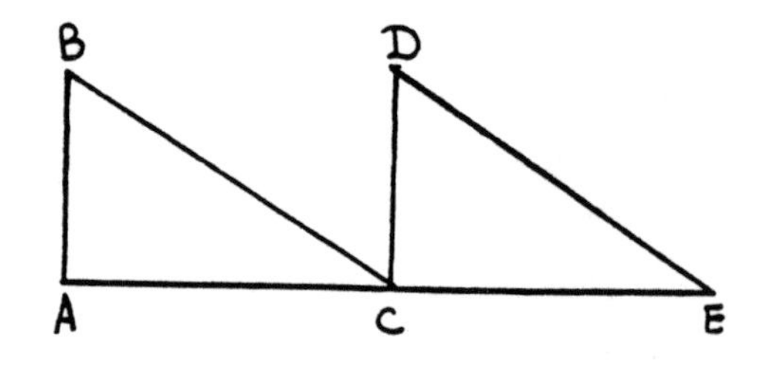

$\triangle$ABC and $\triangle$CDE are not overlapping.

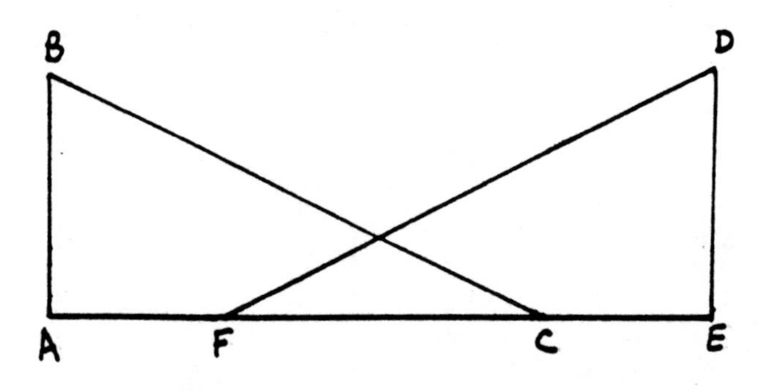

ΔABC and ΔEDF <u>are</u> overlapping at segment $\overline{FC}$.

Study each of the following figures and indicate which two triangles are overlapping.

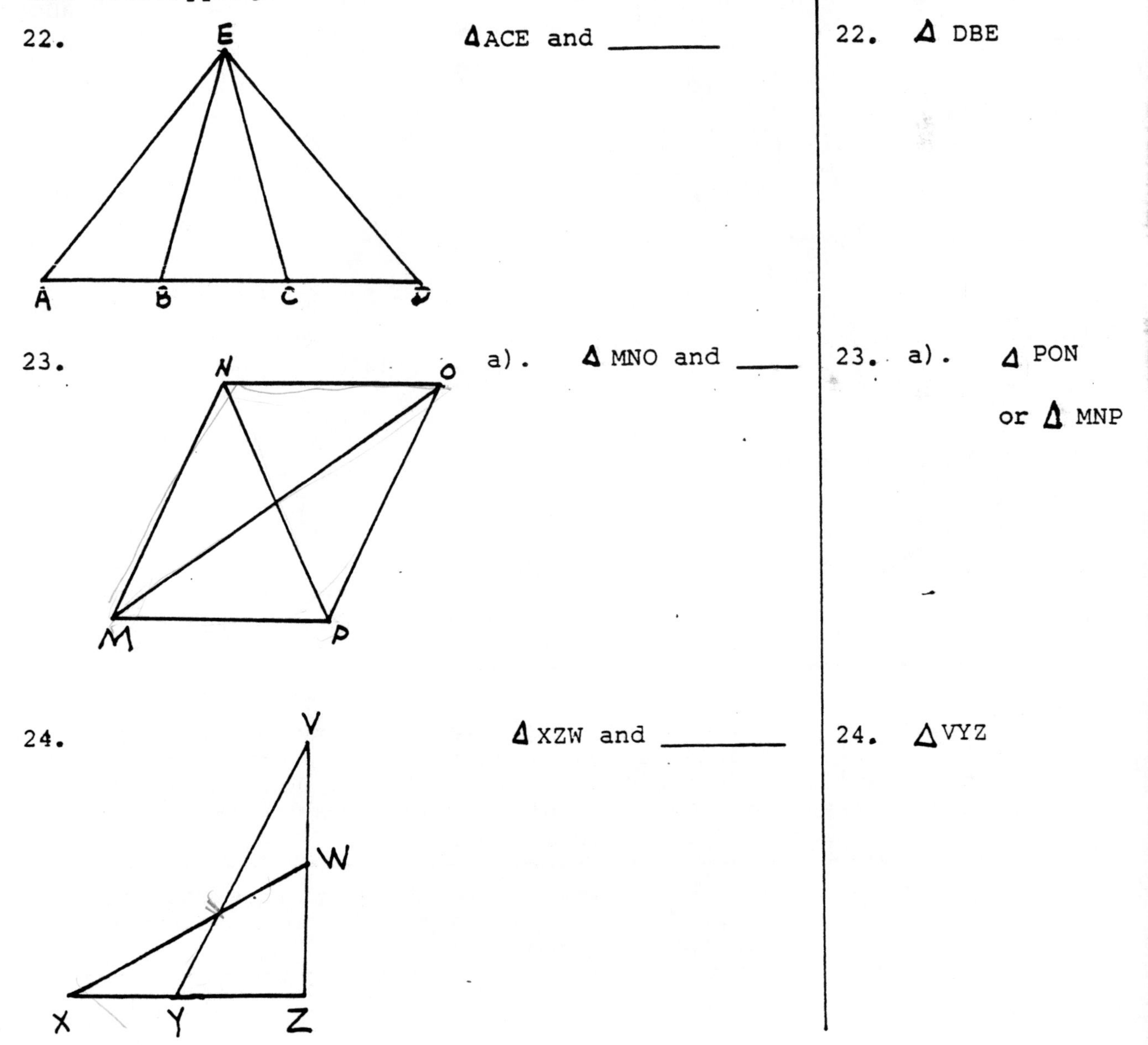

22. ΔACE and ________

23. a). Δ MNO and ____

24. Δ XZW and ________

22. Δ DBE

23. a). Δ PON or Δ MNP

24. ΔVYZ

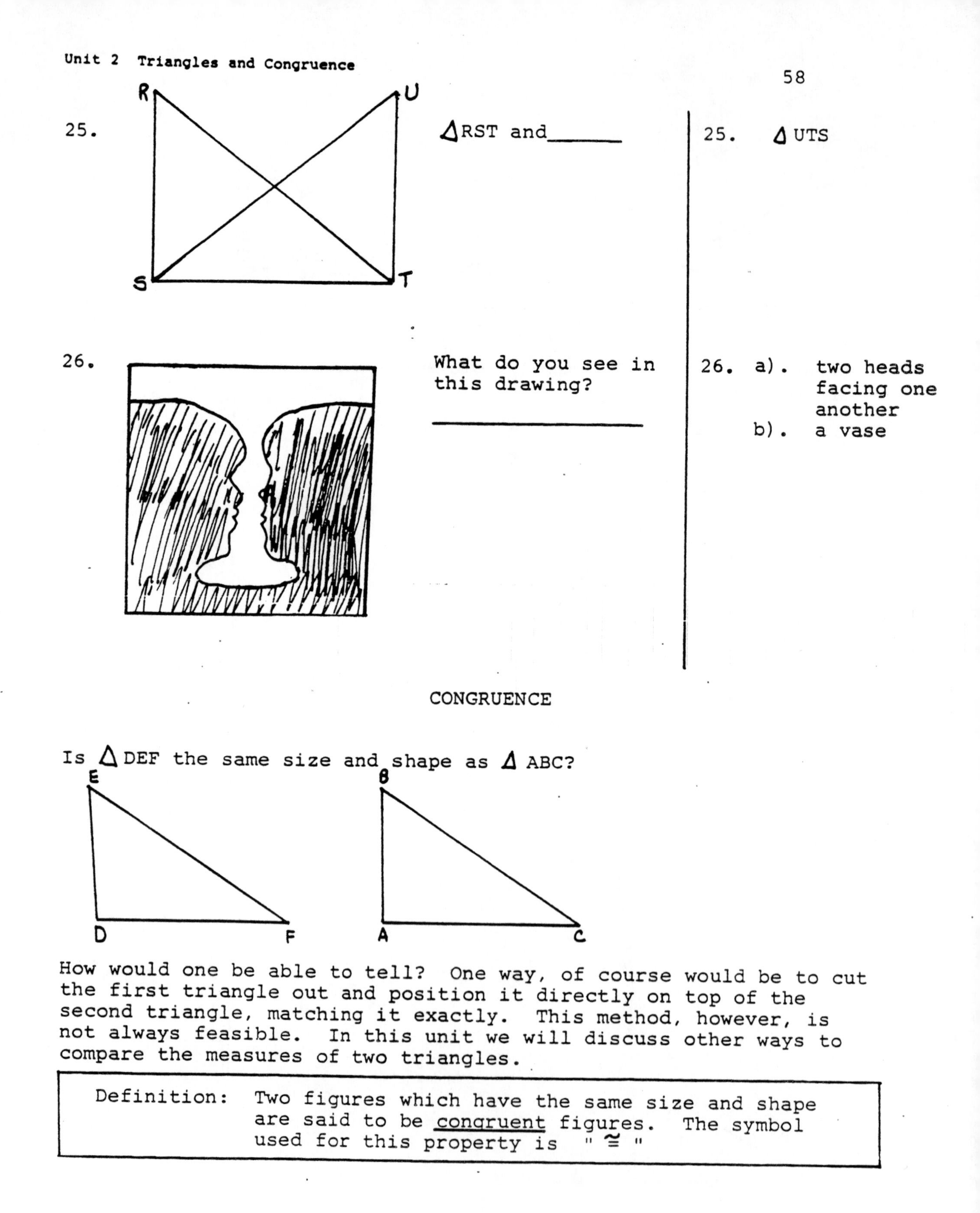

25. ΔRST and______

26. What do you see in this drawing?

____

25. Δ UTS

26. a). two heads facing one another
b). a vase

CONGRUENCE

Is Δ DEF the same size and shape as Δ ABC?

How would one be able to tell? One way, of course would be to cut the first triangle out and position it directly on top of the second triangle, matching it exactly. This method, however, is not always feasible. In this unit we will discuss other ways to compare the measures of two triangles.

Definition:	Two figures which have the same size and shape are said to be <u>congruent</u> figures. The symbol used for this property is " $\cong$ "

The question arises, why is the concept of congruence studied? In the physical world, there are many examples of congruent figures. A machine that seals cans in a mass production line must use circular tops that are all congruent. A paper press that cuts and prints notebook paper will make each sheet exactly the same. All 9 x 10 glass window panes are produced in the same shape in order that each may fit a standard window.

In considering industry and its output of appliances, furnishings, paper and machine supplies, food production, transport vehicles and all other processes, it is no wonder why "congruence" is considered the single most outstanding geometric principle permeating every facet of each of our daily lives.

This chapter will limit its discussions of congruence to comparing two triangles. Below are illustrated $\triangle RST$ and $\triangle ABC$.

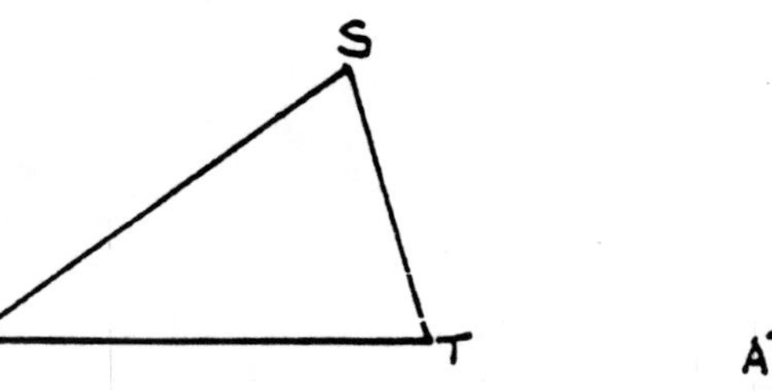

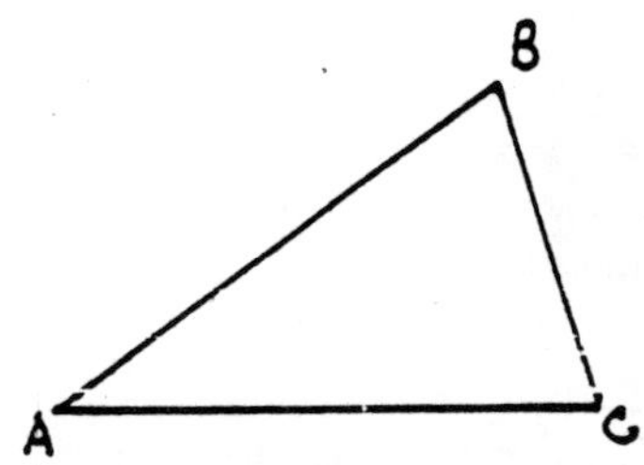

$\angle A$ and $\angle R$ are said to be corresponding parts of these triangles. Why?

Answer:

They are in the same relative positions with regard to the sides of the compared triangles. Likewise $\angle S$ corresponds to $\angle B$ and $\angle T$ corresponds to $\angle C$. The same relationship exists between the sides. $\overline{RS}$ corresponds to $\overline{AB}$, $\overline{ST}$ corresponds to $\overline{BC}$ and $\overline{RT}$ to $\overline{AC}$.

27. In $\triangle PQR$ and $\triangle XYZ$, give the corresponding parts.

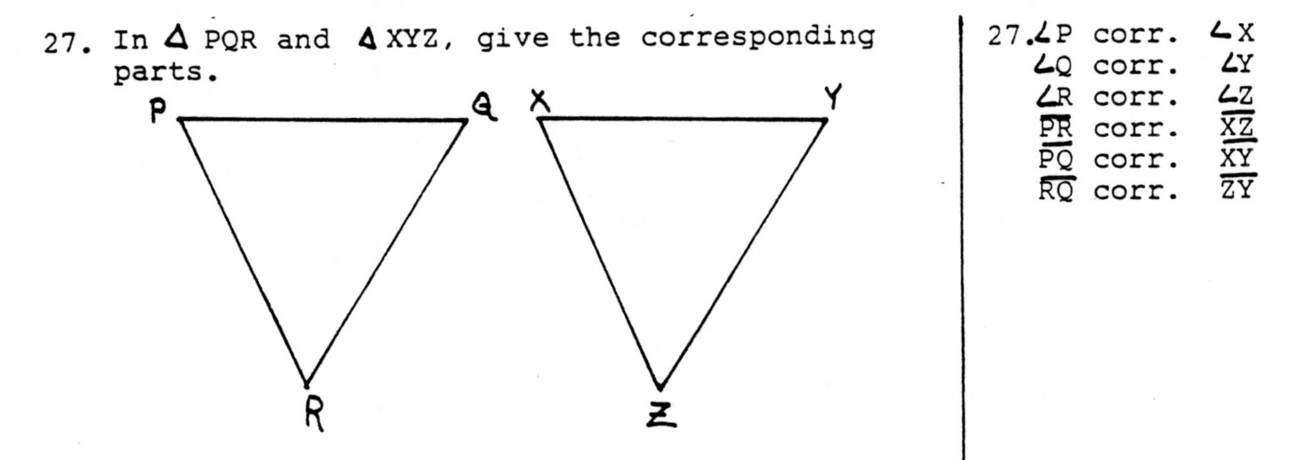

27. $\angle P$ corr. $\angle X$
$\angle Q$ corr. $\angle Y$
$\angle R$ corr. $\angle Z$
$\overline{PR}$ corr. $\overline{XZ}$
$\overline{PQ}$ corr. $\overline{XY}$
$\overline{RQ}$ corr. $\overline{ZY}$

Definition: Two triangles are <u>congruent</u> if and only if all corresponding parts are of equal measure.

If $\Delta LMN \cong \Delta PAC$,

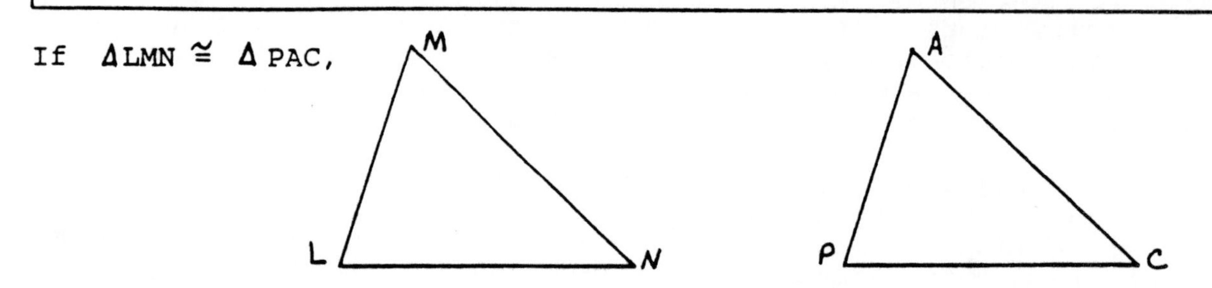

then the corresponding angles are equal ($\angle L = \angle P$, $\angle M = \angle A$, $\angle N = \angle C$) and the corresponding sides are equal. ($\overline{LM} = \overline{PA}$, $\overline{LN} = \overline{PC}$, $\overline{MN} = \overline{AC}$). Note the position of the letters as listed in $\Delta LMN \cong \Delta PAC$. The triangles are named in the order of their corresponding verticies. For example, $\angle L$ corresponds to $\angle P$ so if letter L is first in the listing of one triangle's name then P must be first in the listing of the second triangles's name. Next, if M is listed then A would correspond and so on. The congruence can be indicated as $\Delta MNL \cong \Delta ACP$ or $\Delta NLM \cong \Delta CPA$ as well as $\Delta LMN \cong \Delta PAC$.

28. In each of the following, record the congruence symbolically for the given triangle.

a.

b.

28.

a. $\Delta ABC \cong \Delta XYZ$
or $\Delta BCA \cong \Delta YZX$
or $\Delta CAB \cong \Delta ZXY$

b. $\Delta PQR \cong \Delta XYZ$
or $\Delta QRP \cong \Delta YZX$
or $\Delta RPQ \cong \Delta ZXY$

Exercise 3

1. If $\triangle MNO \cong \triangle PNO$, complete the following statements.

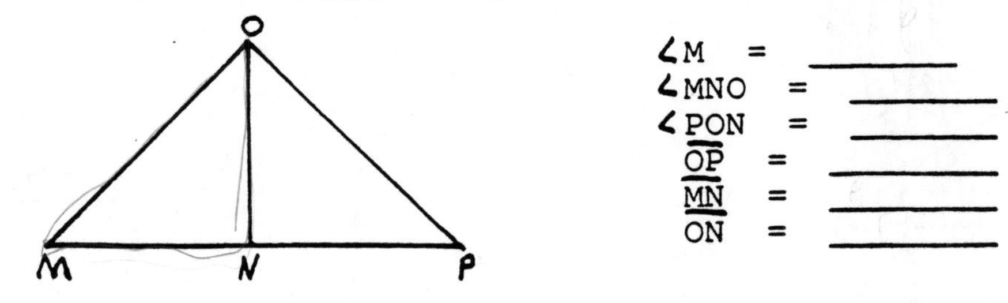

$\angle M$ = ________
$\angle MNO$ = ________
$\angle PON$ = ________
$\overline{OP}$ = ________
$\overline{MN}$ = ________
ON = ________

2. List the six pairs of corresponding congruent parts for the following congruences. Use a sketch if necessary.
   a. $\triangle PRQ \cong \triangle ABC$

   b. $\triangle LMN \cong \triangle CUT$

3. Write the congruence from studying each of the six pairs of corresponding congruent parts.

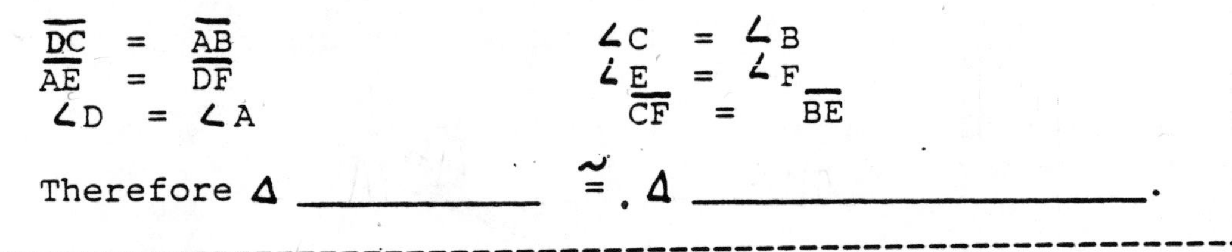

$\overline{DC} = \overline{AB}$ | $\angle C = \angle B$
$\overline{AE} = \overline{DF}$ | $\angle E = \angle F$
$\angle D = \angle A$ | $\overline{CF} = \overline{BE}$

Therefore $\triangle$ ____________ $\cong$ $\triangle$ ____________.

Markings can be used to indicate corresponding parts of congruent triangles. For example, if $\triangle ABC \cong \triangle WHN$,

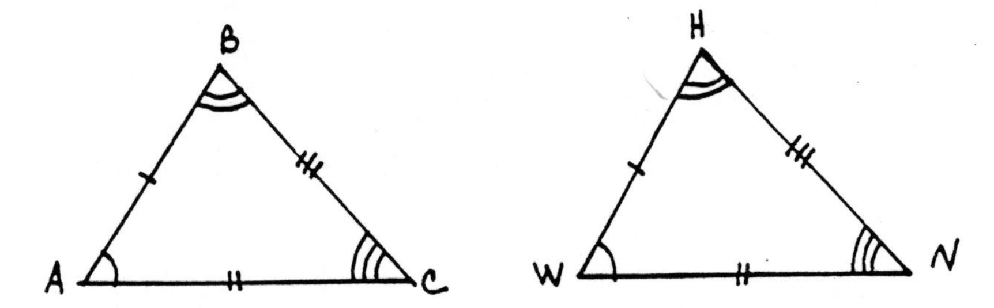

then $\angle A = \angle W$. To indicate this a single mark could be used; $\angle B = \angle H$, a double mark; and, $\angle C = \angle N$, a triple mark. This same process applies to the corresponding sides.

29. If $\triangle DEF \cong \triangle GHI$, mark the triangles to indicate the equal parts.

29. One possible marking may be.

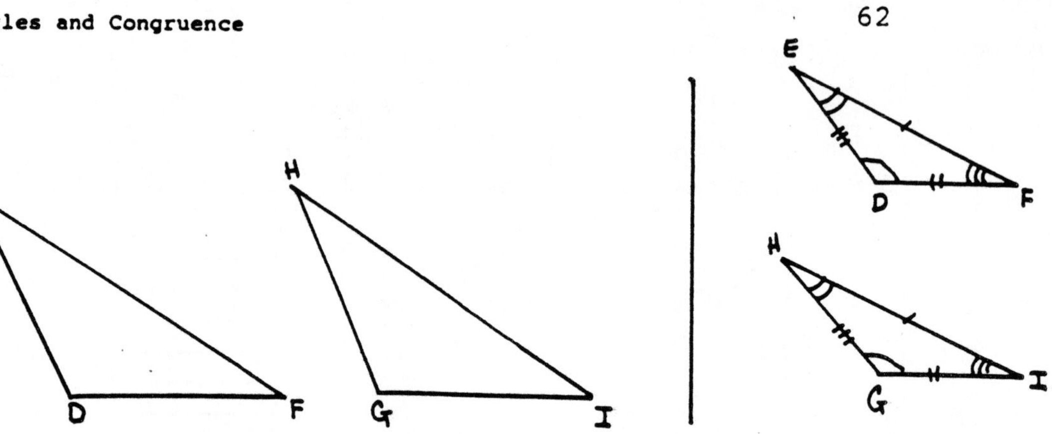

Markings can also be helpful in writing the congruence relationship. For example, study the diagram below with the indicated markings, $\angle J = \angle M$, $\angle K = \angle N$, $\angle L = \angle O$, $\overline{JK} = \overline{MN}$, $\overline{KL} = \overline{NO}$, and $\overline{JL} = \overline{MO}$.

Since all the corresponding sides and angles of both figures are equal then $\triangle JKL \cong \triangle MNO$.

30. From the given diagram, write the congruence relationship.

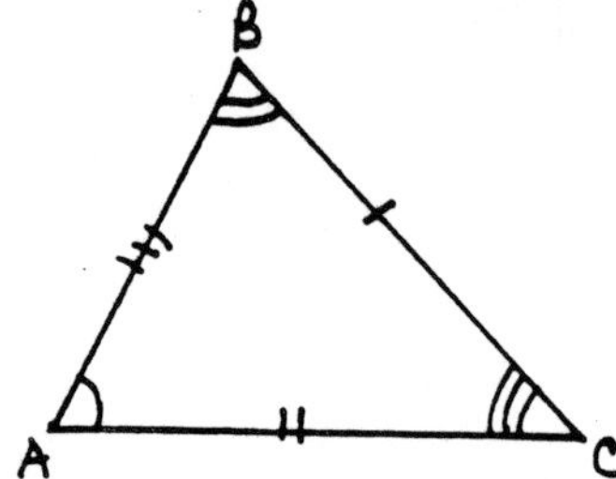

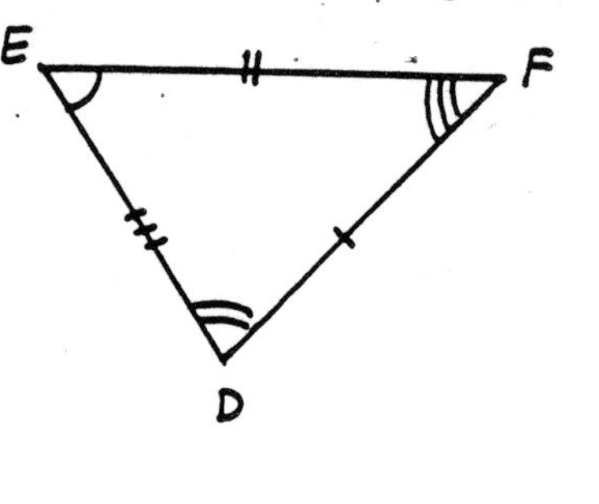

30. $\triangle ABC \cong \triangle EDF$
same as
$\triangle BCA \cong \triangle DFE$
$\triangle CAB \cong \triangle FED$

Study this figure!

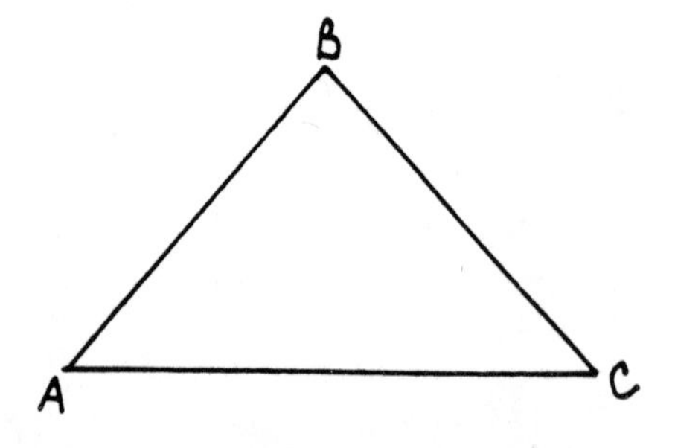

$\angle A$ is referred to as the included angle between sides $\overline{AB}$ and $\overline{AC}$ because sides $\overline{AB}$ and $\overline{AC}$ intersect to form $\angle A$. $\angle B$ is the included angle between sides $\overline{AB}$ and $\overline{BC}$.

31. $\angle C$ is the included angle between which two sides? __________ __________	31. $\overline{BC}$ and $\overline{AC}$

Below is a marked diagram to indicate an included angle between two sides.

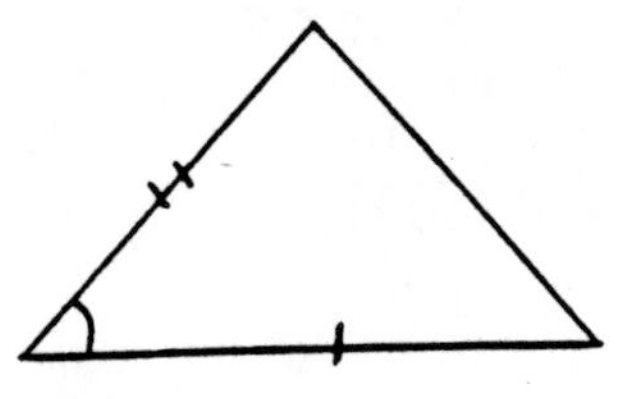

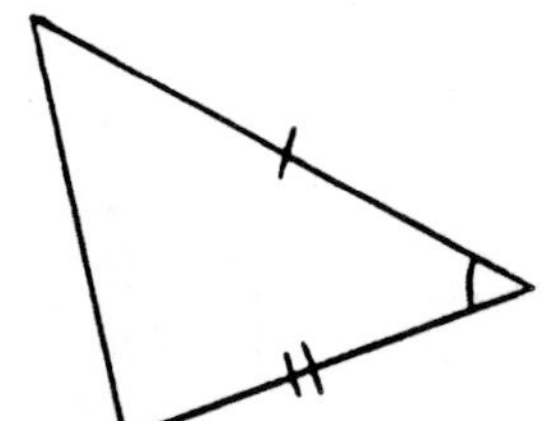

The marked diagram below indicates an angle which is <u>not</u> an included angle.

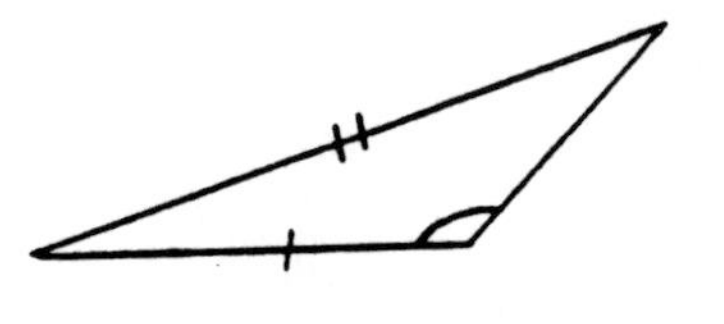

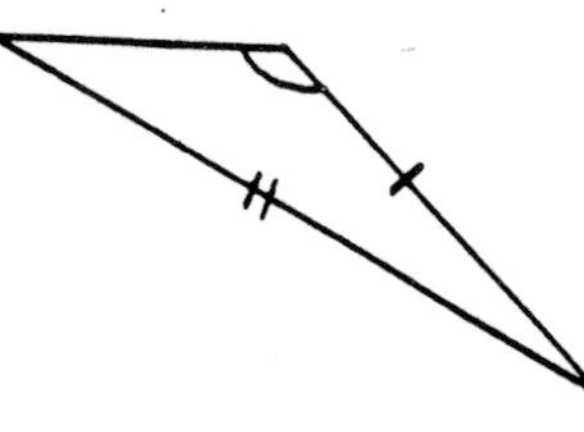

Refer to the diagram on the bottom of page 62. Similarly, $\overline{AB}$ is the included side between $\angle A$ and $\angle B$ and $\overline{BC}$ is the included side between $\angle B$ and $\angle C$.

32. $\overline{AC}$ is the included side between which two angles? _____ _____	32. $\angle A$ and $\angle C$

Below is a marked diagram to indicate an included side between two angles.

The marked diagram below indicates a side which is <u>not</u> an included side.

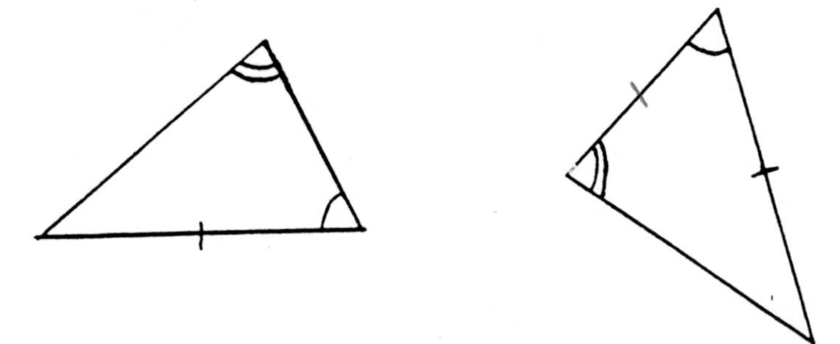

Exercise 4

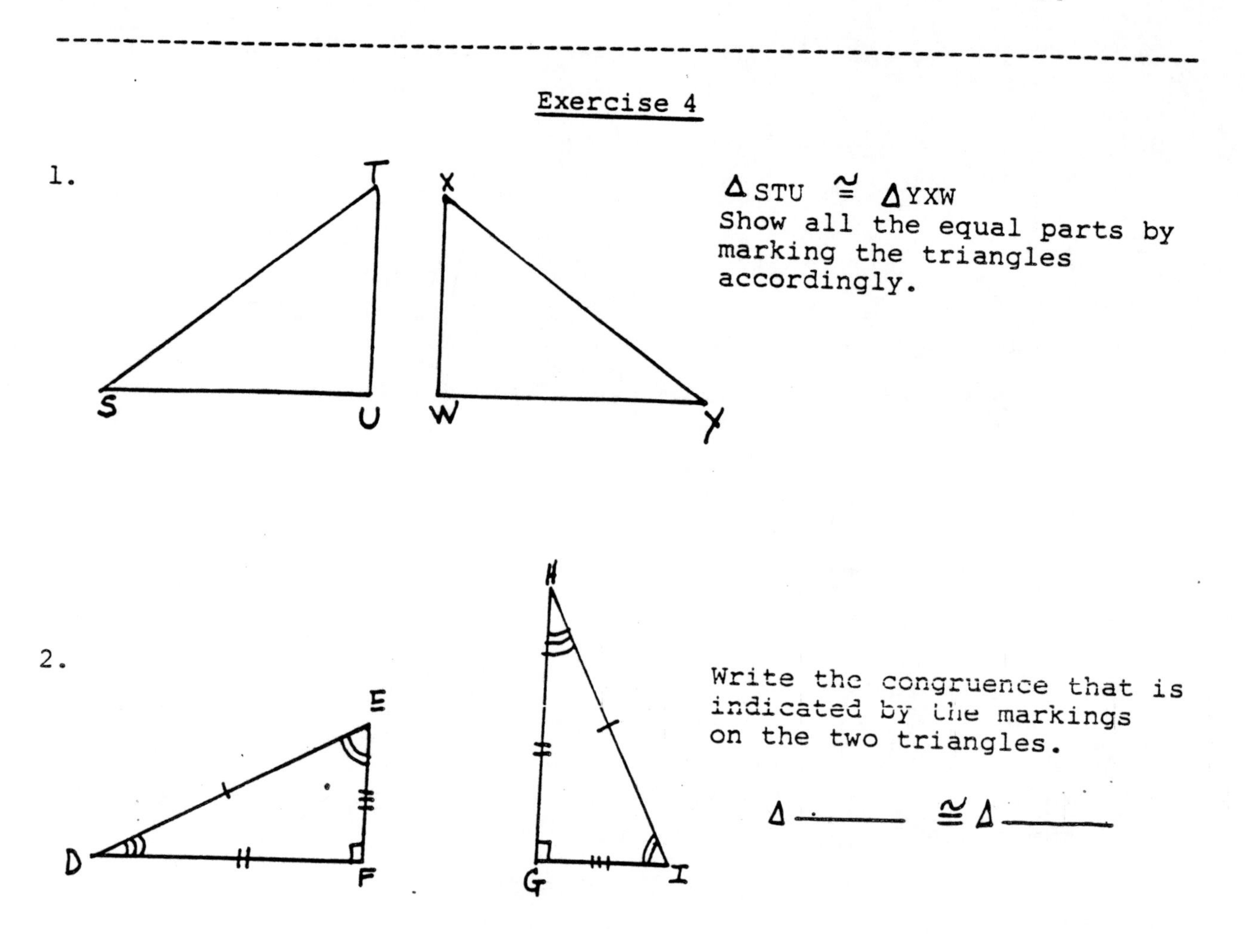

1. △STU ≅ △YXW
Show all the equal parts by marking the triangles accordingly.

2. Write the congruence that is indicated by the markings on the two triangles.

Δ ______ ≅ Δ ______

3. Which of the following markings indicate <u>included</u> <u>angle</u>? Write "yes" or "no" to each illustration.

a. ______

c. ______

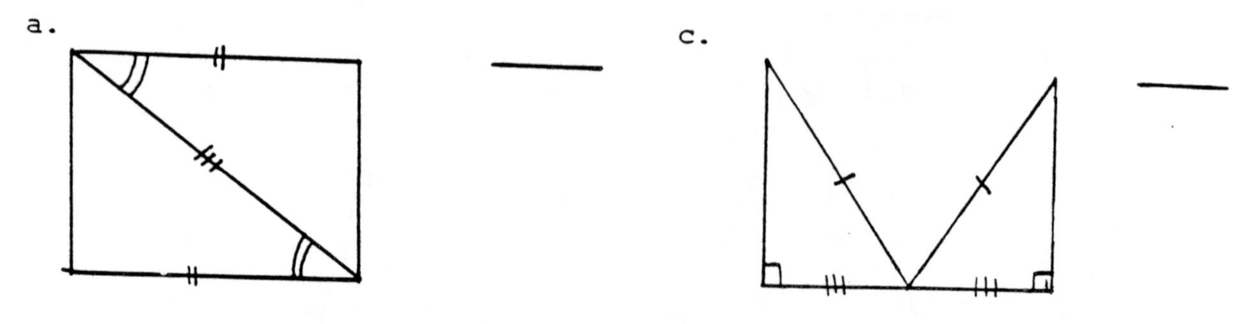

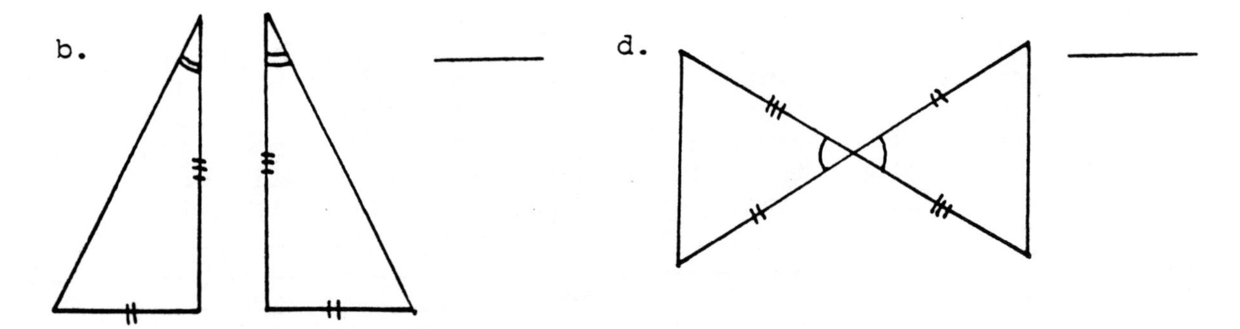

4. Which of the following markings indicate <u>included</u> <u>side</u>? Write "yes" or "no" to each illustration.

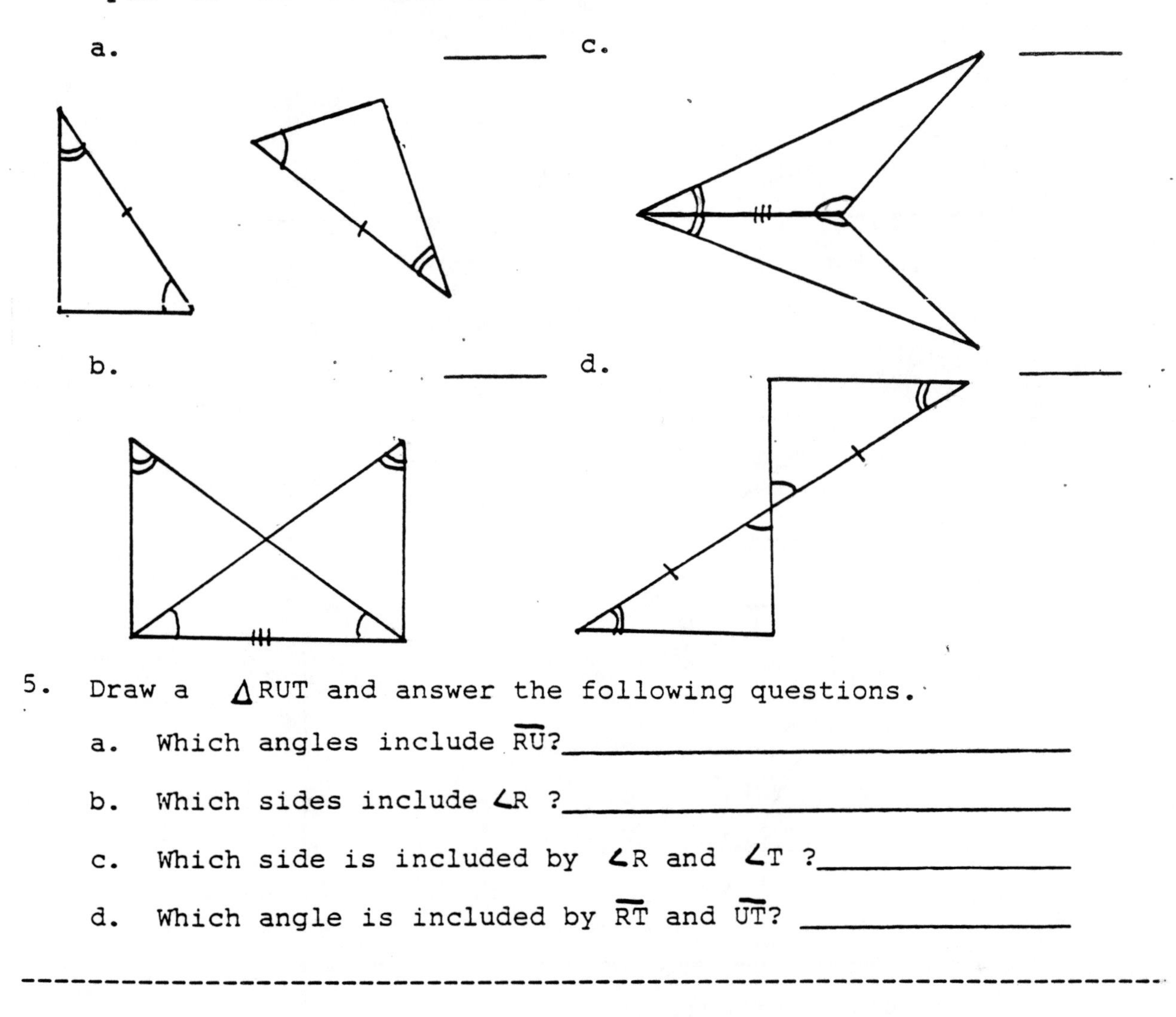

5. Draw a $\triangle RUT$ and answer the following questions.

   a. Which angles include $\overline{RU}$? ______________________

   b. Which sides include $\angle R$? ______________________

   c. Which side is included by $\angle R$ and $\angle T$? ______________

   d. Which angle is included by $\overline{RT}$ and $\overline{UT}$? ______________

In order for triangles to be congruent, must all corresponding parts be known to be equal? No. There are certain conditions that would force a triangle to retain an exact shape. So, congruence can be shown without knowing all parts of one triangle equal to all parts of another triangle. These conditions will be stated in the form of postulates. As you recall, a postulate is a statement whose truth is accepted without proof.

Postulate 1:	If three sides of one triangle are equal to three corresponding sides of a second triangle, the triangles are said to be congruent. (Abbreviated SSS = SSS)

Examples:

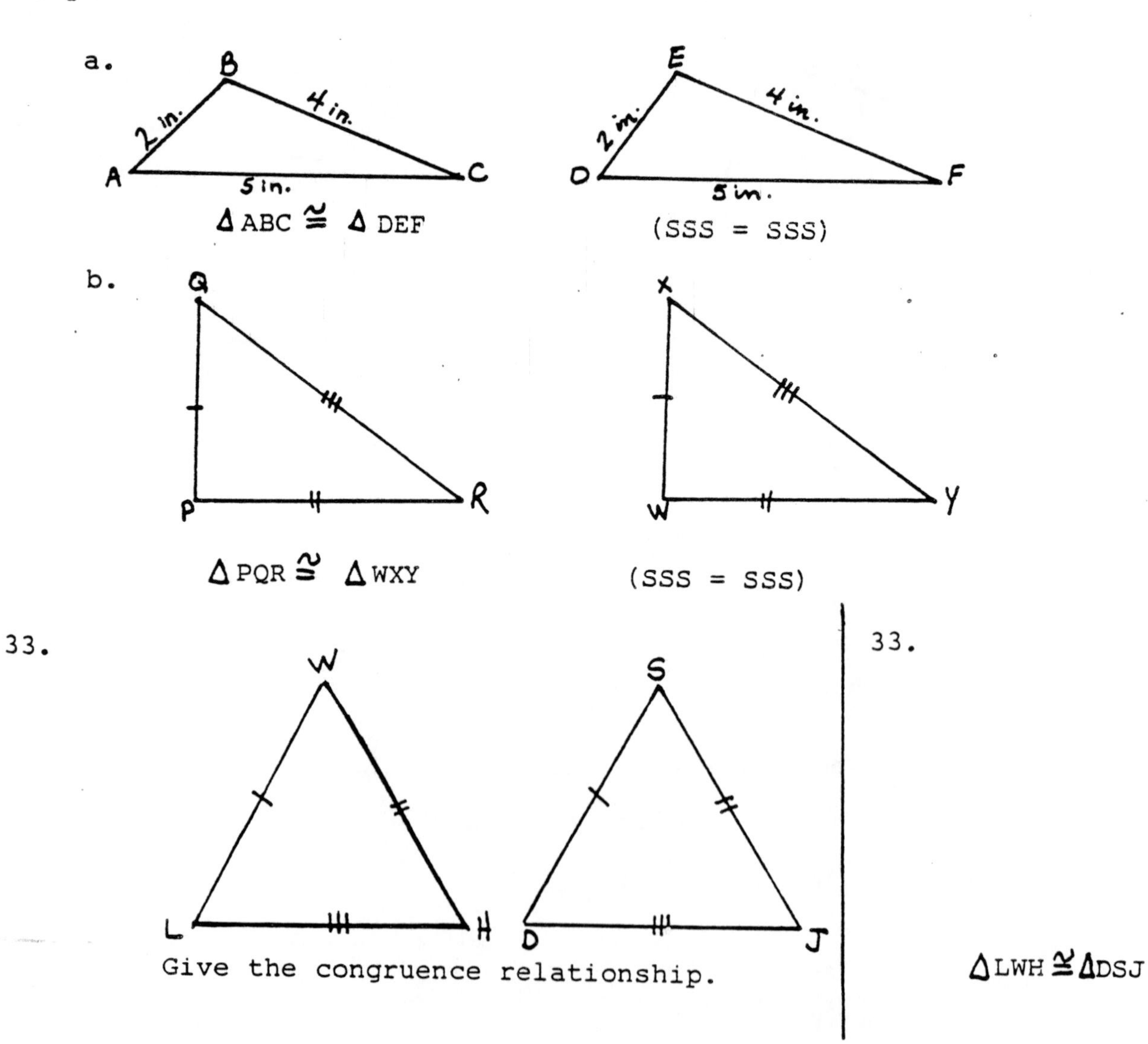

a. $\Delta ABC \cong \Delta DEF$ (SSS = SSS)

b. $\Delta PQR \cong \Delta WXY$ (SSS = SSS)

33. Give the congruence relationship.

33. $\Delta LWH \cong \Delta DSJ$

Postulate 2:	If two sides and an included angle of one triangle are equal to two sides and an included angle of a second triangle, the triangles are congruent. (SAS = SAS)

Examples:

a.

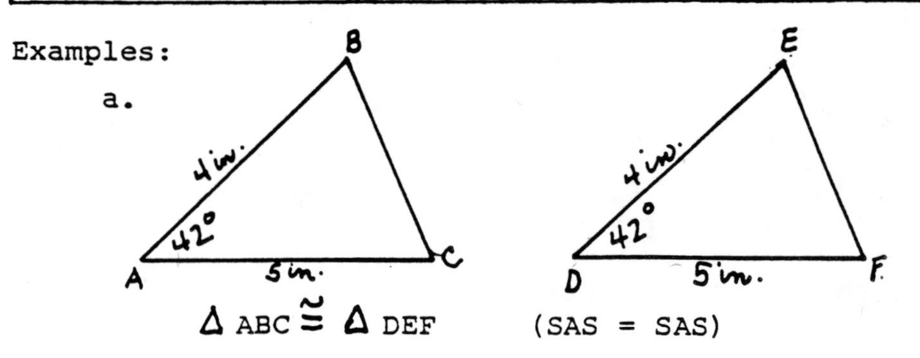

$\triangle$ ABC $\cong$ $\triangle$ DEF (SAS = SAS)

b.

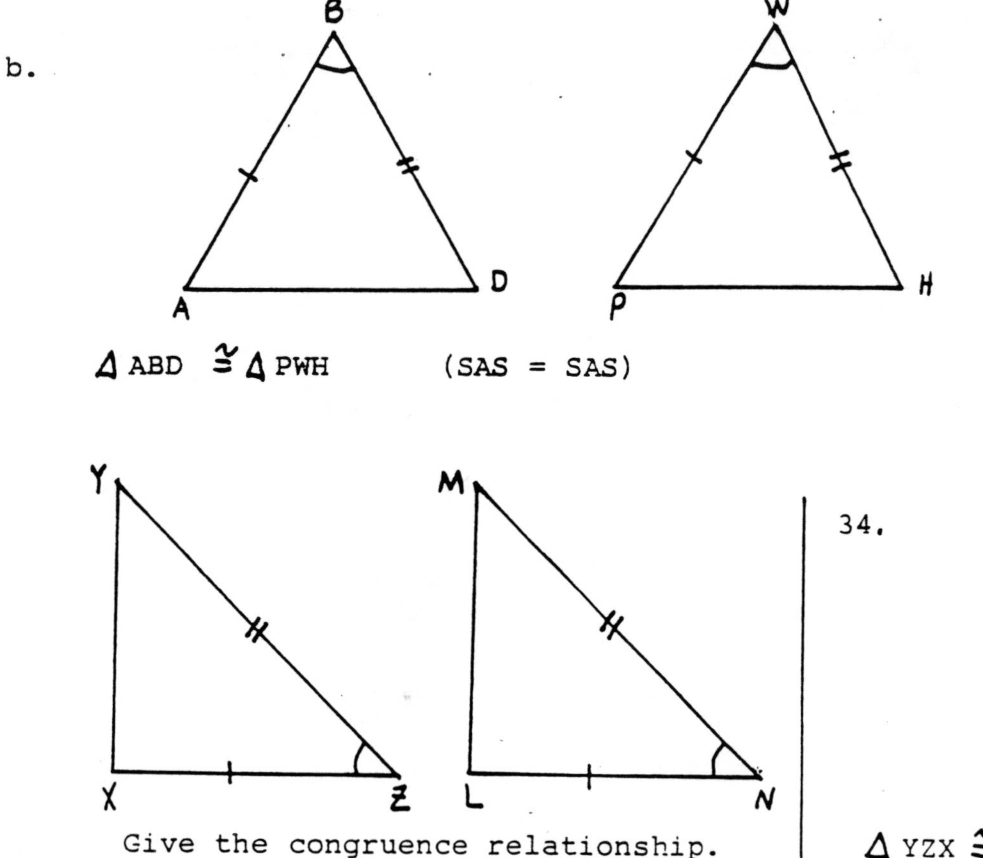

$\triangle$ ABD $\cong$ $\triangle$ PWH (SAS = SAS)

34.

Give the congruence relationship.

34. $\triangle$ YZX $\cong$ $\triangle$ MNL

Postulate 3:	If two angles and an included side of one triangle are equal to two angles and an included side of a second triangle, the triangles are congruent. (ASA = ASA)

Examples:

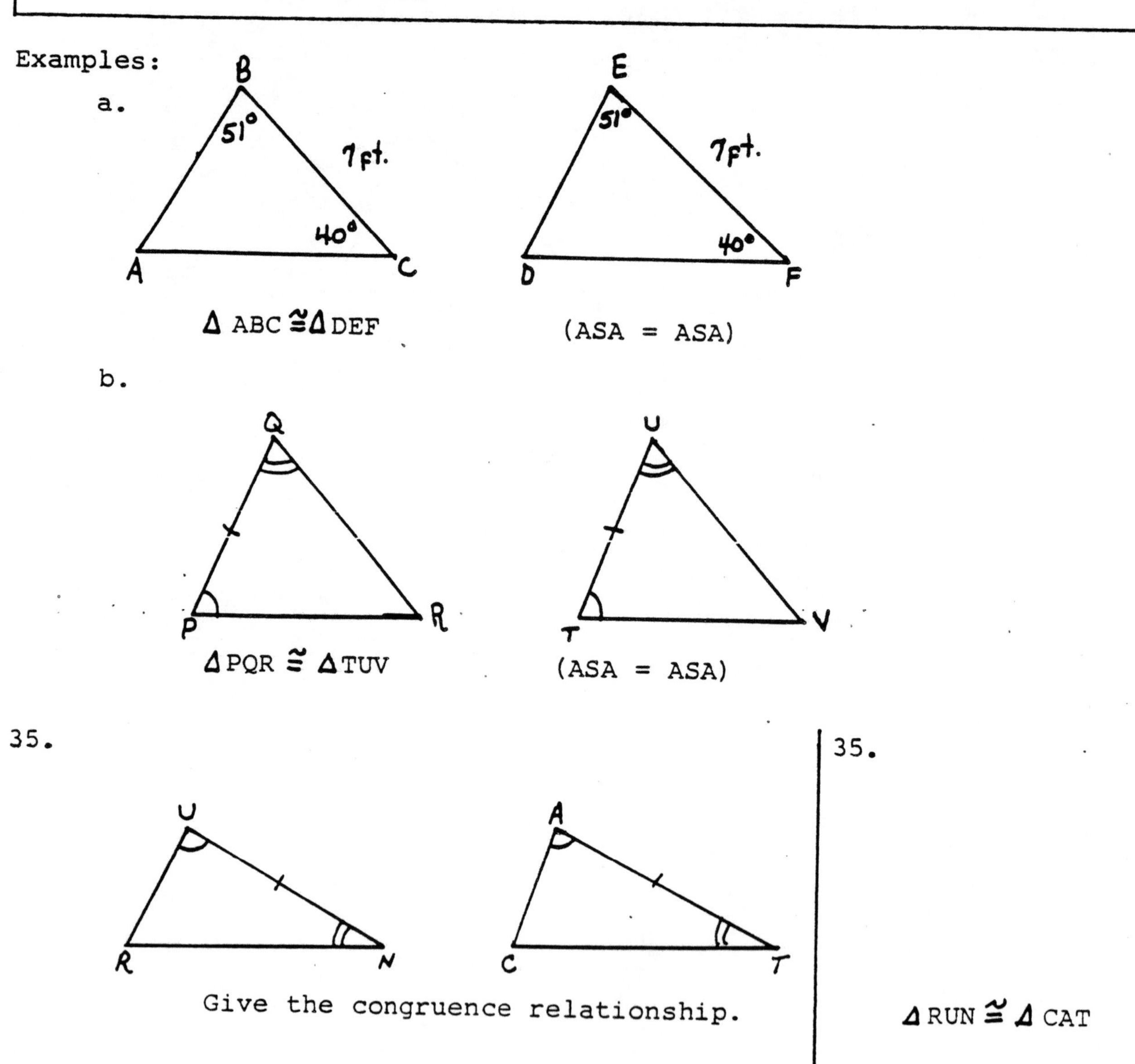

Give the congruence relationship.

The three postulates were illustrated separately and with examples that gave the appropriate data. Study the following examples.

Examples:

a.

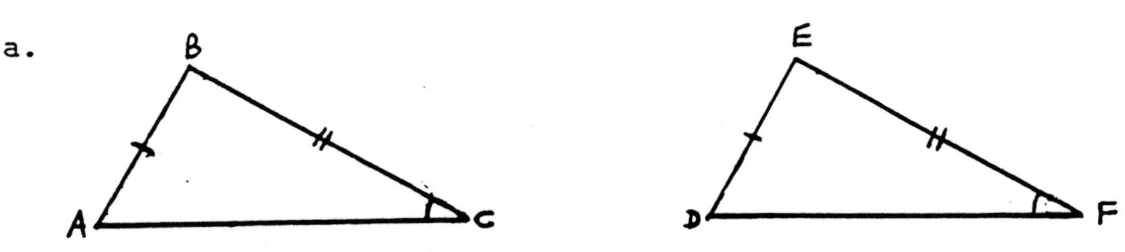

Can we conclude these triangles are congruent? No. Why not? The two triangles are not congruent because two sides and an angle of one triangle are equal to two sides and an angle of the second but the known angle is <u>not the included angle</u>. Actually, when this set of conditions is known, the triangles may not even appear the same as illustrated below.

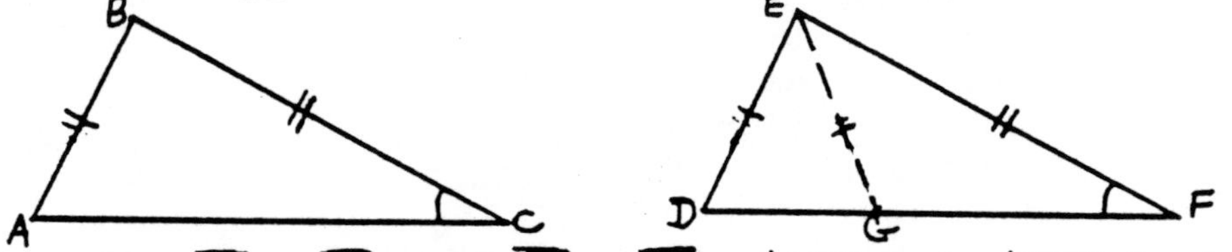

$\angle C = \angle F$, $\overline{BC} = \overline{EF}$, and $\overline{AB} = \overline{DE}$. ΔABC and ΔDEF look alike and may be congruent. However, observe that the two sides and given angle may also occur as follows: $\angle C = \angle F$, $\overline{BC} = \overline{EF}$, and $\overline{AB} = \overline{EG}$. ΔABC is an acute triangle and ΔGEF is an obtuse triangle. Obviously, ΔABC and ΔGEF are not congruent. Consequently, the conditions, SSA = SSA, do not give congruence in all cases and cannot be used as a reason for showing congruence.

b.

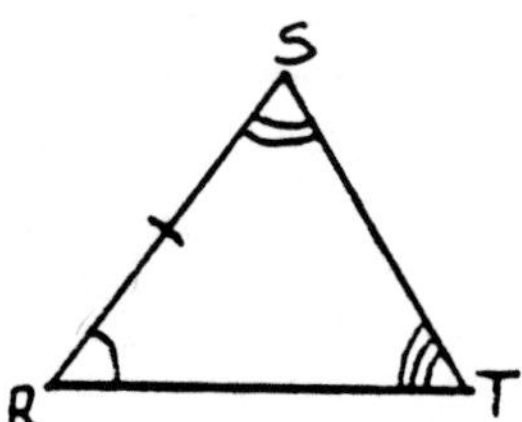

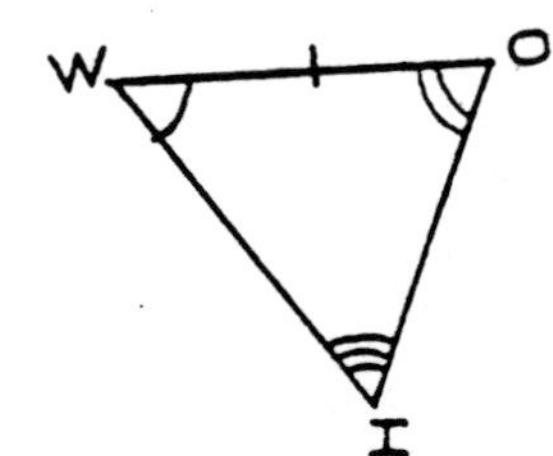

Can we conclude these triangles are congruent? Yes. $\angle R = \angle W$, $\angle S = \angle O$, and $\overline{RS} = \overline{WO}$ so $\Delta RST \cong \Delta WOI$ by ASA = ASA. The information that $\angle T = \angle I$ was not needed.

c.

In this example the triangles are not congruent. Three angles of one triangle equal the corresponding three angles of a second triangle but nothing is known about any of its sides. We do <u>not</u> have an AAA = AAA postulate.

36.

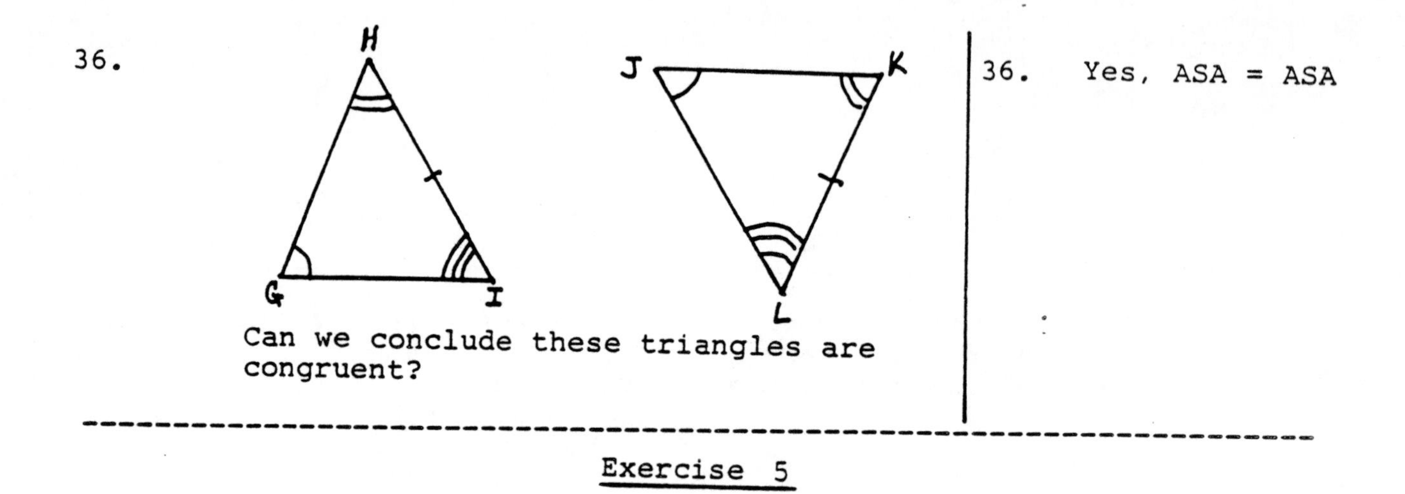

Can we conclude these triangles are congruent?

36. Yes, ASA = ASA

Exercise 5

State whether or not you have sufficient information to determine if the triangles are congruent by Postulate 1, 2, or 3. If the triangles are congruent, state which postulate.

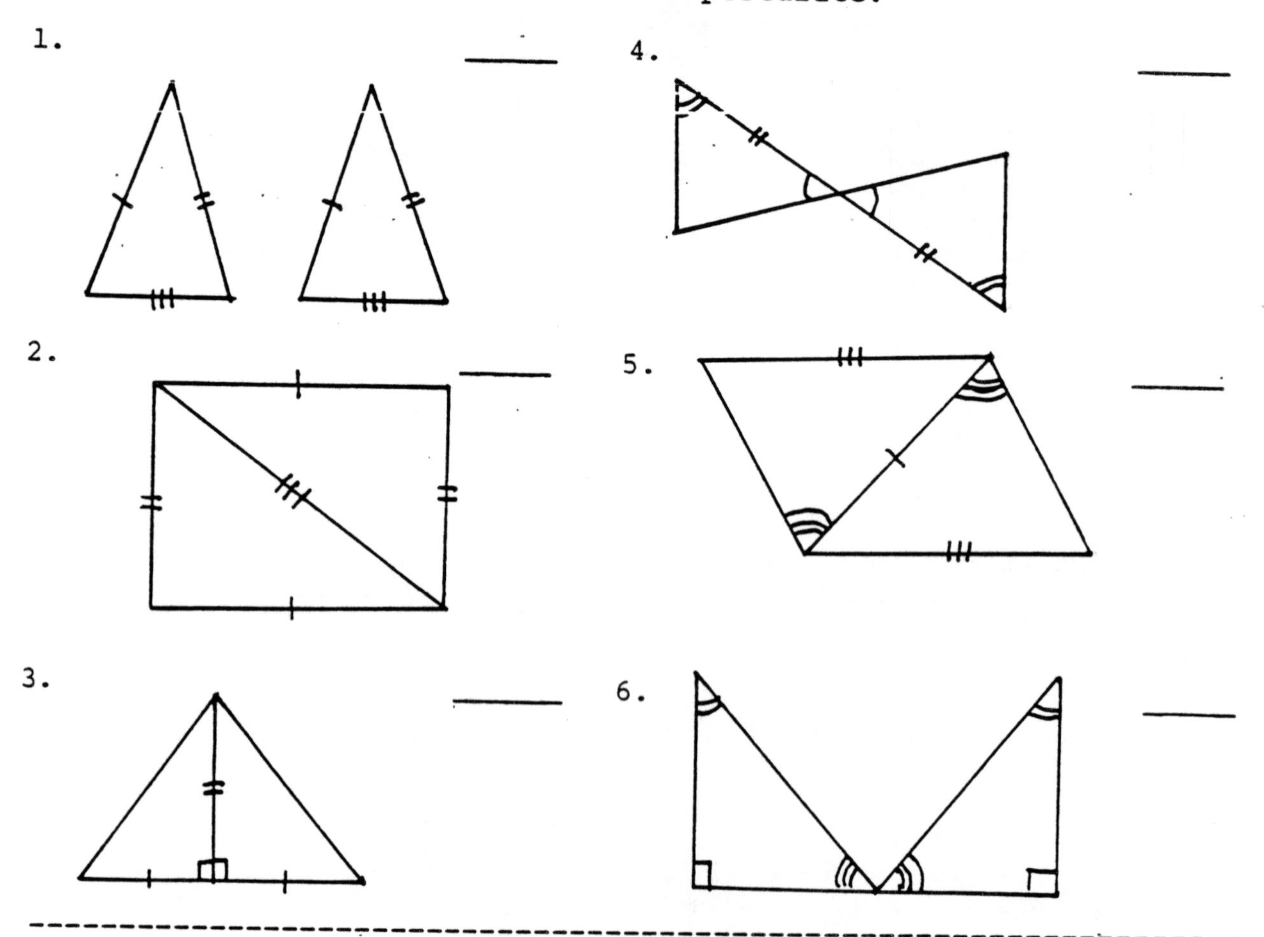

Unit 2 Review

1. Define the following:

triangle
obtuse triangle
perimeter
isosceles triangle
equilateral triangle
scalene triangle
right triangle

acute triangle
equiangular triangle
hypotenuse
altitude
median
congruent
" ≅ "

2. Which of the following pairs of triangles are congruent by the SAS , ASA, or SSS postulate? Indicate whether or not the triangles are congruent and why.

a. ____

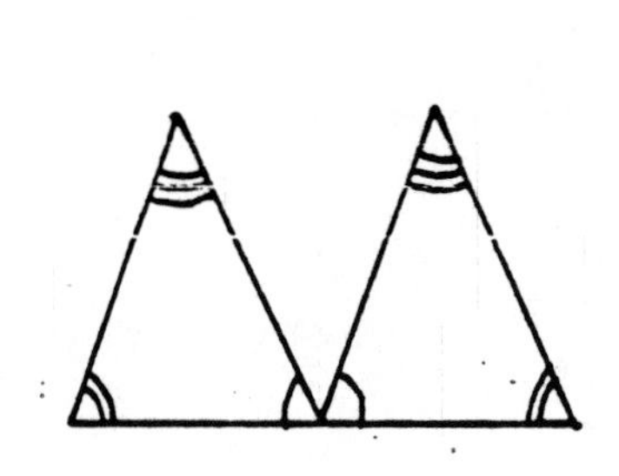

b. ____

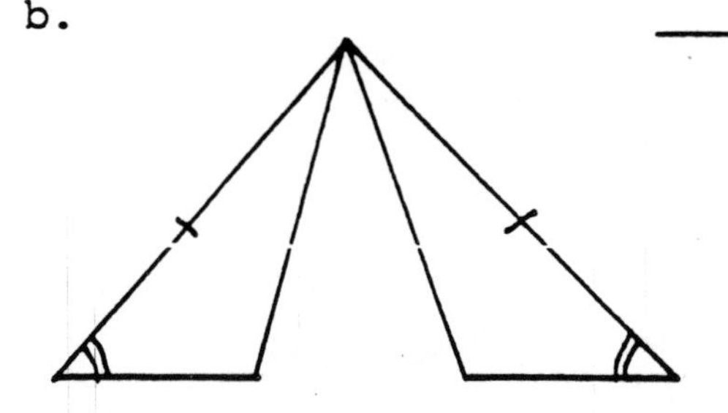

c. ____

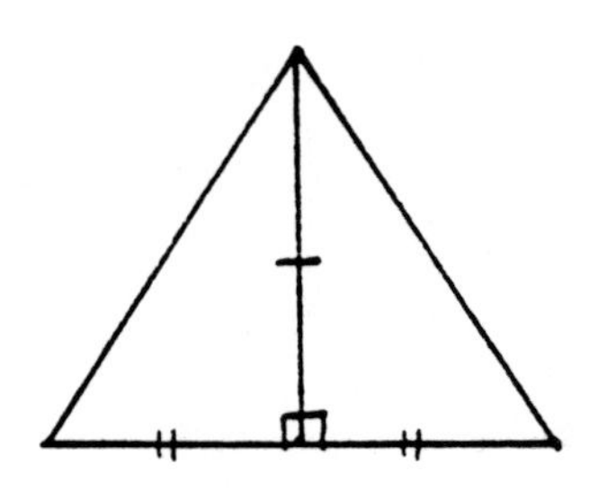

d. ____

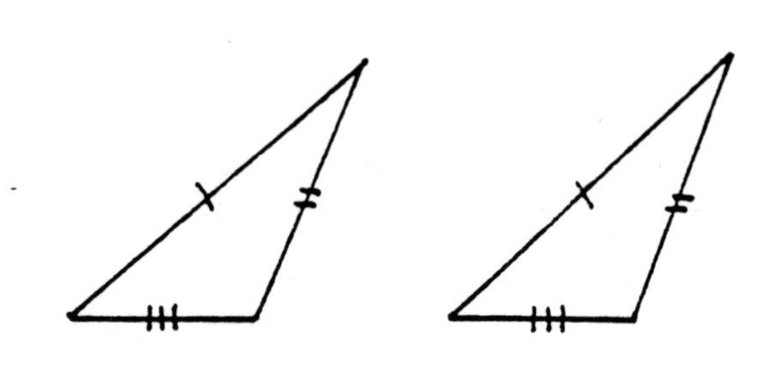

e. ____

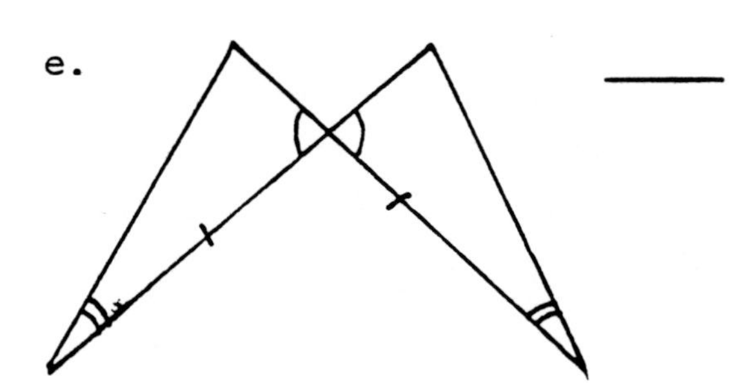

f. ____

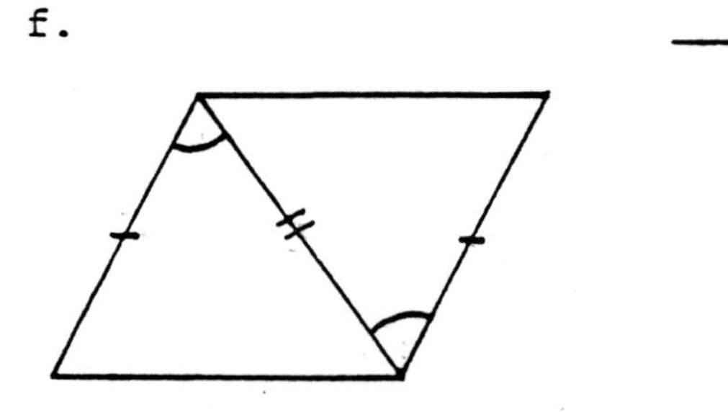

3. Classify each of the following triangles by the lengths of their sides or the measure of their angles as indicated in the figures.

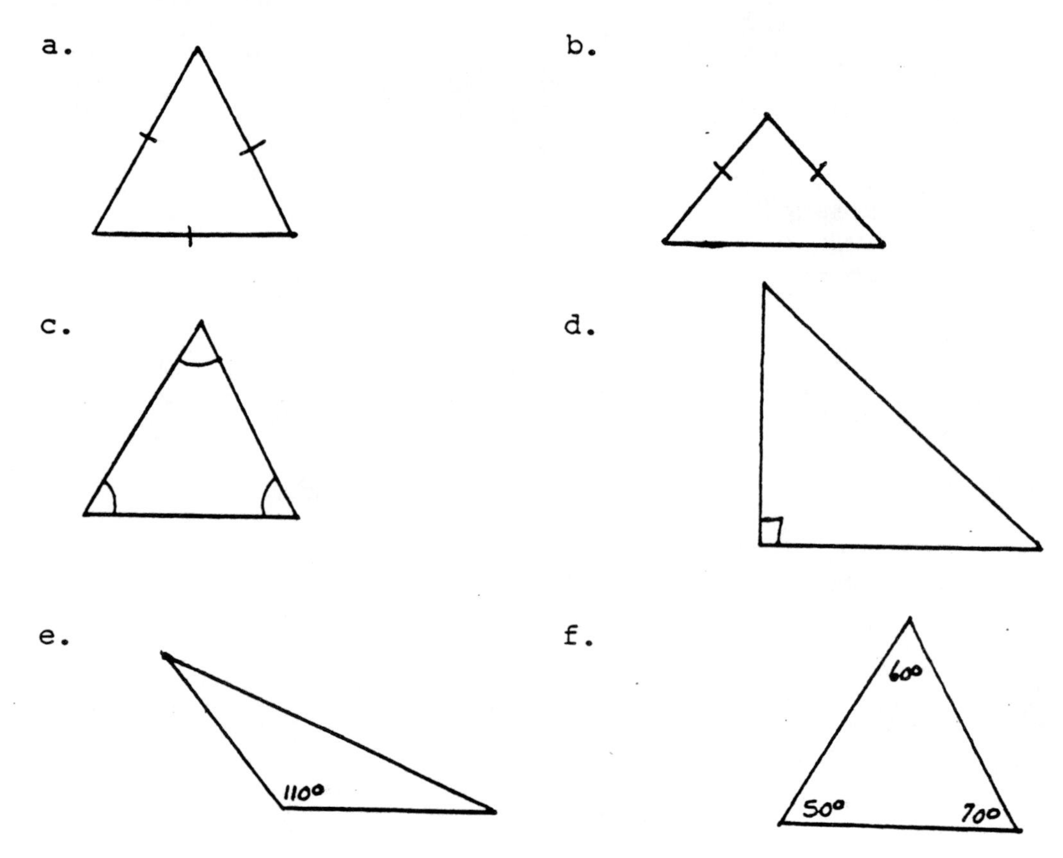

4. Identify the congruence for each pair of triangles by using the proper notation.

a. D C A B

figure 1

$\triangle ACD \cong \triangle BDC$

equal sides	equal angles
______	______
______	______
______	______

b. Q P T S R

figure 2

$\triangle PQT \cong \triangle RQS$

equal sides	equal angles
______	______
______	______
______	______

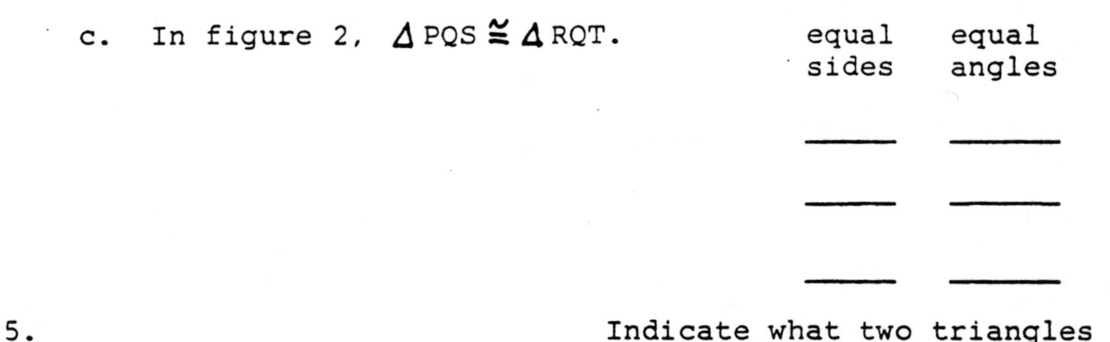

c. In figure 2, $\triangle PQS \cong \triangle RQT$.

equal sides	equal angles
_____	_____
_____	_____
_____	_____

5. Indicate what two triangles are congruent using the proper correspondence. ________________

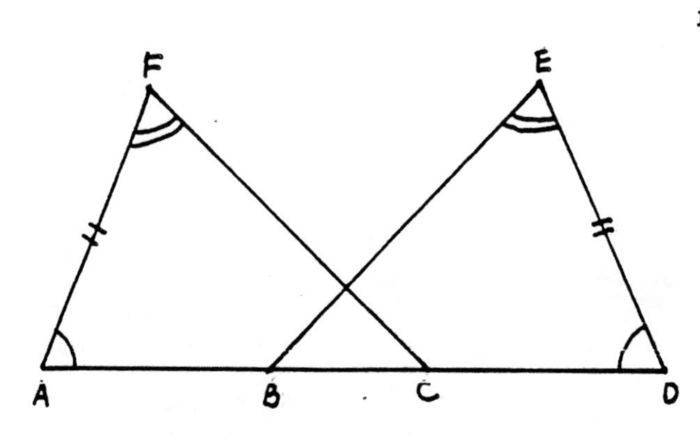

6. The base of an isosceles triangle measures 5 inches. Find the length of one of the equal sides of the triangle if the perimeter equals 23 inches. ________________________

7. This triangular garden spot has 59 feet of fencing surrounding it. What are the lengths of each side of the plot? _____ _____, _____.

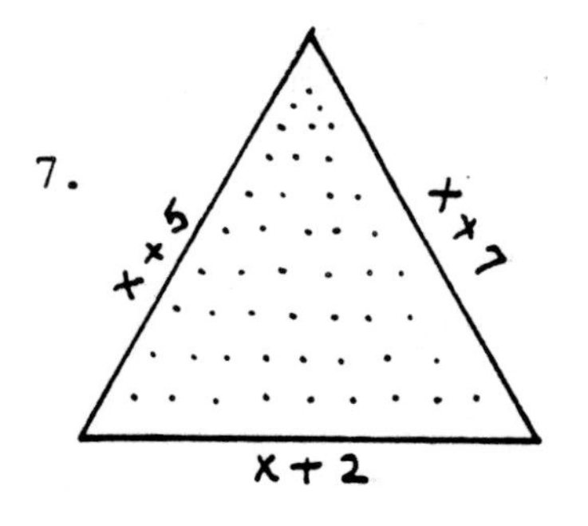

8. Construct the altitude and median to base $\overline{AB}$ in triangle ABC.

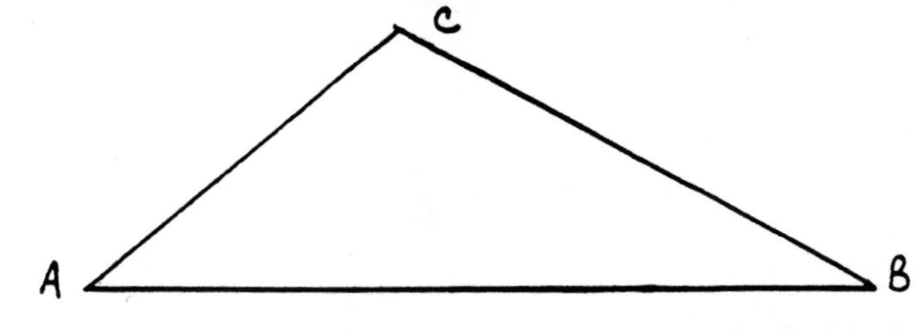

9. Study each of the following figures and indicate which two triangles are overlapping.

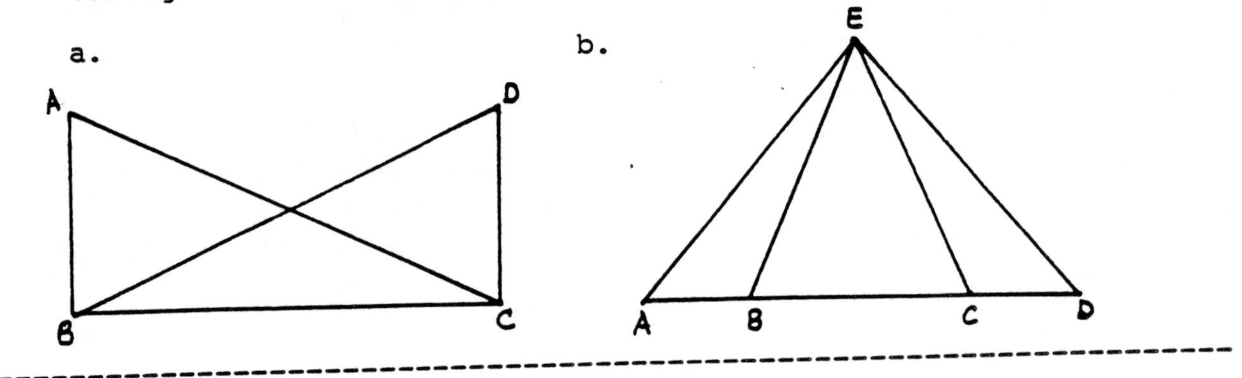

Unit 3

Simple Proofs

Learning objectives:

1. The student will differentiate between the hypothesis and conclusion of a conditional statement.

2. The student will demonstrate his mastery of the converse, inverse, and contrapositive by writing these statements given a conditional statement.

3. The student will review and use in simple proofs statements of equality about angles and line segments.

4. The student will demonstrate his mastery of three congruence postulates by applying them to simple proofs.

5. The student will demonstrate his mastery of the following theorems by writing them, by applying them to selected problems and by using them in simple proofs:

    a. If two sides of a triangle are equal, then the angles opposite these sides are equal.

    b. If two angles of a triangle are equal, then the sides opposite these angles are equal.

6. The student will demonstrate the concept of corresponding parts of congruent triangles are equal (abbreviated CPCTE) by applying it to selected problems and by using it in simple proofs.

Simple Proofs

There are two major reasons for understanding and writing simple proofs. One reason is to strengthen your understanding of the concepts involved in the previous units. The second reason, even more important, is to demonstrate that common sense is fallible and that there is a real need to use proofs in mathematics.

Study the pictures below and answer the following questions:

1.

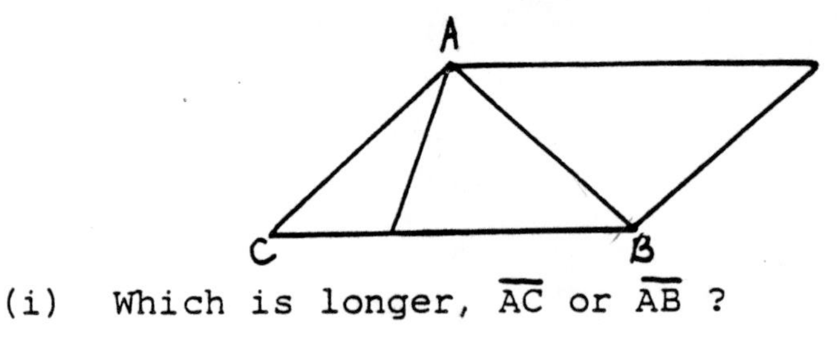

(i) Which is longer, $\overline{AC}$ or $\overline{AB}$?

(ii) Are $\overline{XZ}$ and $\overline{MY}$ the same length? (See drawing to the right.)

M
X Y Z

1. (i). neither, they are the same length.
(ii). yes

No two people will always arrive at exactly the same conclusions when they are depending solely upon their senses. As illustrated in example 1, some senses can be fooled simply by the use of optical illusions.

Even basic common sense can be fallible and can also lead you to incorrect conclusions. Study the examples following.

2. Use only your common sense to identify which of the following algebraic expressions are prime?
(i). 13
(ii). $x^2 + 9$
(iii). $5x^2 - 3x - 14$

2. (i). prime
(ii). prime
(iii). not prime $(5x+7)(x-2)$

3. (i). The sum of two numbers is 10. The product of these two numbers is 24. What are the numbers?

(ii). The distance between two hospitals is 20 miles more than one-fourth of the distance between them. What is the distance between them?

3. (i) 4,6

(ii) d = 26 2/3

IF p THEN q

You were asked to only use common sense to arrive at your answers. Problems 2 and 3 involved a few examples that could be worked out quite easily. However, the more difficult examples required algebraic manipulations and could not be solved by simple opinion or emotion. In geometry we must eliminate personal emotions and depend upon proofs as a basis for drawing conclusions. These proofs are derived from a logical sequence of steps. This sequence of steps is not innate and most students will have to learn to think logically.

From Unit 1, we discussed that the system of logic that we employ in geometry is called deductive reasoning. The four elements of deductive reasoning are postulates, theorems, defined terms, and undefined terms. Recall that in deductive reasoning you are given very general statements to begin with. You then draw upon the four elements in developing a very specific conclusion. Deductive reasoning involves a logical connection between "what is given" and "what is to be proven".

Logical Connection

What is given ⟶ What is to be proven

To start you must be given some information. This given information is the "if" part of the conditional statement and is called the hypothesis. From this given hypothesis as well as by using previously proven theorems, postulates, defined terms, and undefined terms we develop a valid chain of reasoning to reach a logical conclusion. The conclusion is the "then" part of the statement. If we succeed at constructing a logical connection between the hypothesis and the conclusion, then the IF-THEN statement has been proven. This type of "if-then" statement is a conditional statement.

The following is an example of an if-then statement.

If today is December 10th, then the quarter will end soon.

If __________ (p), then __________ (q)

1. p represents what part of the statement?

   ____

2. q represents what part of the statement?

   ____

3. In the following statement identify the hypothesis and the conclusion.

   If I study hard, then I will pass geometry.

Answers:

1. "if" or hypothesis
2. "then" or conclusion

(a) If p ________________
(b) then q ________________

3. (a) If I study hard
(b) I will pass geometry

Other relationships can be formed between the hypothesis and conclusion by introducing the converse, inverse, and contrapositive of statements. Study the relationships below.

Statement: If p, then q.

Converse: If q, then p.

Inverse: If not p, then not q.

Contrapositive: If not q, then not p.

Use this statement in answering question 4:

If Sam is a man, then Sam is a human being.
(p) (q)

4. The converse is started below. Please complete the statement using -- If q then p.

If Sam is a human being, then (q)

________________________. (p)

4. Sam is a man

5. The inverse is started below. Please complete the statement using -- If not p then not q.

If Sam is not a man, then ________. (not p) (not q)

5. Sam is not a human being

6. Complete the contrapositive statement using -- If not q, then not p.

If Sam is not a human being, (not q)
then ________________________. (not p)

6. Sam is not a man

Exercise 1

Find the converse, contrapositive and inverse for each of the following statements. (The statements may or may not be true statements.)

1. If you live in Virginia, then you live in the United States.

2. If you operate a car, then you can fly an airplane.

3. If a triangle is not scalene, then it is equilateral.

4. If you are blond, then you are not intelligent.

You have undoubtedly discovered that some statements may or may not be valid. For example, let's discuss the statement given on page 3.

If Sam is a man, then Sam is a human being.

The statement is truthful. It is valid in its logic because you can not possibly be a man and not be human at the same time.

The contrapositive is also valid. If Sam is not at all human then he could not possibly be a man.

Statements and their contrapositives are, indeed, logical equivalents of one another. If the statement is true, so follows the contrapositive. If the statement is false, the contrapositive will also be false. Therefore, it follows that once you've proven a theorem to be true, you need not do a second proof for the contrapositive. A theorem and its contrapositive are logical equivalents.

However, the same does not apply to the converse. A converse of a statement is not always true even though the original statement may be valid. For example, in the statement above, the converse indicated that if Sam is a human being then Sam is a man. I'm sure that the female protion of the human race will have something to say about the gross oversight implied in the statement.

If a theorem is proven to be true, its converse is not necessarily true and must be proven with a second proof.

Review of Basic Concepts

The previous units have provided a sturdy foundation upon which you will build your understanding of simple proofs.

Let's review some concepts about equality that you will be using in writing simple proofs.

CPCTE

Equality

A. You can prove that two angles are equal if:

1. They are equal to the same or equal angles.
2. They are supplements of the same or equal angles.
3. They are complements of the same or equal angles.
4. They are right angles. All right angles are equal.
5. They are vertical angles. Pairs of vertical angles are equal.
6. They are the same angle. Any angle is equal to itself. (Reflexive)
7. They are parts of a bisected angle.
8. CORRESPONDING ANGLES OF ≅ △s

IF 2 SIDES OF A TRIANGLE ARE EQUAL THAN THE ANGLES OPPSITE THOSE SIDES ARE EQUAL.

B. You can prove that two line segments are equal if:

1. They are equal to the same or equal line segments.
2. They are parts of a bisected line segment.
3. They are the equal sides of an isosceles triangle. (Definition of an isosceles triangle.)
4. They are the same line segment. Any line segment is equal to itself. (Reflexive)
5. They are the results of adding or subtracting equal line segments.
6. They are multiples or halves of equal line segments.
7. CORRESPONDING SIDES OF ≅ △s

Given the following information, prove that the angles or line segments in each problem are equal. State why they are equal using any of the previous statements that would logically apply.

7.

B

A C

Given: $\triangle ABC$ is isosceles, $\overline{AC}$ is the base.

Prove: $\overline{AB} = \overline{BC}$

7. Part B-#3 $\overline{AB}$ and $\overline{BC}$ are the equal sides of an isosceles triangle.

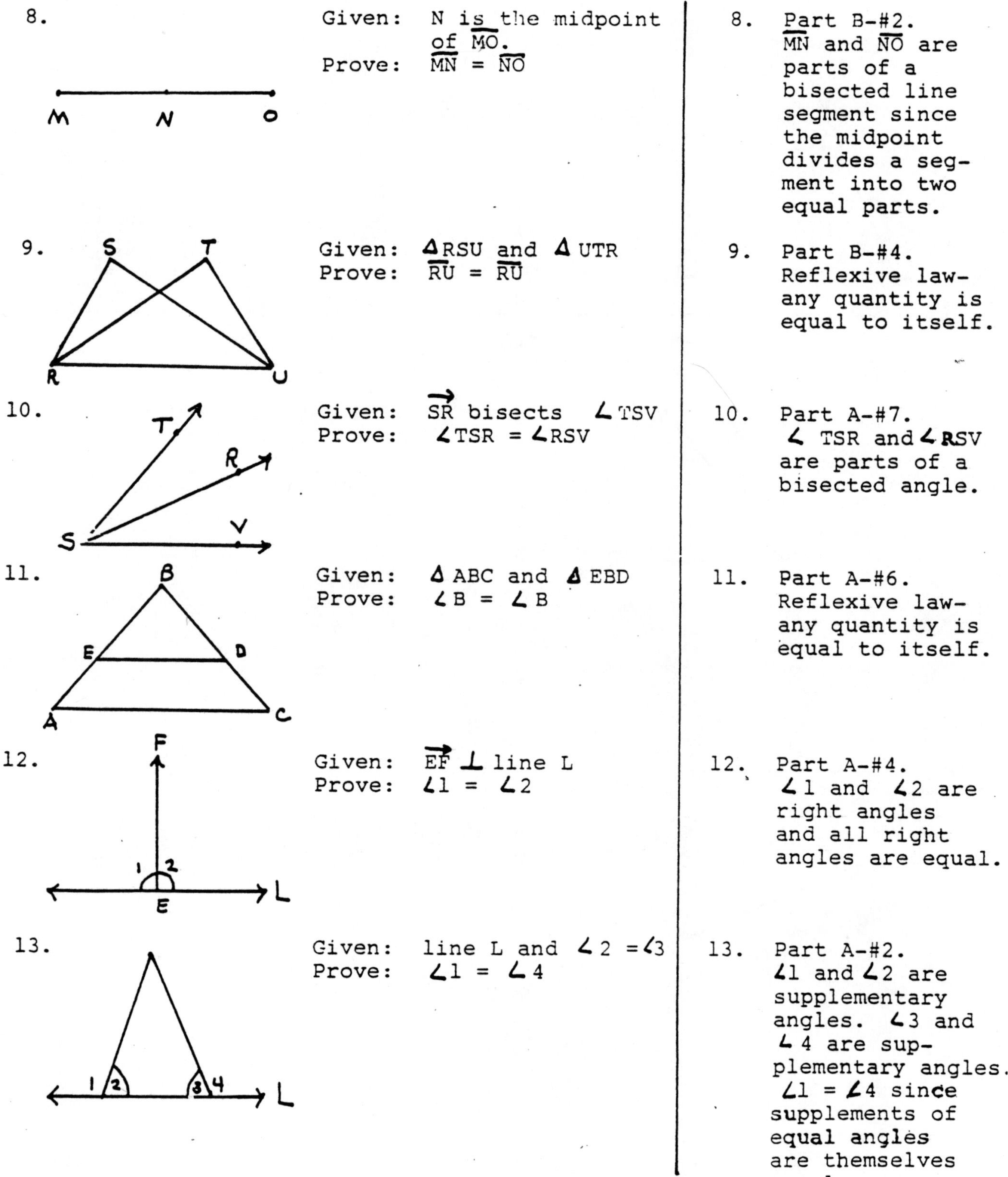

8. Given: N is the midpoint of $\overline{MO}$.
Prove: $\overline{MN} = \overline{NO}$

8. Part B-#2.
$\overline{MN}$ and $\overline{NO}$ are parts of a bisected line segment since the midpoint divides a segment into two equal parts.

9. Given: ΔRSU and ΔUTR
Prove: $\overline{RU} = \overline{RU}$

9. Part B-#4.
Reflexive law- any quantity is equal to itself.

10. Given: $\overrightarrow{SR}$ bisects $\angle$TSV
Prove: $\angle$TSR = $\angle$RSV

10. Part A-#7.
$\angle$ TSR and $\angle$RSV are parts of a bisected angle.

11. Given: Δ ABC and Δ EBD
Prove: $\angle$B = $\angle$B

11. Part A-#6.
Reflexive law- any quantity is equal to itself.

12. Given: $\overrightarrow{EF} \perp$ line L
Prove: $\angle$1 = $\angle$2

12. Part A-#4.
$\angle$1 and $\angle$2 are right angles and all right angles are equal.

13. Given: line L and $\angle$2 = $\angle$3
Prove: $\angle$1 = $\angle$4

13. Part A-#2.
$\angle$1 and $\angle$2 are supplementary angles. $\angle$3 and $\angle$4 are supplementary angles. $\angle$1 = $\angle$4 since supplements of equal angles are themselves equal.

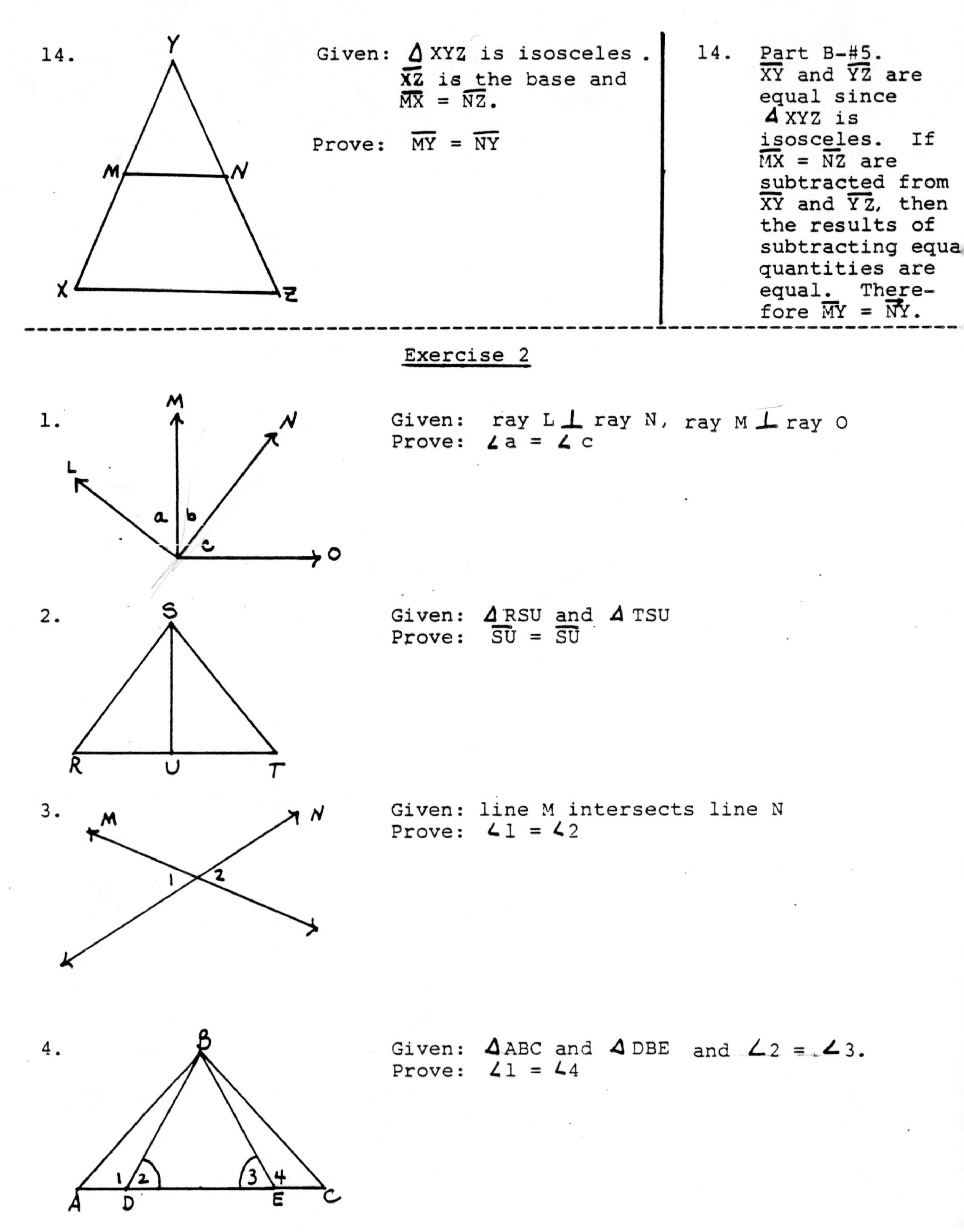

14. Given: Δ XYZ is isosceles. $\overline{XZ}$ is the base and $\overline{MX} = \overline{NZ}$.

Prove: $\overline{MY} = \overline{NY}$

14. Part B-#5. $\overline{XY}$ and $\overline{YZ}$ are equal since Δ XYZ is isosceles. If $\overline{MX} = \overline{NZ}$ are subtracted from $\overline{XY}$ and $\overline{YZ}$, then the results of subtracting equa[l] quantities are equal. Therefore $\overline{MY} = \overline{NY}$.

Exercise 2

1. Given: ray L $\perp$ ray N, ray M $\perp$ ray O

   Prove: $\angle a = \angle c$

2. Given: Δ RSU and Δ TSU

   Prove: $\overline{SU} = \overline{SU}$

3. Given: line M intersects line N

   Prove: $\angle 1 = \angle 2$

4. Given: Δ ABC and Δ DBE and $\angle 2 = \angle 3$.

   Prove: $\angle 1 = \angle 4$

5.

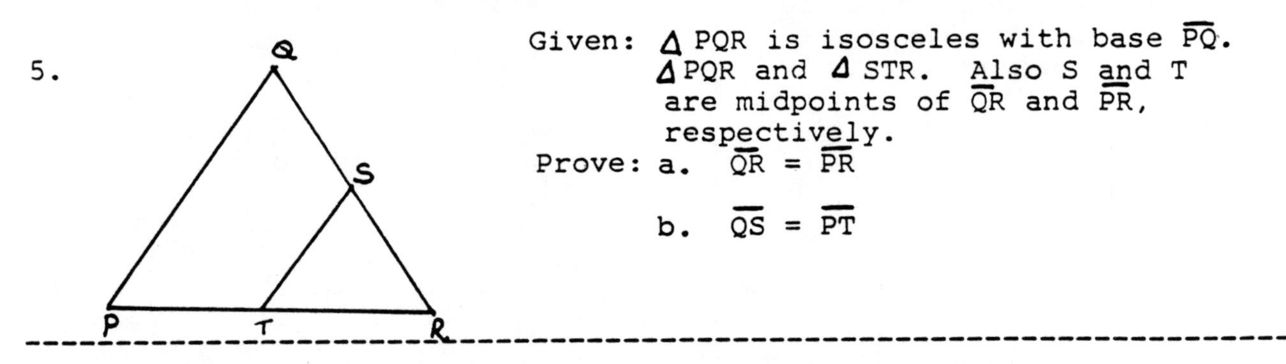

Given: $\triangle$PQR is isosceles with base $\overline{PQ}$. $\triangle$PQR and $\triangle$STR. Also S and T are midpoints of $\overline{QR}$ and $\overline{PR}$, respectively.

Prove: a. $\overline{QR} = \overline{PR}$

b. $\overline{QS} = \overline{PT}$

In most cases the problems that you have just completed could be proven in one or two steps. Those problems were merely an introduction to the next set of proofs that involve more than one step to arrive at the conclusion. So, what does a complete proof look like?

Anatomy of a Proof

For this section you will need to review the unit on congruence. Three ways to prove triangles congruent are SSS = SSS, SAS = SAS, and ASA = ASA.

(In the proof of statements, the use of approved symbols and abbreviations is acceptable.)

Procedure:	Example of a proof:
I. Divide the problem up into what is given and what is to be proven.	I. Given: $\triangle$ABC with $\overline{AB} = \overline{BC}$, $\overrightarrow{BD}$ bisects $\angle$B and $\angle A = \angle C$ Prove: $\triangle ABD \cong \triangle CBD$
II. Use the given to mark your diagram.	II. B A D C

III. Present a plan to prove your hypothesis.	III. Plan: Since two triangles can be proven congruent only by ASA = ASA, SAS = SAS, or SSS= SSS, you must select the one that best fits the information that you were given. You were given two angles and an included side. Therefore you can not use SSS or SAS. The ASA postulate would be the one that you would select.

	Statements	Reasons
IV. Set up your t-table. Provide logical statements on the left. Provide valid reasons for each statement to the right of the statement.	IV.	
	1. $\overline{AB} = \overline{BC}$ (side)	1. Given
	2. $\overrightarrow{BD}$ bisects $\angle B$	2. Given
	3. $\angle ABD = \angle CBD$ (angle)	3. Def. of bisect
	4. $\angle A = \angle C$ (angle)	4. Given
	5. $\triangle ABD \cong \triangle CBD$	5. Angle side angle ASA = ASA

Using postulates, definitions, and theorems, examine the following proofs. In some you will be asked to supply the reasons.

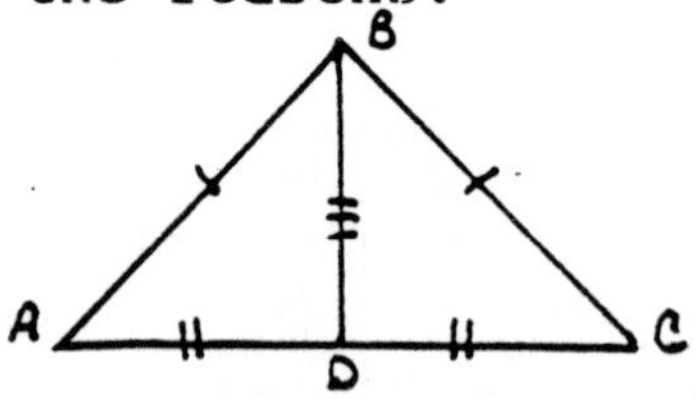

15. Given: $\triangle ABC$ with $\overline{AB} = \overline{BC}$ and $\overline{AD} = \overline{DC}$

Prove: $\triangle ABD \cong \triangle CBD$

Statements	Reasons
1. $\overline{AB} = \overline{BC}$ $\overline{AD} = \overline{DC}$	1. Given
2. $\overline{BD} = \overline{BD}$	2. Reflexive Law
3. $\triangle ABD \cong \triangle CBD$	3. SSS = SSS

Note: Markings generally help in demonstrating the given and pointing out what is still needed. Examine the markings on the figure above. In statement 1, $\overline{AB} = \overline{BC}$. On the diagram notice each of these has a single dash. In statement 1, $\overline{AD} = \overline{DC}$. On the diagram notice each of these has a double dash. Then in examining the two triangles, congruence can be obtained by SSS = SSS. This is only one of many methods of marking. Colored pencils may also be beneficial.

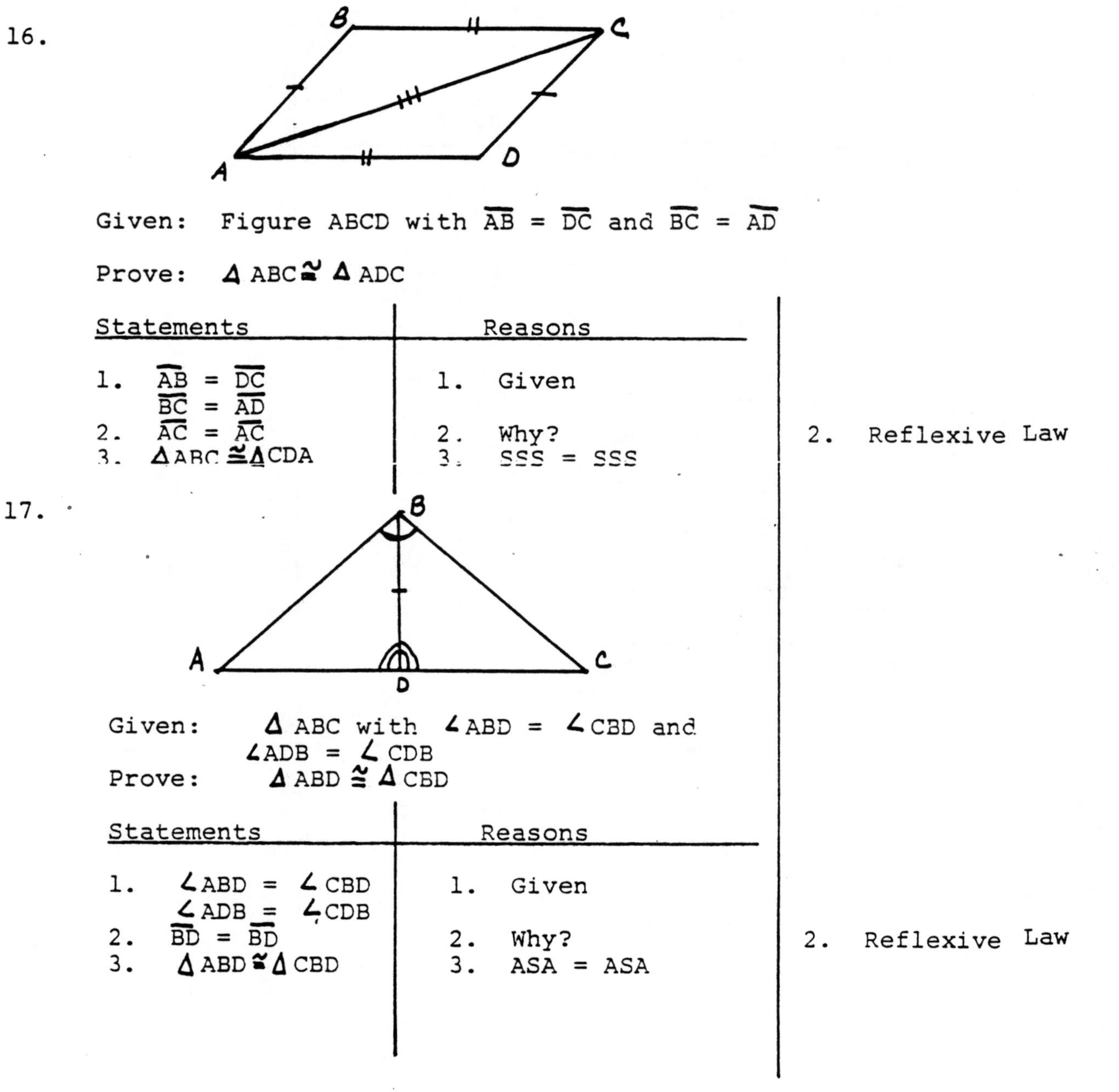

16.

Given: Figure ABCD with $\overline{AB} = \overline{DC}$ and $\overline{BC} = \overline{AD}$

Prove: $\Delta ABC \cong \Delta ADC$

Statements	Reasons	
1. $\overline{AB} = \overline{DC}$ $\overline{BC} = \overline{AD}$	1. Given	
2. $\overline{AC} = \overline{AC}$	2. Why?	2. Reflexive Law
3. $\Delta ABC \cong \Delta CDA$	3. SSS = SSS	

17.

Given: ΔABC with $\angle ABD = \angle CBD$ and $\angle ADB = \angle CDB$

Prove: $\Delta ABD \cong \Delta CBD$

Statements	Reasons	
1. $\angle ABD = \angle CBD$ $\angle ADB = \angle CDB$	1. Given	
2. $\overline{BD} = \overline{BD}$	2. Why?	2. Reflexive Law
3. $\Delta ABD \cong \Delta CBD$	3. ASA = ASA	

18.

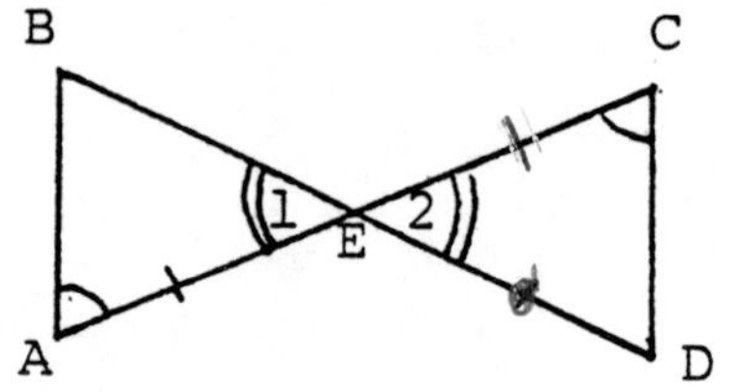

Given: $\overline{BD}$ bisects $\overline{AC}$ at E, $\angle A = \angle C$

Prove: $\triangle ABE \cong \triangle CDE$

Statements	Reasons
1. $\overline{BD}$ bisects $\overline{AC}$ at E, $\angle A = \angle C$	1. Given
2. $\overline{AE} = \overline{EC}$	2. Why?
3. $\angle 1$ and $\angle 2$ are vertical angles.	3. Def. of vertical angles.
4. $\angle 1 = \angle 2$	4. Pairs of vertical angles are equal.
5. $\triangle ABE \cong \triangle CDE$	5. Why?

18.

2. Def. of bisector

5. ASA = ASA

19.

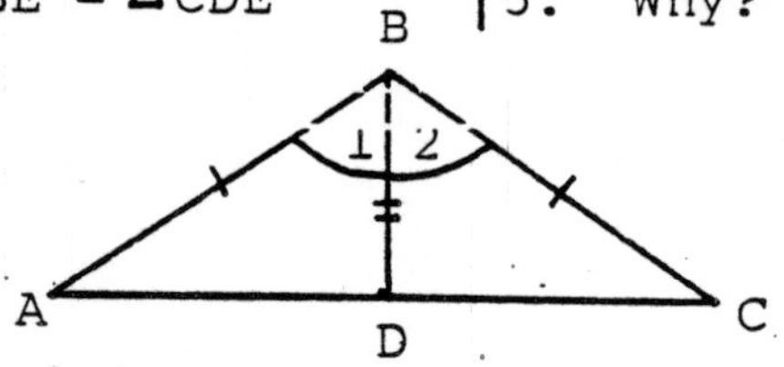

Given: $\triangle ABC$ is isosceles with $\overline{AB} = \overline{BC}$, $\angle 1 = \angle 2$

Prove: $\triangle ABD \cong \triangle CBD$

Statements	Reasons
1. $\triangle ABC$ is isosceles with $\overline{AB} = \overline{BC}$, $\angle 1 = \angle 2$	1. Given
2. $\overline{BD} = \overline{BD}$	2. Reflexive Law
3. $\triangle ABD \cong \triangle CBD$	3. Why?

19.

3. SAS = SAS

20.

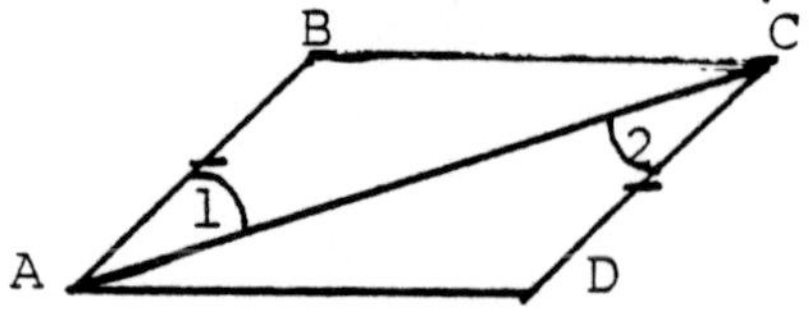

Given: $\overline{AB} = \overline{DC}$, $\angle 1 = \angle 2$

Prove: $\triangle ABC \cong \triangle CDA$

20.

Statements	Reasons	
1. $\overline{AB} = \overline{DC}$, $\angle 1 = \angle 2$	1. Given	
2. $\overline{AC} = \overline{AC}$	2. Why?	2. Reflexive
3. $\triangle ABC \cong \triangle CDA$	3. Why?	3. SAS = SAS

Exercise 3

Using the concepts on page 80 and procedures outlined on pages 83 and 84, set up your t-table and prove the following exercises.

1.

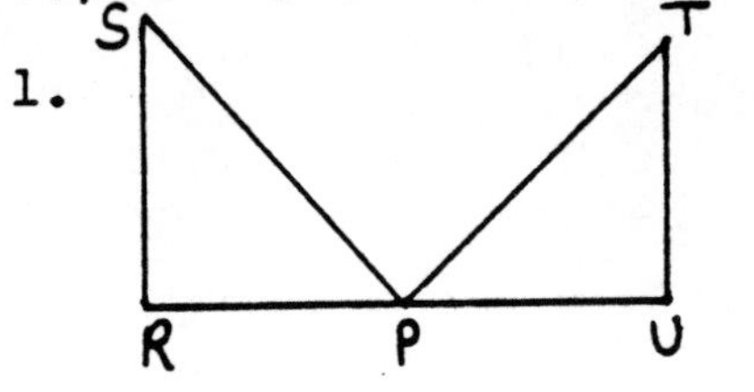

Given: $\angle S = \angle T$, $\overline{SR} = \overline{TU}$ and $\overline{SR} \perp \overline{RU}$ and $\overline{TU} \perp \overline{RU}$

Prove: $\triangle SRP \cong \triangle TUP$

2. Given: $\overline{MO}$ and $\overline{NP}$ bisect one another at their point of intersection.

Prove: $\triangle NMQ \cong \triangle POQ$

3.

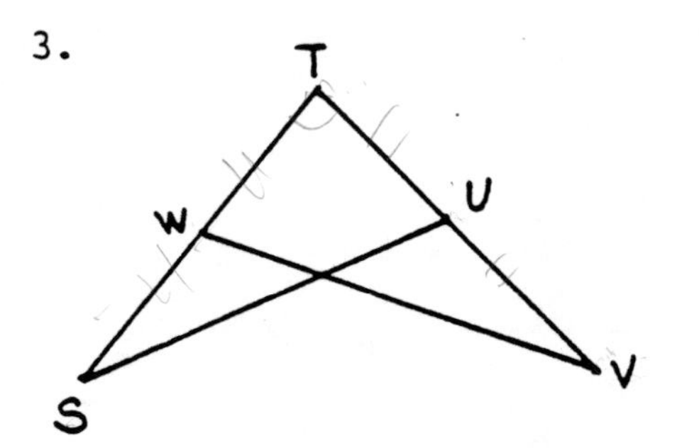

Given: $\overline{WT} = \overline{UT}$, $\overline{ST} = \overline{TV}$

Prove: $\triangle STU \cong \triangle VTW$

4.

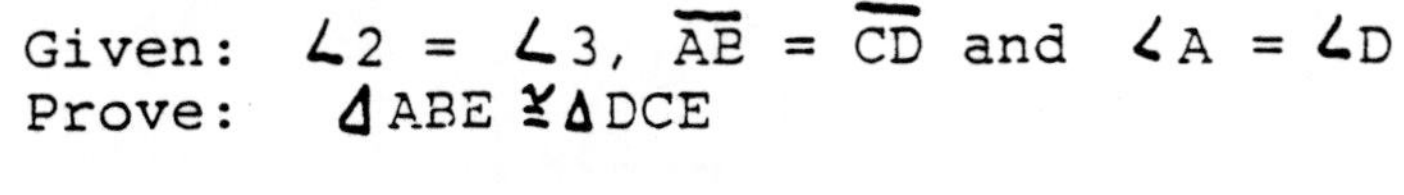

Given: $\angle 2 = \angle 3$, $\overline{AB} = \overline{CD}$ and $\angle A = \angle D$

Prove: $\triangle ABE \cong \triangle DCE$

5.

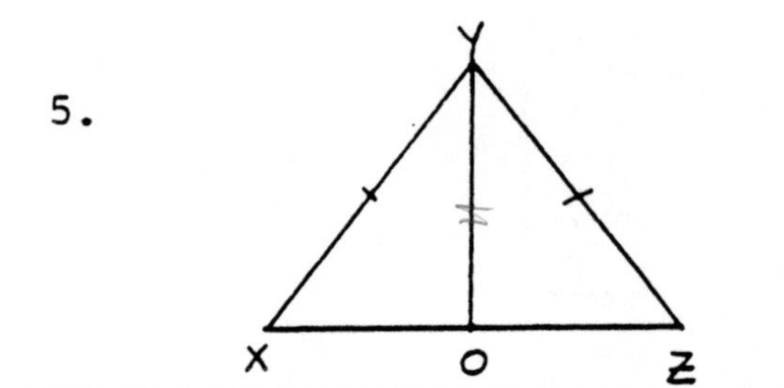

Given: ΔXYZ is isosceles with base $\overline{XZ}$ and $\overline{XO} = \overline{OZ}$
Prove: $\Delta XYO \cong \Delta ZYO$

In proving two triangles congruent, you have proven that these two figures have the same size and shape. Therefore, each pair of corresponding sides and each pair of corresponding angles are also equal. <u>Corresponding parts of congruent triangles are equal</u>. This complete statement may be abbreviated as C.P.C.T.E. Remember that you must <u>first</u> prove the triangles congruent before you can use CPCTE. Study the following examples.

Example:

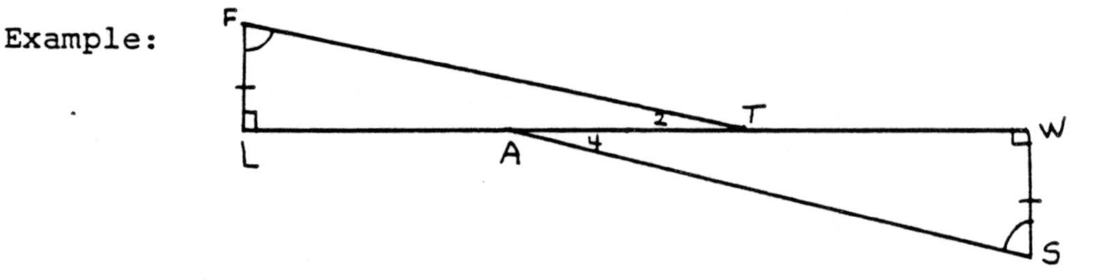

Given: $\overline{FL} \perp \overline{LT}$, $\overline{WS} \perp \overline{WA}$, $\angle F = \angle S$ and $\overline{FL} = \overline{WS}$
Prove: $\angle 2 = \angle 4$

Statements	Reasons
1. $\overline{FL} \perp \overline{LT}$, $\overline{WS} \perp \overline{WA}$, $\overline{WS} = \overline{FL}$, $\angle F = \angle S$	1. Given
2. $\angle L$ and $\angle W$ are right angles.	2. Definition of perpendicular lines.
3. $\angle L = \angle W$	3. All right angles are equal.
4. $\Delta FLT \cong \Delta SWA$	4. ASA = ASA
5. $\angle 2 = \angle 4$	5. CPCTE

Example:

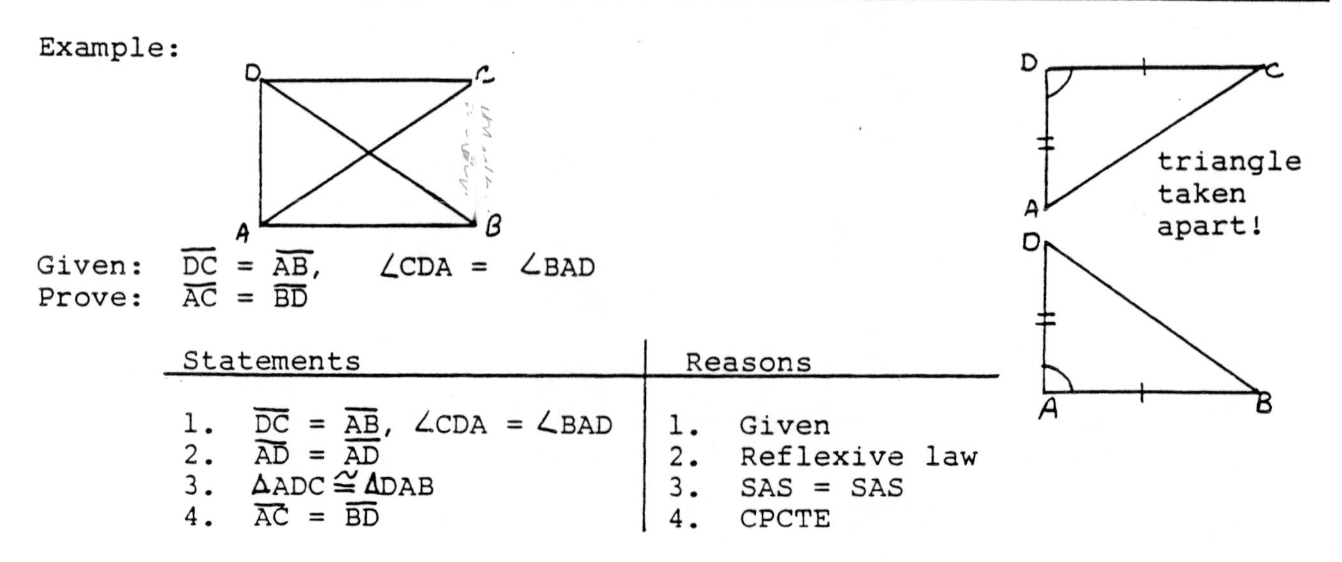

Given: $\overline{DC} = \overline{AB}$, $\angle CDA = \angle BAD$
Prove: $\overline{AC} = \overline{BD}$

Statements	Reasons
1. $\overline{DC} = \overline{AB}$, $\angle CDA = \angle BAD$	1. Given
2. $\overline{AD} = \overline{AD}$	2. Reflexive law
3. $\Delta ADC \cong \Delta DAB$	3. SAS = SAS
4. $\overline{AC} = \overline{BD}$	4. CPCTE

Proof of a Theorem

> Theorem 1: If two sides of a triangle are equal, then the angles opposite these sides are equal.

Remember the ____IF____ is the given, and the ____THEN____ is the part to prove.

"if" part
"then" part

So, if we use a triangle ABC with $\overline{AB} = \overline{BC}$ as the two equal sides, then we would need to prove which two angles are equal? The "given" is $\overline{AB} = \overline{BC}$ and the "to prove" is $\angle A = \angle C$.

This proof will be slightly different from the others. Before beginning, construct the bisector of $\angle B$. This can be done without altering the given or doing anything to what is to be proven. Constructing the bisector of an angle is one of our basic constructions. It is very important to remember that this can be done only if the "given" and the "to prove" is not altered.

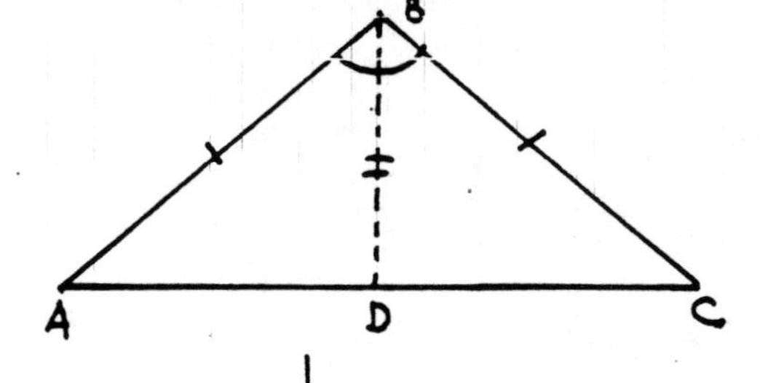

Statements	Reasons
1. $\overline{AB} = \overline{BC}$	1. Given
2. $\overline{BD}$ bisects $\angle B$	2. By construction
3. $\angle ABD = \angle CBD$	3. Def. of angle bisector
4. $\overline{BD} = \overline{BD}$	4. Reflexive Law
5. $\triangle ABD \cong \triangle CBD$	5. SAS = SAS
6. $\angle A = \angle C$	6. CPCTE

Looking at the converse of this theorem, we have:

> Theorem 2: If two angles of a triangle are equal, then the sides opposite these angles are equal.

Note: The converse of a theorem should also be proven. The proof of this theorem is not included since it is more difficult and would require more background to develop the logical steps. We will, however, use this theorem in proofs further along in the unit.

These two theorems are very useful in proving relationships between other triangles .

21. For example, in $\triangle ABC$ if $\angle 1 = \angle 2$ and $\overline{AE} = \overline{DC}$, prove $\overline{AB} = \overline{CB}$.

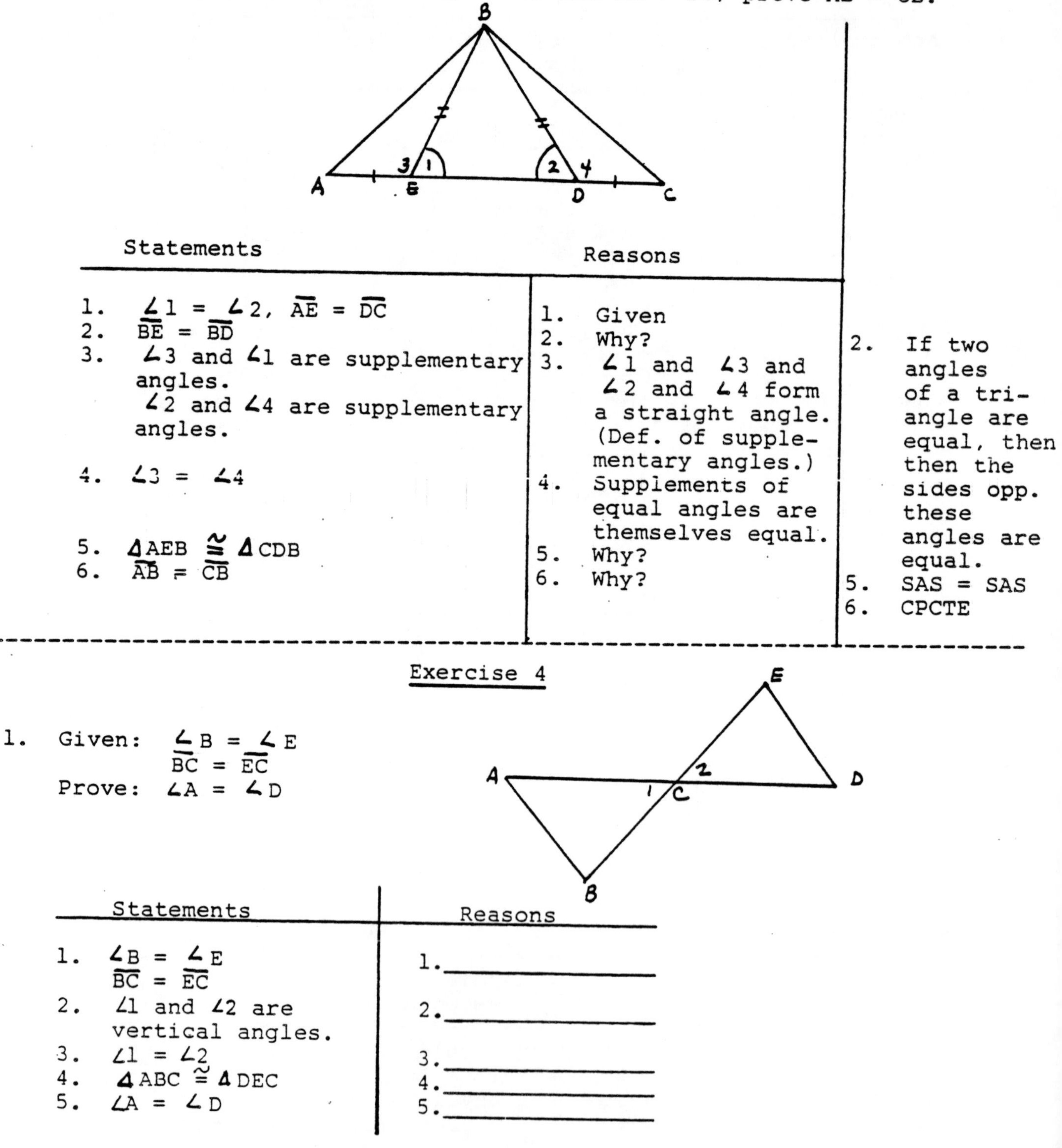

Statements	Reasons	
1. $\angle 1 = \angle 2$, $\overline{AE} = \overline{DC}$	1. Given	
2. $\overline{BE} = \overline{BD}$	2. Why?	2. If two angles of a triangle are equal, then then the sides opp. these angles are equal.
3. $\angle 3$ and $\angle 1$ are supplementary angles. $\angle 2$ and $\angle 4$ are supplementary angles.	3. $\angle 1$ and $\angle 3$ and $\angle 2$ and $\angle 4$ form a straight angle. (Def. of supplementary angles.)	
4. $\angle 3 = \angle 4$	4. Supplements of equal angles are themselves equal.	
5. $\triangle AEB \cong \triangle CDB$	5. Why?	5. SAS = SAS
6. $\overline{AB} = \overline{CB}$	6. Why?	6. CPCTE

Exercise 4

1. Given: $\angle B = \angle E$
   $\overline{BC} = \overline{EC}$
   Prove: $\angle A = \angle D$

Statements	Reasons
1. $\angle B = \angle E$, $\overline{BC} = \overline{EC}$	1.____________
2. $\angle 1$ and $\angle 2$ are vertical angles.	2.____________
3. $\angle 1 = \angle 2$	3.____________
4. $\triangle ABC \cong \triangle DEC$	4.____________
5. $\angle A = \angle D$	5.____________

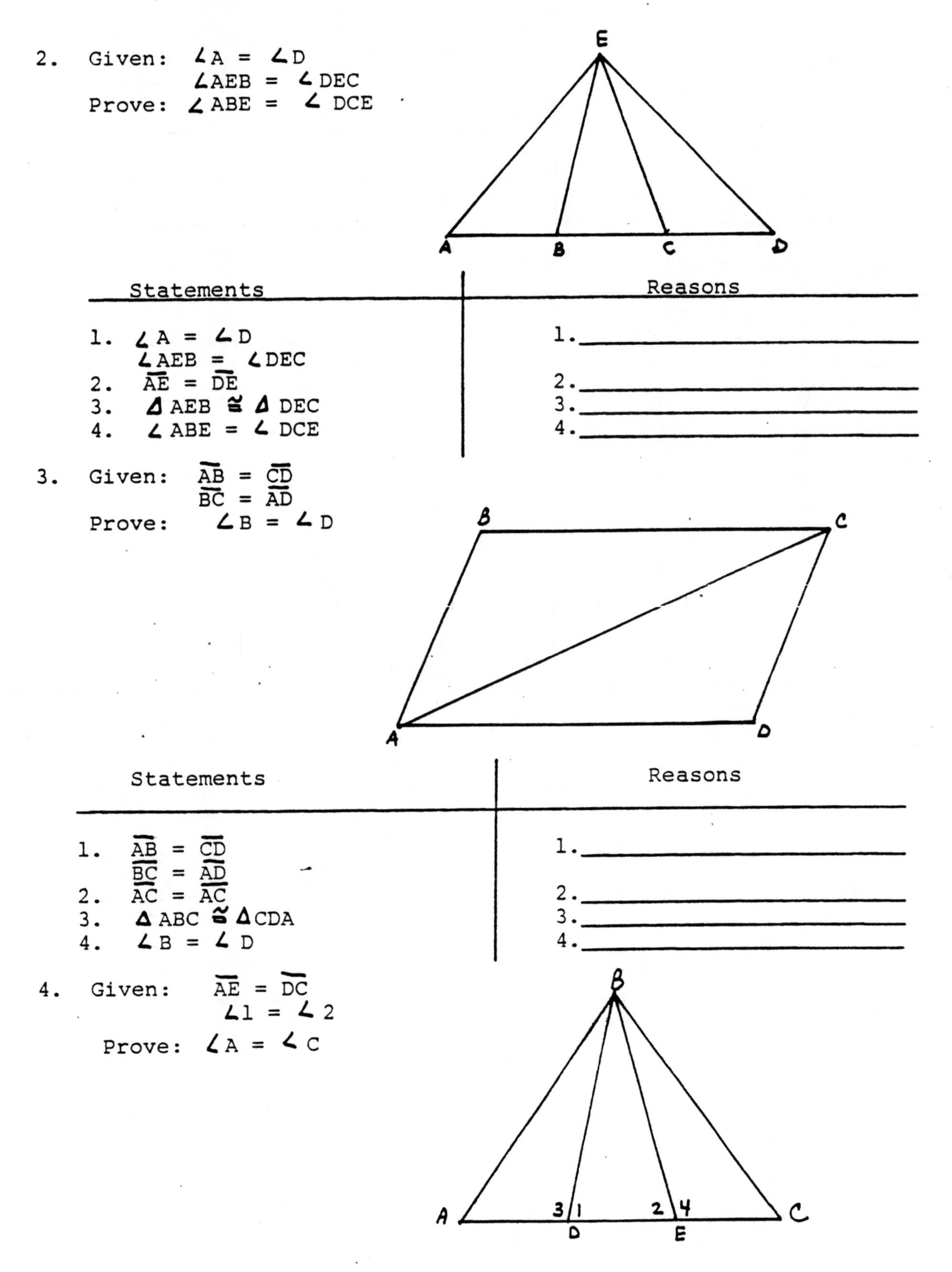

2. Given: $\angle A = \angle D$
$\angle AEB = \angle DEC$
Prove: $\angle ABE = \angle DCE$

Statements	Reasons
1. $\angle A = \angle D$ $\angle AEB = \angle DEC$	1.____________
2. $\overline{AE} = \overline{DE}$	2.____________
3. $\triangle AEB \cong \triangle DEC$	3.____________
4. $\angle ABE = \angle DCE$	4.____________

3. Given: $\overline{AB} = \overline{CD}$
$\overline{BC} = \overline{AD}$
Prove: $\angle B = \angle D$

Statements	Reasons
1. $\overline{AB} = \overline{CD}$ $\overline{BC} = \overline{AD}$	1.____________
2. $\overline{AC} = \overline{AC}$	2.____________
3. $\triangle ABC \cong \triangle CDA$	3.____________
4. $\angle B = \angle D$	4.____________

4. Given: $\overline{AE} = \overline{DC}$
$\angle 1 = \angle 2$
Prove: $\angle A = \angle C$

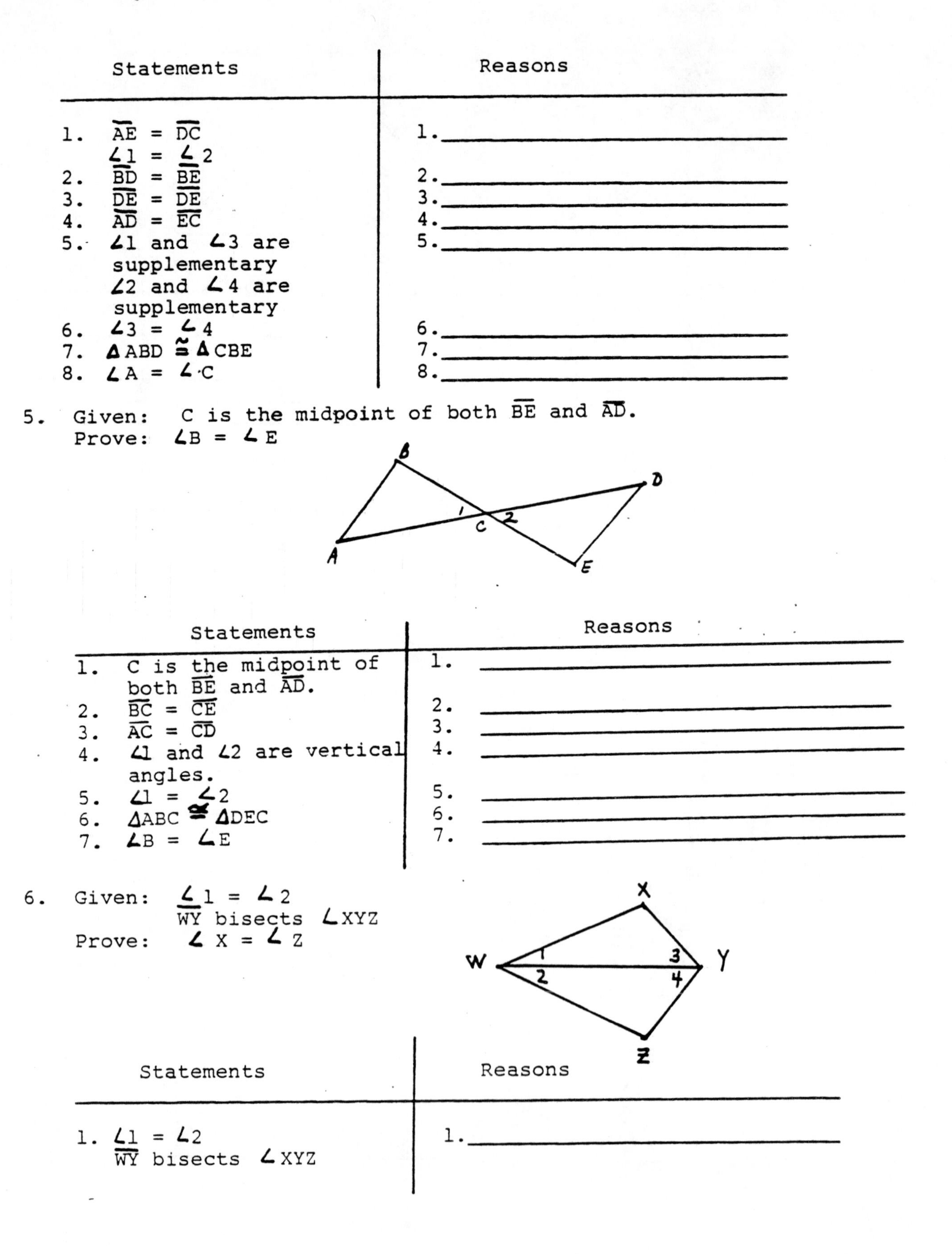

Statements	Reasons
1. $\overline{AE} = \overline{DC}$ $\angle 1 = \angle 2$	1.____________________
2. $\overline{BD} = \overline{BE}$	2.____________________
3. $\overline{DE} = \overline{DE}$	3.____________________
4. $\overline{AD} = \overline{EC}$	4.____________________
5. $\angle 1$ and $\angle 3$ are supplementary $\angle 2$ and $\angle 4$ are supplementary	5.____________________
6. $\angle 3 = \angle 4$	6.____________________
7. $\triangle ABD \cong \triangle CBE$	7.____________________
8. $\angle A = \angle C$	8.____________________

5. Given: C is the midpoint of both $\overline{BE}$ and $\overline{AD}$.
   Prove: $\angle B = \angle E$

Statements	Reasons
1. C is the midpoint of both $\overline{BE}$ and $\overline{AD}$.	1. ____________________
2. $\overline{BC} = \overline{CE}$	2. ____________________
3. $\overline{AC} = \overline{CD}$	3. ____________________
4. $\angle 1$ and $\angle 2$ are vertical angles.	4. ____________________
5. $\angle 1 = \angle 2$	5. ____________________
6. $\triangle ABC \cong \triangle DEC$	6. ____________________
7. $\angle B = \angle E$	7. ____________________

6. Given: $\angle 1 = \angle 2$
   $\overline{WY}$ bisects $\angle XYZ$
   Prove: $\angle X = \angle Z$

Statements	Reasons
1. $\angle 1 = \angle 2$ $\overline{WY}$ bisects $\angle XYZ$	1.____________________

2. $\angle 3 = \angle 4$	2. ______________
3. $\overline{WY} = \overline{WY}$	3. ______________
4. $\triangle WXY \cong \triangle WZY$	4. ______________
5. $\angle X = \angle Z$	5. ______________

7. Given: $\overline{BD}$ is the perpendicular bisector of $\overline{AC}$.

   Prove: $\angle ABD = \angle CBD$

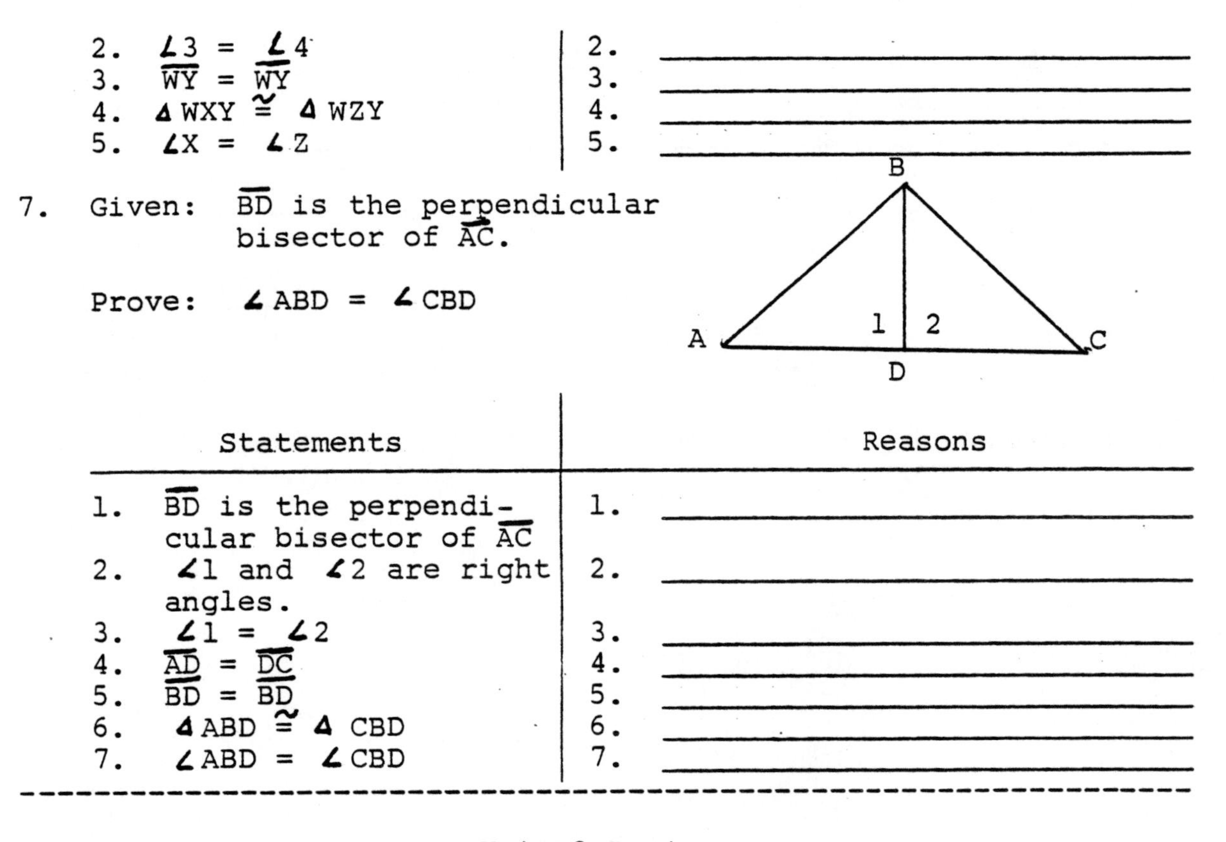

Statements	Reasons
1. $\overline{BD}$ is the perpendicular bisector of $\overline{AC}$	1. ______________
2. $\angle 1$ and $\angle 2$ are right angles.	2. ______________
3. $\angle 1 = \angle 2$	3. ______________
4. $\overline{AD} = \overline{DC}$	4. ______________
5. $\overline{BD} = \overline{BD}$	5. ______________
6. $\triangle ABD \cong \triangle CBD$	6. ______________
7. $\angle ABD = \angle CBD$	7. ______________

Unit 3 Review

1. The "IF" part of a conditional statement is called the ______________ and the "THEN" part is called the __________.

2. Write the converse, inverse, and contrapositive of the following conditional statement:

   "If I study hard, then I will pass geometry."

3. Does the converse of a theorem need to be proven if the theorem has already been proven? Why or why not?

4. Three ways to prove two triangles ar congruent are ________, ________, and ________.

5. Given: $\overline{AG} = \overline{FE}$, $\overline{BF} = \overline{DG}$, $\angle DGE = \angle BFA$

   Prove: $\triangle ABF \cong \triangle EDG$

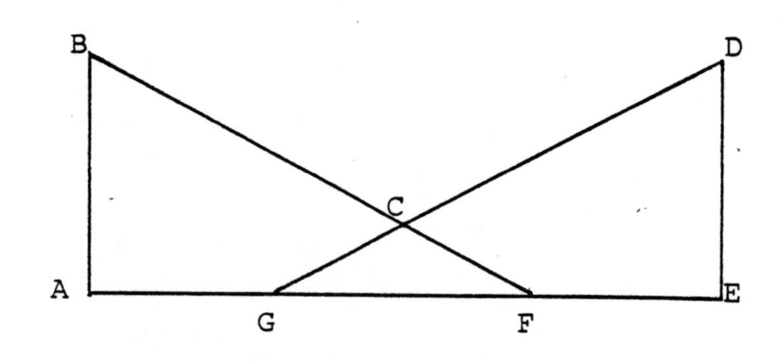

Statements	Reasons
1. $\overline{AG} = \overline{FE}$ $\overline{BF} = \overline{DG}$ $\angle DGE = \angle BFA$	1.________________
2. $\overline{GF} = \overline{GF}$	2.________________
3. $\overline{AF} = \overline{GE}$	3.________________
4. $\triangle ABF \cong \triangle EDG$	4.________________

6. Given: $\angle 1 = \angle 4$, $\angle 2 = \angle 3$
   Prove: $\triangle ABD \cong \triangle CDB$

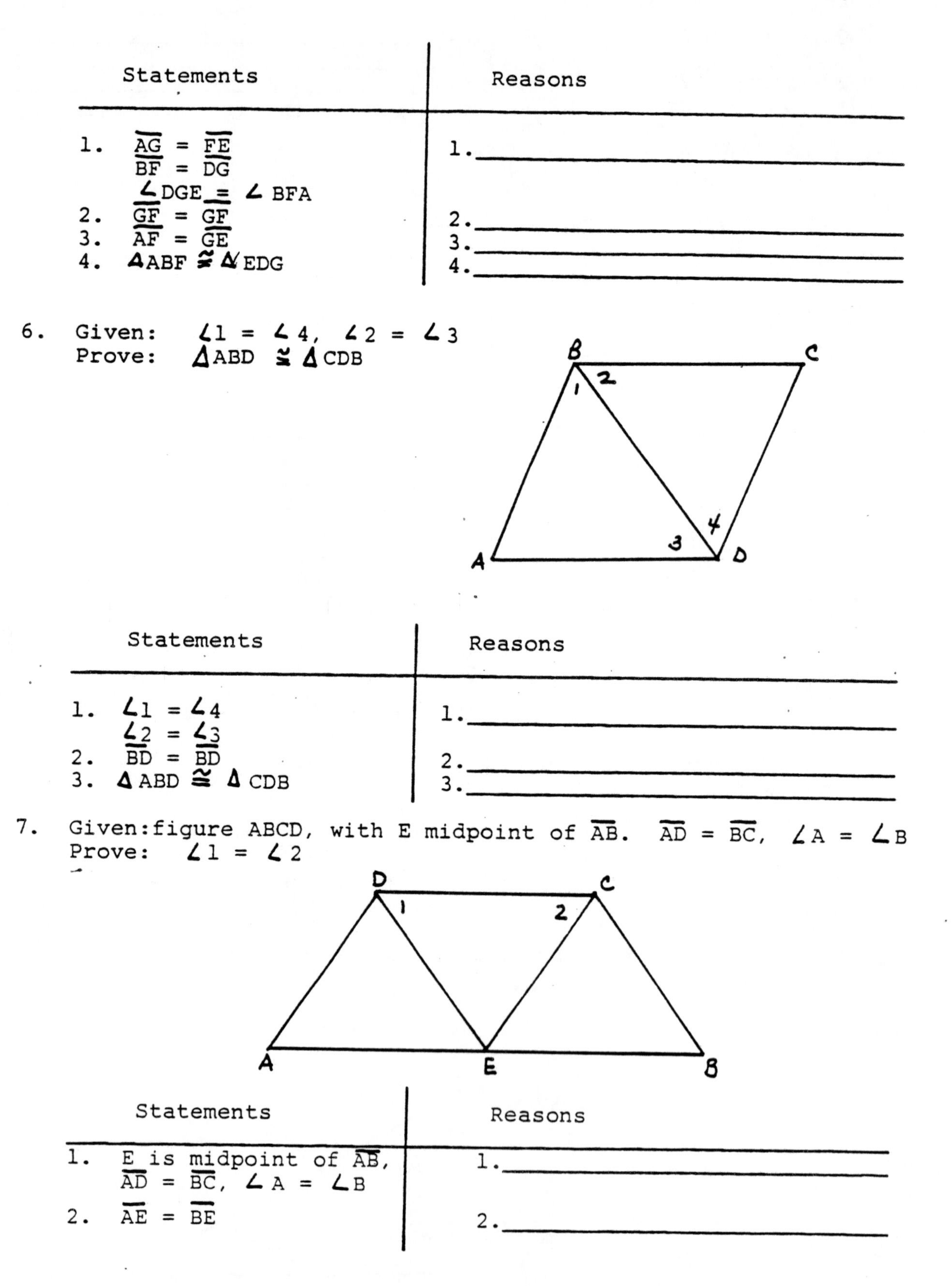

Statements	Reasons
1. $\angle 1 = \angle 4$ $\angle 2 = \angle 3$	1.________________
2. $\overline{BD} = \overline{BD}$	2.________________
3. $\triangle ABD \cong \triangle CDB$	3.________________

7. Given: figure ABCD, with E midpoint of $\overline{AB}$. $\overline{AD} = \overline{BC}$, $\angle A = \angle B$
   Prove: $\angle 1 = \angle 2$

Statements	Reasons
1. E is midpoint of $\overline{AB}$, $\overline{AD} = \overline{BC}$, $\angle A = \angle B$	1.________________
2. $\overline{AE} = \overline{BE}$	2.________________

Statements	Reasons
3. $\triangle ADE \cong \triangle BCE$	3.__________
4. $\overline{DE} = \overline{CE}$	4.__________
5. $\angle 1 = \angle 2$	5.__________

8. Given: $\overline{CD}$ bisects $\overline{AB}$; $\overline{CD} \perp \overline{AB}$
   Prove: $\triangle ABC$ is isosceles.

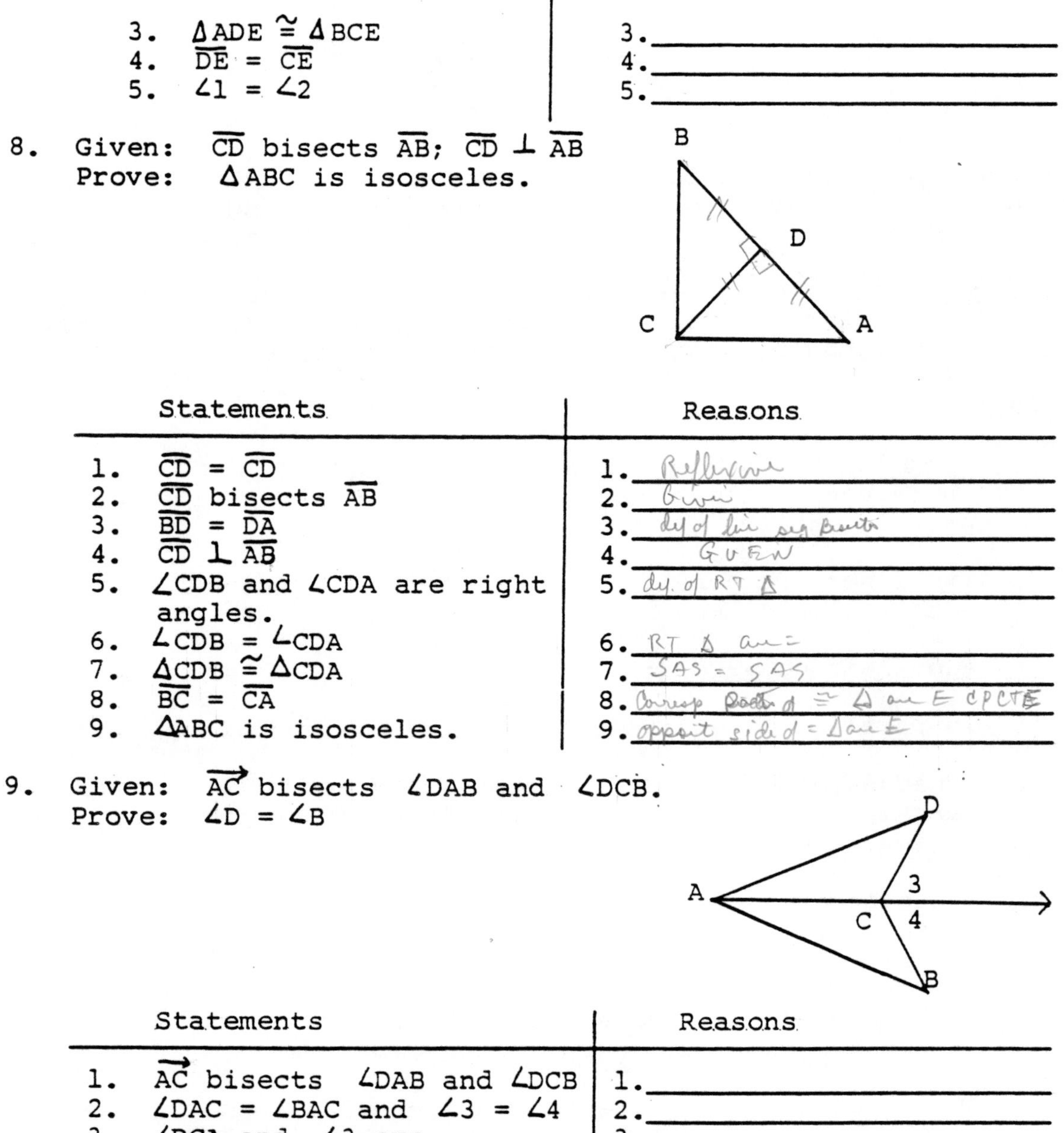

Statements	Reasons
1. $\overline{CD} = \overline{CD}$	1. Reflexive
2. $\overline{CD}$ bisects $\overline{AB}$	2. Given
3. $\overline{BD} = \overline{DA}$	3. def of line seg Bisector
4. $\overline{CD} \perp \overline{AB}$	4. GIVEN
5. $\angle CDB$ and $\angle CDA$ are right angles.	5. def. of RT ∆
6. $\angle CDB = \angle CDA$	6. RT ∆ are =
7. $\triangle CDB \cong \triangle CDA$	7. SAS = SAS
8. $\overline{BC} = \overline{CA}$	8. Corresp parts of ≅ ∆ are = CPCTE
9. $\triangle ABC$ is isosceles.	9. opposite sides of = ∆ are =

9. Given: $\overrightarrow{AC}$ bisects $\angle DAB$ and $\angle DCB$.
   Prove: $\angle D = \angle B$

D
A 3
C 4
B

Statements	Reasons
1. $\overrightarrow{AC}$ bisects $\angle DAB$ and $\angle DCB$	1.__________
2. $\angle DAC = \angle BAC$ and $\angle 3 = \angle 4$	2.__________
3. $\angle DCA$ and $\angle 3$ are supplementary angles. $\angle BCA$ and $\angle 4$ are supplementary angles.	3.__________
4. $\angle DCA = \angle BCA$	4.__________
5. $\overline{AC} = \overline{AC}$	5.__________
6. $\triangle ACD \cong \triangle ACB$	6.__________
7. $\angle D = \angle B$	7.__________

Unit 4

Parallel Lines

Learning objectives:

The student will demonstrate his mastery of the following definitions, postulates, and theorems by writing them, by applying them to solutions of selected problems, or using them as reasons in a proof.

<u>Definitions</u>: parallel lines, skew lines, transversal, alternate interior angles, alternate exterior angles, corresponding angles, corollary, and exterior angle.

<u>Postulates</u>:

1. Parallel Postulate. Through any point, P, not on line l, there can be constructed one and only one line m that is parallel to l.

2. Two lines are parallel if they are both perpendicular to a third line.

<u>Theorems and their converses</u>:

If two parallel lines are cut by a transversal, then

1. Nonadjacent angles on the opposite sides of the transversal but interior to the two lines (alternate interior angles) are equal.

2. Nonadjacent angles on the opposite sides of the transversal but exterior to the two lines (alternate exterior angles) are equal.

3. Angles on the same side of the transversal but interior to the two lines are supplementary.

4. Angles on the same side of the transversal and in the same position relative to the two lines (corresponding angles) are equal.

<u>Theorems</u>:

5. The sum of the interior angles of a triangle is 180°.

6. If a line joins the midpoints of two sides of a triangle, that line is parallel to and equal to one-half the measure of the third side.

Corollaries:

1. In a right triangle, the two acute angles are complementary.

2. Each exterior angle of a triangle equals the sum of its non-adjacent interior angles.

The following constructions are required:

1. Construct parallel lines using alternate interior angles equal.

2. Construct parallel lines using Postulate 2.

Have you ever noticed that the horizontal lines on a piece of notebook paper never intersect or that teeth on a comb never cross one another? The environment is abundant with examples of parallel lines. The rows of mortar between bricks, railroad tracks, the horizontal bars in a five-barred gate, and the center and outer edge markings on an interstate highway provide a few illustrations of parallel lines.

Definition:	Lines that are (1) in the same plane and (2) do not intersect are called parallel lines.

A plane can be described as a flat surface that extends indefinitely in every direction.

The first relationship of parallel lines refers to the fact that these lines must lie totally within the same plane.

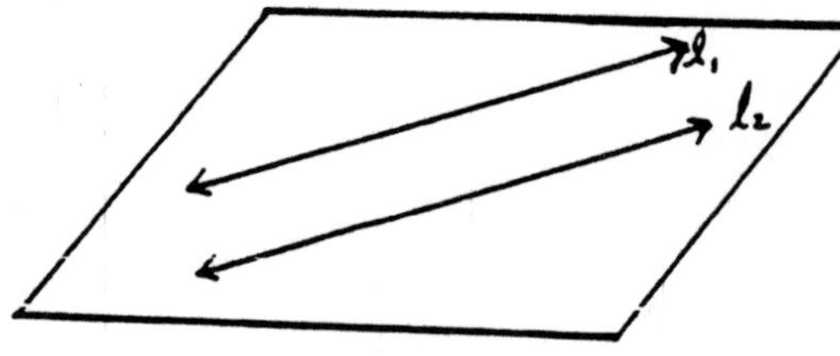

plane

Figure 4.1

Line l_1 is parallel to l_2.

The symbol for parallel is $\parallel$.

Therefore, $l_1 \parallel l_2$.

That is, we should be able to lay a piece of paper flat against both lines. If one line passes through the paper and the other lies on the paper, these lines are not contained in the same plane and, therefore, are not parallel.

Definition:	Lines that are (1) not in the same plane and (2) do not intersect are called skew.

Skew lines are illustrated below:

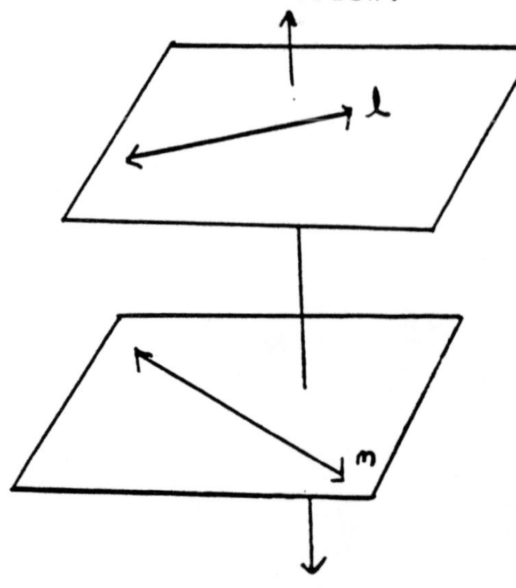

Figure 4.2

Lines m and l are not in the same plane and do not intersect.

Lines that are not parallel but are in the same plane must intersect at some point within that plane. See l_3 and l_4 in Figure 4.3.

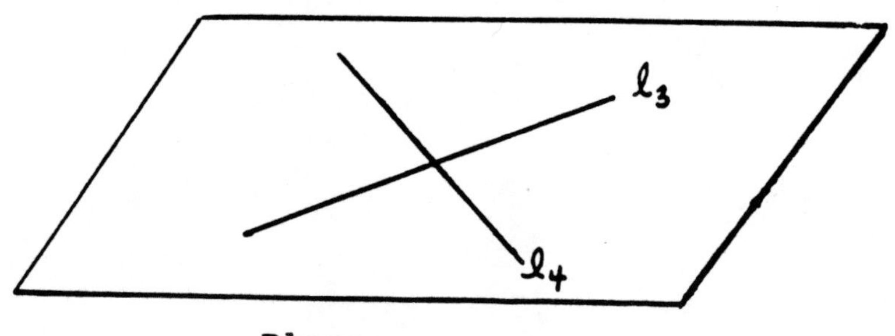

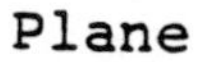
Plane

Figure 4.3

1. Which of the following lines appear to be intersecting lines and which ones are || ?

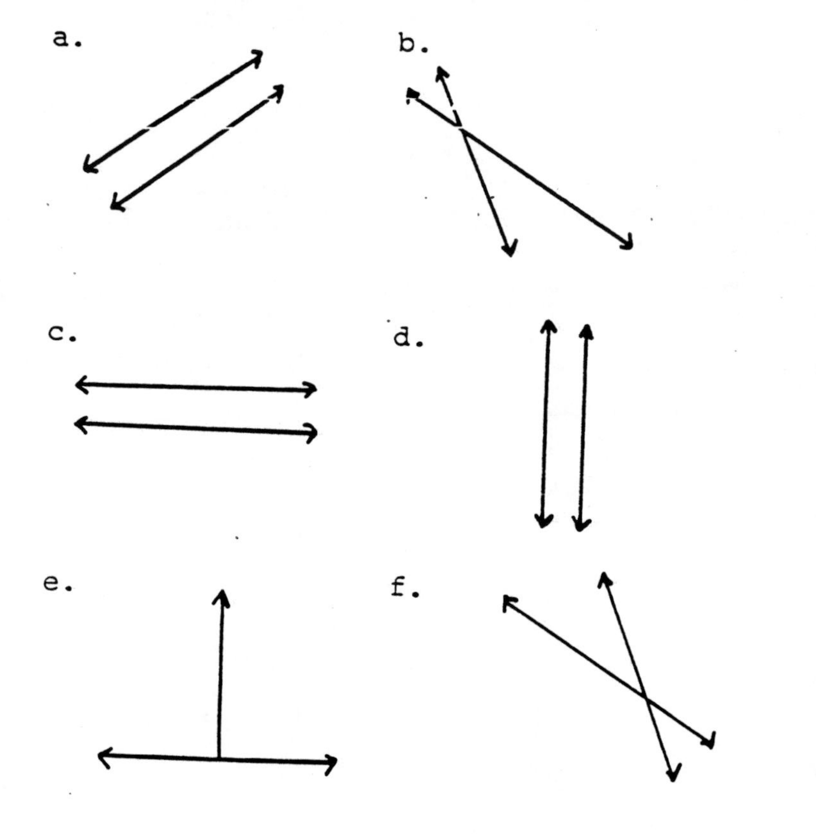

1. parallel a, c, d

   intersecting b,e,f

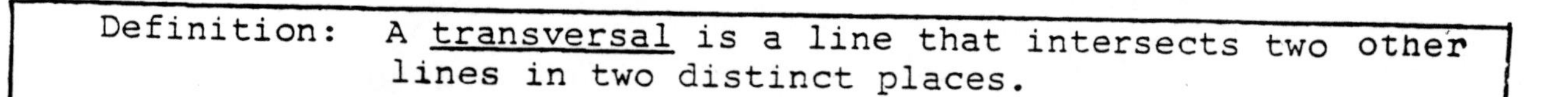

Definition: A <u>transversal</u> is a line that intersects two other lines in two distinct places.

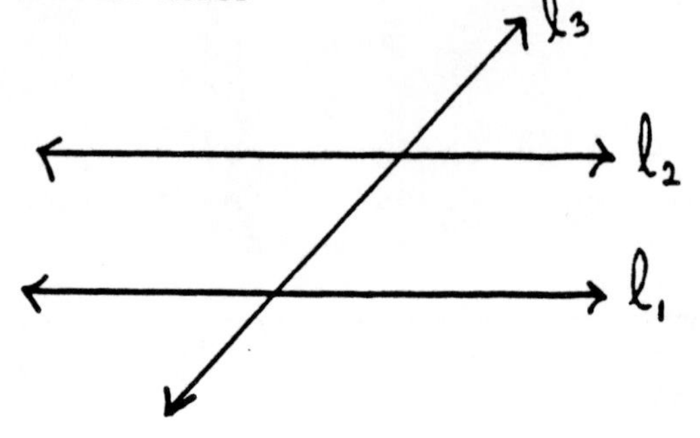

$l_1 \parallel l_2$ and l_3 is transversal.

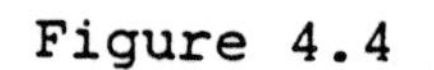

Figure 4.4

Is line l_3 in Figure 4.5 a transversal?

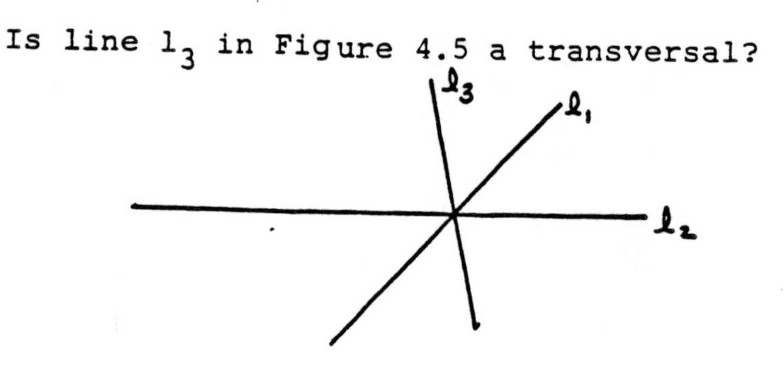

No, l_3 is not a transversal. It is a line but it does not intersect two other lines in <u>two</u> distinct places. It intersects l_1 and l_2 in only <u>one</u> point.

Figure 4.5

ANGLES RELATED TO TWO LINES INTERSECTED BY A TRANSVERSAL

Definition:	<u>Alternate interior angles</u> are nonadjacent angles on opposite sides of the transversal but "interior" to the two lines.

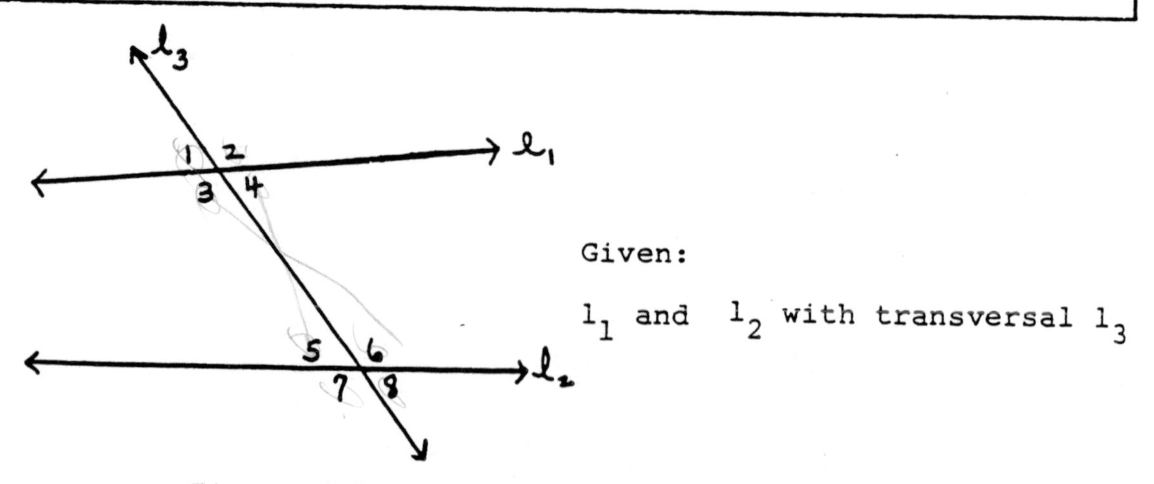

Given:

l_1 and l_2 with transversal l_3

Figure 4.6

The pair of angles ∠3 and ∠6; ∠4 and ∠5 are called alternate interior.

> Definition: <u>Alternate exterior angles</u> are nonadjacent angles on opposite sides of the transversal but "exterior" to the two lines.

In Figure 4.6, the pair of angles that fit this description are angles ∠1 and ∠8; ∠2 and ∠7.

> Definition: <u>Corresponding angles</u> are angles on the same side of the transversal and in the same relative position to the given lines.

In Figure 4.6, ∠1 and ∠5; ∠2 and ∠6; ∠3 and ∠7; ∠4 and ∠8 are all pairs of corresponding angles.

Using a protractor, measure the alternate interior angles, the alternate exterior angles, and the corresponding angles in Figure 4.6. You will discover that if two lines are parallel, the pairs of alternate interior angles, alternate exterior angles and corresponding angles are all <u>equal</u>.

Example: Given l || m and ∠1 = 48°. With the use of your protractor, find the other numbered angles.

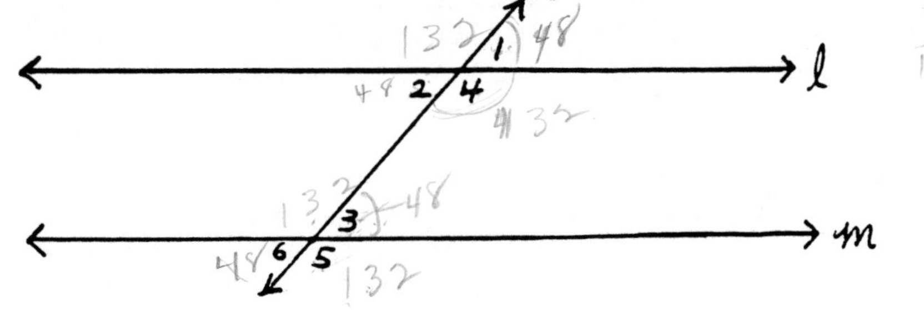

∠1 = 48°

∠2 = 48°

∠3 = 48°

∠4 = 132°

∠5 = 132°

∠6 = 48°

Observe here, ∠3 = 48°, ∠4 = 132° and $48^\circ + 132^\circ = 180^\circ$. This equation indicates that ∠3 and ∠4 are <u>supplementary</u>.

2.

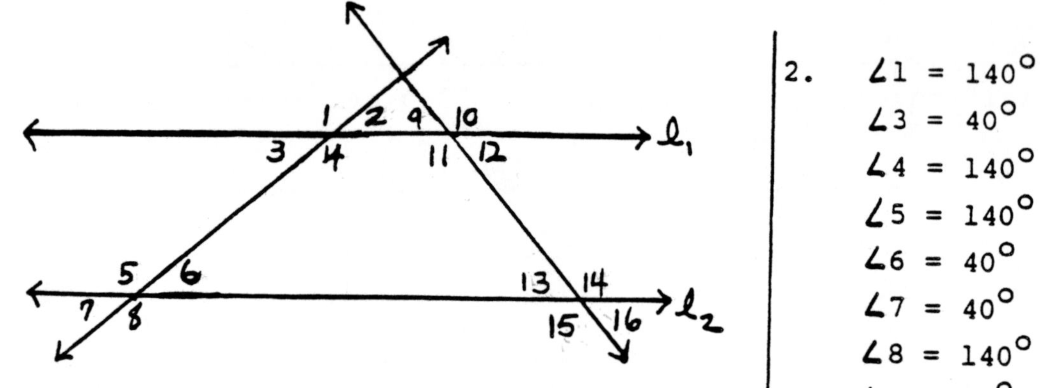

$l_1 \parallel l_2$ and $\angle 2 = 40^\circ$ and $\angle 13 = 50^\circ$. By using your protractor, find the following measures:

$\angle 1$ = ________	$\angle 8$ = ________
$\angle 3$ = ________	$\angle 9$ = ________
$\angle 4$ = ________	$\angle 10$ = ________
$\angle 5$ = ________	$\angle 11$ = ________
$\angle 6$ = ________	$\angle 12$ = ________
$\angle 7$ = ________	$\angle 14$ = ________
	$\angle 15$ = ________
	$\angle 16$ = ________

2. $\angle 1 = 140^\circ$
$\angle 3 = 40^\circ$
$\angle 4 = 140^\circ$
$\angle 5 = 140^\circ$
$\angle 6 = 40^\circ$
$\angle 7 = 40^\circ$
$\angle 8 = 140^\circ$
$\angle 9 = 50^\circ$
$\angle 10 = 130^\circ$
$\angle 11 = 130^\circ$
$\angle 12 = 50^\circ$
$\angle 14 = 130^\circ$
$\angle 15 = 130^\circ$
$\angle 16 = 50^\circ$

By using your protractor in answering the steps in problem 2, you should have discovered the following four theorems:

"If two parallel lines are cut by a transversal, then ..."

<u>Theorem (1)</u>: Nonadjacent angles on the opposite sides of the transversal but interior to the two lines (alternate interior angles) are equal.

<u>Theorem (2)</u>: Nonadjacent angles on the opposite sides of the transversal but exterior to the two lines (alternate exterior angles) are equal.

<u>Theorem (3)</u>: Angles on the <u>same</u> <u>side</u> of the transversal but interior to the two lines are supplementary.

<u>Theorem (4)</u>: Angles on the same side of the transversal and in the same position relative to the two lines (corresponding angles) are equal.

Now, re-work problem 2, page 102. This time do not use your protractor. Use the four theorems for parallel lines and other definitions and theorems about angles to find the measure of each angle. One possible set of reasons for the angle measures follows:

$\angle 1 = 140°$ Supplementary to $\angle 2$
$\angle 3 = 40°$ Vertical to $\angle 2$
$\angle 4 = 140°$ Vertical to $\angle 1$
$\angle 5 = 140°$ Alternate interior to $\angle 4$
$\angle 6 = 40°$ Alternate interior to $\angle 3$
$\angle 7 = 40°$ Corresponding to $\angle 3$
$\angle 8 = 140°$ Corresponding to $\angle 4$
$\angle 9 = 50°$ Corresponding to $\angle 13$
$\angle 10 = 130°$ Supplementary to $\angle 9$
$\angle 11 = 130°$ Vertical to $\angle 10$
$\angle 12 = 50°$ Vertical to $\angle 9$
$\angle 14 = 130°$ Corresponding to $\angle 10$
$\angle 15 = 130°$ Corresponding to $\angle 11$
$\angle 16 = 50°$ Corresponding to $\angle 12$

Algebraic expressions may be used in the evaluation of angles and parallel lines.

Example: Given $p \parallel q$. Find the value of x and the measures of each angle as shown in the figure.

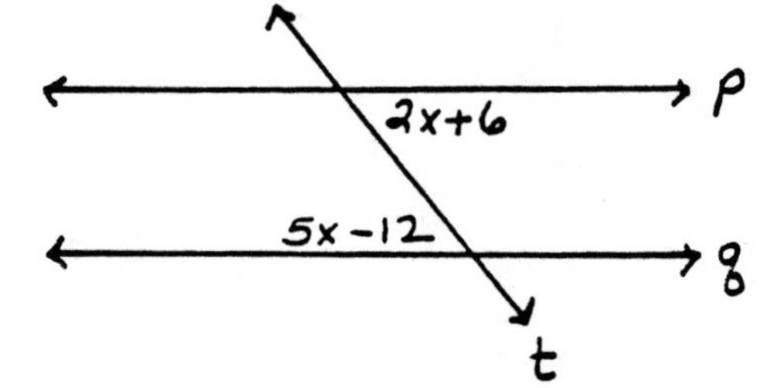

Since alternate interior angles are equal,

then $2x + 6 = 5x - 12$

Now, solve.

$$\begin{aligned} 2x + 6 &= 5x - 12 \\ -3x + 6 &= -12 \\ -3x &= -18 \\ x &= 6 \end{aligned}$$

By substituting 6 into each expression, we find:

$2x + 6$	(or)	$5x - 12$
$2(6) + 6$		$5(6) - 12$
$12 + 6$		$30 - 12$
18°		18°

The value of x is 6 and the angles both measure 18°.

3. You solve the following:

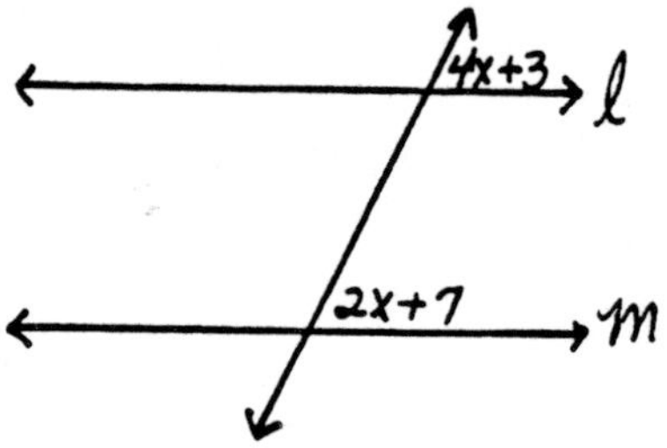

Given l ∥ m. Find x and the measure of the angles.

(a) The two angles are called ________ and are equal.
(b) $4x + 3 =$ ________.

Simplify to

(c) $x =$ ________
(d) One angle is $4x + 3$ which equals ________.

3.

(a) corresponding angles
(b) $4x+3=2x+7$
$2x=4$
(c) $x = 2$
(d) $4(2) + 3$
$8 + 3$
11°

4.

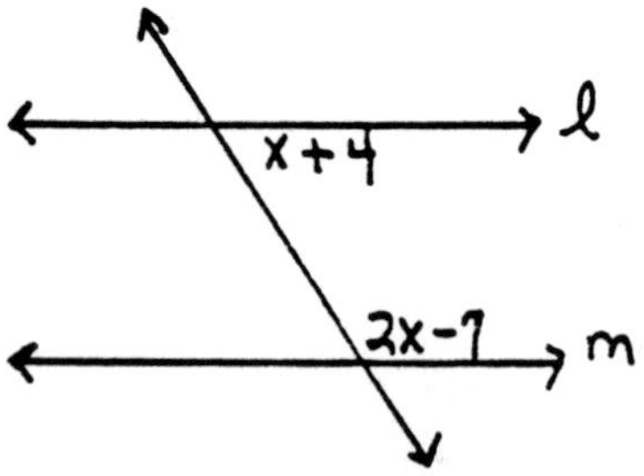

Given l ∥ m. Find x and the measure of the angles.

(a) The two angles are called ________.
(b) $(x + 4) + (2x - 7) =$ ________

Simplify to
(c) $x =$ ________
(d) One angle is $x + 4$ which equals____.
(e) The other angle equals ________.

4.

(a) supple-mentary
(b) $(x+4)+(2x-7)=180^\circ$
$3x - 3 = 180$
$3x = 183$
(c) $x = 61$
(d) $(61) + 4 = 65^\circ$
(e) $2(61)-7 = 115^\circ$

Parallel lines give us four new concepts (Theorems, page 102) that can be applied to simple proofs.

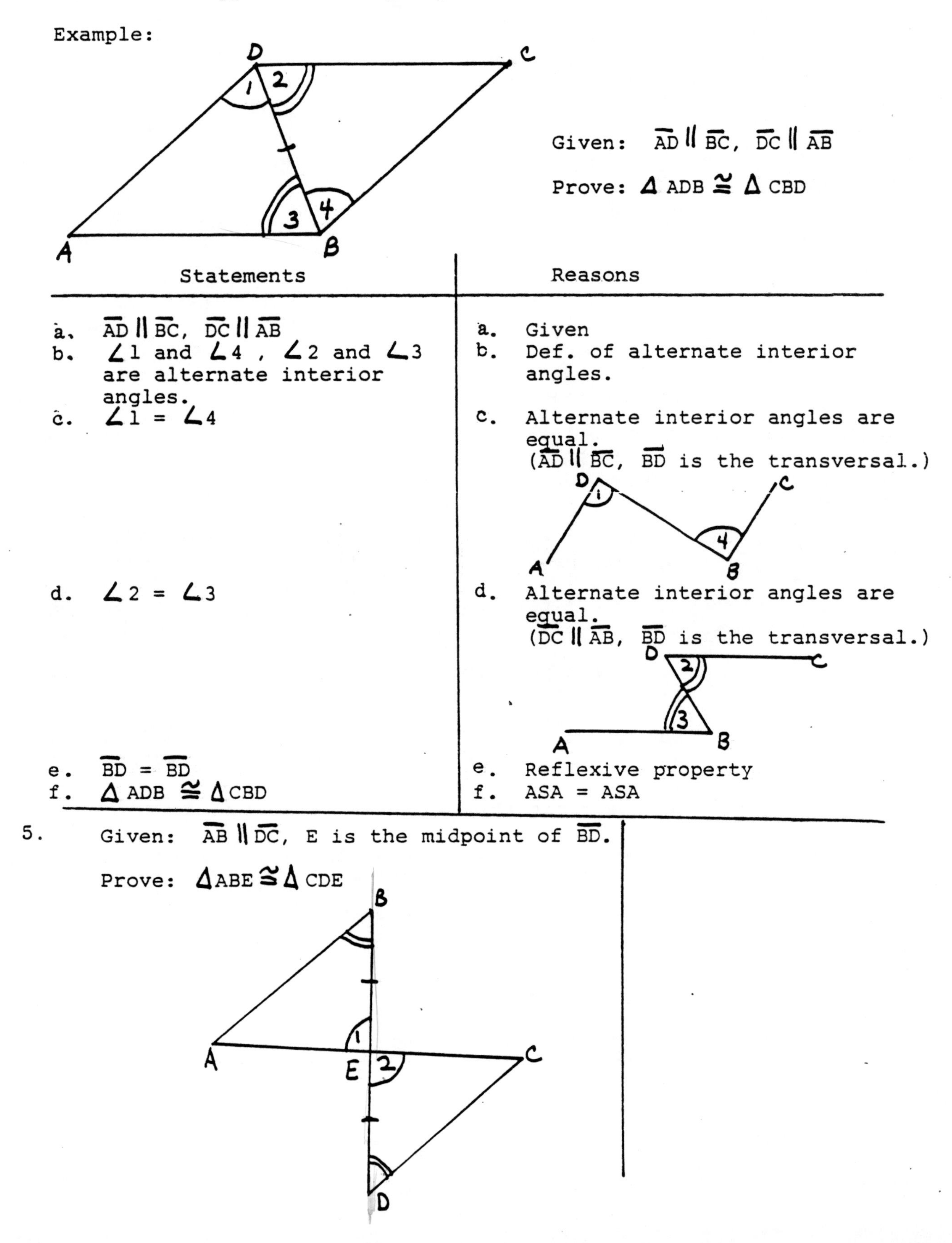

Example:

Given: $\overline{AD} \parallel \overline{BC}$, $\overline{DC} \parallel \overline{AB}$

Prove: $\triangle ADB \cong \triangle CBD$

Statements	Reasons
a. $\overline{AD} \parallel \overline{BC}$, $\overline{DC} \parallel \overline{AB}$	a. Given
b. $\angle 1$ and $\angle 4$, $\angle 2$ and $\angle 3$ are alternate interior angles.	b. Def. of alternate interior angles.
c. $\angle 1 = \angle 4$	c. Alternate interior angles are equal. ($\overline{AD} \parallel \overline{BC}$, $\overline{BD}$ is the transversal.)
d. $\angle 2 = \angle 3$	d. Alternate interior angles are equal. ($\overline{DC} \parallel \overline{AB}$, $\overline{BD}$ is the transversal.)
e. $\overline{BD} = \overline{BD}$	e. Reflexive property
f. $\triangle ADB \cong \triangle CBD$	f. ASA = ASA

5. Given: $\overline{AB} \parallel \overline{DC}$, E is the midpoint of $\overline{BD}$.

Prove: $\triangle ABE \cong \triangle CDE$

	Statements		Reasons	5.	
a.	$\overline{AB} \parallel \overline{DC}$, E is the midpoint of $\overline{BD}$.	a.	Given	a.	-----
b.	$\angle 1$ and $\angle 2$ are vertical angles.	b.	Why?	b.	Def. of vertical angles.
c.	$\angle 1 = \angle 2$	c.	Why?	c.	Vertical angles are equal.
d.	$\overline{BE} = \overline{ED}$	d.	Why?	d.	Def. of midpoint.
e.	$\angle B$ and $\angle D$ are alternate interior angles.	e.	Why?	e.	Def. of alternate interior angles.
f.	$\angle B = \angle D$	f.	Why?	f.	Alternate interior angles are equal.
g.	$\triangle ABE \cong \triangle CDE$	g.	Why?	g.	ASA = ASA

Converses of conditional statements have been mentioned earlier. Recall that the converse interchanges the hypothesis and conclusion of a conditional statement. The four theorems studied in this chapter (page 102)have true converses. The theorems are as follows:

(1) If two lines are parallel, the alternate interior angles formed by a transversal cutting these lines are equal.

(2) If two lines are parallel, the alternate exterior angles formed by a transversal cutting these lines are equal.

(3) If two lines are parallel, the corresponding angles formed by a transversal cutting these lines are equal.

(4) If two lines are parallel, angles on the same side of the transversal but interior to the two lines are supplementary.

6. The converse of the first theorem reads as follows:

(i) If two lines cut by a transversal form equal alternate interior angles, then the lines are parallel.

Write the converses of the other three theorems below:

(ii) ____

____.

(iii)____.

____.

(iv) ______________________________

____

7.

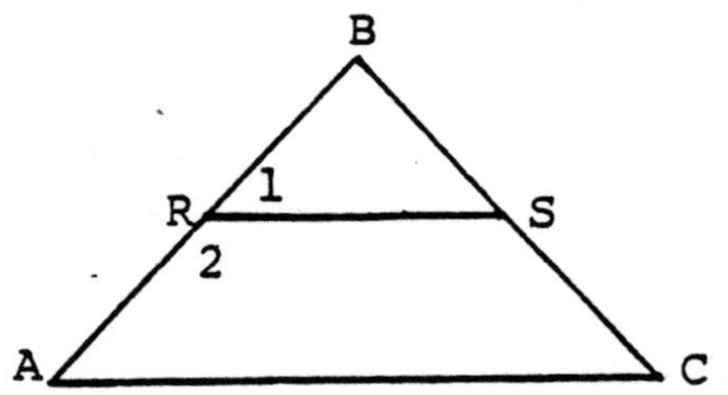

Let's consider the following conditions:

(a) If $\angle 1 = \angle A$, is $\overline{RS} \parallel \overline{AC}$? Why?______
(Hint: Converse (iii))

(b) If $\angle 1 = \angle C$ and $\overline{AB} = \overline{BC}$, is $\overline{RS} \parallel \overline{AC}$?
______
Why? ______
(Hint: Converse (iii))

(c) If $\angle RSC = 103^\circ$ and $\angle C = 77^\circ$, is $\overline{RS} \parallel \overline{AC}$? ______
Why?______
(Hint: Converse (iv))

7.

(a) Yes.
If corresponding angles are equal, then the lines are parallel.

(b) Yes, if $\overline{AB} = \overline{BC}$ then $\angle A = \angle C$. (Angles opposite equal sides of a triangle are equal.) By substitution, $\angle 1 = \angle A$ plus $\angle 1$ and $\angle A$ are corresponding angles; hence, the lines are parallel.

(c) Yes, $\angle RSC$ and $\angle C$ are supplementary; therefore, $\overline{RS} \parallel \overline{AC}$.

Exercise 1

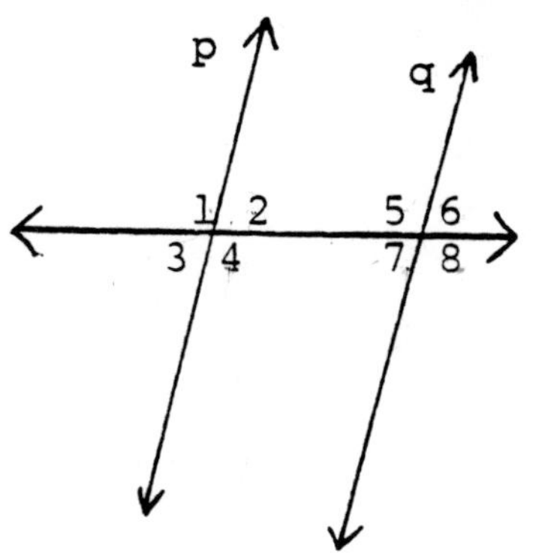

Lines p $\parallel$ q. $\angle 2 = 57^\circ$ Find the measure of each angle and give a reason for each answer.

2. In problem 1, find x and the measure of each angle if $\angle 4 = 2x$ and $\angle 7 = x + 4$.

3. In problem 1, find x and the measure of each angle if $\angle 2 = 3x - 2$ and $\angle 7 = x + 4$.

4. In problem 1, find x and the measure of each angle if $\angle 3 = 2x - 10$ and $\angle 7 = x + 4$.

5. Lines l and m are not in the same plane. (See illustration.) They are called what type of lines?

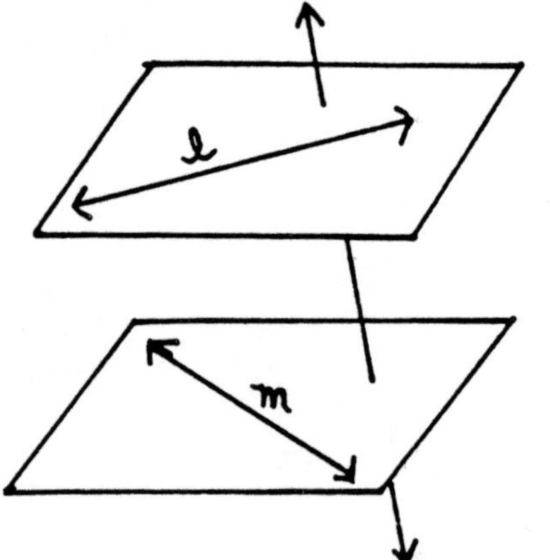

6. If two lines are parallel and are cut by a transversal, what pairs of equal angles are formed? (Name three types.)

7. If two lines are parallel and are cut by a transversal, what pairs of angles are supplementary? (Demonstrate your answer with examples.)

8.

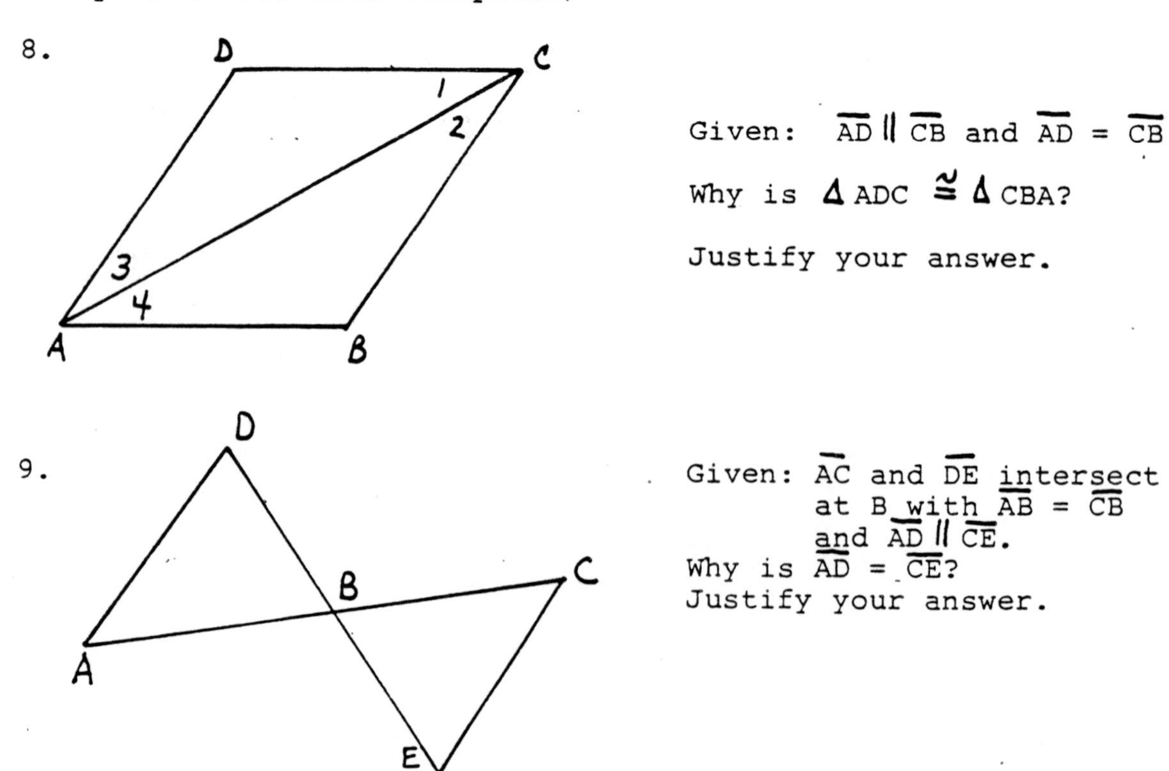

Given: $\overline{AD} \parallel \overline{CB}$ and $\overline{AD} = \overline{CB}$

Why is $\triangle ADC \cong \triangle CBA$?

Justify your answer.

9. Given: $\overline{AC}$ and $\overline{DE}$ intersect at B with $\overline{AB} = \overline{CB}$ and $\overline{AD} \parallel \overline{CE}$.
Why is $\overline{AD} = \overline{CE}$?
Justify your answer.

TRIANGLES AND PARALLEL LINES

Theorem 5: The sum of the interior angles of a triangle is 180°.

The proof of this theorem requires the use of parallel lines and their related theorems.

Given: $\triangle ABC$ and line $l \parallel \overline{AC}$.

Prove: $\angle A + \angle 2 + \angle C = 180^\circ$

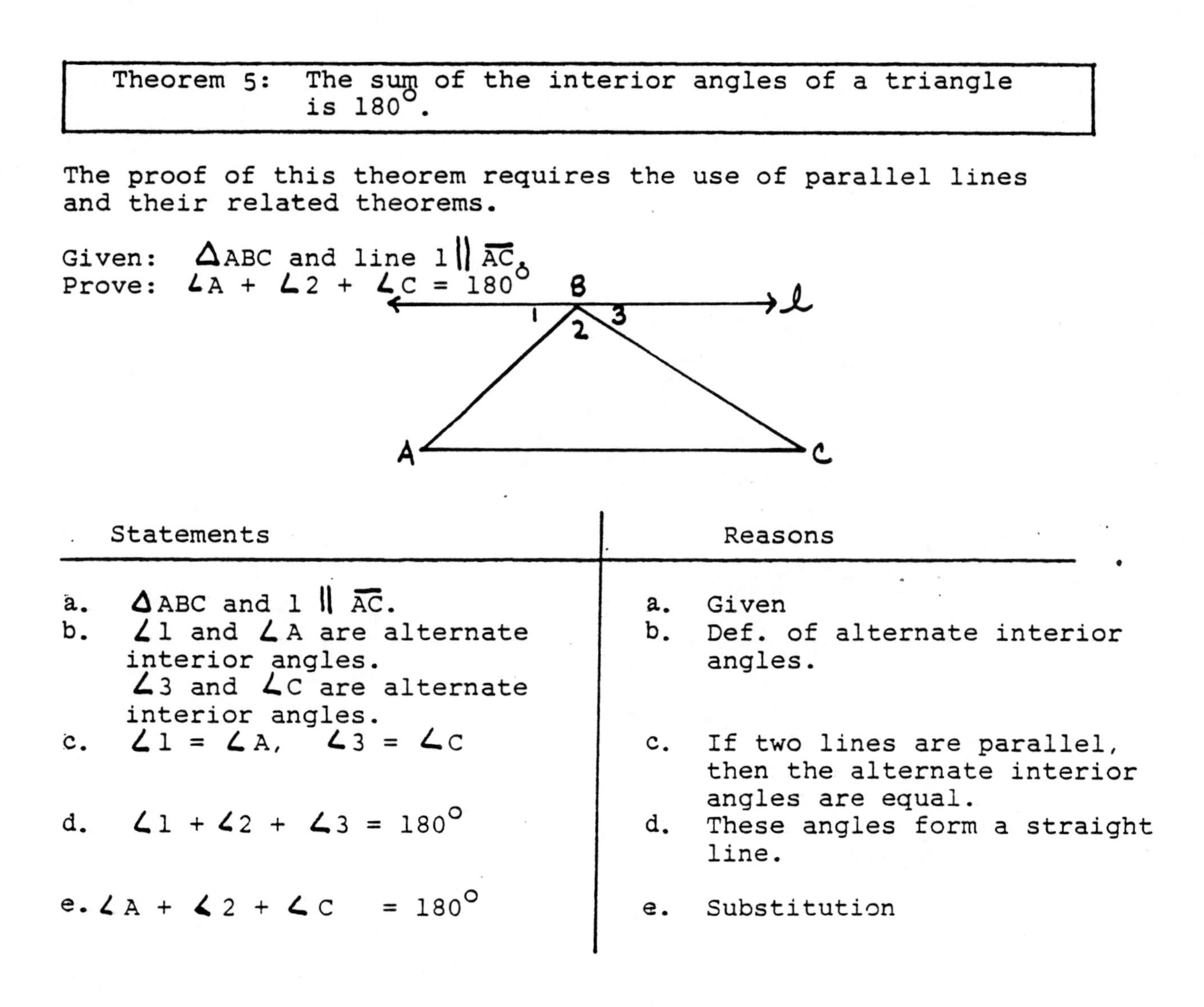

Statements	Reasons
a. $\triangle ABC$ and $l \parallel \overline{AC}$.	a. Given
b. $\angle 1$ and $\angle A$ are alternate interior angles. $\angle 3$ and $\angle C$ are alternate interior angles.	b. Def. of alternate interior angles.
c. $\angle 1 = \angle A$, $\angle 3 = \angle C$	c. If two lines are parallel, then the alternate interior angles are equal.
d. $\angle 1 + \angle 2 + \angle 3 = 180^\circ$	d. These angles form a straight line.
e. $\angle A + \angle 2 + \angle C = 180^\circ$	e. Substitution

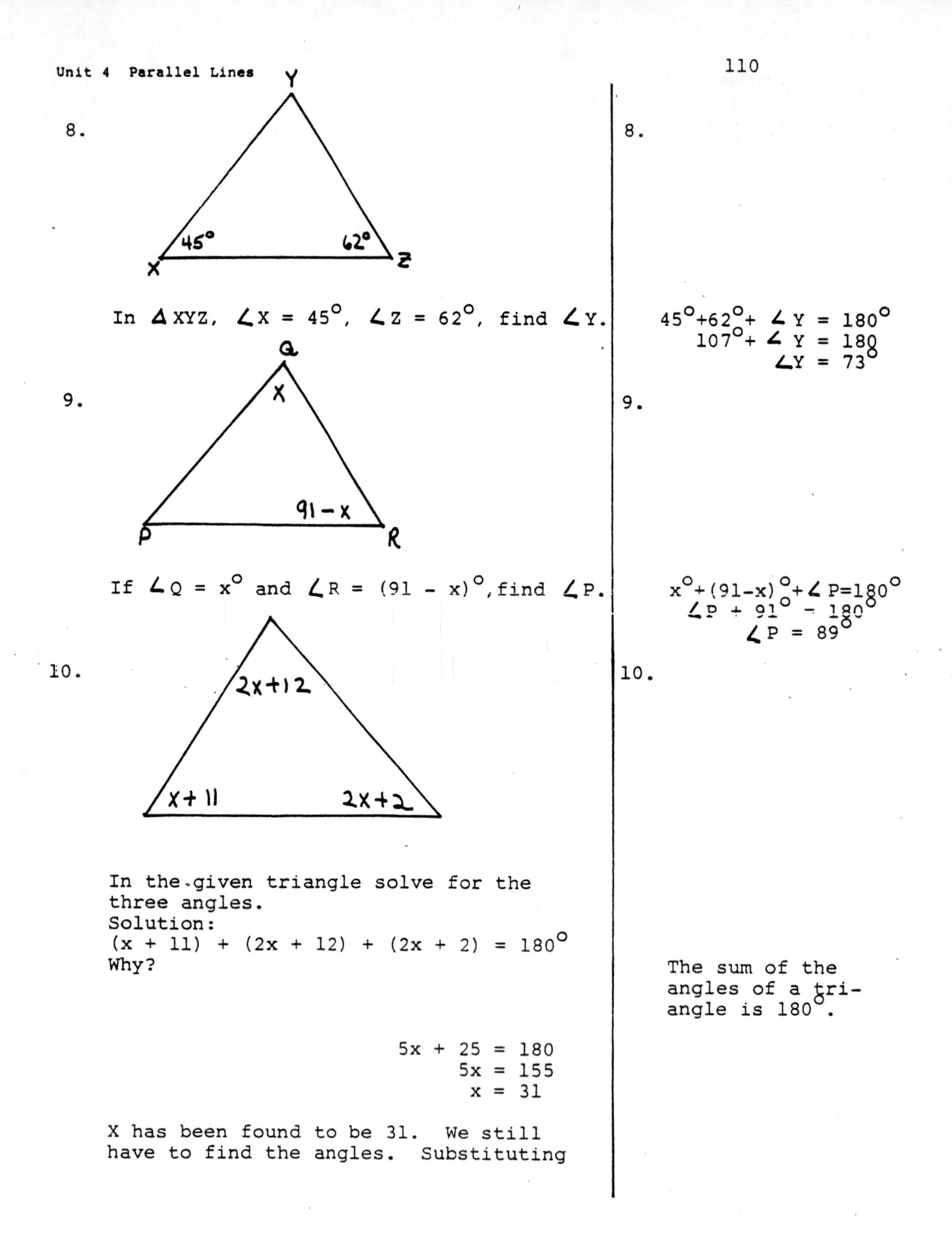

8.

In $\triangle XYZ$, $\angle X = 45^\circ$, $\angle Z = 62^\circ$, find $\angle Y$.

8.

$45^\circ + 62^\circ + \angle Y = 180^\circ$
$107^\circ + \angle Y = 180^\circ$
$\angle Y = 73^\circ$

9.

If $\angle Q = x^\circ$ and $\angle R = (91 - x)^\circ$, find $\angle P$.

9.

$x^\circ + (91 - x)^\circ + \angle P = 180^\circ$
$\angle P + 91^\circ = 180^\circ$
$\angle P = 89^\circ$

10.

In the given triangle solve for the three angles.
Solution:
$(x + 11) + (2x + 12) + (2x + 2) = 180^\circ$
Why?

10.

The sum of the angles of a triangle is 180°.

$5x + 25 = 180$
$5x = 155$
$x = 31$

X has been found to be 31. We still have to find the angles. Substituting

$x = 31$ into $\underline{x + 11} = 31 + 11 = 42°$ and $\underline{2x + 12} = 2(31) + 12 = 62 + 12 = 74°$ and $\underline{2x + 2} = 2(31) + 2 = 62 + 2 = 64°$

11. Given $\triangle ABC$ with the angles labeled. Find x.

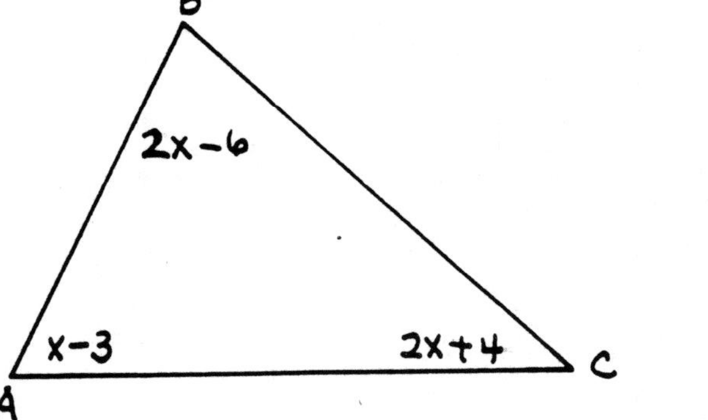

11.

$(2x-6)+(x-3)+(2x+4)=180°$
$5x - 5 = 180$
$x = 37$

$2x-6=2(37)-6$
$= 74-6=68°$

$x-3=37-3=34$

$2x+4=2(37)+4$
$=74+4=78°$

Example: One angle of a triangle is five more than twice a second. A third is ten less than twice the second angle. Find the value of the angles of the triangle.

x = second angle

five more than twice a second
$5 + 2 \cdot x$ = first angle

ten less than twice a second
$2 \cdot x - 10$ = third angle

Since the sum of the angles of a triangle is 180°, we obtain our equation:

$$(x) + (5 + 2x) + (2x - 10) = 180$$
$$5x - 5 = 180$$
$$5x = 185$$
$$x = 37$$

Now, $5 + 2x = 5 + 2(37)$
$= 5 + 74$
$= 79°$

and $2x - 10 = 2(37) - 10$
$= 74 - 10$
$= 64°$

12. Given the following problem, find the solution:
In a triangle one angle is three times a second. The third angle is six less than twice the second. Find the angles.

12.

Let x = second angle
_______________ = first angle
_______________ = third angle
The equation is _____________________.

The angles are _________, ___________,
______________.

13.

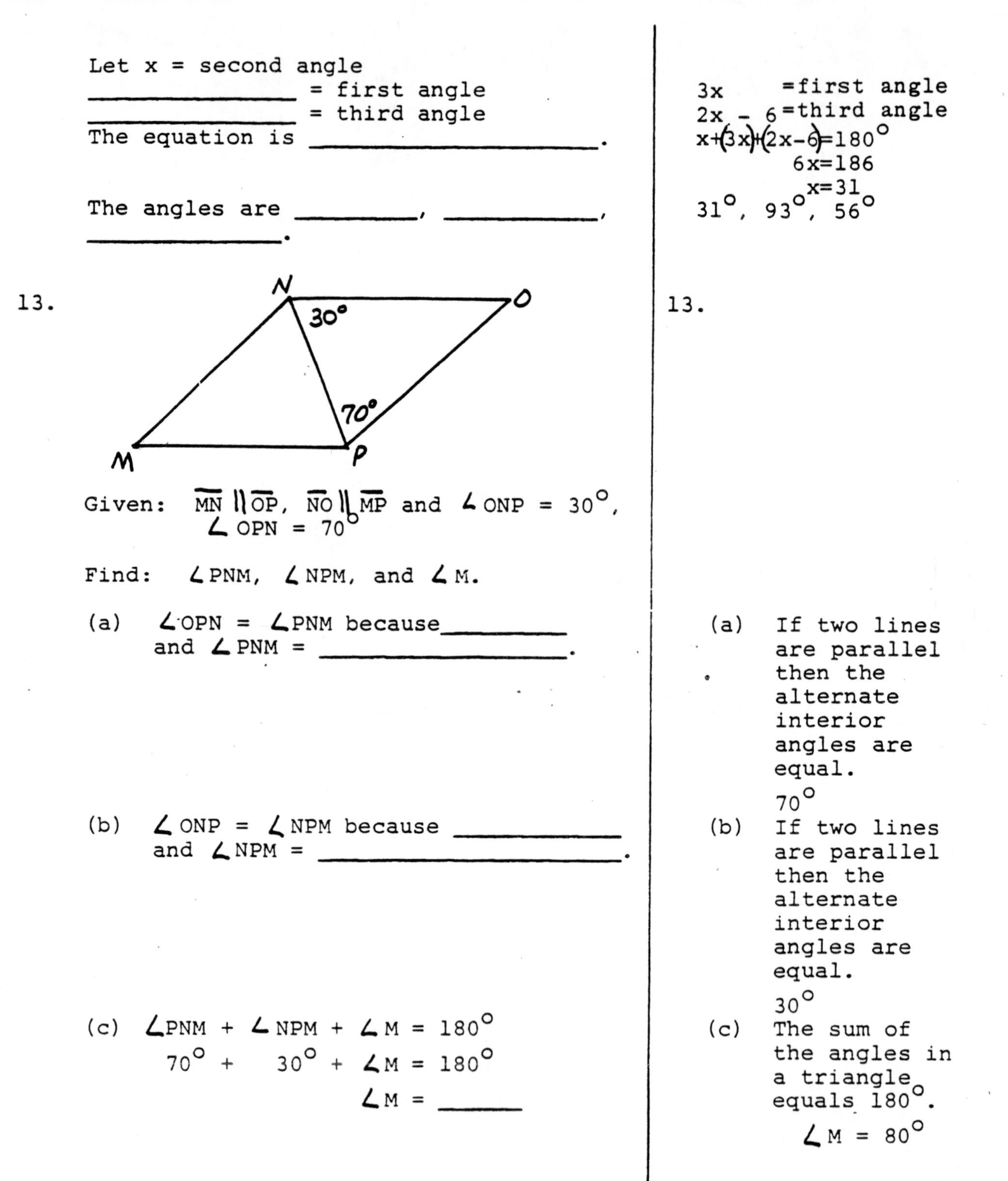

Given: $\overline{MN} \parallel \overline{OP}$, $\overline{NO} \parallel \overline{MP}$ and $\angle ONP = 30^\circ$, $\angle OPN = 70^\circ$

Find: $\angle PNM$, $\angle NPM$, and $\angle M$.

(a) $\angle OPN = \angle PNM$ because__________ and $\angle PNM$ = ____________________.

(b) $\angle ONP = \angle NPM$ because ____________ and $\angle NPM$ = ______________________.

(c) $\angle PNM + \angle NPM + \angle M = 180^\circ$

$70^\circ + 30^\circ + \angle M = 180^\circ$

$\angle M$ = ______

$3x$ = first angle
$2x - 6$ = third angle
$x + (3x) + (2x - 6) = 180^\circ$
$6x = 186$
$x = 31$
31°, 93°, 56°

13.

(a) If two lines are parallel then the alternate interior angles are equal.
70°

(b) If two lines are parallel then the alternate interior angles are equal.
30°

(c) The sum of the angles in a triangle equals 180°.
$\angle M = 80^\circ$

Theorem 5 has several interesting conclusions which can be drawn with minimal effort. Statements which are theorems but are easily obtained as an extension of another theorem are called <u>corollaries</u>.

Corollary 1: In a right triangle, the two acute angles are complementary.

Let triangle ABC be a right triangle with $\angle C$ a right angle.

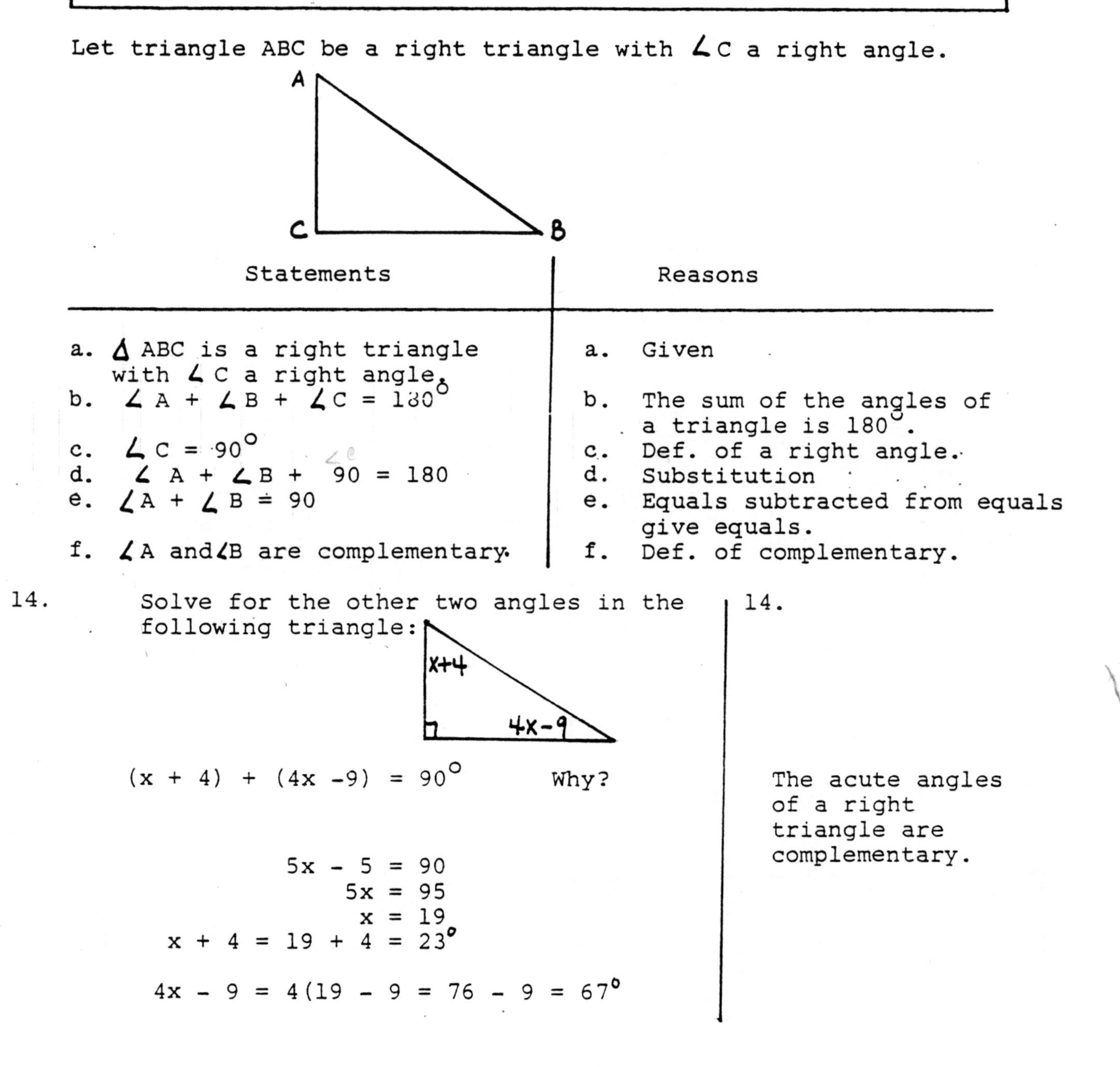

Statements	Reasons
a. $\triangle$ ABC is a right triangle with $\angle C$ a right angle.	a. Given
b. $\angle A + \angle B + \angle C = 180^\circ$	b. The sum of the angles of a triangle is 180°.
c. $\angle C = 90^\circ$	c. Def. of a right angle.
d. $\angle A + \angle B + 90 = 180$	d. Substitution
e. $\angle A + \angle B = 90$	e. Equals subtracted from equals give equals.
f. $\angle A$ and $\angle B$ are complementary.	f. Def. of complementary.

14. Solve for the other two angles in the following triangle:

$(x + 4) + (4x - 9) = 90^\circ$ Why?

$5x - 5 = 90$

$5x = 95$

$x = 19$

$x + 4 = 19 + 4 = 23^\circ$

$4x - 9 = 4(19 - 9 = 76 - 9 = 67^\circ$

14. The acute angles of a right triangle are complementary.

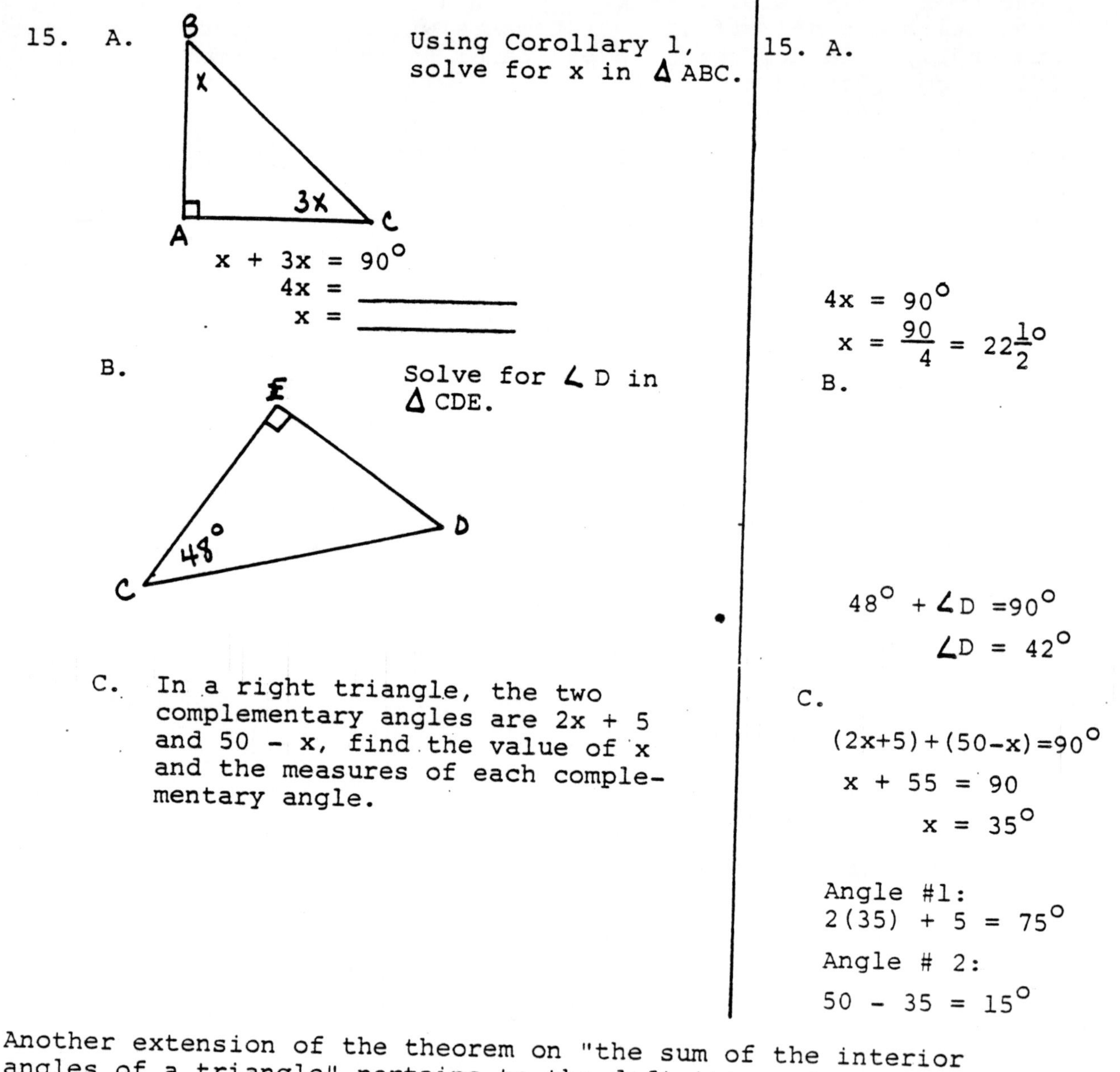

15. A. Using Corollary 1, solve for x in $\triangle ABC$.

$x + 3x = 90^\circ$
$4x =$ ________
$x =$ ________

B. Solve for $\angle D$ in $\triangle CDE$.

C. In a right triangle, the two complementary angles are $2x + 5$ and $50 - x$, find the value of x and the measures of each complementary angle.

15. A.

$4x = 90^\circ$

$x = \frac{90}{4} = 22\frac{1}{2}^\circ$

B.

$48^\circ + \angle D = 90^\circ$

$\angle D = 42^\circ$

C.

$(2x+5)+(50-x)=90^\circ$

$x + 55 = 90$

$x = 35^\circ$

Angle #1:
$2(35) + 5 = 75^\circ$

Angle # 2:
$50 - 35 = 15^\circ$

Another extension of the theorem on "the sum of the interior angles of a triangle" pertains to the definition of exterior angles.

Definition:	An <u>exterior angle</u> of a triangle is formed whenever one of the sides of a triangle is extended through its vertex. An exterior angle is adjacent to an interior angle of a triangle.

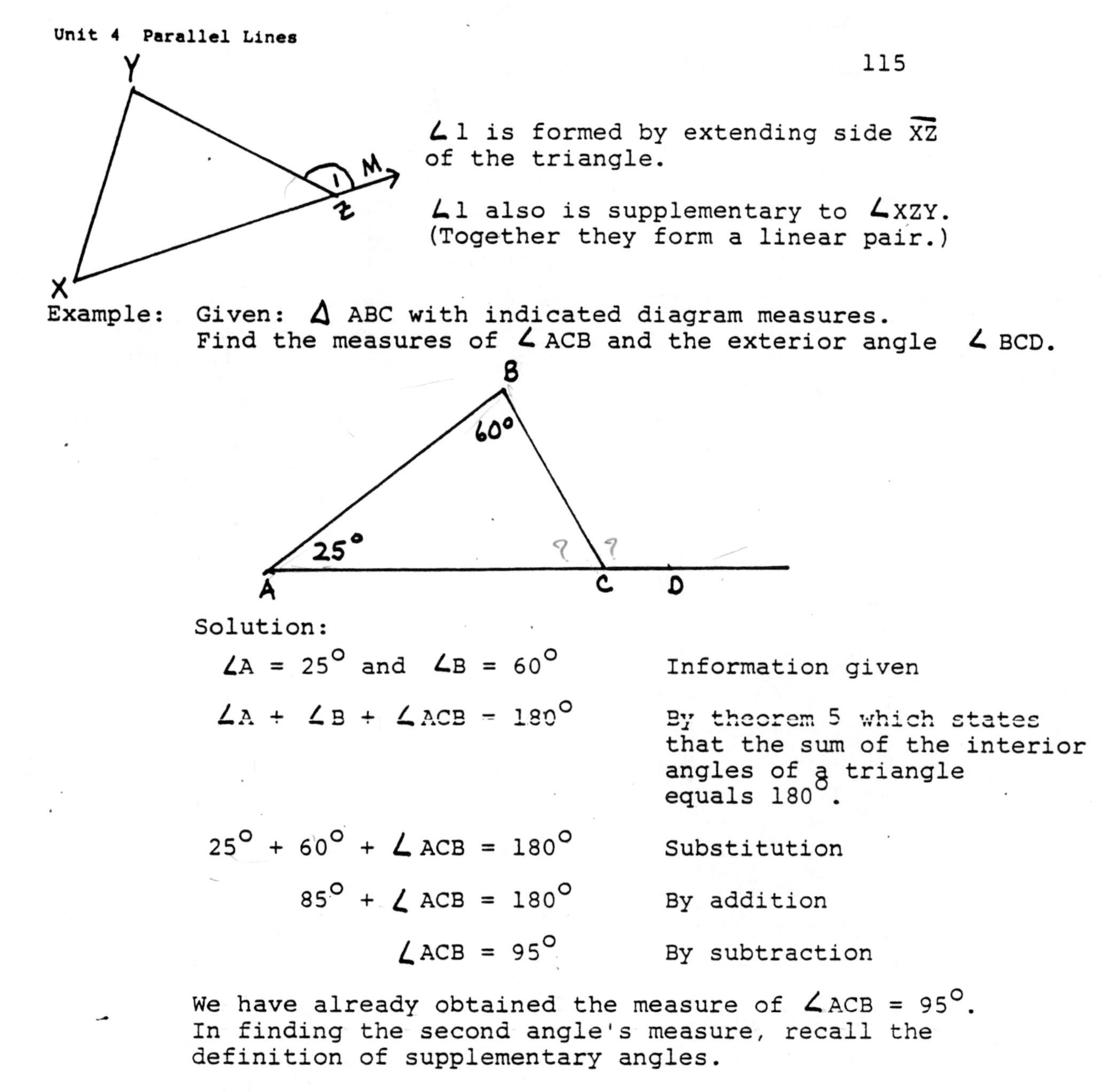

$\angle 1$ is formed by extending side $\overline{XZ}$ of the triangle.

$\angle 1$ also is supplementary to $\angle XZY$. (Together they form a linear pair.)

Example: Given: $\triangle$ ABC with indicated diagram measures. Find the measures of $\angle ACB$ and the exterior angle $\angle BCD$.

Solution:

$\angle A = 25^\circ$ and $\angle B = 60^\circ$	Information given
$\angle A + \angle B + \angle ACB = 180^\circ$	By theorem 5 which states that the sum of the interior angles of a triangle equals 180°.
$25^\circ + 60^\circ + \angle ACB = 180^\circ$	Substitution
$85^\circ + \angle ACB = 180^\circ$	By addition
$\angle ACB = 95^\circ$	By subtraction

We have already obtained the measure of $\angle ACB = 95^\circ$. In finding the second angle's measure, recall the definition of supplementary angles.

$\angle BCD$ is supplementary to $\angle ACB$.

$\angle BCD + \angle ACB = 180^\circ$

$\angle BCD + 95^\circ = 180^\circ$

$\therefore \angle BCD = 85^\circ$

$\angle BCD$ is called an <u>exterior</u> angle to $\triangle$ ABC which is also supplementary to $\angle ACB$.

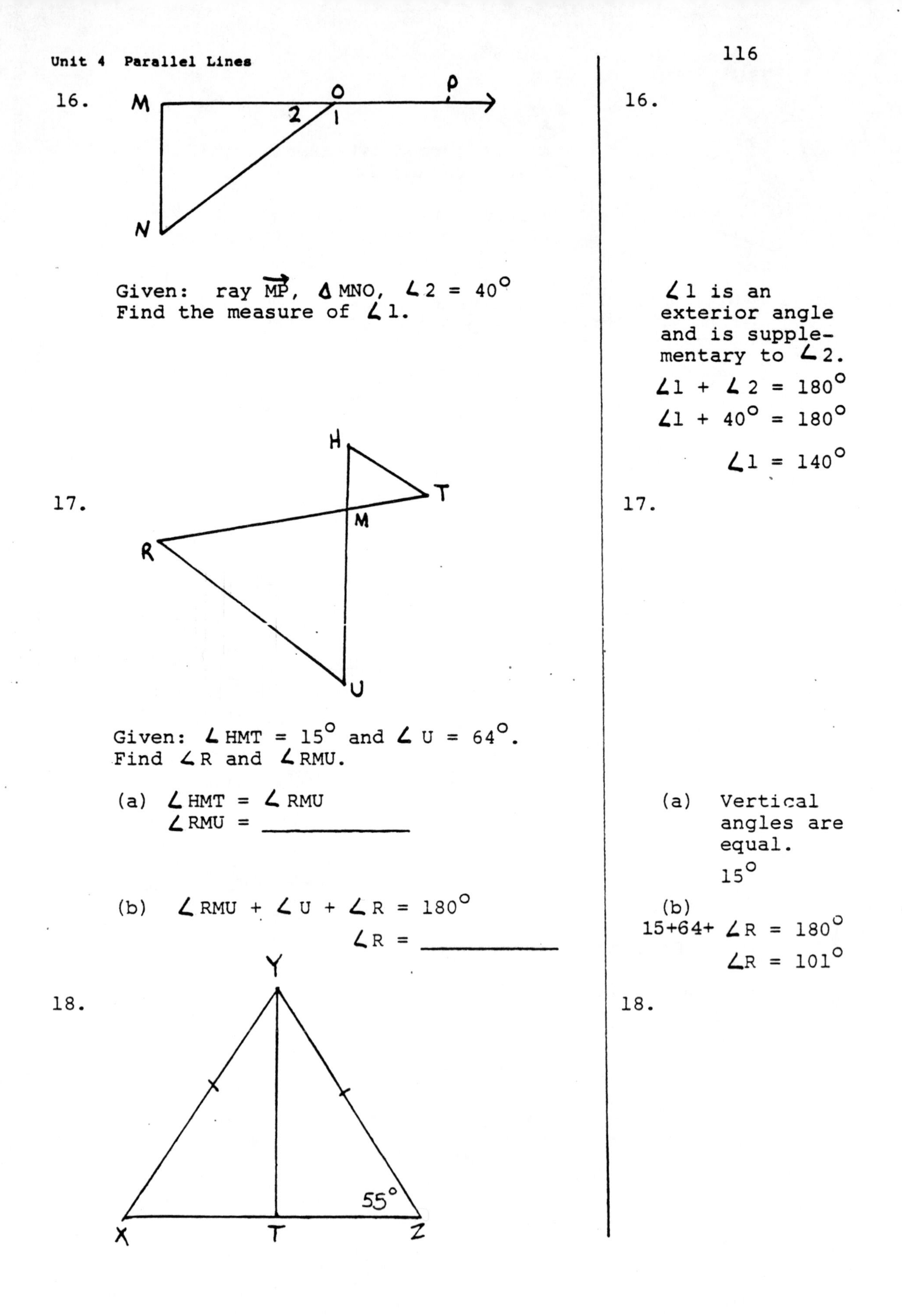

16.

Given: ray $\overrightarrow{MP}$, $\triangle MNO$, $\angle 2 = 40^\circ$
Find the measure of $\angle 1$.

16.

$\angle 1$ is an exterior angle and is supplementary to $\angle 2$.

$\angle 1 + \angle 2 = 180^\circ$

$\angle 1 + 40^\circ = 180^\circ$

$\angle 1 = 140^\circ$

17.

Given: $\angle HMT = 15^\circ$ and $\angle U = 64^\circ$.
Find $\angle R$ and $\angle RMU$.

(a) $\angle HMT = \angle RMU$
$\angle RMU =$ ____________

(b) $\angle RMU + \angle U + \angle R = 180^\circ$
$\angle R =$ ____________

17.

(a) Vertical angles are equal.
15°

(b)
$15 + 64 + \angle R = 180^\circ$
$\angle R = 101^\circ$

18.

18.

Given: ΔXYZ is isosceles and $\overline{YT}$ is an altitude of ΔXYZ.

Find: $\angle$X, $\angle$XYT, and $\angle$YTX

(a) $\angle X = \angle Z$

$\angle X =$ ______

(a) Angles opposite equal sides in a triangle are equal.

55°

(b) $\angle YTX = 90^\circ$ because ____________ ______________________________.

(b) $\overline{YT} \perp \overline{XZ}$ by the def. of altitude and perpendicular lines form right angles.

(c) $\angle X + \angle YTX + \angle XYT = 180^\circ$

$\angle XYT =$ ______

(c) 35°

19.

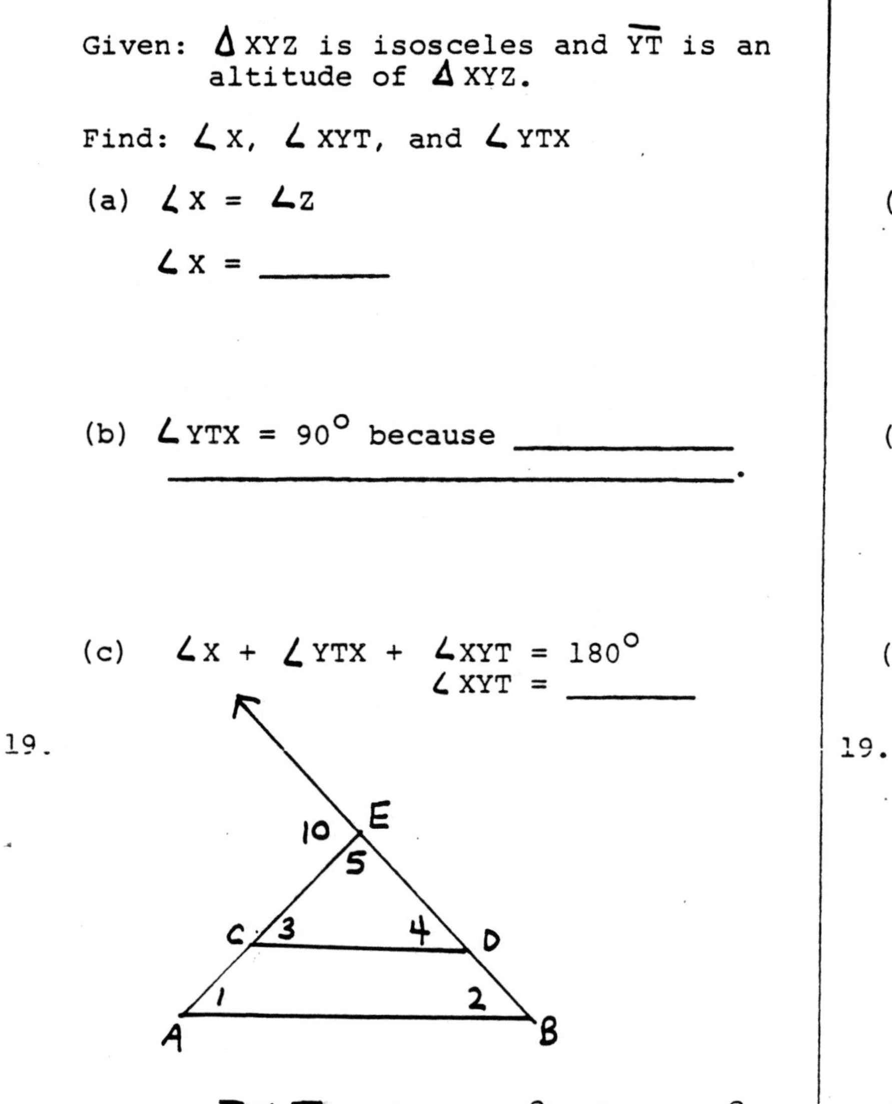

Given: $\overline{CD} \parallel \overline{AB}$, $\angle 2 = 70^\circ$, $\angle 3 = 81^\circ$
Find the measures of the other angles.

19.

$\angle 1 = \angle 3 = 81^\circ$
If two lines are parallel, then the corresponding angles are equal.

$\angle 2 = \angle 4 = 70^\circ$

corresponding angles are equal.

$\angle 3 + \angle 4 + \angle 5 = 180^\circ$

The sum of the angles of a triangle equals 180°.

$81^\circ + 70^\circ + \angle 5 = 180^\circ$
$151^\circ + \angle 5 = 180^\circ$
$\angle 5 = 29^\circ$

(answer continued)

$\angle 10$ is exterior to $\angle 5$.
$\angle 10 = 151^\circ$

In exercise 19, note that the measure of $\angle 10$ equals the sum of the measures of $\angle 3$ and $\angle 4$.

$$\angle 10 = \angle 3 + \angle 4$$

$$\angle 10 = 81^\circ + 70^\circ$$

$$\angle 10 = 151^\circ$$

Corollary 2:	Each exterior angle of a triangle equals the sum of its two non-adjacent interior angles.

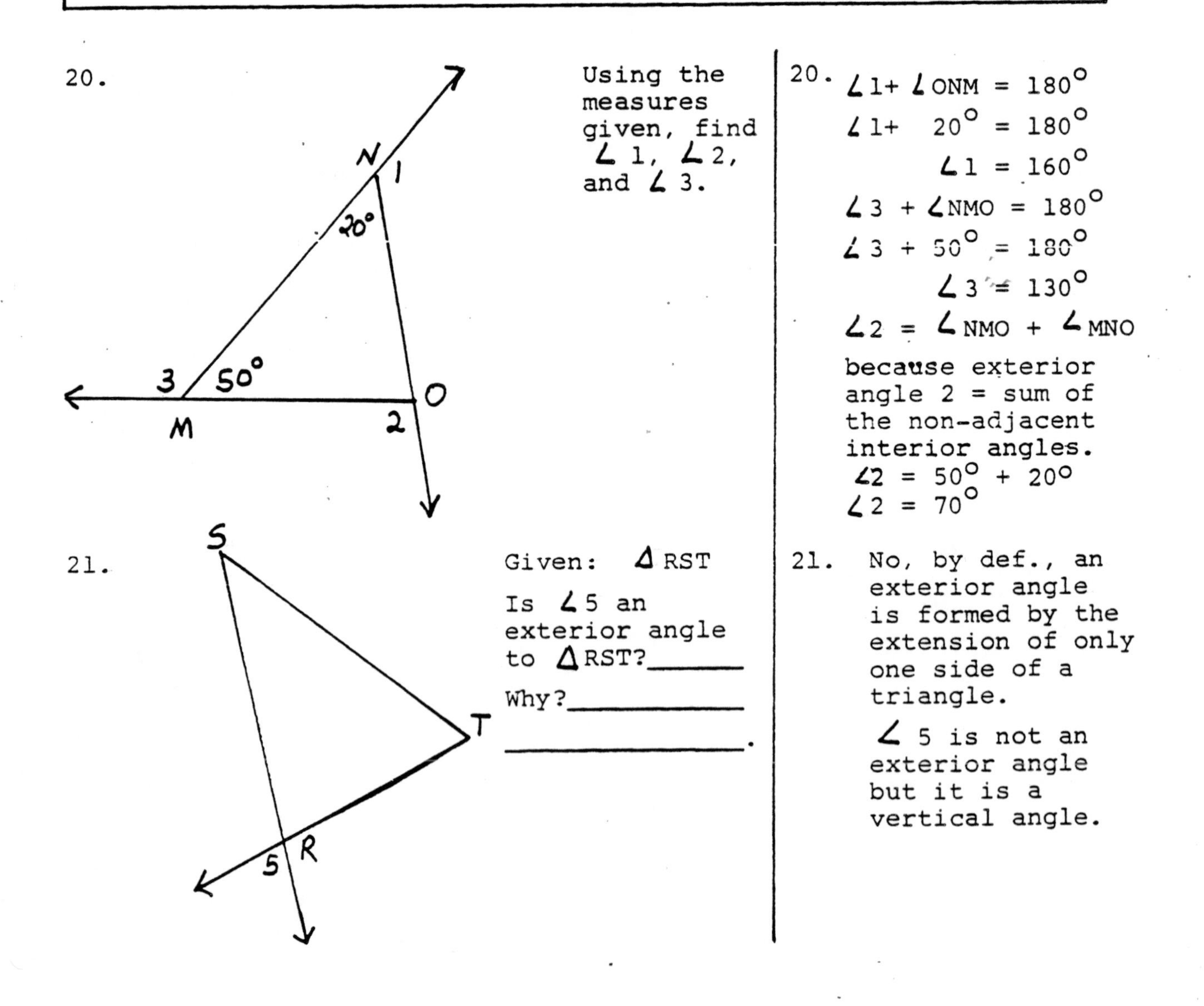

20. Using the measures given, find $\angle 1$, $\angle 2$, and $\angle 3$.

20. $\angle 1 + \angle ONM = 180^\circ$
$\angle 1 + 20^\circ = 180^\circ$
$\angle 1 = 160^\circ$
$\angle 3 + \angle NMO = 180^\circ$
$\angle 3 + 50^\circ = 180^\circ$
$\angle 3 = 130^\circ$
$\angle 2 = \angle NMO + \angle MNO$
because exterior angle 2 = sum of the non-adjacent interior angles.
$\angle 2 = 50^\circ + 20^\circ$
$\angle 2 = 70^\circ$

21. Given: ΔRST

Is $\angle 5$ an exterior angle to ΔRST?______

Why?______________

______________.

21. No, by def., an exterior angle is formed by the extension of only one side of a triangle.

$\angle 5$ is not an exterior angle but it is a vertical angle.

Exercise 2

1. Find the unknown angle in the following triangle.

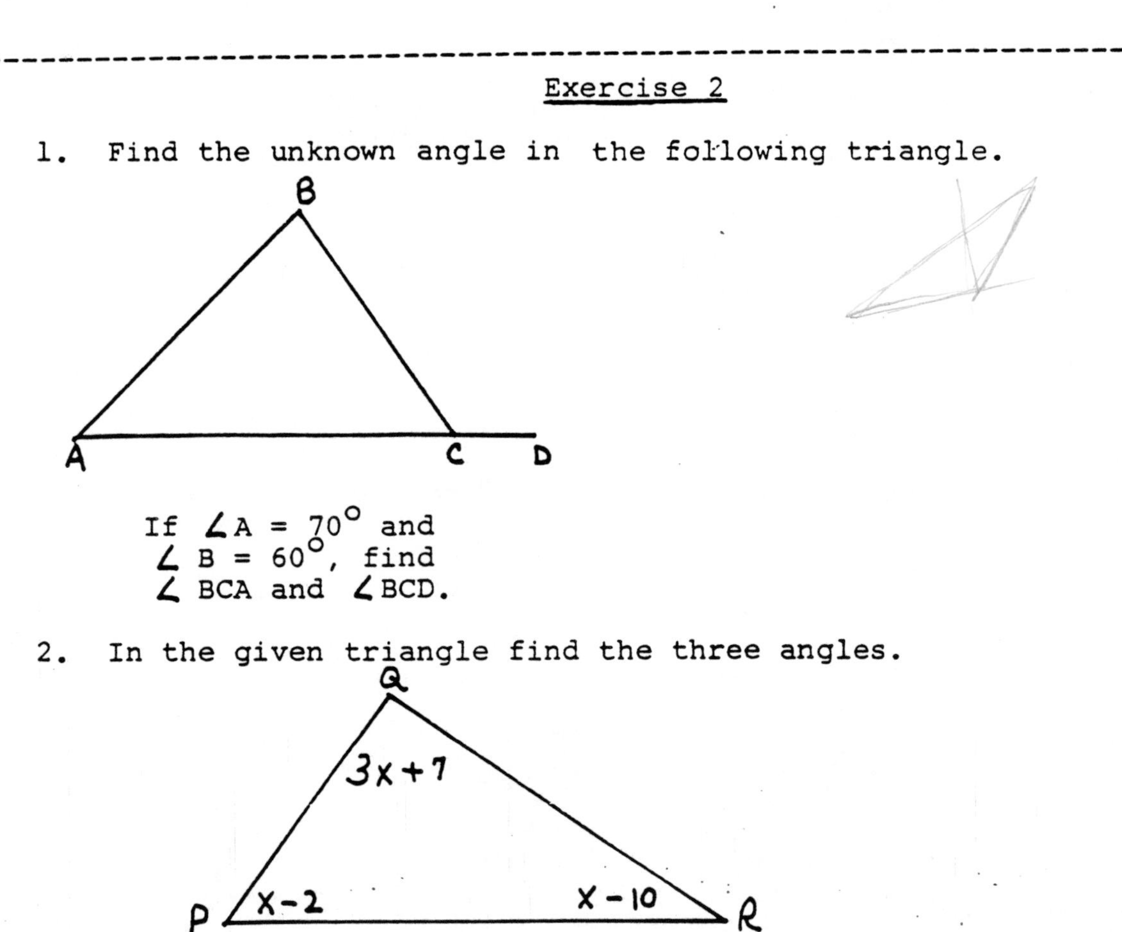

If $\angle A = 70^\circ$ and $\angle B = 60^\circ$, find $\angle BCA$ and $\angle BCD$.

2. In the given triangle find the three angles.

3. One angle is twice a second in a triangle. A third is ten less than six times the second. Find the angles of this triangle.

4. Solve for the other two angles in the following triangle.

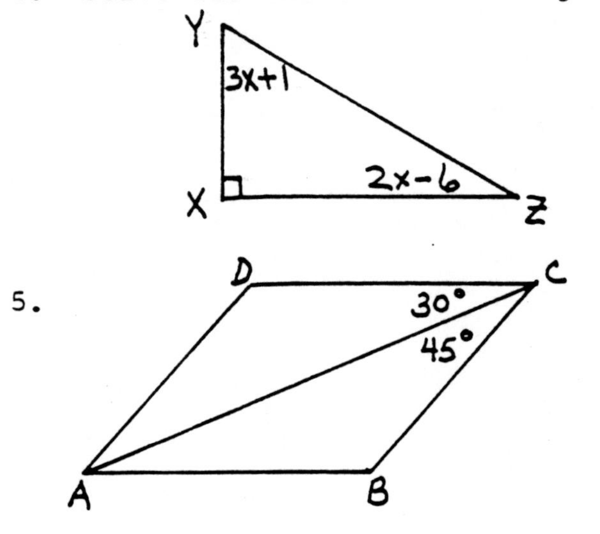

5. Given: $\overline{AD} \parallel \overline{BC}$, $\overline{AB} \parallel \overline{DC}$, $\angle DCA = 30^\circ$, $\angle ACB = 45^\circ$
Find: $\angle DAC$, $\angle CAB$, $\angle D$ and $\angle B$.
Justify your answer.

6.

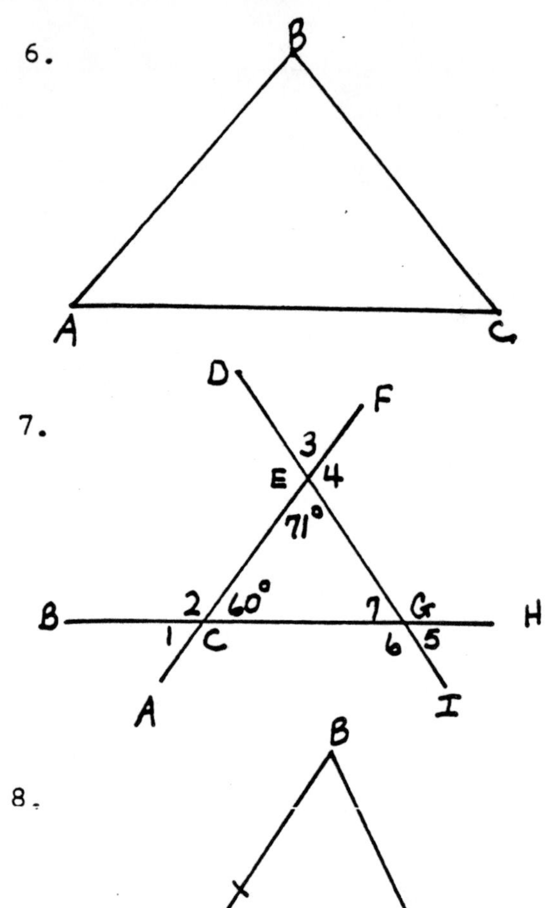

Given: $\angle A = 43^\circ$, $\overline{AB}$ = 6 in., $\overline{BC}$ = 6 in.
Find: $\angle C$ and $\angle B$
What kind of triangle is $\triangle ABC$?

7.

Find the measures of all the numbered angles. Name all the angles exterior to $\triangle CGE$.

8.

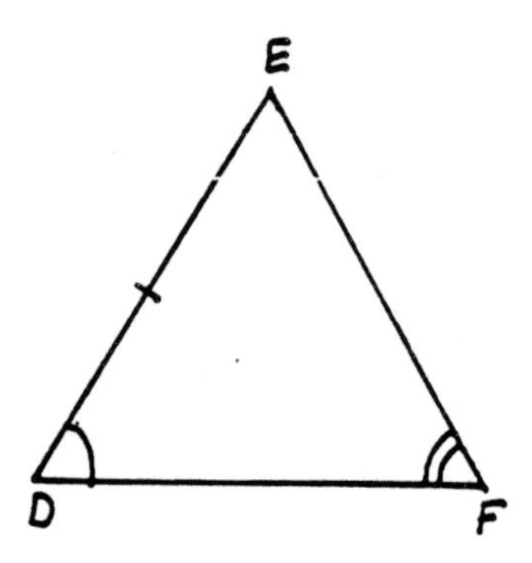

Given: $\angle A = \angle D$, $\angle C = \angle F$, $\overline{AB} = \overline{DE}$
Prove: $\triangle ABC \cong \triangle DEF$

Fill in the appropriate reasons for each statement listed below:

Statements	Reasons
a. $\angle A = \angle D$, $\angle C = \angle F$, $\overline{AB} = \overline{DE}$	a.
b. $\angle A + \angle B + \angle C = 180^\circ$ $\angle D + \angle E + \angle F = 180^\circ$	b.
c. $\angle A + \angle B + \angle C = \angle D + \angle E + \angle F$	c.
d. $\angle B = \angle E$	d.
e. $\triangle ABC \cong \triangle DEF$	e. *

* Note: Recall the three methods discussed to prove triangles congruent were ASA = ASA, SAS = SAS, and SSS= SSS. This exercise is actually another method to prove congruence. This is AAS = AAS. If two angles

of a triangle are given, the third angle can always be found. When any side is given, enough data is known to obtain congruence by ASA = ASA.

Theorem 6: If a line joins the midpoints of two sides of a triangle, that line is parallel to and equal to one-half the measure of the third side.

Example:

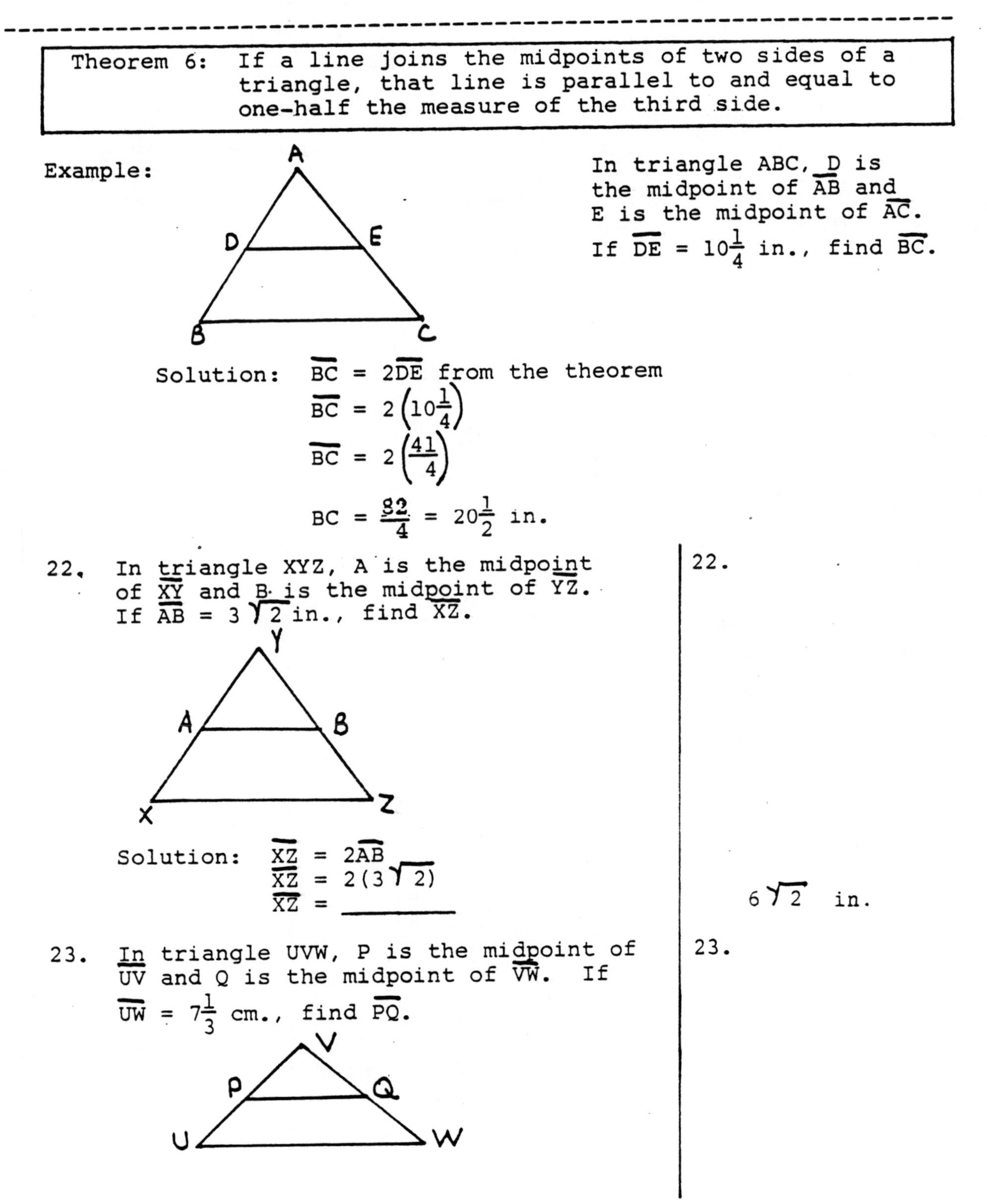

In triangle ABC, D is the midpoint of $\overline{AB}$ and E is the midpoint of $\overline{AC}$. If $\overline{DE} = 10\frac{1}{4}$ in., find $\overline{BC}$.

Solution: $\overline{BC} = 2\overline{DE}$ from the theorem

$\overline{BC} = 2\left(10\frac{1}{4}\right)$

$\overline{BC} = 2\left(\frac{41}{4}\right)$

$BC = \frac{82}{4} = 20\frac{1}{2}$ in.

22. In triangle XYZ, A is the midpoint of $\overline{XY}$ and B is the midpoint of $\overline{YZ}$. If $\overline{AB} = 3\sqrt{2}$ in., find $\overline{XZ}$.

Solution: $\overline{XZ} = 2\overline{AB}$

$\overline{XZ} = 2(3\sqrt{2})$

$\overline{XZ} =$ ________

22. $6\sqrt{2}$ in.

23. In triangle UVW, P is the midpoint of $\overline{UV}$ and Q is the midpoint of $\overline{VW}$. If $\overline{UW} = 7\frac{1}{3}$ cm., find $\overline{PQ}$.

23.

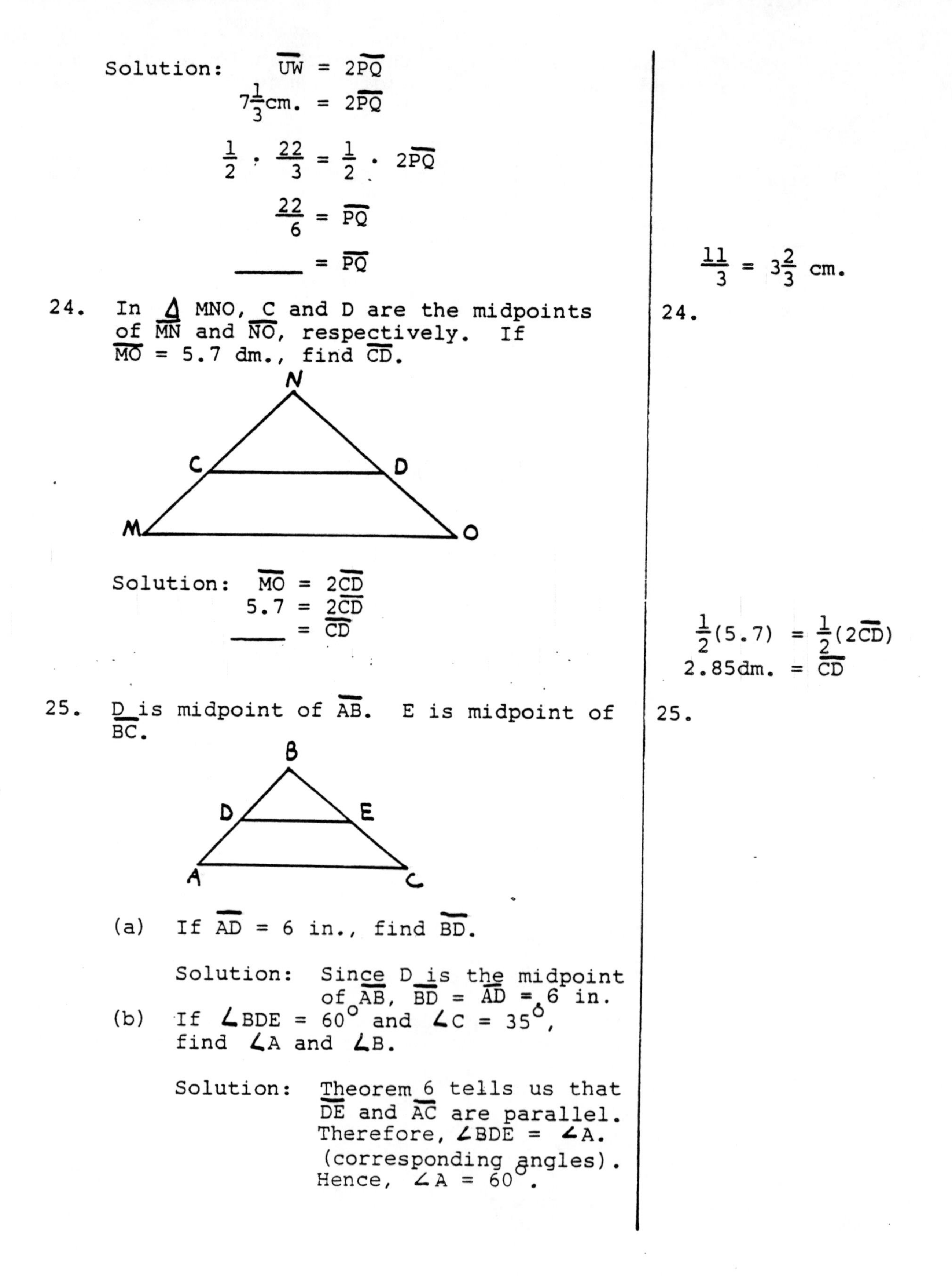

Solution:

$$\overline{UW} = 2\overline{PQ}$$

$$7\tfrac{1}{3}\text{cm.} = 2\overline{PQ}$$

$$\frac{1}{2} \cdot \frac{22}{3} = \frac{1}{2} \cdot 2\overline{PQ}$$

$$\frac{22}{6} = \overline{PQ}$$

$$\_\_\_\_\_ = \overline{PQ}$$

$$\frac{11}{3} = 3\tfrac{2}{3}\text{ cm.}$$

24. In $\triangle$ MNO, C and D are the midpoints of $\overline{MN}$ and $\overline{NO}$, respectively. If $\overline{MO}$ = 5.7 dm., find $\overline{CD}$.

Solution:

$$\overline{MO} = 2\overline{CD}$$

$$5.7 = 2\overline{CD}$$

$$\_\_\_\_\_ = \overline{CD}$$

24.

$$\frac{1}{2}(5.7) = \frac{1}{2}(2\overline{CD})$$

$$2.85\text{dm.} = \overline{CD}$$

25. D is midpoint of $\overline{AB}$. E is midpoint of $\overline{BC}$.

(a) If $\overline{AD}$ = 6 in., find $\overline{BD}$.

Solution: Since D is the midpoint of $\overline{AB}$, $\overline{BD} = \overline{AD} = 6$ in.

(b) If $\angle BDE = 60^\circ$ and $\angle C = 35^\circ$, find $\angle A$ and $\angle B$.

Solution: Theorem 6 tells us that $\overline{DE}$ and $\overline{AC}$ are parallel. Therefore, $\angle BDE = \angle A$. (corresponding angles). Hence, $\angle A = 60^\circ$.

25.

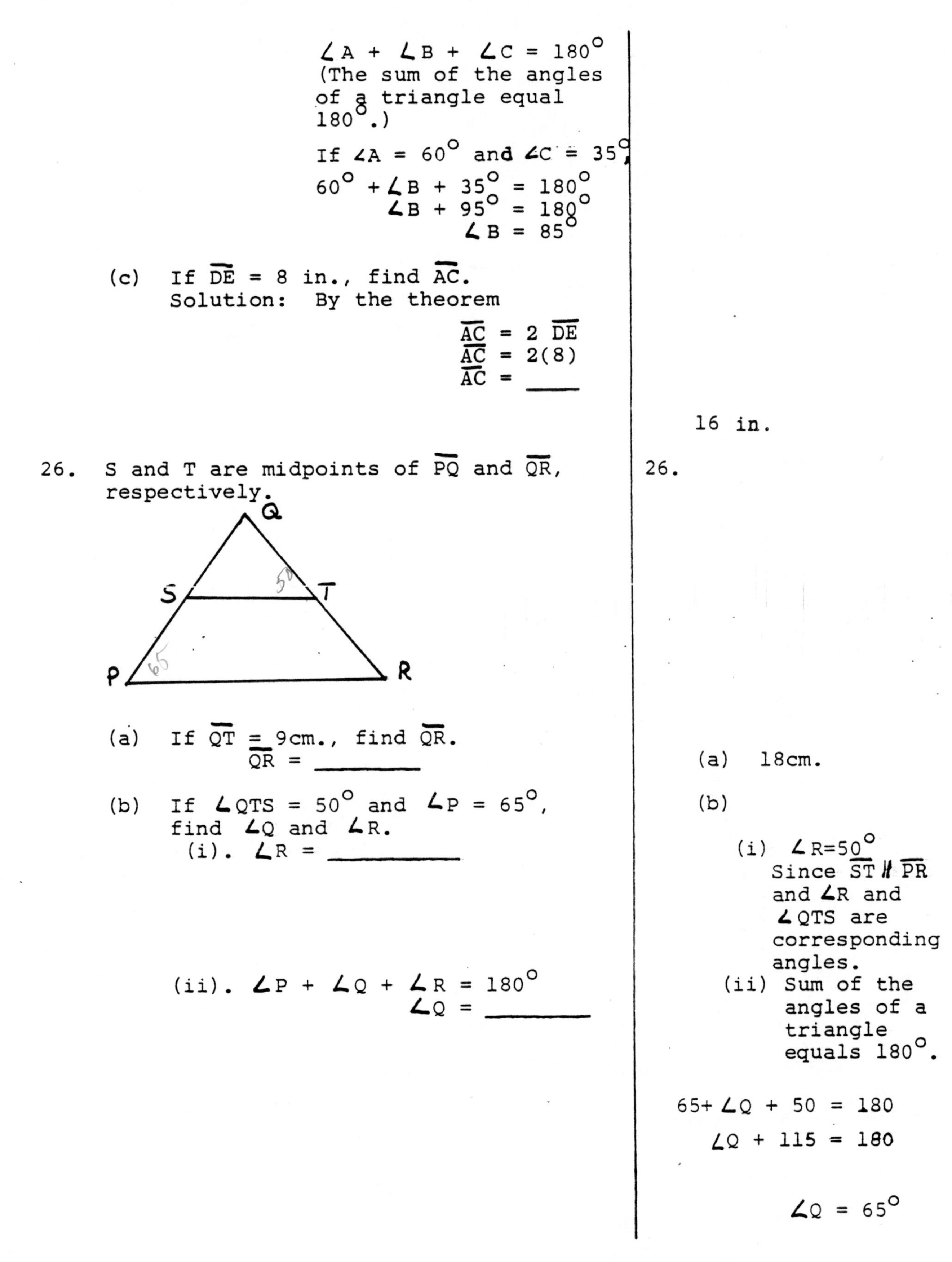

$\angle A + \angle B + \angle C = 180^\circ$
(The sum of the angles of a triangle equal 180°.)

If $\angle A = 60^\circ$ and $\angle C = 35^\circ$,

$60^\circ + \angle B + 35^\circ = 180^\circ$
$\angle B + 95^\circ = 180^\circ$
$\angle B = 85^\circ$

(c) If $\overline{DE}$ = 8 in., find $\overline{AC}$.
Solution: By the theorem

$\overline{AC} = 2\ \overline{DE}$
$\overline{AC} = 2(8)$
$\overline{AC}$ = ____

16 in.

26. S and T are midpoints of $\overline{PQ}$ and $\overline{QR}$, respectively.

(a) If $\overline{QT}$ = 9cm., find $\overline{QR}$.
$\overline{QR}$ = ________

(b) If $\angle QTS = 50^\circ$ and $\angle P = 65^\circ$, find $\angle Q$ and $\angle R$.
(i). $\angle R$ = _________

(ii). $\angle P + \angle Q + \angle R = 180^\circ$
$\angle Q$ = ________

26.

(a) 18cm.

(b)

(i) $\angle R = 50^\circ$
Since $\overline{ST} \parallel \overline{PR}$ and $\angle R$ and $\angle QTS$ are corresponding angles.

(ii) Sum of the angles of a triangle equals 180°.

$65 + \angle Q + 50 = 180$

$\angle Q + 115 = 180$

$\angle Q = 65^\circ$

(c) If $\overline{ST} = 4\frac{3}{5}$ ft., find $\overline{PR}$.

$\overline{PR} = 2\overline{ST}$

$\overline{PR} = 2\left(4\frac{3}{5}\right)$

$\overline{PR} = 2(\quad)$

$\overline{PR} =$ ______

(c) $\overline{PR} = 2\left(\frac{23}{5}\right) = \frac{46}{5}$

$\overline{PR} = 9\frac{1}{5}$ ft.

Exercise 3

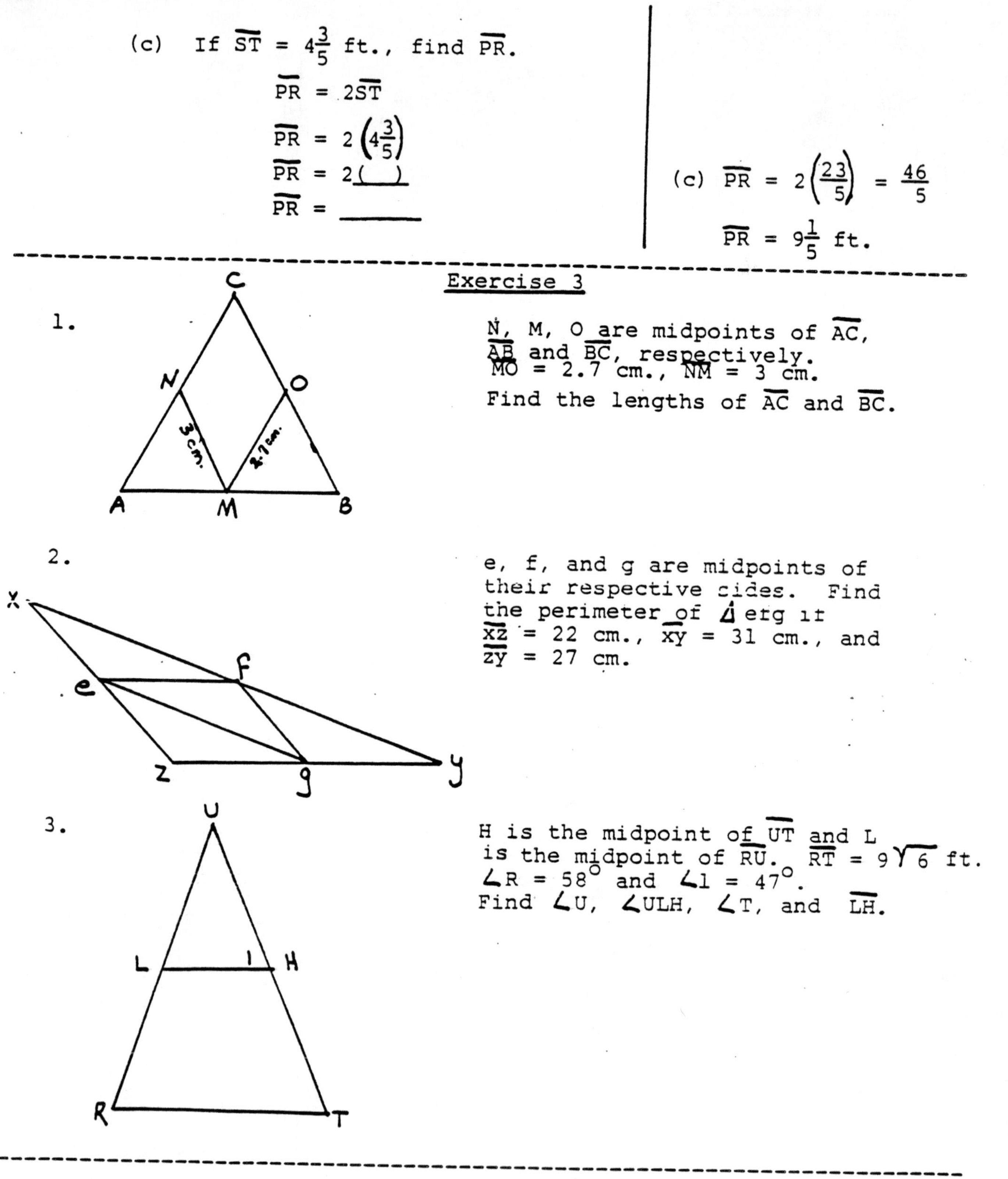

1. N, M, O are midpoints of $\overline{AC}$, $\overline{AB}$ and $\overline{BC}$, respectively. $\overline{MO} = 2.7$ cm., $\overline{NM} = 3$ cm. Find the lengths of $\overline{AC}$ and $\overline{BC}$.

2. e, f, and g are midpoints of their respective sides. Find the perimeter of $\triangle$ efg if $\overline{xz} = 22$ cm., $\overline{xy} = 31$ cm., and $\overline{zy} = 27$ cm.

3. H is the midpoint of $\overline{UT}$ and L is the midpoint of $\overline{RU}$. $\overline{RT} = 9\sqrt{6}$ ft. $\angle R = 58^\circ$ and $\angle 1 = 47^\circ$. Find $\angle U$, $\angle ULH$, $\angle T$, and $\overline{LH}$.

CONSTRUCTIONS

Euclid, a famous ancient mathematician, put forth a theoretical principle presently called the <u>Parallel Postulate</u>. This postulate has become well known in math circles and much discussion has centered around its attempted proof.

However, it is a postulate and we need not attempt its proof. We will merely demonstrate the parallel postulate by construction.

Postulate 1:	Parallel Postulate Through any point, P, not on line 1, there can be constructed one and only one line m that is parallel to 1.

Construction # 1

We will use this postulate and converse (i) in the construction below:

Given:

Step 1: Through P draw a line T that also intersects line 1. Notice that the intersection of lines T and 1 form $\angle 1$.

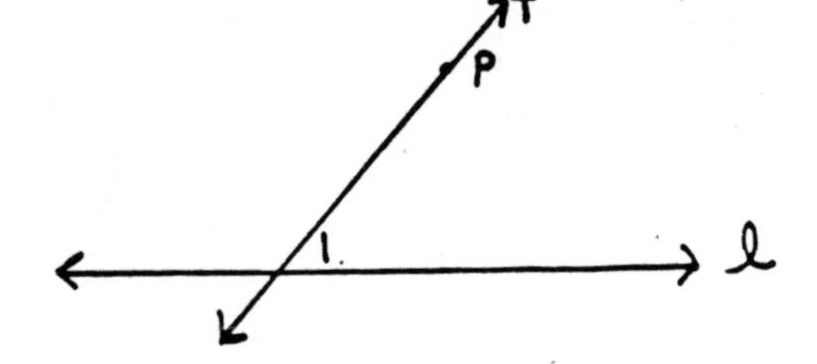

Step 2: Copy angle $\angle 1$ on the opposite side of the transversal using P as the vertex and T as one side of the new angle. Draw line m through P and the arc intersection as shown.

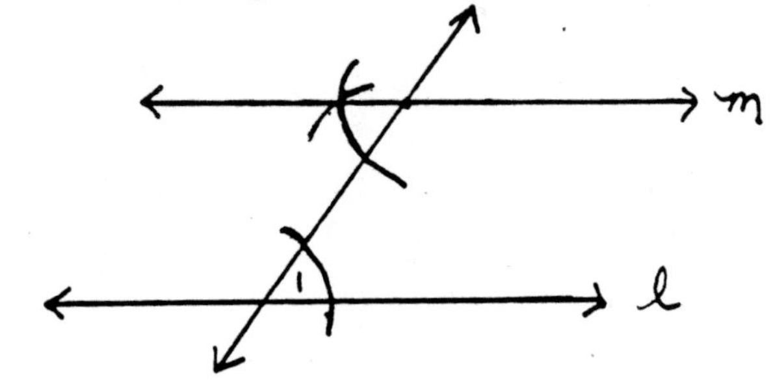

Since the new angle and $\angle 1$ are constructed as equal alternate interior angles, then line 1 is indeed parallel to m.

In construction #1, the alternate interior angles were drawn as acute angles. If the alternate interior angles were both 90°, they would be equal;and, therefore, would form two parallel lines.

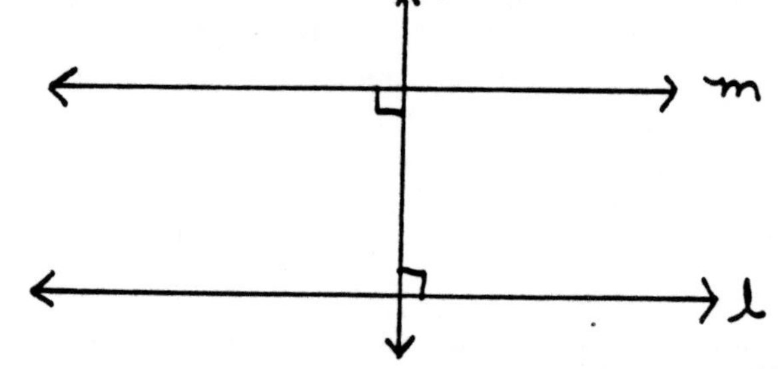

Postulate 2:	Two lines are parallel if they are both perpendicular to a third line.

Since $T \perp m$ and $T \perp l$, then $m \parallel l$.

Construction # 2

We will use this postulate 2 to construct two lines that are parallel to one another and perpendicular to the same line.

Step 1: Draw line l and select two distinct points, P and Q, on l.

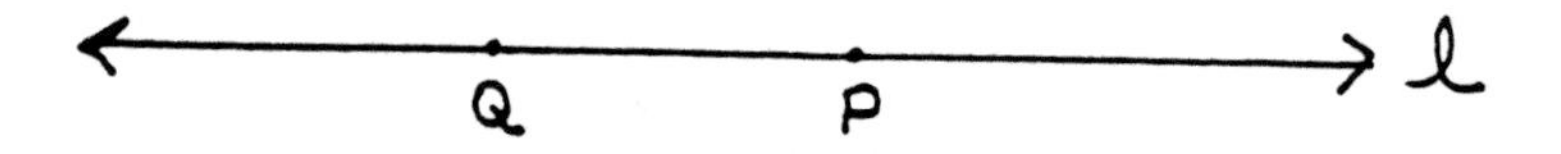

Step 2. Select a radius on your compass. Using Q then P as centers, make arcs on each side of each point.

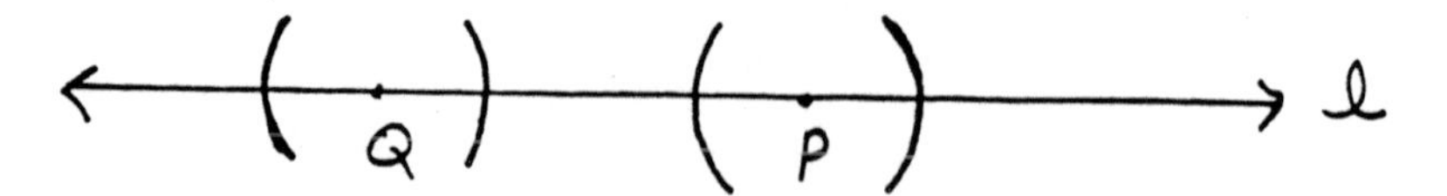

Step 3: With Q as the midpoint of the segment, $\overline{XY}$, formed at Q, construct a perpendicular bisector of $\overline{XY}$. Do the same construction for the segment, $\overline{RS}$, formed at P.

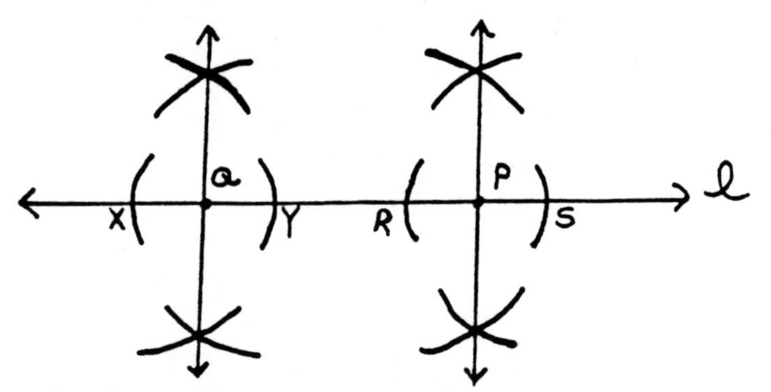

The two lines formed are perpendicular to the same line; and, therefore, are themselves parallel.

Exercise 4

1. How many lines can you construct through S that will be parallel to the given line? Demonstrate your answer by construction.

• S

2. How do you know that lines m and n are parallel?

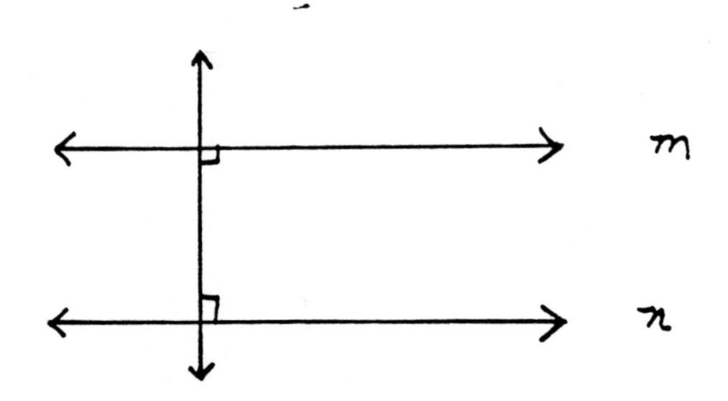

3. Construct a line that passes through H and is parallel to line n by using the steps outlined in construction #1.

H.

n

Review Unit 4

1. If two lines are cut by a transversal, then name three ways you can prove that two lines are parallel.

2\.

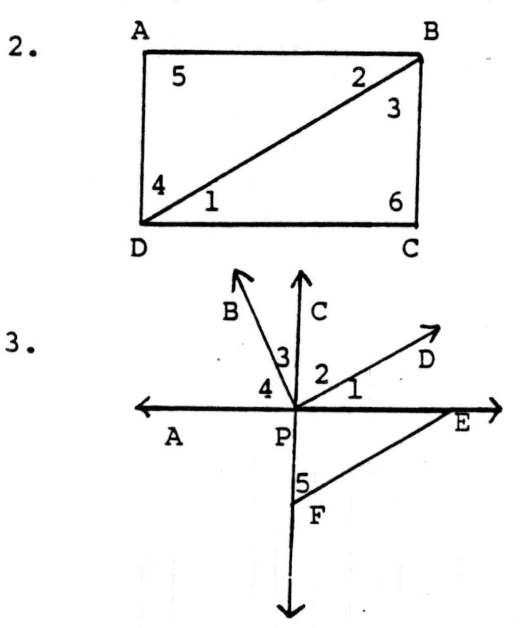

Given: $\overline{AB} \parallel \overline{DC}$, $\overline{AD} \parallel \overline{BC}$

Find the measures of all the angles if $\angle 1 = 30^\circ$ and $\angle 3 = 60^\circ$.

$\angle 2$ = ______ $\angle 4$ = ______

$\angle 5$ = ______ $\angle 6$ = ______

3\. If $\overrightarrow{PD} \parallel \overline{FE}$, $\overleftrightarrow{CF} \perp \overleftrightarrow{AE}$, $\overrightarrow{PD} \perp \overrightarrow{PB}$ and $\angle 5 = 30^\circ$, find the other numbered angles.

4. Define the following terms: parallel lines, skew lines, alternate interior angles, alternate exterior angles, corresponding angles, transversal and exterior angle.

5. Draw a line and select two distinct points on the line. Construct a line through each of these two points to obtain parallel lines using postulate 2.

6\.

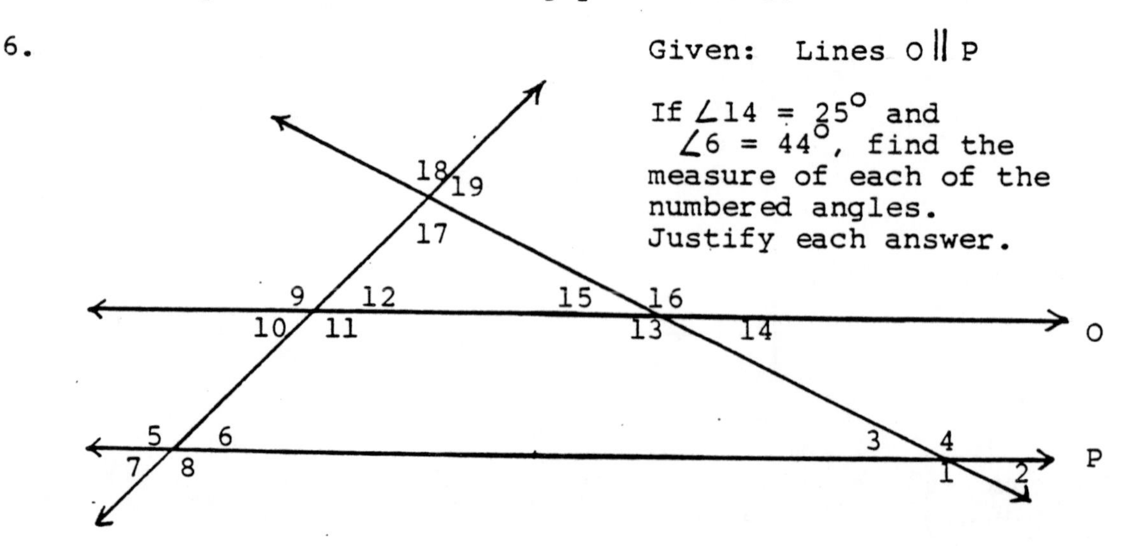

Given: Lines O $\parallel$ P

If $\angle 14 = 25^\circ$ and $\angle 6 = 44^\circ$, find the measure of each of the numbered angles. Justify each answer.

7.

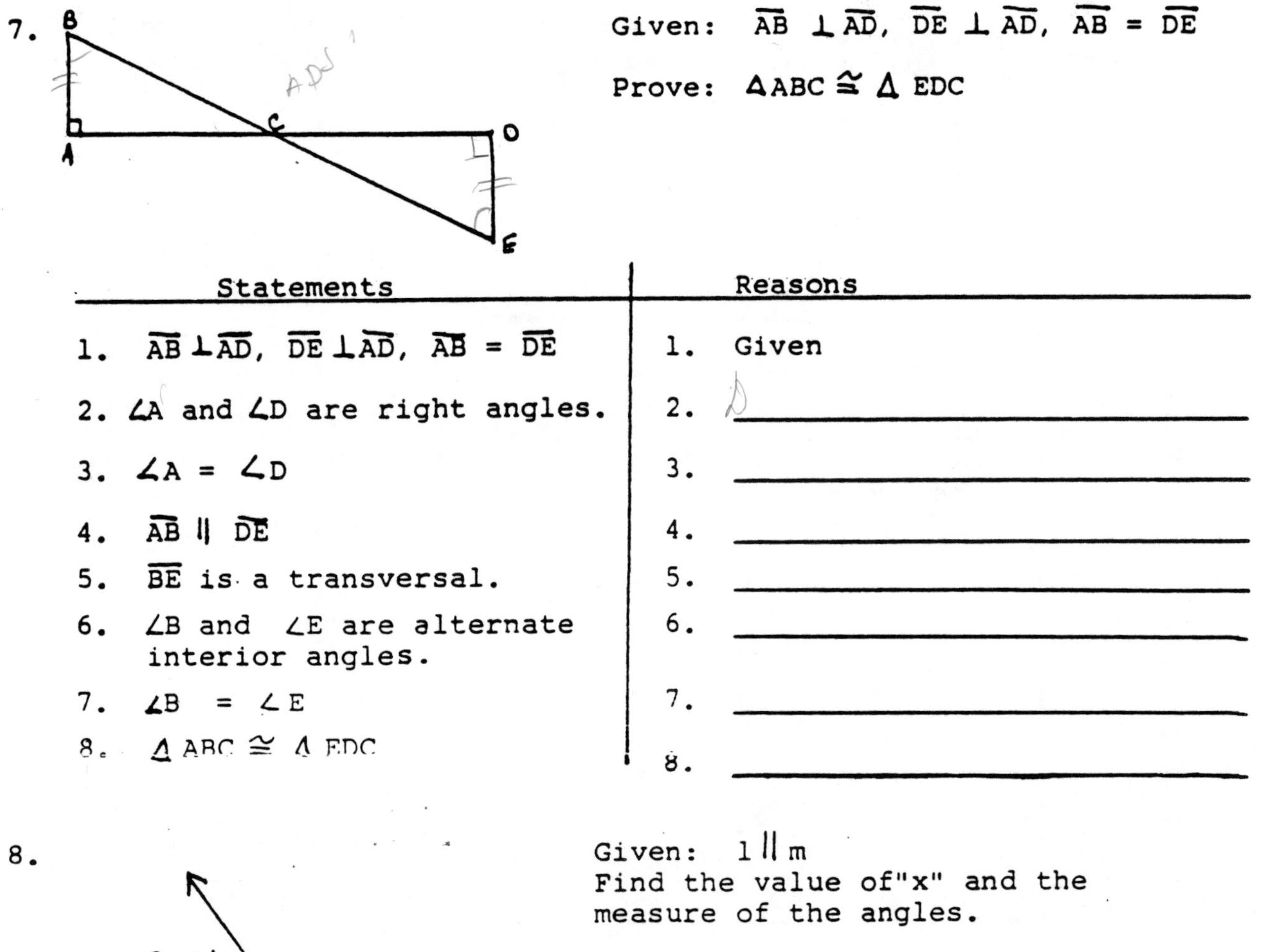

Given: $\overline{AB} \perp \overline{AD}$, $\overline{DE} \perp \overline{AD}$, $\overline{AB} = \overline{DE}$

Prove: $\triangle ABC \cong \triangle EDC$

Statements	Reasons
1. $\overline{AB} \perp \overline{AD}$, $\overline{DE} \perp \overline{AD}$, $\overline{AB} = \overline{DE}$	1. Given
2. $\angle A$ and $\angle D$ are right angles.	2. ________
3. $\angle A = \angle D$	3. ________
4. $\overline{AB} \parallel \overline{DE}$	4. ________
5. $\overline{BE}$ is a transversal.	5. ________
6. $\angle B$ and $\angle E$ are alternate interior angles.	6. ________
7. $\angle B = \angle E$	7. ________
8. $\triangle ABC \cong \triangle EDC$	8. ________

8.

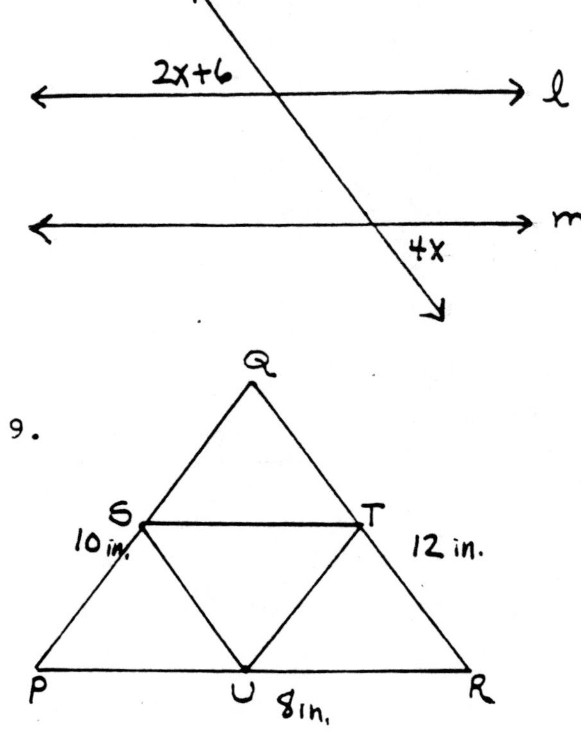

Given: $l \parallel m$
Find the value of "x" and the measure of the angles.

9.

S, T, U are midpoints of $\overline{PQ}$, $\overline{QR}$, and $\overline{PR}$, respectively. Find the perimeter of $\triangle STU$.

10. Using postulate 1, construct a line through point M that is parallel to the given line.

. M

⟵⟶

11. In a triangle, the first angle is eleven more than three times the second. The third is one less than twice the second. Find the angles.

12. In $\triangle ABC$, $\angle A$ is a right angle. Find $\angle B$ and $\angle C$.

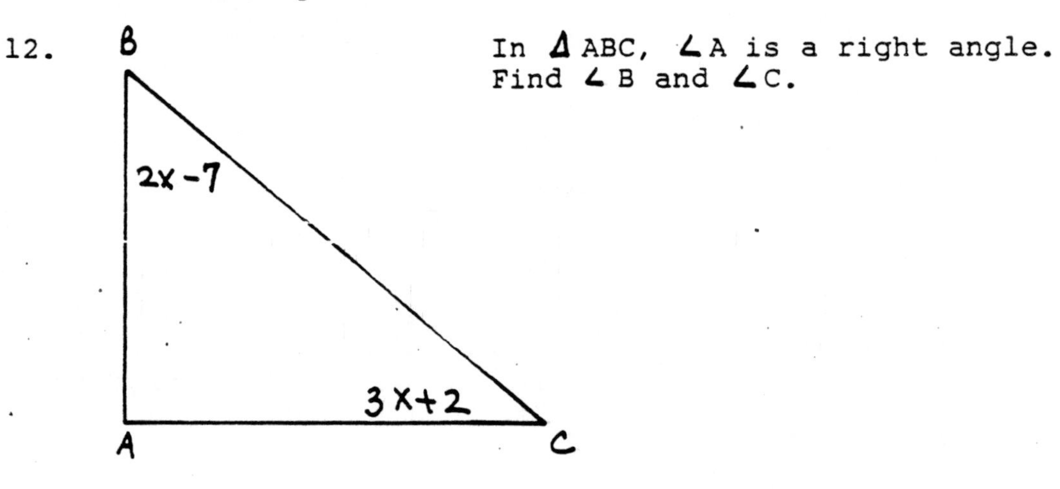

13. In the figure, line $0 \perp l$, $0 \perp m$ and $l \parallel m \parallel n$ and $\angle 9 = 130^\circ$. Find the measure of each of the other angles and state your reason why.

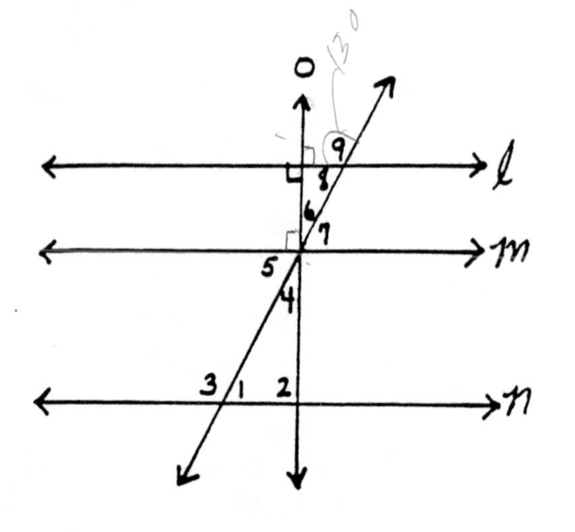

Unit 5

Quadrilaterals

Learning objectives:

The student will demonstrate his mastery of the following definitions and theorems by writing them and by applying them to the solution of selected problems:

<u>Definitions</u>: quadrilateral, convex, verticies, sides, consecutive, opposite, diagonal, parallelogram, rectangle, square, rhombus, trapezoid, base, median of a trapezoid, isosceles trapezoid

<u>Theorems</u>:

(1) If a quadrilateral is a parallelogram, then both pairs of opposite sides are equal.

(2) If the quadrilateral is a parallelogram, then both pairs of opposite angles of a quadrilateral are equal.

(3) Consecutive angles of a parallelogram are supplementary.

(4) The median of a trapezoid is parallel to the bases and is equal to one-half of the sum of the bases.

QUADRILATERALS

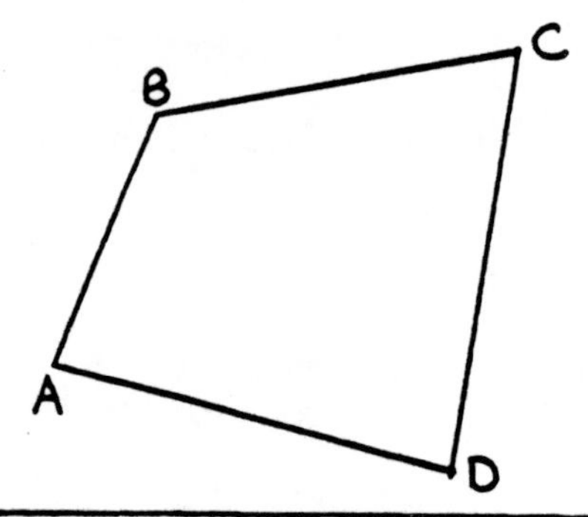

Definition:	If A, B, C, and D are points, no three of which are collinear and if sides $\overline{AB}$, $\overline{BC}$, $\overline{CD}$, and $\overline{AD}$ intersect only at these points, then the figure so formed is called a quadrilateral.

The illustrations below are two examples of four-sided closed figures (quadrilaterals).

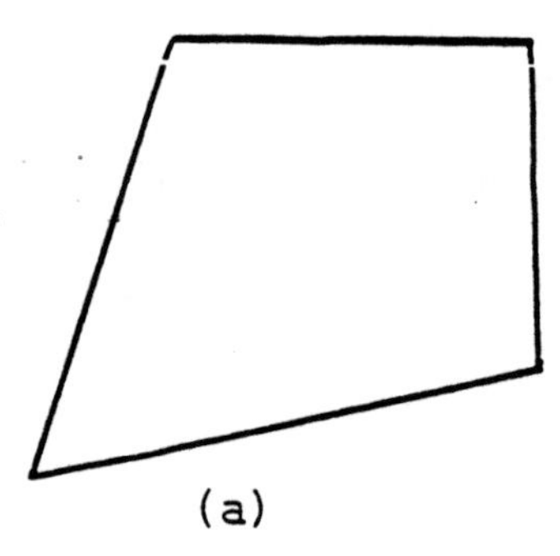

(a)

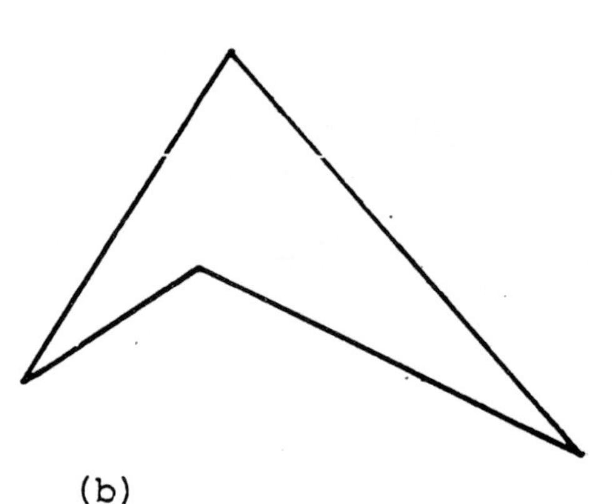

(b)

Figure (a) represents a convex quadrilateral. Figure (b) is not a convex figure but is a quadrilateral.

Definition:	A quadrilateral is convex if one may connect any two points belonging to the figure without any portion of the line lying in the exterior of the figure.

(a)

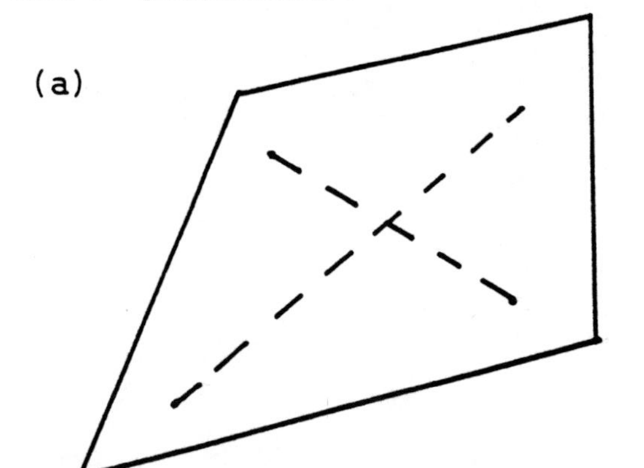

Convex - dotted lines connecting any two points in this figure are also contained within the figure.

(b)

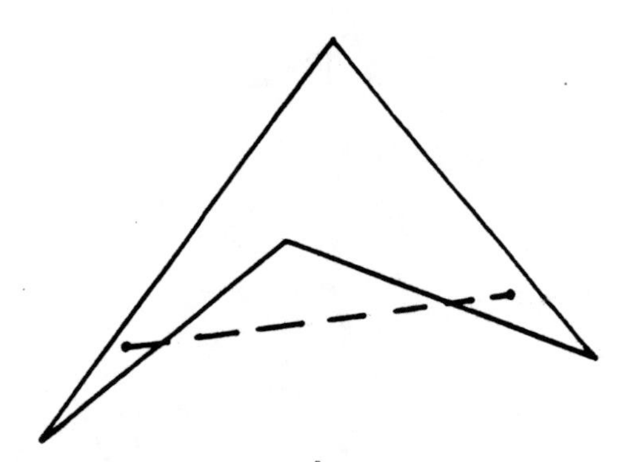

Not convex - dotted lines connecting the two points in this figure are not contained within the figure.

1. Which of the following are convex quadrilaterals? Tell why or why not.

a.

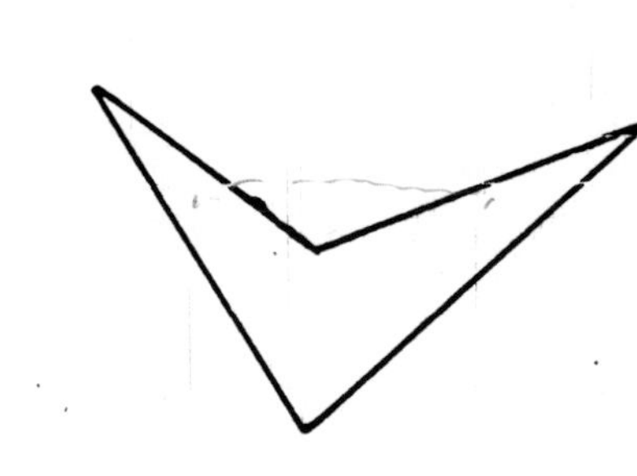

b.

c.

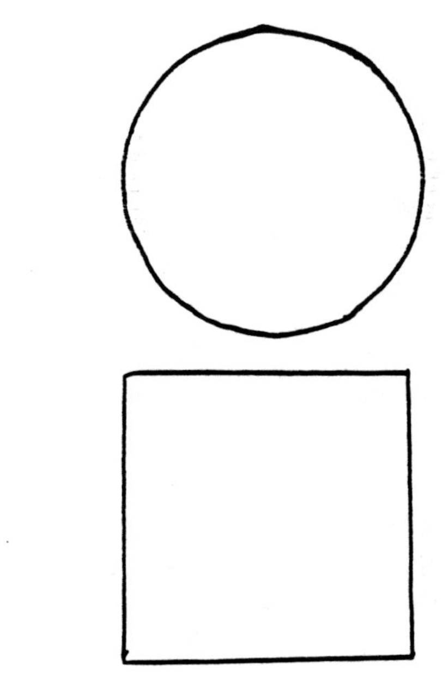

1.

a. not convex

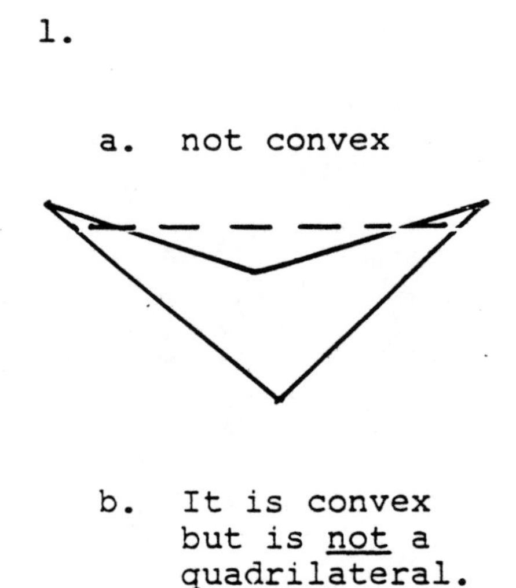

b. It is convex but is not a quadrilateral.

c. convex

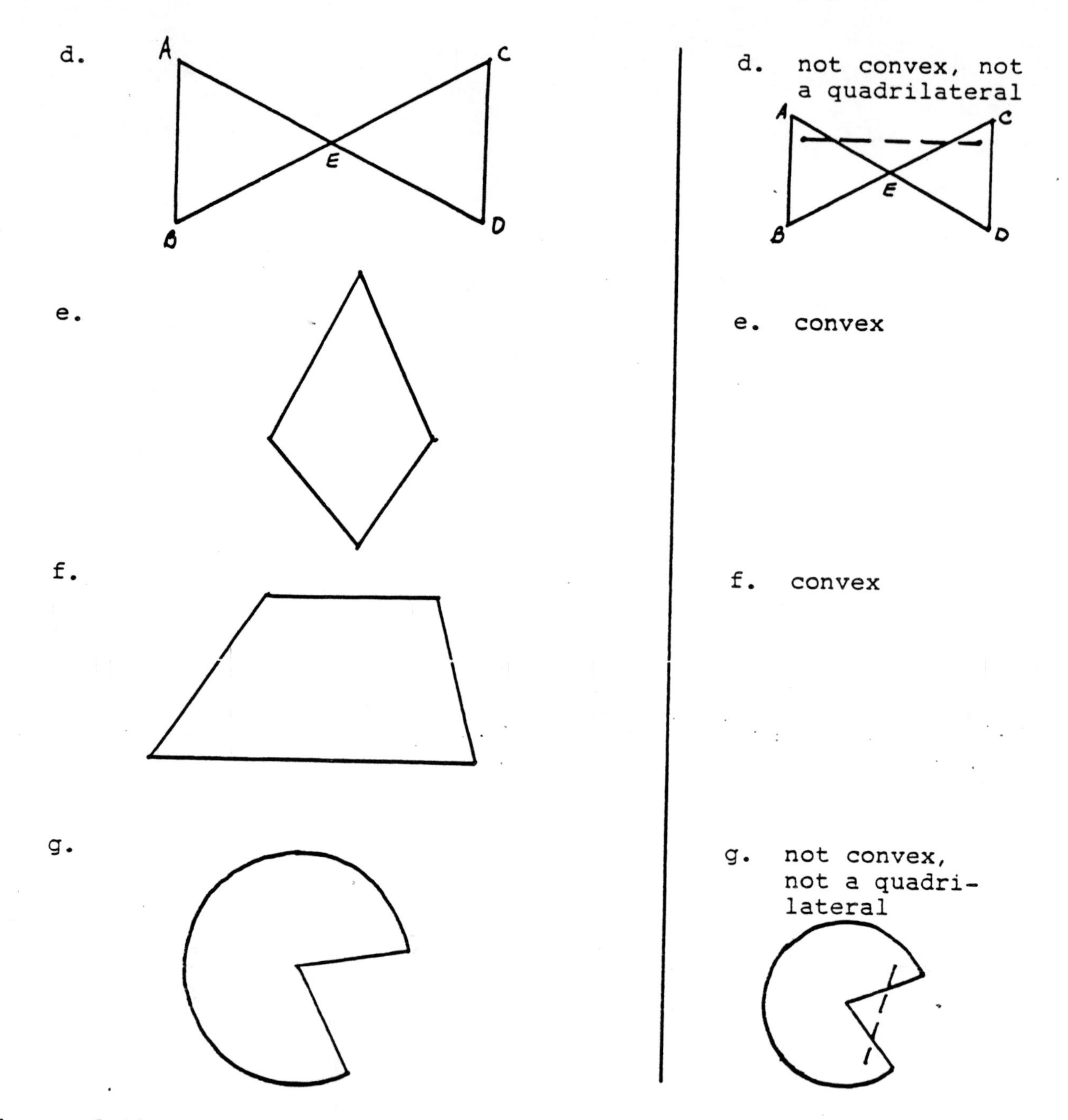

The quadrilaterals that we will be discussing are all closed, four-sided convex figures.

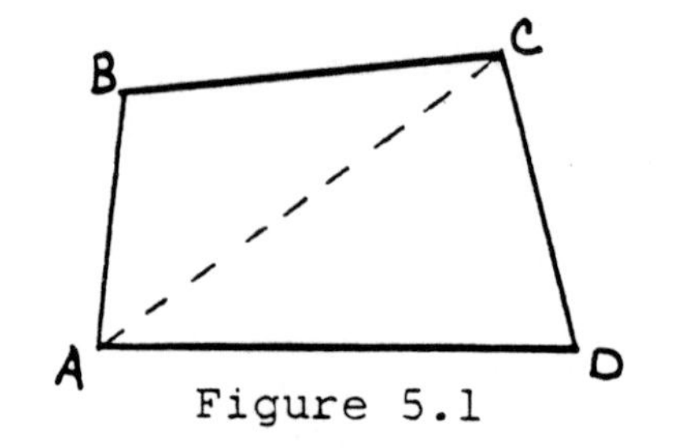

Figure 5.1

Figure ABCD is one example of this type of quadrilateral. Points A, B, C, and D are called the verticies of of the quadrilateral and the line segments are called the sides. The sides that intersect one another are called consecutive. Two angles are consecutive if they have a side in common. $\angle A$ and $\angle B$ are examples of consecutive angles. Sides $\overline{CD}$ and $\overline{AD}$ are examples of consecutive sides. Angles A and C are called opposite angles because they do not share a common side. $\overline{BC}$ and $\overline{AD}$ are opposite sides because they do not intersect.

The diagonal of a quadrilateral is a line segment that connects pairs of opposite verticies. See the dotted line segment in Figure 5.1.

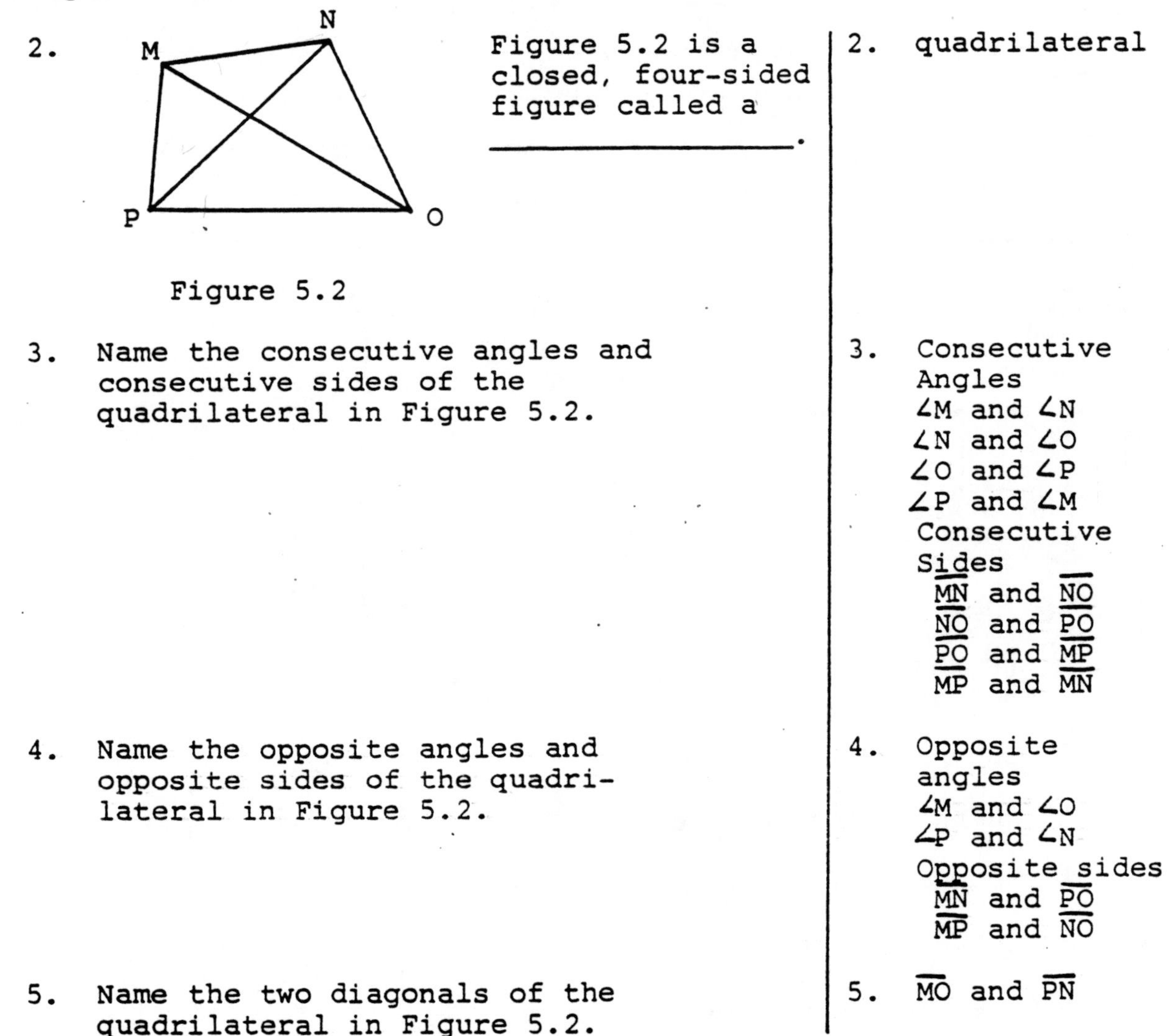

Figure 5.2

2. Figure 5.2 is a closed, four-sided figure called a ________________.

Answer 2. quadrilateral

3. Name the consecutive angles and consecutive sides of the quadrilateral in Figure 5.2.

Answer 3. Consecutive Angles
$\angle M$ and $\angle N$
$\angle N$ and $\angle O$
$\angle O$ and $\angle P$
$\angle P$ and $\angle M$
Consecutive Sides
$\overline{MN}$ and $\overline{NO}$
$\overline{NO}$ and $\overline{PO}$
$\overline{PO}$ and $\overline{MP}$
$\overline{MP}$ and $\overline{MN}$

4. Name the opposite angles and opposite sides of the quadrilateral in Figure 5.2.

Answer 4. Opposite angles
$\angle M$ and $\angle O$
$\angle P$ and $\angle N$
Opposite sides
$\overline{MN}$ and $\overline{PO}$
$\overline{MP}$ and $\overline{NO}$

5. Name the two diagonals of the quadrilateral in Figure 5.2.

Answer 5. $\overline{MO}$ and $\overline{PN}$

The most common types of quadrilaterals are defined as follows:

Parallelogram - is a quadrilateral in which each pair of opposite sides are parallel to one another.

Rectangle - is a parallelogram having one right angle.

Square - is a rectangle all of whose sides are equal.

Rhombus -	is a parallelogram all of whose sides are equal.
Trapezoid -	is a quadrilateral having a single pair of opposite sides that are parallel. These sides are called bases.

The schematic diagram illustrates the relationships of these common types of quadrilaterals.

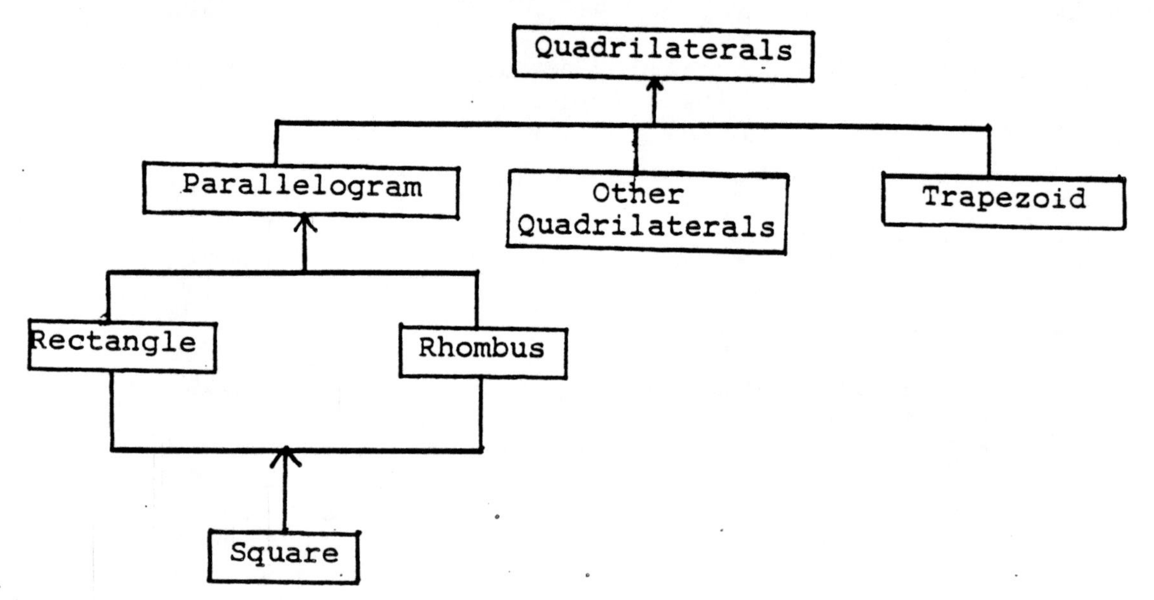

6. Using the diagram, answer the following statements about quadrilaterals. (Use yes or no.)
   a. A rectangle is a trapezoid.
   b. A square is a parallelogram.
   c. A rhombus is a square.
   d. A square is a rectangle.
   e. A rectangle is a square.
   f. A square is a rhombus.
   g. A trapezoid is a quadrilateral.
   h. A rhombus is a parallelogram.

6.
   a. no
   b. yes
   c. no
   d. yes
   e. no
   f. yes
   g. yes
   h. yes

		7.
7.	Using the definitions, which quadrilaterals fit these given conditions?	
a.	Has one right angle and is a parallelogram.	a. rectangle
b.	Has all the sides equal and is a parallelogram.	b. rhombus
c.	Has a right angle, all sides equal, and is a parallelogram.	c. square
d.	Has both pairs of opposite sides parallel.	d. parallelogram
e.	Has only one pair of parallel sides.	e. trapezoid

PARALLELOGRAMS

Each of the following statements concerning quadrilaterals is a theorem. Write out the converses of these theorems as an exercise.

> Theorem 1: If a quadrilateral is a parallelogram, then both pairs of opposite sides are equal.

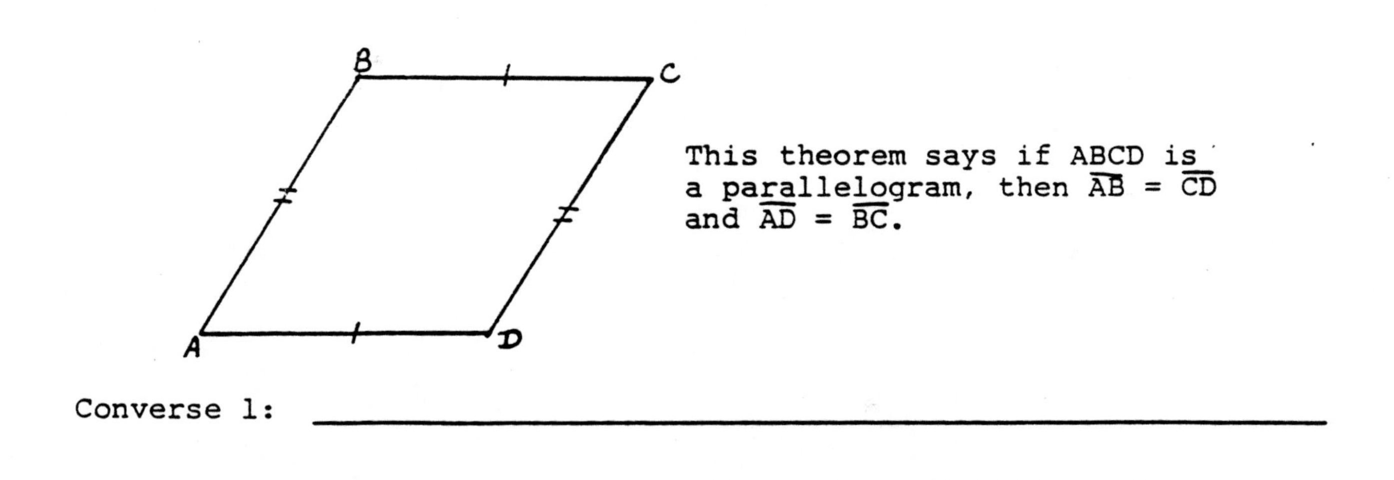

This theorem says if ABCD is a parallelogram, then $\overline{AB} = \overline{CD}$ and $\overline{AD} = \overline{BC}$.

Converse 1: ____

____

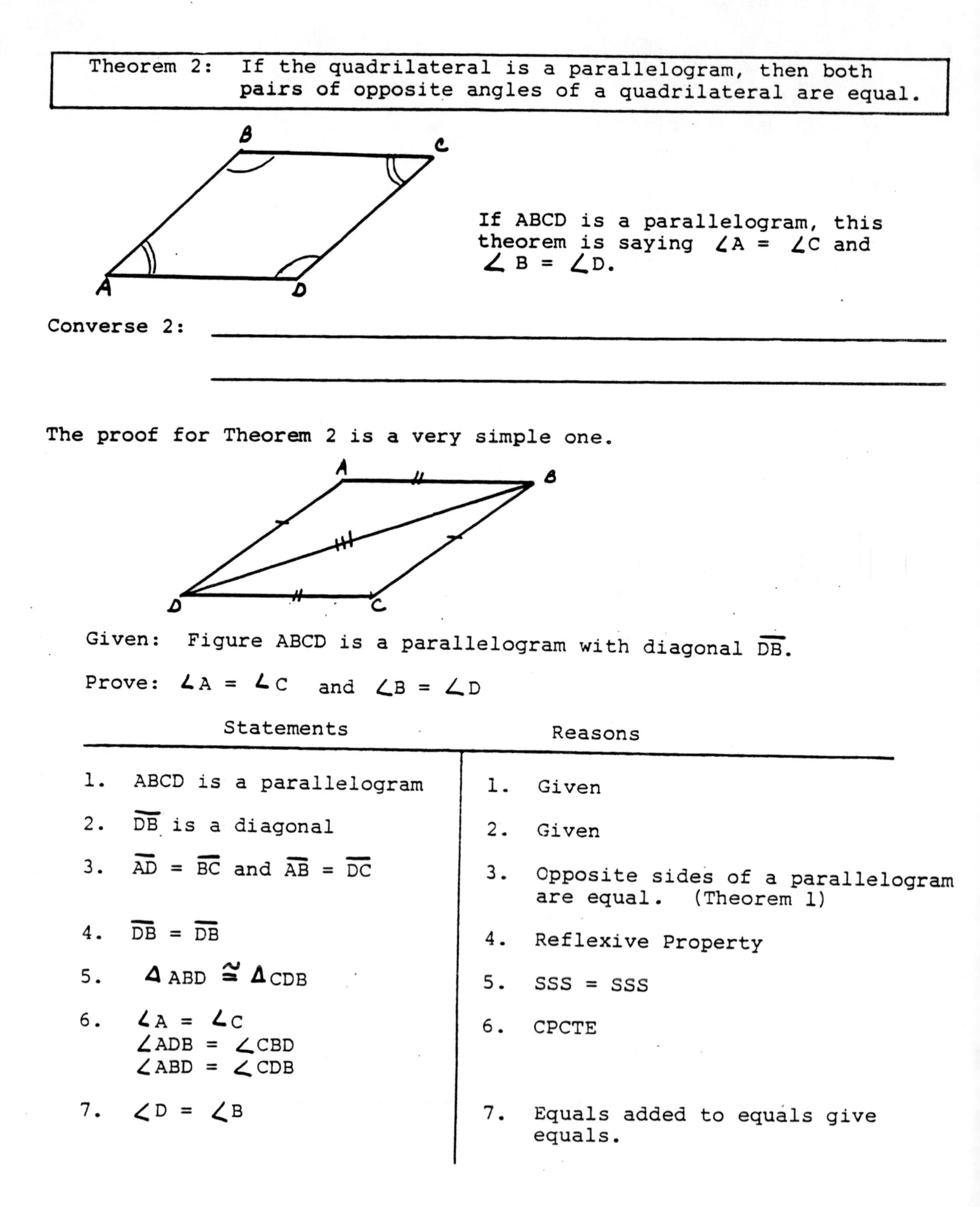

Theorem 2:	If the quadrilateral is a parallelogram, then both pairs of opposite angles of a quadrilateral are equal.

If ABCD is a parallelogram, this theorem is saying $\angle A = \angle C$ and $\angle B = \angle D$.

Converse 2: ____

____

The proof for Theorem 2 is a very simple one.

Given: Figure ABCD is a parallelogram with diagonal $\overline{DB}$.

Prove: $\angle A = \angle C$ and $\angle B = \angle D$

Statements	Reasons
1. ABCD is a parallelogram	1. Given
2. $\overline{DB}$ is a diagonal	2. Given
3. $\overline{AD} = \overline{BC}$ and $\overline{AB} = \overline{DC}$	3. Opposite sides of a parallelogram are equal. (Theorem 1)
4. $\overline{DB} = \overline{DB}$	4. Reflexive Property
5. $\triangle ABD \cong \triangle CDB$	5. SSS = SSS
6. $\angle A = \angle C$ $\angle ADB = \angle CBD$ $\angle ABD = \angle CDB$	6. CPCTE
7. $\angle D = \angle B$	7. Equals added to equals give equals.

From the proof, we discovered that a diagonal of a parallelogram divides the parallelogram into two congruent triangles. The corresponding parts of these triangles are equal. Therefore, the opposite angles of a parallelogram are indeed equal.

8. Use the following steps to prove that consecutive angles of a parallelogram are supplementary. The statements are written in, you provide the reasons.

Given: MNOP is a parallelogram.

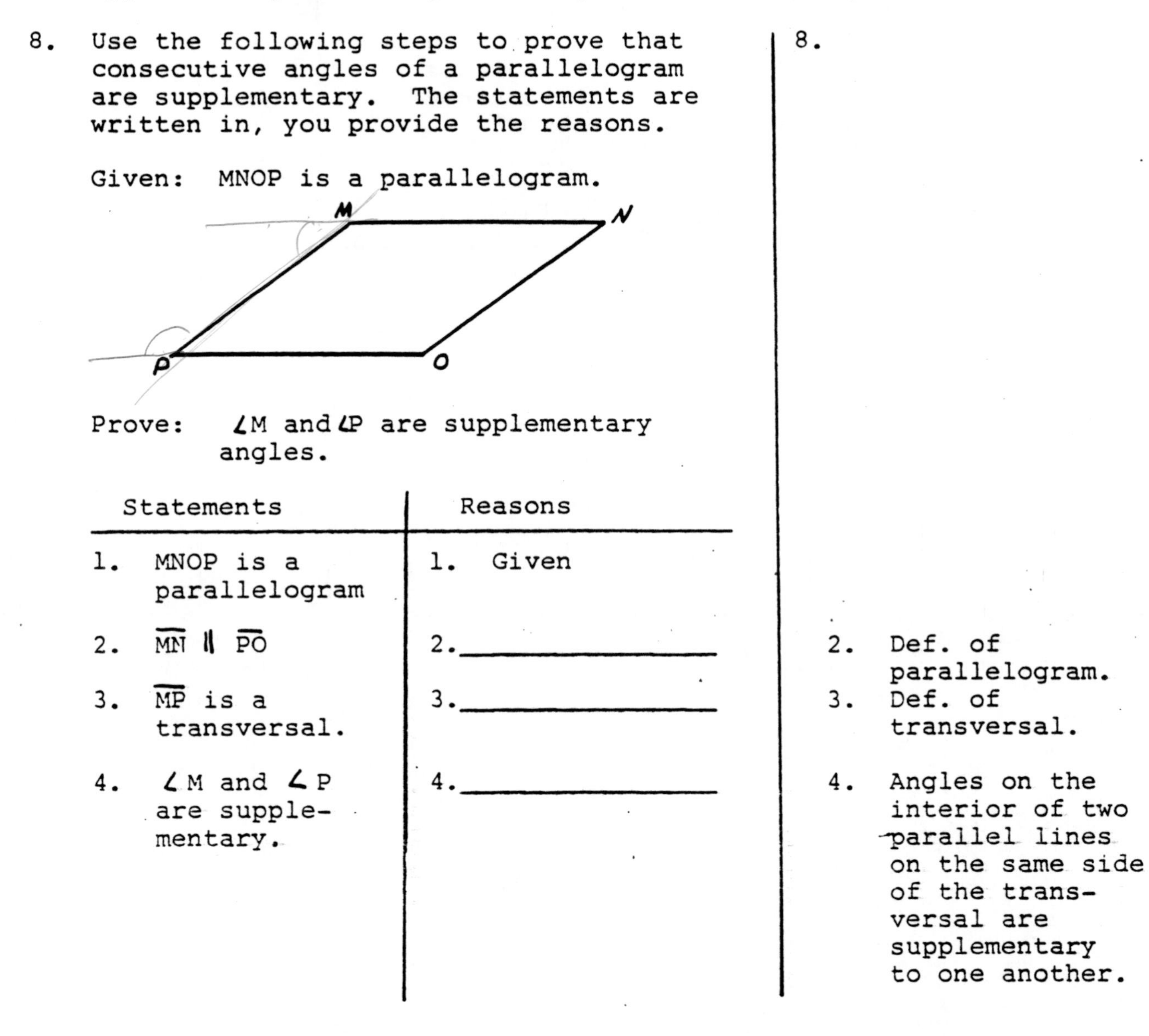

Prove: ∠M and ∠P are supplementary angles.

Statements	Reasons
1. MNOP is a parallelogram	1. Given
2. $\overline{MN} \parallel \overline{PO}$	2.____________
3. $\overline{MP}$ is a transversal.	3.____________
4. ∠M and ∠P are supplementary.	4.____________

8.

2. Def. of parallelogram.
3. Def. of transversal.
4. Angles on the interior of two parallel lines on the same side of the transversal are supplementary to one another.

We have just proven a theorem for parallelograms.

Theorem 3: Consecutive angles of a parallelogram are supplementary.

9. If the measure of one angle of a rhombus is 40°, find the measures of the other angles.

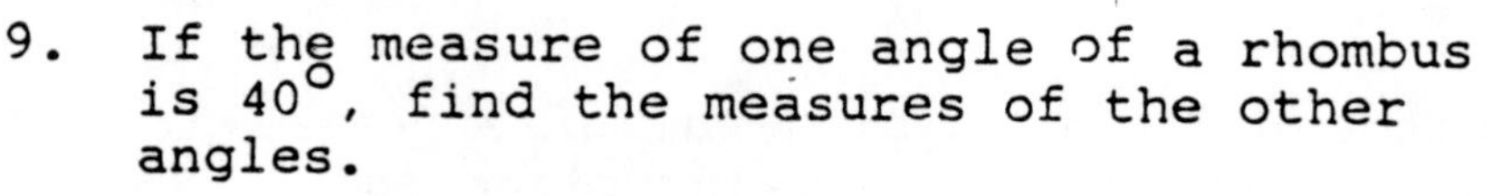

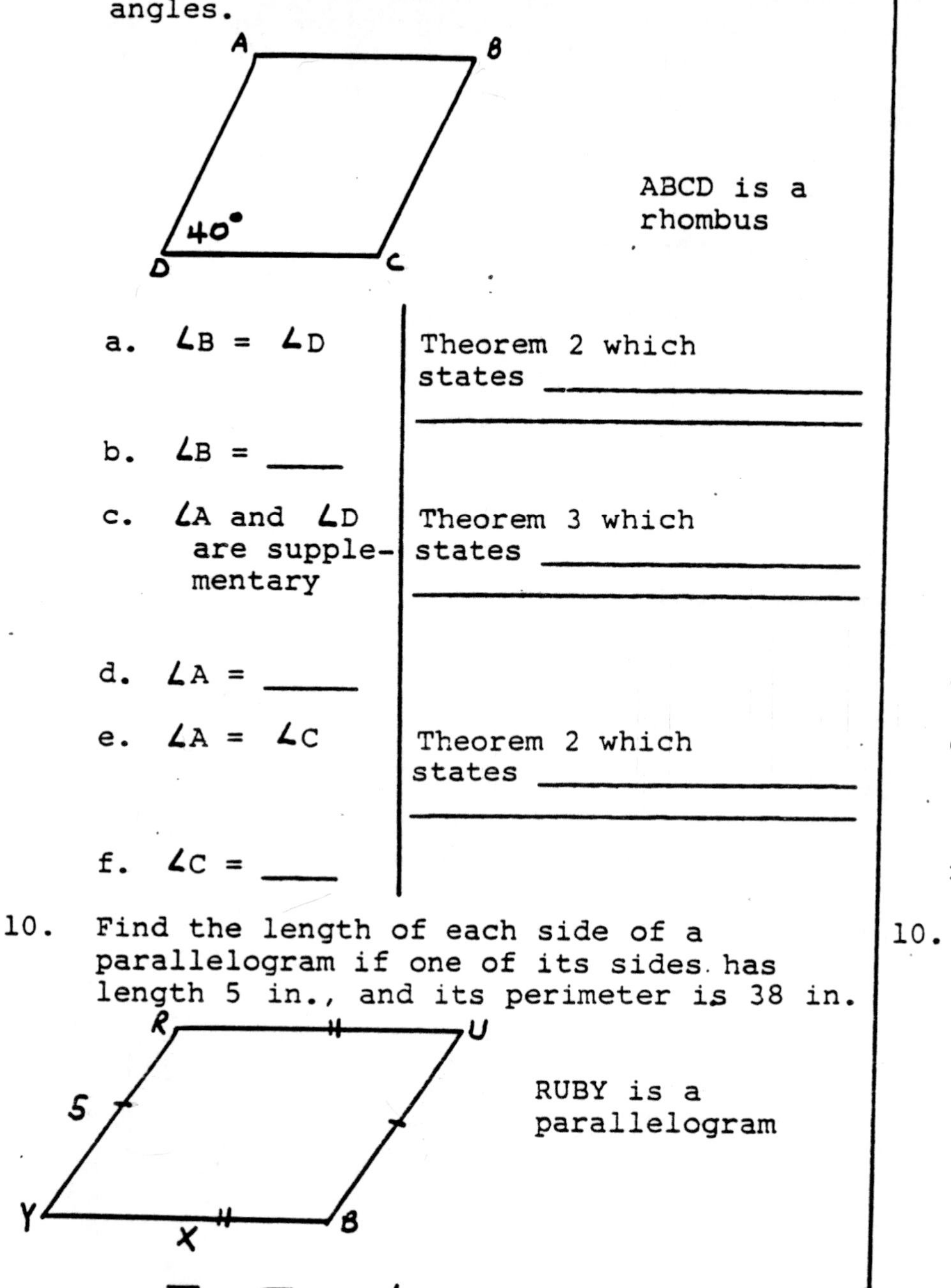

ABCD is a rhombus

a. ∠B = ∠D — Theorem 2 which states ______________________

b. ∠B = ____

c. ∠A and ∠D are supplementary — Theorem 3 which states ______________________

d. ∠A = ____

e. ∠A = ∠C — Theorem 2 which states ______________________

f. ∠C = ____

10. Find the length of each side of a parallelogram if one of its sides has length 5 in., and its perimeter is 38 in.

R U Y B X 5

RUBY is a parallelogram

a. $\overline{RY} = \overline{UB}$, $\overline{RU} = \overline{YB}$ — Theorem 1 which states ______________________

b. $\overline{UB}$ = ____

$\overline{RU}$ = ____

9.

a. Opposite angles of a parallelogram are equal.

b. 40°

c. Consecutive angles of a parallelogram are supplementary.

d. 140°

e. Opposite angles of a parallelogram are equal.

f. 140°

10.

a. Both pairs of opposite sides of a parallelogram are equal.

b. 5 in.

x

c. $38 = 2x + 2(5)$ because__________

d. Solve for x; _____

11.

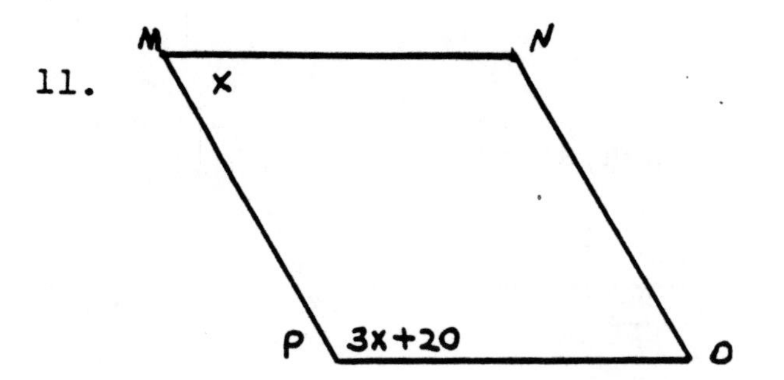

Find the value of x and the measure of the two angles shown in the parallelogram.

$\angle M$ and $\angle P$ are supplementary.

Therefore $\angle M + \angle P = 180^\circ$

and $x + (3x + 20) = 180$

$4x + 20 = 180$

$4x = 160$

$x = 40^\circ$

Therefore $3x + 20 =$ __________

12.

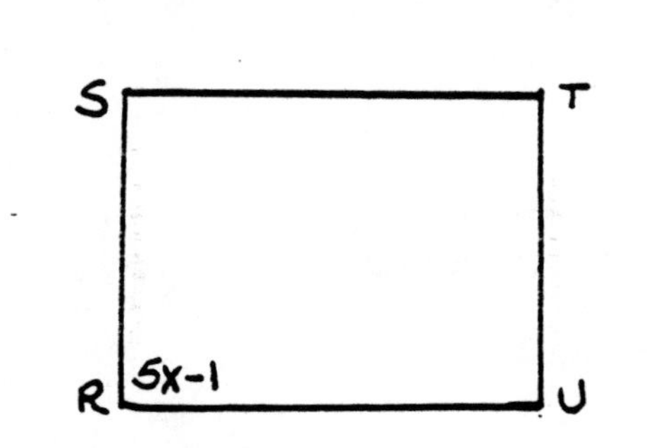

RSTU is a rectangle.

Solve for x.

$\angle R$ is a right angle since RSTU is a rectangle.

Therefore, $5x - 1 = 90^\circ$

$5x = 91$

$x =$ _____

c. Def. of perimeter.

d. $38 = 2x + 10$

$28 = 2x$

14 in. $= x$

11. 140°

12. $18\frac{1}{5}^\circ$

13.

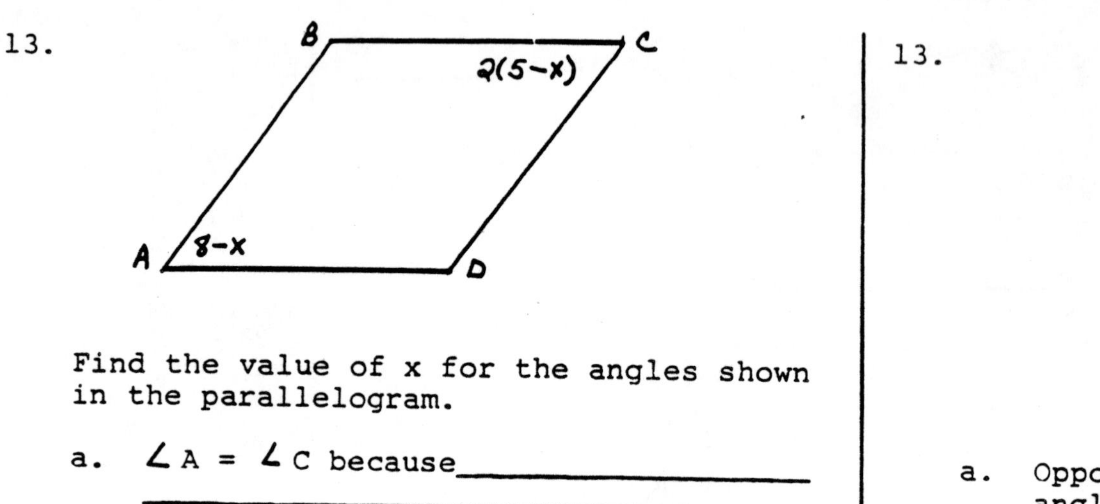

Find the value of x for the angles shown in the parallelogram.

a. $\angle A = \angle C$ because ________________________

b. Therefore, $(8 - x) = 2(5 - x)$
$x =$ ____

13.

a. Opposite angles of a parallelogram are equal.

b. $x = 2^{\circ}$

Exercise 1

1. If the measure of one angle of a rhombus is 86°, find the measures of the other angles.

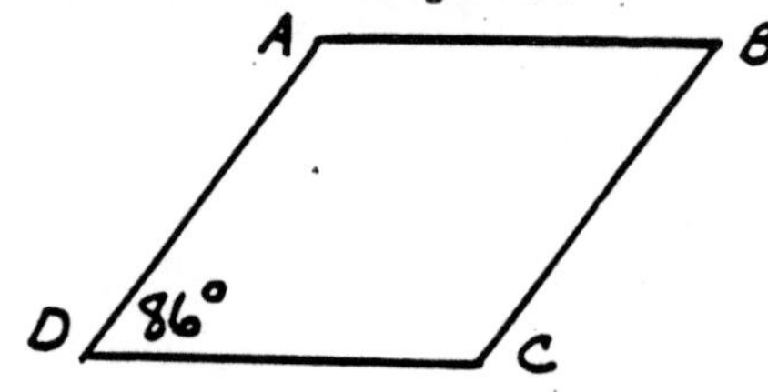

2. Find the length of each side of a parallelogram if one of its sides has length 8 in. and its perimeter is 46 in.

3. Given ABCD is a parallelogram, find x.

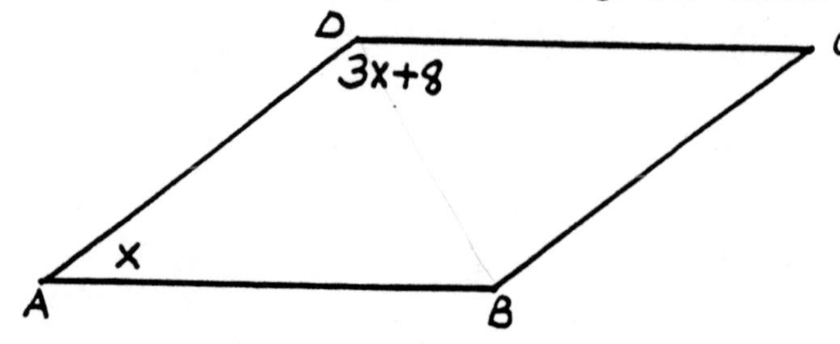

4. Given ABCD is a square with perimeter 75 m., find the length of a side and also the value of x.

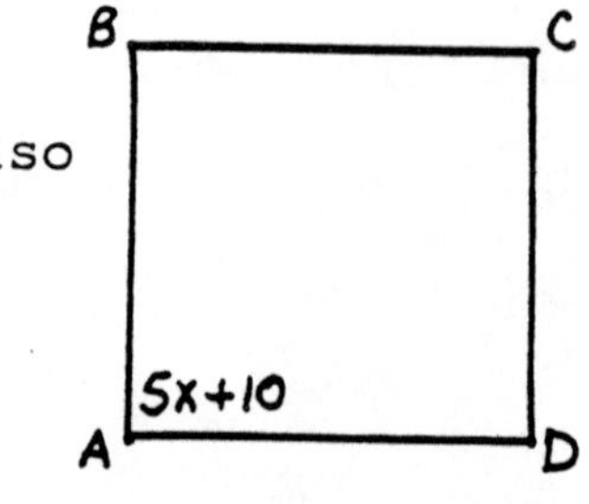

5. Given ABCD is a parallelogram, find the value of each angle.

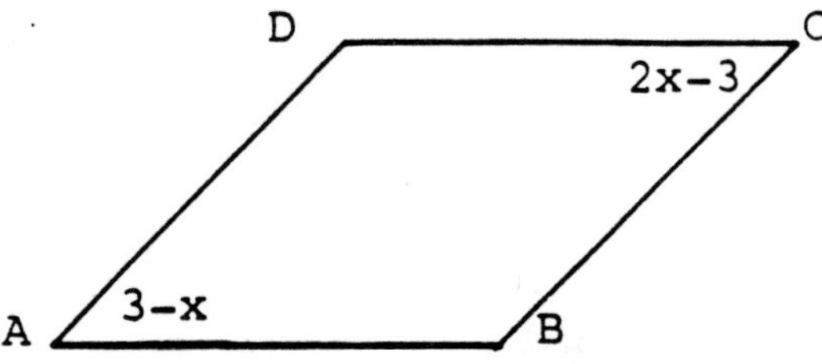

6. Supply the reasons for the following proof of Theorem 1.

Theorem 1:	If a quadrilateral is a parallelogram, then both pairs of opposite sides are equal.

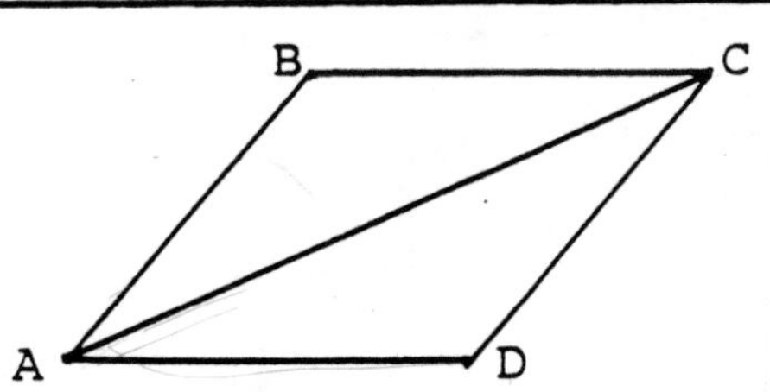

Given: Figure ABCD is a parallelogram with diagonal $\overline{AC}$.

Prove: $\overline{AB} = \overline{CD}$ and $\overline{AD} = \overline{BC}$

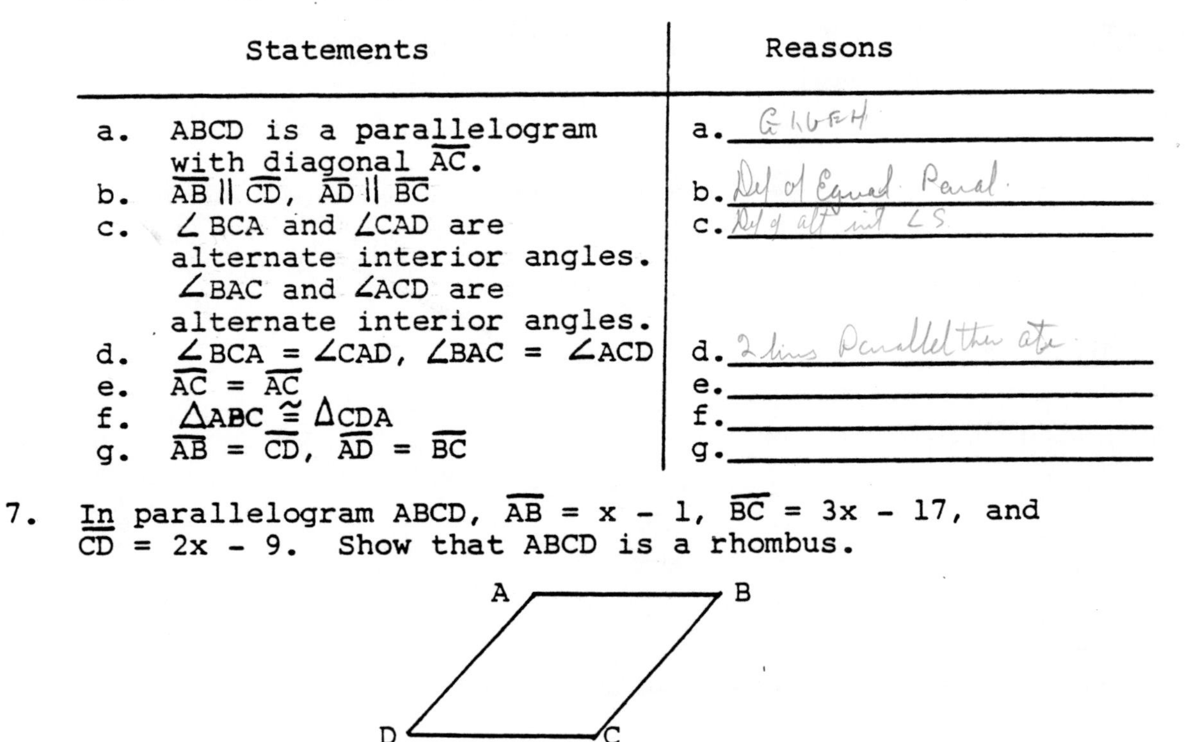

Statements	Reasons
a. ABCD is a parallelogram with diagonal $\overline{AC}$.	a. GIVEN
b. $\overline{AB} \parallel \overline{CD}$, $\overline{AD} \parallel \overline{BC}$	b. Def of Equal Paral
c. $\angle BCA$ and $\angle CAD$ are alternate interior angles. $\angle BAC$ and $\angle ACD$ are alternate interior angles.	c. Def of alt int ∠s
d. $\angle BCA = \angle CAD$, $\angle BAC = \angle ACD$	d. 2 lines Parallel then alt
e. $\overline{AC} = \overline{AC}$	e.
f. $\triangle ABC \cong \triangle CDA$	f.
g. $\overline{AB} = \overline{CD}$, $\overline{AD} = \overline{BC}$	g.

7. In parallelogram ABCD, $\overline{AB} = x - 1$, $\overline{BC} = 3x - 17$, and $\overline{CD} = 2x - 9$. Show that ABCD is a rhombus.

A B

D C

TRAPEZOID

The trapezoid is a quadrilateral that does not come under the heading of parallelogram. (See diagram, page 136.)

A trapezoid is a quadrilateral that has only <u>one pair</u> of opposite sides that are parallel. These parallel sides are called <u>bases</u>. The non-parallel sides of a trapezoid are called <u>legs</u>.

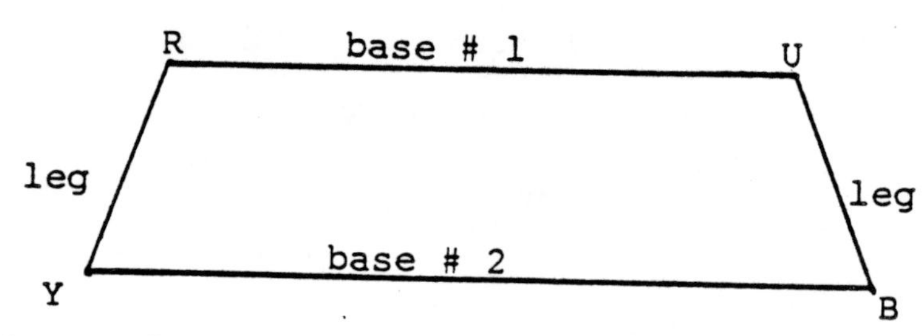

$\overline{RY}$ and $\overline{UB}$ are legs of the figure. $\overline{RU}$ and $\overline{YB}$ are bases.

A special type of trapezoid is called an isosceles trapezoid. An <u>isosceles trapezoid</u> is a trapezoid whose legs are equal.

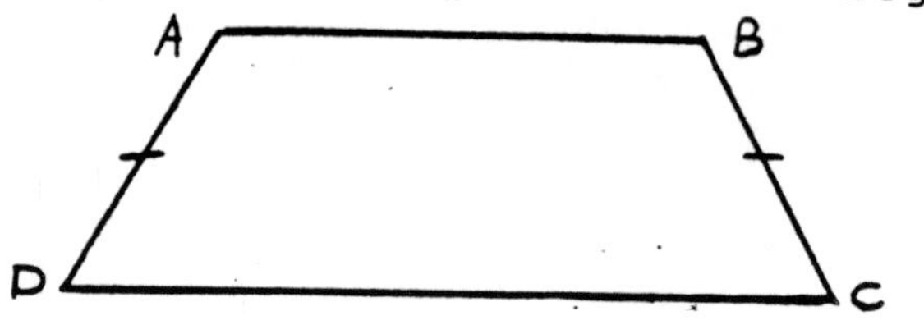

Definition:	The line that joins the midpoints of the legs of any trapezoid is called the <u>median of a trapezoid</u>.

Theorem 4:	The median of a trapezoid is parallel to the bases and is equal to one-half of the sum of the two bases.

14. Given trapezoid ABCD with median $\overline{MN}$.

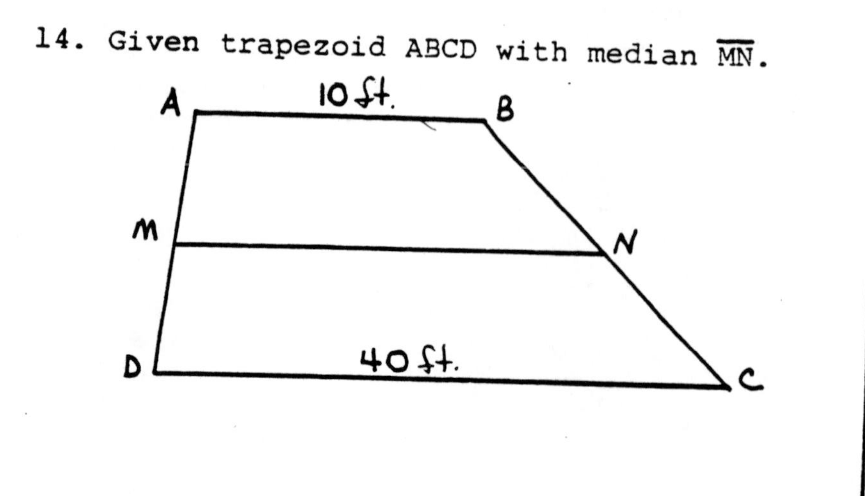

If $\overline{AB}$ = 10 ft. and $\overline{CD}$ = 40 ft., find the measure of $\overline{MN}$.

Median = $\frac{1}{2}$(the sum of the bases)

The equation can be written as

(1) $\overline{MN} = \frac{1}{2}(\overline{AB} + \overline{DC})$

or, by eliminating fractions, as

(2) $2\overline{MN} = \overline{AB} + \overline{DC}$

Remember, $\overline{AB}$ = 10 ft. and $\overline{CD}$ = 40 ft.

Substituting these values into the second equation gives:

$2\overline{MN} = 10 + 40$

$2\overline{MN} = 50$

$\overline{MN}$ = ______

14. 25 ft.

15.

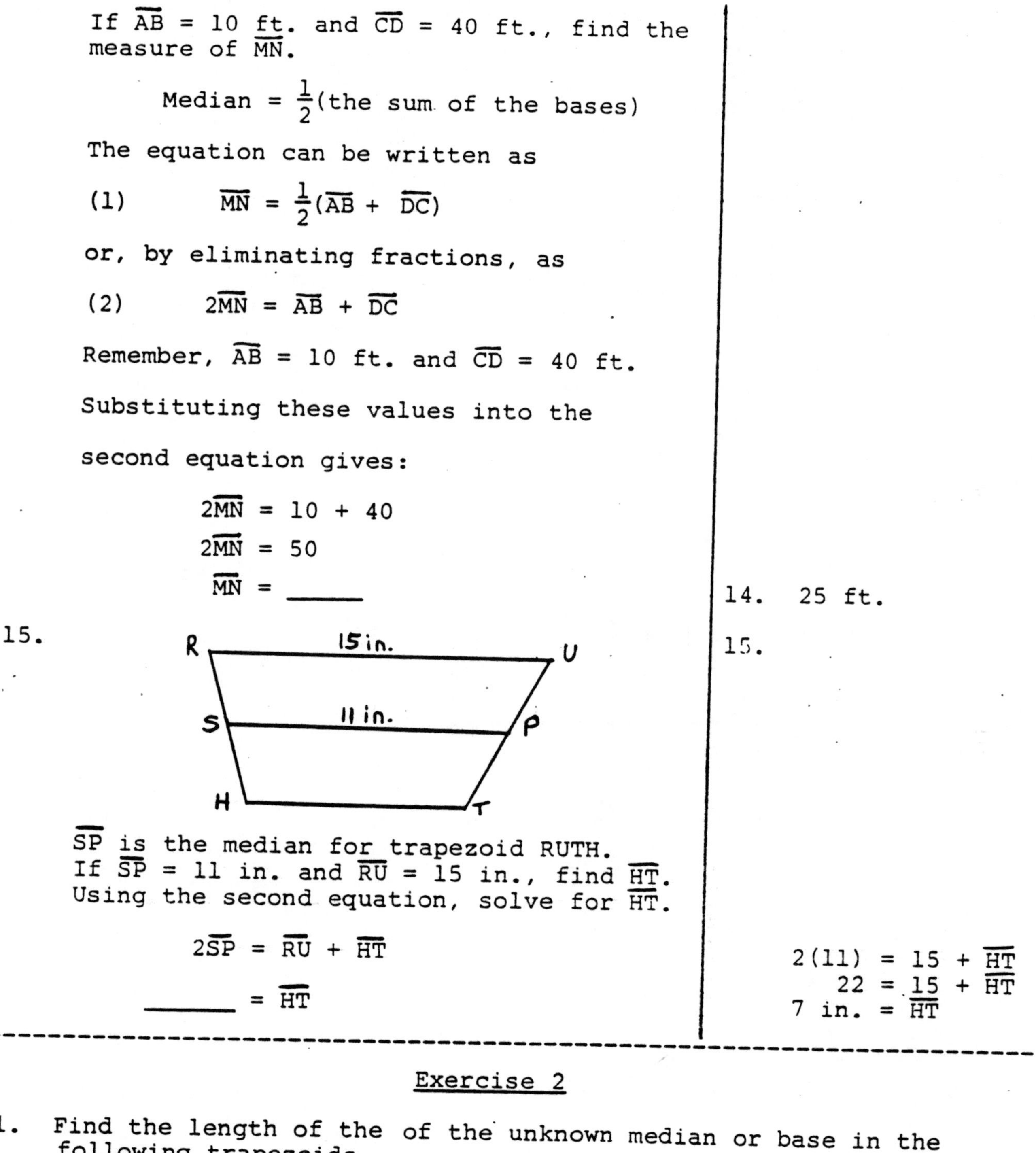

$\overline{SP}$ is the median for trapezoid RUTH.
If $\overline{SP}$ = 11 in. and $\overline{RU}$ = 15 in., find $\overline{HT}$.
Using the second equation, solve for $\overline{HT}$.

$2\overline{SP} = \overline{RU} + \overline{HT}$

______ = $\overline{HT}$

15.

$2(11) = 15 + \overline{HT}$
$22 = 15 + \overline{HT}$
$7 \text{ in.} = \overline{HT}$

<u>Exercise 2</u>

1. Find the length of the of the unknown median or base in the following trapezoids.

a.

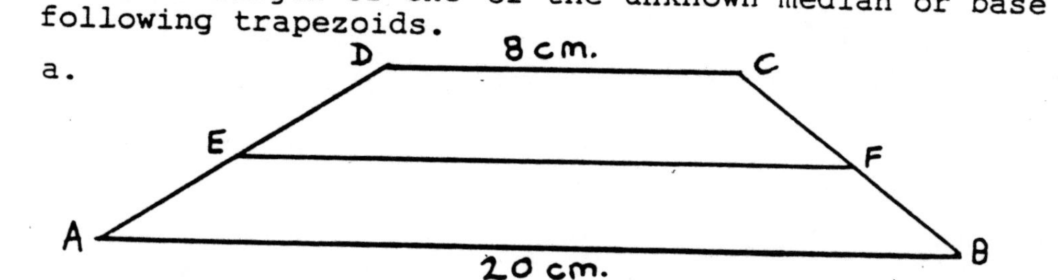

b.

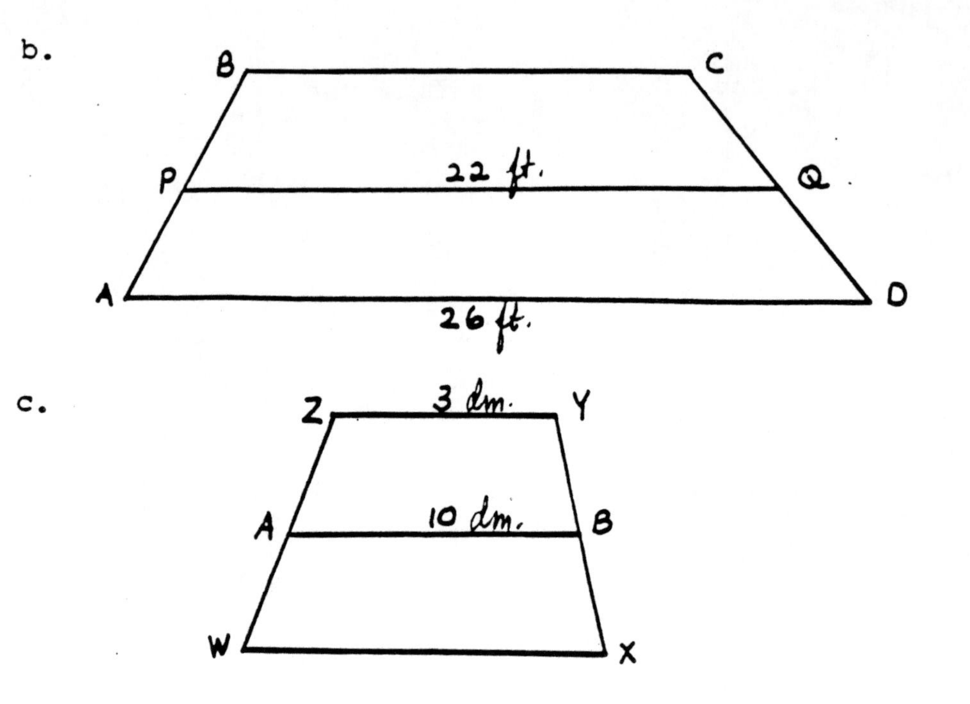

c.

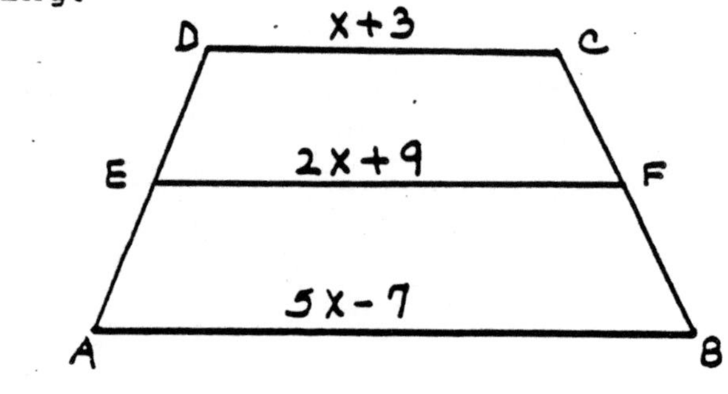

2. Given trapezoid ABCD with median $\overline{EF}$, find x in each of the following.

a.

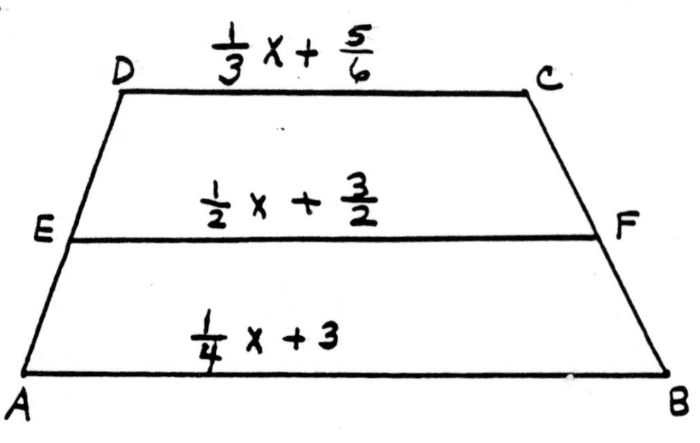

b.

c.

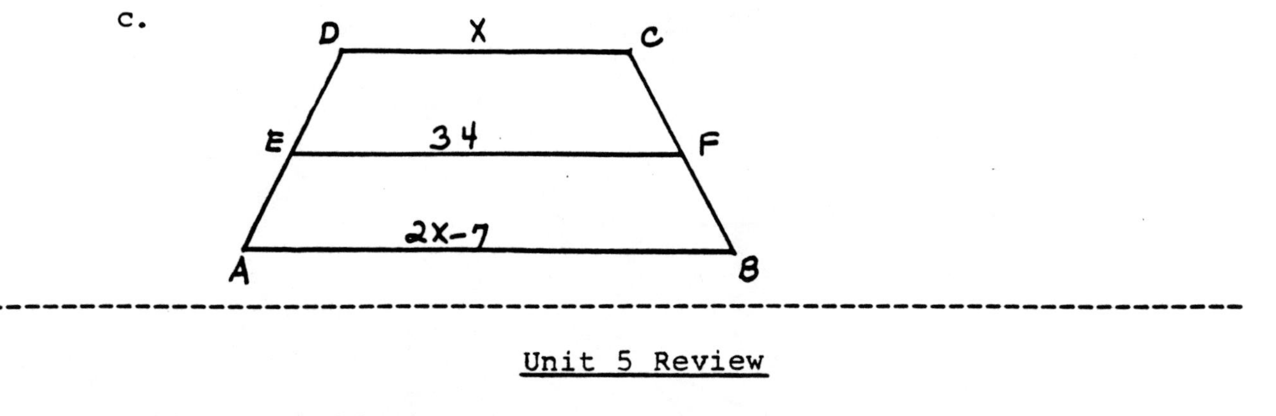

Unit 5 Review

I. Fill in the blanks with the appropriate word or phrase as studied in this unit.

1. A ________________ is a closed, four-sided convex figure.

2. In quadrilateral D C A B the verticies are ________, __________, ___________, and __________. The sides are ______, ________, _________, and ___________. ∠A and ∠C are called ___________ angles. ∠A and ∠D are called __________ angles.

3. A ___________ is a quadrilateral in which each pair of opposite sides is parallel.

4. A __________ is a parallelogram having at least one right angle.

5. A __________ is a rectangle all of whose sides are equal.

6. A _________ is a parallelogram all of whose sides are equal.

7. A __________ is a quadrilateral having only one pair of parallel sides which are called ____________.

8. An _________ ____________ is a trapezoid whose legs are equal.

9. The line that joins the midpoints of the legs of a trapezoid is called the ___________ of a trapezoid.

II. Solve the following.

a.

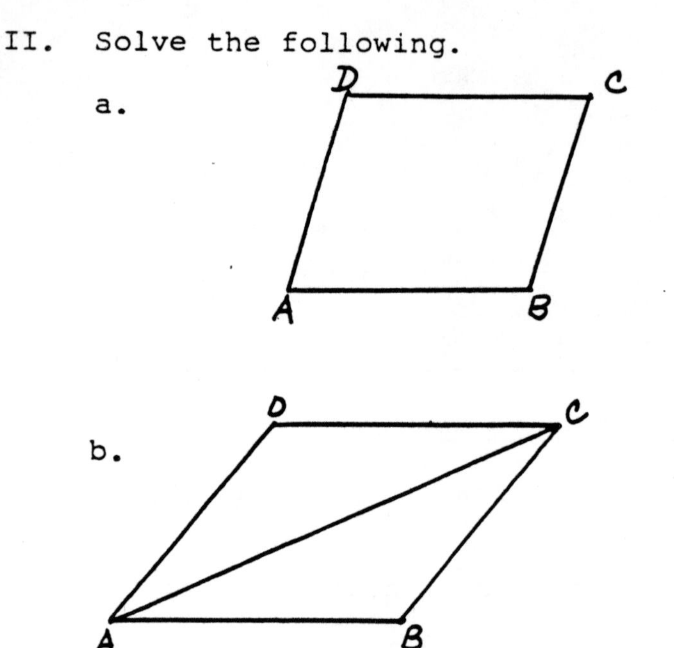

a. In parallelogram ABCD, $\overline{AB} = 3x - 7$ and $\overline{DC} = 2x + 30$. Find x.

b. In parallelogram ABCD, $\angle CAB = 2x - 3$ and $\angle DCA = 3x - 7$. Find x. If $\angle D = 105^\circ$, find $\angle B$ and $\angle DAB$.

c. Find the length of each side of the parallelogram below if one of its sides has length 8 dm. and its perimeter is 60 dm.

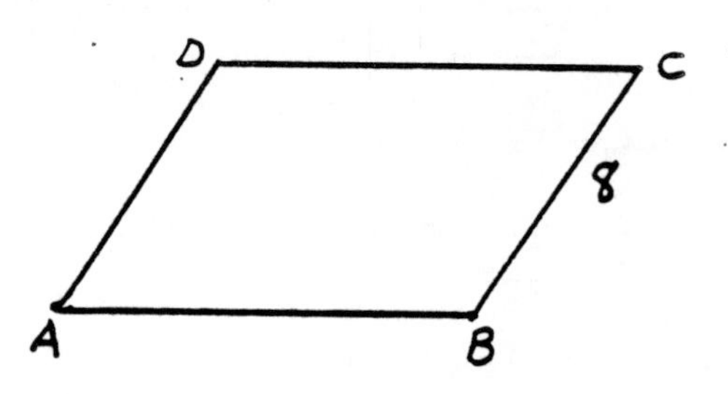

d. Find the value of x and the measure of the angles shown in the parallelogram.

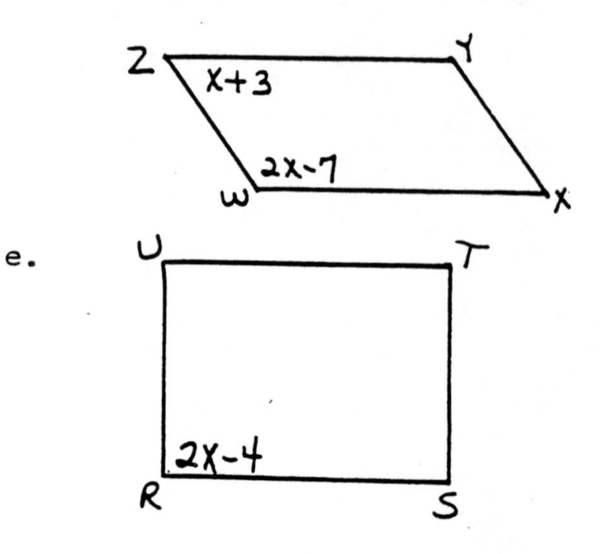

e. RSTU is a rectangle. Solve for x.

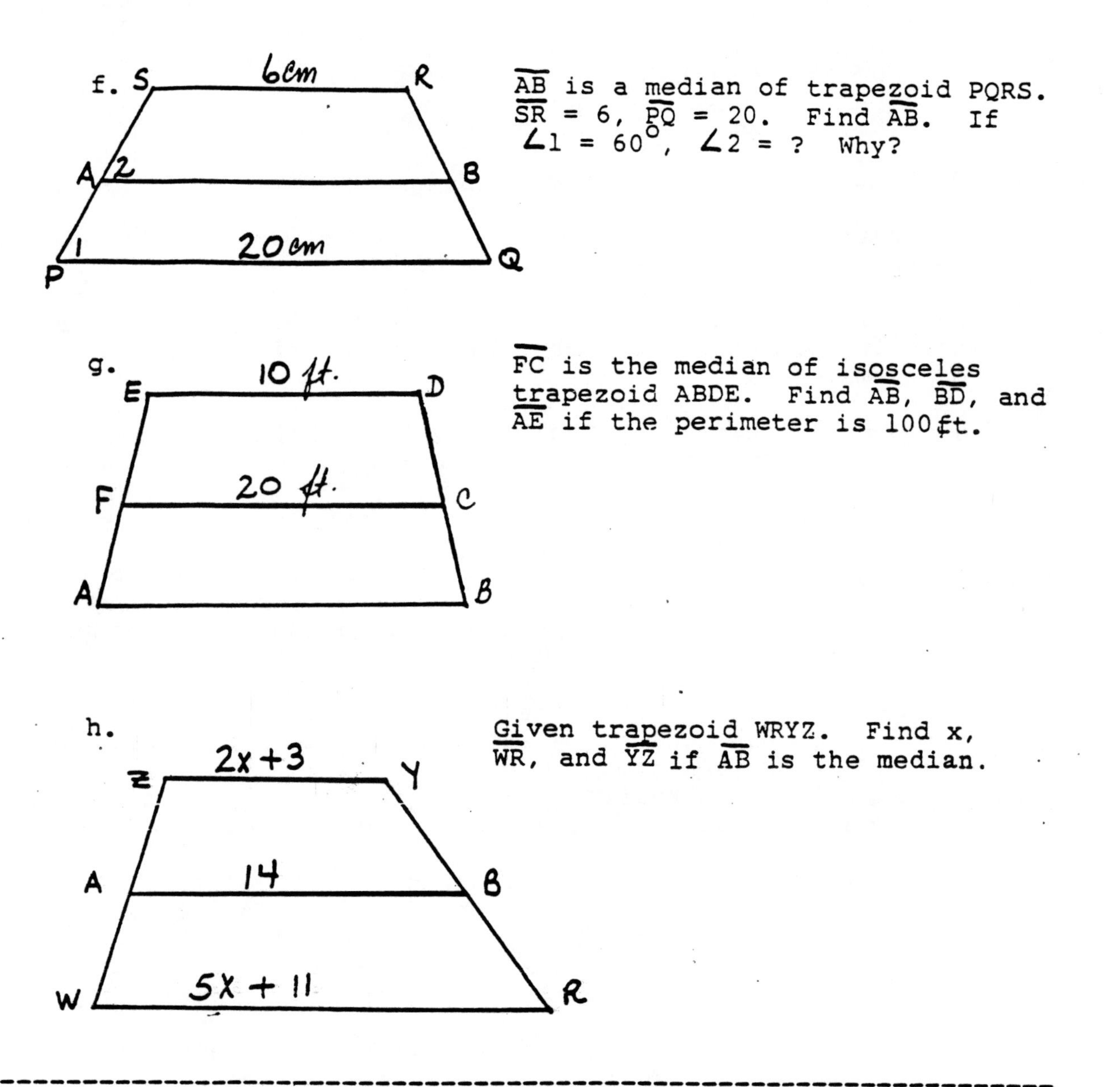

f. $\overline{AB}$ is a median of trapezoid PQRS. $\overline{SR} = 6$, $\overline{PQ} = 20$. Find $\overline{AB}$. If $\angle 1 = 60^\circ$, $\angle 2 = ?$ Why?

g. $\overline{FC}$ is the median of isosceles trapezoid ABDE. Find $\overline{AB}$, $\overline{BD}$, and $\overline{AE}$ if the perimeter is 100 ft.

h. Given trapezoid WRYZ. Find x, $\overline{WR}$, and $\overline{YZ}$ if $\overline{AB}$ is the median.

Unit 6

Inequalities

Learning Objectives:

1. The student will write the meanings for " $>$ " and " $<$ ".

2. The student will write an algebraic open sentence for an English sentence involving inequalities.

3. The student will use the following properties to solve linear, certain quadratic and some cubic inequalities and other selected problems:

   a. Order Property I: The Three Possibilities Postulate (Trichotomy Law). In comparing any two numbers only one of the following can exist at any one time. Either $a = b$, $a > b$ or $a < b$.

   b. Order Property II: Transitive Law. In comparing two real numbers, with a third real number c, the following order will exist: If a is greater than b and b is greater than c, then a is greater than c.

   c. Order Property III: The Addition Property.
   Part 1: If equals are added to unequals, the sums are unequal in the same order.
   Part 2: If unequals are added to unequals of the same order, the sums are unequal in the same order.

   d. Order Property IV: The Subtraction Property.
   Part 1: If equals are subtracted from unequals, the differences are unequal in the same order.
   Part 2: If unequals are subtracted from equals, the differences are unequal in the opposite order.

   e. Order Property V: The Multiplication Property.
   Part 1: If unequals are multiplied by the same positive number, the products are unequal in the same order.
   Part 2: If unequals are multiplied by the same negative number, the products are unequal in the opposite order.

4. Using definitions, postulates, and previously proven theorems, the student will apply the following theorems to the solution of related problems:

   a. The whole is greater than any of its parts.

   b. If two sides of a triangle are unequal, the angles opposite them are unequal and the larger angle is opposite the longer side.

c. If two angles of a triangle are unequal, the sides opposite them are unequal and the longer side is opposite the larger angle.

INEQUALITIES

. . . -4 -3 -2 -1 0 1 2 3 4 . . .

In comparing numbers on the number line, the furthest number to the right has the greatest value. For example, -2 and 0. The number that has the greater value is zero since it is further to the right on the number line than -2.

Similarly, a number furthest to the left on the number line has the lesser value. For example, 1 and 4. The number that has the lesser value is one since it is further to the left on the number line than 4.

The symbols used for inequality are $>$ for "is greater than" and $<$ for "is less than". These symbols compare any two numbers whose values are not equal.

Example 1: 0 is greater than -2 may be written as $0 > -2$.

Example 2: 1 is less than 4 may be written as $1 < 4$.

You complete the following using the correct symbol.

1. -5 ____ -3
2. -3 ____ 0
3. 1 ____ 0
4. -2 ____ 6
5. -10 ____ -19
6. 3 ____ -3
7. 0 ____ 9

1. $-5 < -3$
2. $-3 < 0$
3. $1 > 0$
4. $-2 < 6$
5. $-10 > -19$
6. $3 > -3$
7. $0 < 9$

Any number greater than zero is <u>positive</u> and may be written as: $x > 0$ (x representing any number greater than zero).

If x were a <u>negative</u> number then it would be expressed as $x < 0$. (x representing any number less than zero).

8.	A negative number, x, may be expressed as ______________.	8.	$x < 0$
9.	4 is less than 6 ______________	9.	$4 < 6$
10.	Is $6 > 4$ a correct response to problem 9? __________ Why or why not? __________	10.	Yes, 6 is greater than 4 has the same meaning as 4 is less than 6.
11.	23 is greater than 22. ______________	11.	$23 > 22$ <u>or</u> $22 < 23$
12.	0 is less than 15. ______________	12.	$0 < 15$ <u>or</u> $15 > 0$
13.	Y is a positive number. ______________	13.	$Y > 0$
14.	Twice a number is greater than 3. ______________	14.	$2x > 3$
15.	The sum of a number and -5 is greater than 2. ______________	15.	$[x + (-5)] > 2$
16.	8z is a negative number. ______________	16.	$8z < 0$

The inequality symbol, in combination with the equal sign, will result in two possible conditions to be considered in studying the example.

Example: $x \leq 0$ means
1. x may be negative <u>or</u>
2. x may be equal to zero.

Hence, x may be negative or zero.

$x \geq 0$ means x may be positive or zero.

Complete these problems:

17.	x is negative	17.	$x < 0$
18.	x is zero.	18.	$x = 0$
19.	x may be negative or zero.	19.	$x \leq 0$
20.	x may be positive or zero.	20.	$x \geq 0$
21.	x may be greater than or equal to 10.	21.	$x \geq 10$
22.	x may be less than or equal to 10.	22.	$x \leq 10$

If one number falls between two other numbers, the expression is written as follows:

Example: 7 falls between 6 and 9 and may be expressed as "7 is greater than 6 and also less than 9".

$$6 \angle 7 \angle 9$$

Problem	Answer
23. Write a number that is between -1 and 15 using the inequality symbol and x.	23. $-1 \angle x \angle 15$
24. A number is less than 1 and greater than 0.	24. $0 \angle x \angle 1$
25. x is greater than 8 and less than or equal to 200.	25. $8 \angle x \leq 200$
26. x is less than -5 and greater than or equal to -9.	26. $-9 \leq x \angle -5$
27. x is a positive number and has a value less than or equal to 2.	27. $0 \angle x \leq 2$

Order Properties of Inequality

Order Property I: The Three Possibilities Postulate (Trichotomy Law)

In comparing any two numbers only one of the following can exist at any one time. Compare two numbers, a and b, real numbers, either

1. a is greater than b__________ $a > b$
2. a is less than b__________ $a \angle b$
3. a is equal to b __________ $a = b$

Problem	Answer
28. Compare 10 and $\frac{20}{2}$. Can 10 equal $\frac{20}{2}$ $\left(10 = \frac{20}{2}\right)$ and can 10 be less than $\frac{20}{2}$ $\left(10 \angle \frac{20}{2}\right)$ at the same time?	28. no
29. Compare x and 0. What are the three possible values of x in relation to zero? Either __________or__________or__________	29. Either $x \angle 0$, or $x > 0$, or $x = 0$.
30. If x is not negative (non-negative) then what are the other two possible values of x? (See problem 29.) ________ __________	30. x can be positive or x can be zero.
31. If x is not positive (non-positive) then what are the other two possible values of x? __________ __________	31. x can be negative or x can be zero.

32. $x \geq 0$ means that x is non-negative. If x is non-positive then how would you write that in symbols?	32. $x \leq 0$
33. If x is not zero then x must be ________ or ________________.	33. $x > 0$ or $x < 0$
34. Write m is non-negative in symbols.	34. $m \geq 0$
35. Write three times z increased by 4 is non-positive.	35. $3z + 4 \leq 0$
36. y is non-positive and greater than -11.	36. $-11 < y \leq 0$

Order Property II: Transitive Law

In comparing two real numbers, a and b, with a third real number c, the following order will exist:

If a is greater than b and b is greater than c, then a is greater than c. In symbols, if $a > b$ and $b > c$, then $a > c$.

Consider the order as it applies to the number line.

. . . -4 -3 -2 -1 0 1 2 3 4 . . .

If 2 is greater than -1 and -1 is greater than -4, then 2 must be greater than -4 also.

If $2 > -1$ and $-1 > -4$, then $2 > -4$.

37. If $12 > 7$ and $7 > -14$, what is the relationship of 12 and -14?	37. $12 > -14$
38. If $0 > -4$ and $-4 > -11$, then ____________.	38. $0 > -11$
39. If $-3 < 6$ and $6 < 39$, then ____________.	39. $-3 < 39$
40. If $x > y$ and $y > 0$, then is x also positive?	40. Yes, because $x > 0$.
41. If $x < y$ and $y < 0$ then ____________.	41. $x < 0$
42. If $x < m$ and $m = 7$ then ____________. (by substitution)	42. $x < 7$

<u>Order Property III</u>: The Addition Property

Part 1: If equals are added to unequals, the sums are unequal in the same order.

In symbols, if $a > b$ and $c = d$, then $a + c > b + d$.

Example: If $5 > 3$ and $2 = 2$, then $(5 + 2) > (3 + 2)$.

Part 2: If unequals are added to unequals of the same order, the sums are unequal in the same order.

In symbols, if $a > b$ and $c > d$, then $a + c > b + d$.

Example: If $5 > 3$ and $1 > 0$, then $5 + 1 > 3 + 0$.

<u>Order Property IV</u>: The Subtraction Property

Part 1: If equals are subtracted from unequals, the differences are unequal in the same order.

In symbols, if $a > b$ and $c = d$, then $a - c > b - d$.

Example: If $5 > 3$ and $2 = 2$, then $5 - 2 > 3 - 2$.

Part 2: If unequals are subtracted from equals, the differences are unequal in the <u>opposite</u> order.

In symbols, if $a > b$ and $c = d$, then $c - a < d - b$. (Note that the symbol is reversed!)

Example: If $5 > 3$ and $2 = 2$, then $2 - 5 < 2 - 3$.

43. If $16 > 5$ and $3 = 3$, then is $16 - 3 > 5 - 3$? Show why or why not.

43. Yes
$16 - 3 > 5 - 3$
$13 > 5 - 3$
$13 > 2$

44. If $16 > 5$ and $3 = 3$ then is $3 - 16 > 3 - 5$? Show why or why not.

44. No, the symbol should be in the opposite order.

$3 - 16 < 3 - 5$

$-13 < 3 - 5$

$-13 < -2$

45. If $16 > 5$ and $3 = 3$ then $16 + 3 > 5 + 3$? Show why or why not.

45. Yes

$16 + 3 > 5 + 3$

$19 > 5 + 3$

$19 > 8$

46. If unequal quantities are added to unequal quantities of the same order, will the sum be unequal in the same order? ________________ Why?________________

46. Yes, addition order property, part 2.

47. If equal quantities are subtracted from unequal quantities, will the difference be unequal in the same order?__________

47. Yes, subtraction order property, part 1.

48. If unequal quantities are subtracted from equal quantities, will the difference be unequal in the opposite order?__________

48. Yes, subtraction order property, part 2.

Order Property V: The Multiplication Property

Part 1: If unequals are multiplied by the same positive number, the products are unequal in the same order.

In symbols, if $a > b$ and $c > 0$, then $ac > bc$.

Example: If $5 > 3$ and 2 is positive then $5 \cdot 2 > 3 \cdot 2$, since $10 > 6$.

Part 2: If unequals are multiplied by the same negative number, the products are unequal in the opposite order.

In symbols, if $a > b$ and $c < 0$ then $ac < bc$.

Example: If $5 > 3$ and -1 is negative, then $5 \cdot -1 < 3 \cdot -1$ since $-5 < -3$.

49. If c is -2 and $11 < 41$ then how do the products $11c$ and $41c$ compare ?

49. The products are unequal in the opposite order.
$11 \cdot -2 > 41 \cdot -2$
$-22 > -82$
Order Property V, Part 2

50. If c is positive 2 and $11 < 41$ then how do the products $11c$ and $41c$ compare ?

50. $11 \cdot 2 < 41 \cdot 2$
$22 < 82$
Order Property V, Part 1

51. If $\frac{1}{4}x > 5$ and 4 is positive then is $x > 20$?

51. Yes
$4\left(\frac{1}{4}x\right) > 5(4)$
$x > 20$

52. If $-2x > 16$ and $\frac{-1}{2}$ is negative then is $x > -8$?

52. No
$\left(\frac{-1}{2}\right)(-2x) < 16\left(\frac{-1}{2}\right)$
$\frac{+2}{2}x < \frac{-16}{2}$
$x < -8$
Multiplication Order Property, Part 2

Exercise 1

1. Identify the order property illustrated in each of the following:

   a. If $6 > 4$ and $4 > 2$, then $6 > 2$.
   b. If $8 > 2$ and $.5 = .5$, then $8 + .5 > 2 + .5$
   c. If $9 > 6$ and $2 = 2$, then $9 - 2 > 6 - 2$.
   d. If $10 > 2$ and 3 is positive, then $30 > 6$.
   e. If $4 < 6$ and -1 is negative, then $-4 > -6$.
   f. If $2x > 3$ and y is negative, then $2xy < 3y$.
   g. If $B < 10$, then B can not equal 10.
   h. If $y > 3$ and $5 = 5$, then $y + 5 > 3 + 5$.

i. If $x > y$ and $y > z$, then $x > z$.

j. If $x < y$ and $7 = 7$, then $7 - x > 7 - y$.

k. If $4 < 6$ and $a < b$, then $4 + a < 6 + b$.

l. If $m < n$ and $c < 0$, then $m \cdot c > n \cdot c$.

m. If $c \neq d$, then $c > d$ or $c < d$.

n. If $y < 8$ and $1 = 1$, then $y - 1 < 8 - 1$.

o. If $7 < m$ and $m < n$, then $7 < n$.

p. If $p > 11$ and 5 is positive, then $p \cdot 5 > 11 \cdot 5$.

2. Fill in with the appropriate word, phrase, or algebraic expression.

   a. If $-6 < 0$ and $0 < 5$, then ________________. (Transitive Law)

   b. If x and y are real numbers, then ___________, ______________, or ______________. (Trichotomy Law)

   c. If $6 < 8$ and -2 is negative, then -12 ___ -16. (Multiplication Property, Part 2)

   d. If $3x + 1 > 0$ and $1 = 1$, then _____________ (Subtraction Property, Part 1)

3. Write an open sentence with the same meaning as the following statements.

   a. Twice x increased by two is positive.

   b. Y is non-negative.

   c. W is non-positive.

   d. Z is positive and less than or equal to 78.

   e. X is between 2 and 10.

   f. The square of x is greater than 4.

   g. X is between -3 and 7 and may equal to 7.

Solving Inequalities

Using the properties on pages 154-157, solve the following inequalities:

Example 1:

$$2x + 5 \angle 4x - 3$$

a) $\underline{-2x} + 2x + 5 \angle \underline{-2x} + 4x - 3$ Property IV, 1

b) $5 \angle 2x - 3$ Simplify

c) $5 \underline{+ 3} \angle 2x - 3 \underline{+ 3}$ Property III, 1

d) $8 \angle 2x$ Simplify

e) $\frac{1}{2}(8) \angle \frac{1}{2}(2x)$ Property V, 1

f) $4 \angle x$ Simplify

Therefore, $2x + 5 \angle 4x - 3$ simplifies to $4 \angle x$. Any value of x greater than 4 will be a solution to the inequality in Example 1.

Example 2:

$$x(x^2 + 5) \angle 8x - 12 + x^3$$

a) $x^3 + 5x \angle 8x - 12 + x^3$ Distributive Law

b) $x^3 \underline{- x^3} + 5x \angle 8x - 12 + x^3 \underline{- x^3}$ Property IV, 1

c) $5x \angle 8x - 12$ Simplify

d) $5x \underline{- 8x} \angle 8x \underline{- 8x} - 12$ Property IV, 1

e) $5x - 8x \angle -12$ Simplify

f) $-3x \angle -12$ Simplify

g) * $\frac{-1}{3}(-3x) > \frac{-1}{3}(-12)$ Property V, 2

h) * $x > 4$ Simplify

* Note the change in ordering! See Property V, 2.

53. Solve by filling in the blank spaces.

$$2(x - 5) > 7x - 2$$

a)	$\_\_\_\_\_\_ > 7x - 2$	Distributive Law
b)	$2x \underline{- 7x} - 10 > 7x \underline{- 7x} - 2$	Why?
c)	$-5x - 10 > -2$	Why?
d)	$\_\_\_\_\_\_ > \_\_\_\_\_\_$	Property III,1
e)	$-5x > \_\_\_\_\_\_$	Simplify
f)	$\frac{-1}{5}(-5x) < \frac{-1}{5}(8)$	Why?
g)	$x < \_\_\_\_\_\_$	Simplfiy

54. Solve: $\frac{5x}{3} < 7 + x$

x has what value? __________

55. Solve: $8x - 4 > \frac{3}{5}x + 2$

x has what value? __________

53.

a) $2x - 10 > 7x - 2$

b) Property IV,1

c) Simplify

d) $-5x-10+10 > -2+10$

e) $-5x > 8$

f) Property V,2

g) $x < \frac{-8}{5}$ or $x < -1\frac{3}{5}$

54.

$$\frac{5x}{3} < 7 + x$$

$$3\left(\frac{5x}{3}\right) < 3(7) + 3(x)$$

$$5x < 21 + 3x$$

$$2x < 21$$

$$x < \frac{21}{2}$$

$$x < 10\frac{1}{2}$$

55.

$$5(8x)-5(4) > 5\left(\frac{3}{5}x\right) +5($$

$$40x - 20 > 3x + 10$$

$$37x - 20 > 10$$

$$37x > 30$$

$$x > \frac{30}{37}$$

Exercise 2

1. Solve the following inequalities. Justify each step.

   a. $11x > -22$

   b. $-3x < 18$

   c. $10x + 3 \geq 9 + 13x$

   d. $2 + 4(x + 3) \geq 2(x + 3)$

   e. $3x + 7(x + 1) \geq -2(x - 7)$

   f. $\frac{3x}{4} - \frac{x}{2} > \frac{3}{8} + \frac{5x}{4}$

   g. $\frac{1}{2}x + \frac{1}{3}x \leq -7$

   h. $4(5 - 2y) - 7 \geq 6(y + 3) - 3(y + 4) + y$

   i. $3x - \frac{x - 2}{2} \geq 5(2 - x)$

   j. $3(x - 2) > -6$

   k. $\frac{-13}{15} + \frac{x + 2}{3} - \frac{4x - 1}{5} < 0$

Theorems of Inequality

Theorem 1:	The whole is greater than any of its parts. In symbols, if $a = b + c$ and $a > 0$, $b > 0$ and $c > 0$, then $a > b$ and $a > c$.

Example: For theorem 1 use these line segments to illustrate the theorem by construction. Add line segment b to c by using only a straightedge and compass.

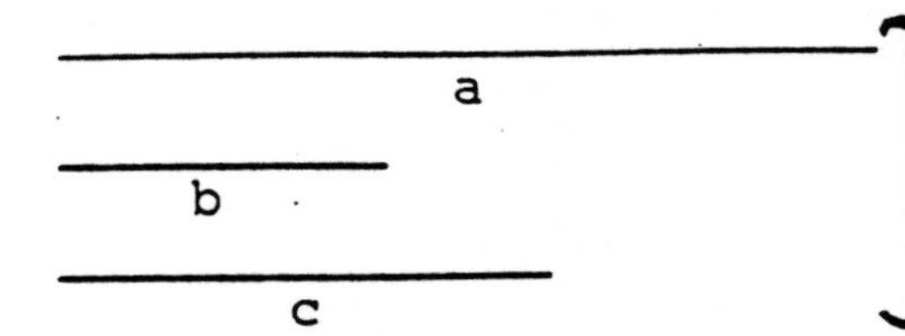

Given these measures for segments a, b, and c.

Copy the above segments to construct the sum of (b + c) equal to a. What are your results?

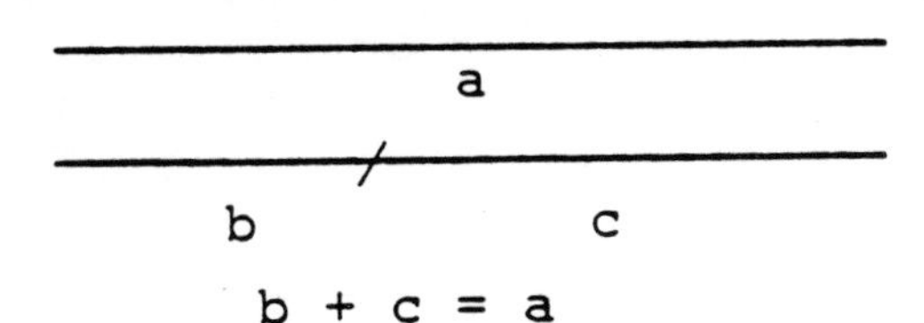

$b + c = a$

Now compare a to each of the original segments.

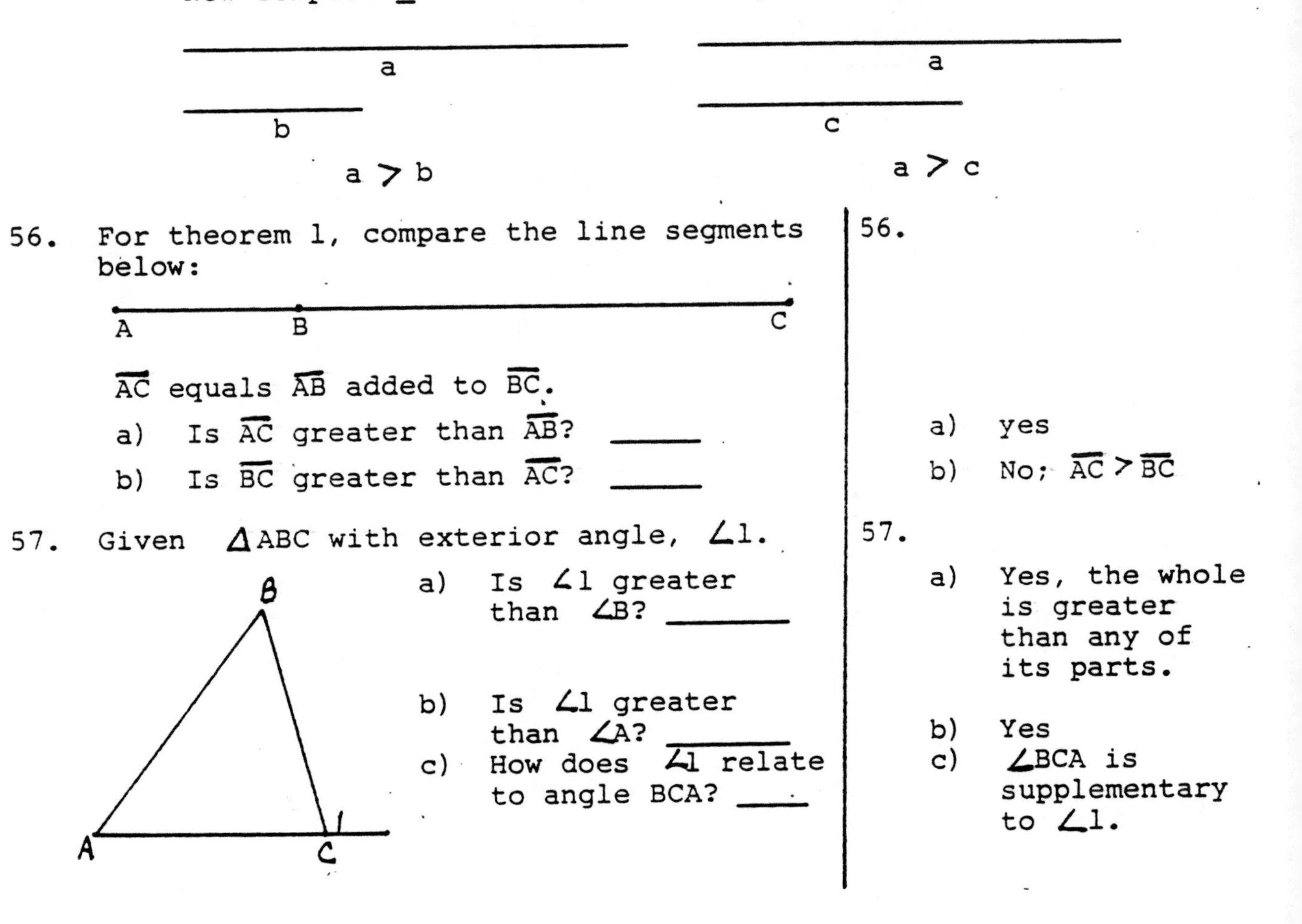

$a > b$ $\qquad$ $a > c$

56. For theorem 1, compare the line segments below:

A B C

$\overline{AC}$ equals $\overline{AB}$ added to $\overline{BC}$.

a) Is $\overline{AC}$ greater than $\overline{AB}$? ______

b) Is $\overline{BC}$ greater than $\overline{AC}$? ______

57. Given $\triangle ABC$ with exterior angle, $\angle 1$.

a) Is $\angle 1$ greater than $\angle B$? ________

b) Is $\angle 1$ greater than $\angle A$? ________

c) How does $\angle 1$ relate to angle BCA? ____

56.

a) yes

b) No; $\overline{AC} > \overline{BC}$

57.

a) Yes, the whole is greater than any of its parts.

b) Yes

c) $\angle BCA$ is supplementary to $\angle 1$.

Theorem 2:	If two sides of a triangle are unequal, the angles opposite them are unequal and the larger angle is opposite the longer side.

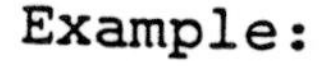

Example:

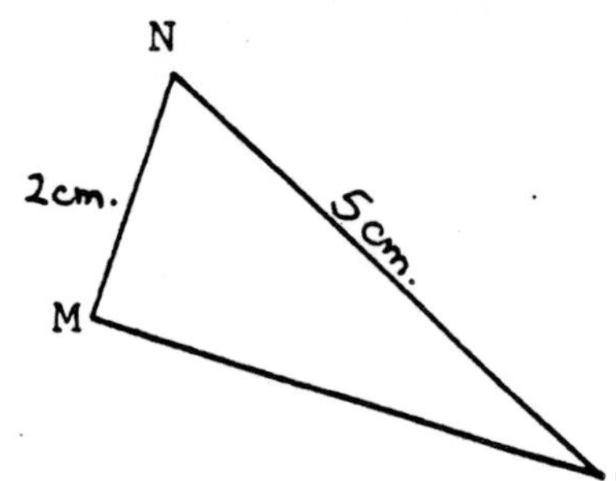

Given $\overline{NM} < \overline{NO}$.

How do $\angle O$ and $\angle M$ compare?

The measure of $\angle O$ is less than the measure of $\angle M$ since it is opposite the shorter side.

Theorem 3:	If two angles of a triangle are unequal, the sides opposite them are unequal and the longer side is opposite the larger angle.

Example:

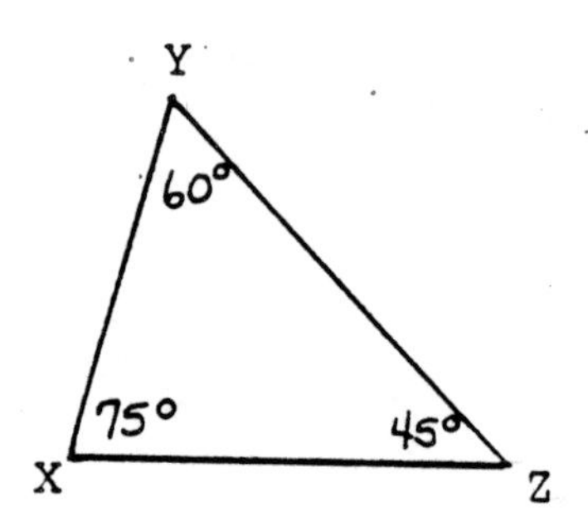

Given: $\angle Y$ measures less than $\angle X$

How does side $\overline{XZ}$ compare to side $\overline{YZ}$?

Side $\overline{XZ}$ is shorter than $\overline{YZ}$ since it is opposite the smaller angle.

58.

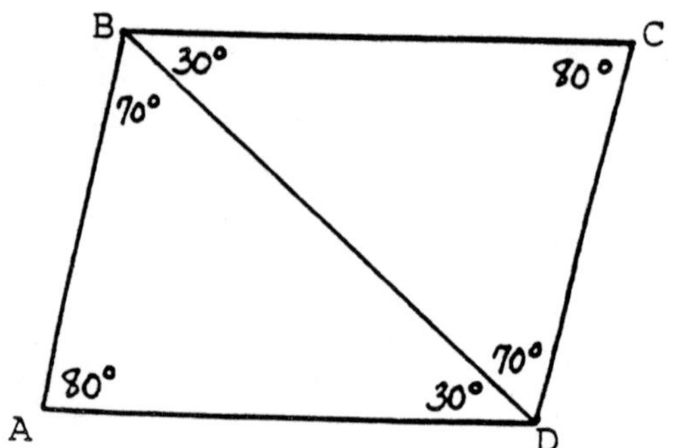

On the basis of the drawing, answer the questions below:

a) In Δ ABD which is the longest side? ______

b) In Δ BCD which is the smallest side? ______

58.

a) $\overline{BD}$

b) $\overline{DC}$

59.

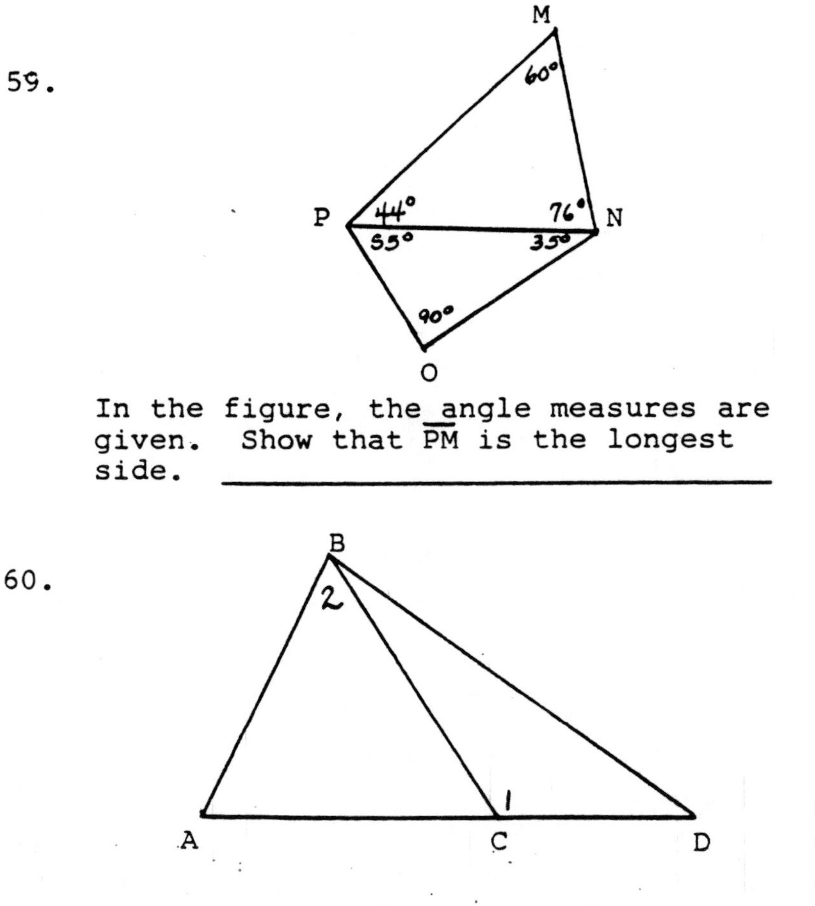

In the figure, the angle measures are given. Show that $\overline{PM}$ is the longest side. ________________________

59. In $\triangle$ PON, side $\overline{PN}$ is the longest but in $\triangle$ PMN $\overline{PN}$ is opposite the 60° angle. Therefore, $\overline{PN}$ is not the longest side in $\triangle$ PMN. The longest side in $\triangle$ PMN is opposite the 76° angle, side $\overline{PM}$.

60.

Given: $\triangle$ ABC with exterior angle, $\angle 1$.

If $\angle 1 = 125^\circ$ and $\angle 2 = 60^\circ$, what side is the longest in $\triangle$ ABC?

60.

$$\angle BCA + \angle 1 = 180^\circ$$
$$\angle BCA + 125^\circ = 180^\circ$$
$$\angle BCA = 55^\circ$$

$$\angle A + \angle 2 = \angle 1$$
$$\angle A + 60^\circ = 125^\circ$$
$$\angle A = 65^\circ$$

$\angle$ A is the largest angle in $\triangle$ ABC. By Theorem 3, $\overline{BC}$ is the longest side.

61.

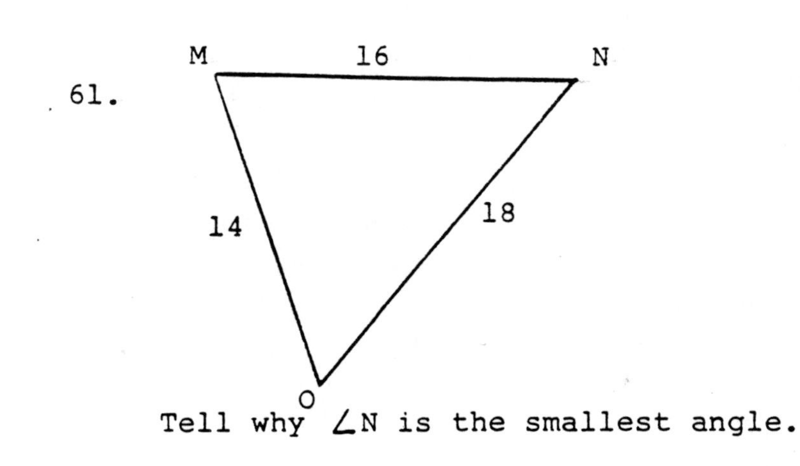

Tell why $\angle$N is the smallest angle.

____

61. By Theorem 2, $\angle$N is opposite the smallest side.

62.

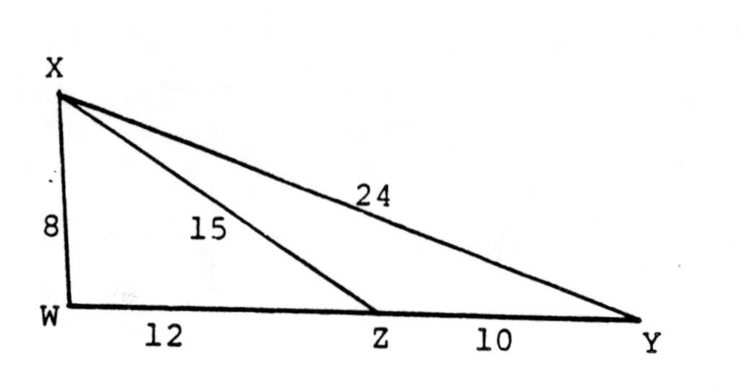

Tell why $\angle WXY$ is not the smallest or the largest angle in $\triangle WXY$.

____

62. In $\triangle WXY$,
$\overline{XY} > \overline{WY} > \overline{XW}$.

By Theorem 2, Angle W > Angle WXY > Angle Y.

In symbols, $\angle W > \angle WXY > \angle Y$

Exercise 3

1.

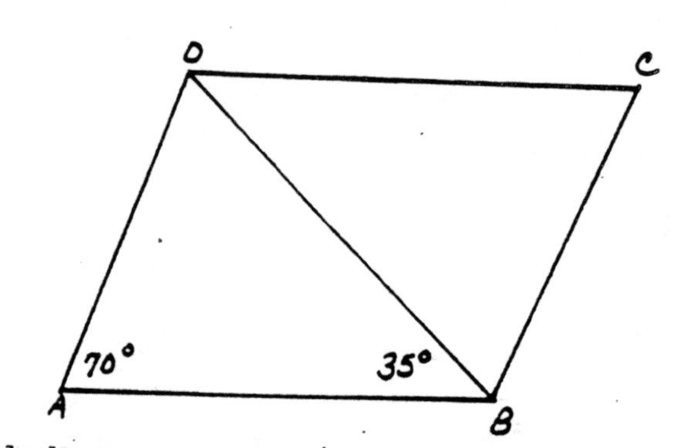

ABCD is a parallelogram. In $\triangle$ DCB which side is the longest? Which side is the shortest? Justify your answer.

2.

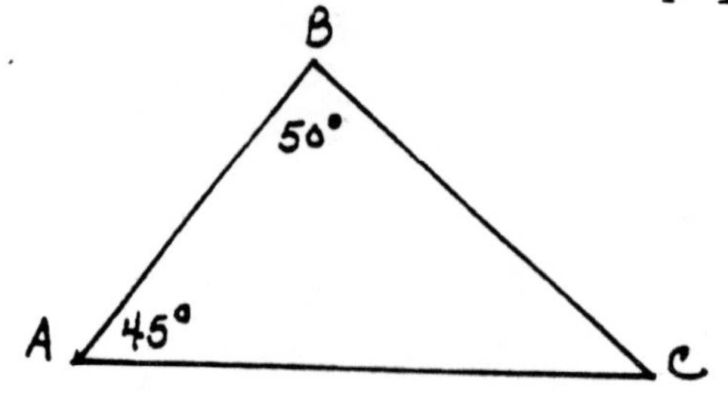

In $\triangle$ ABC which is the longest side? Which is the shortest side?

3. In triangle PQR, find the longest SIDE.

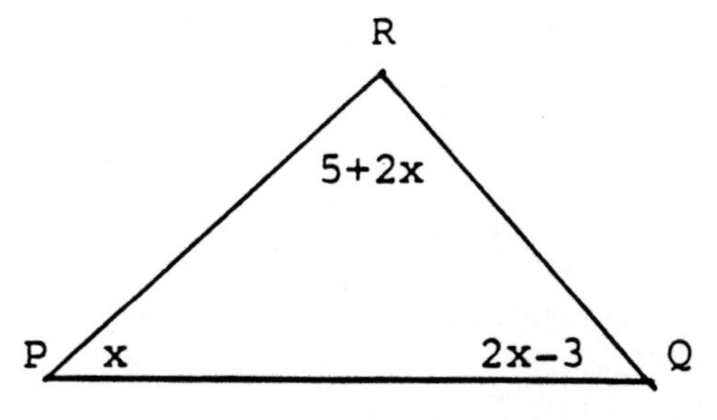

4. $\overline{XY} = 14$. In ΔZNY, which angle is the largest angle? Which is the smallest? Justify your answer.

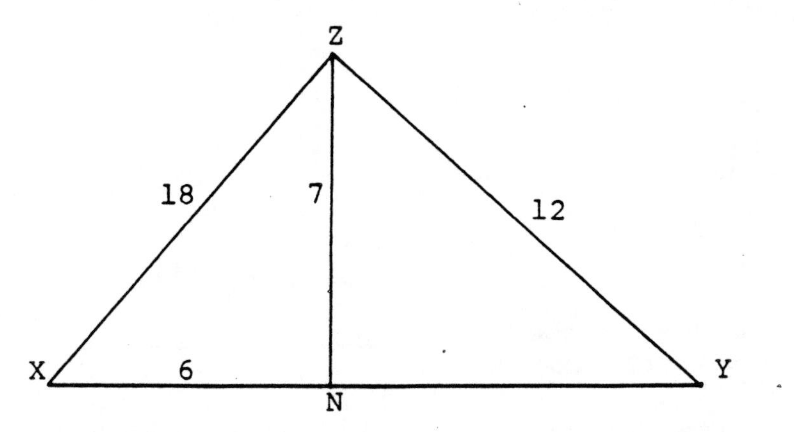

5. Is $\angle CBD$ larger than $\angle A$? Why or why not?

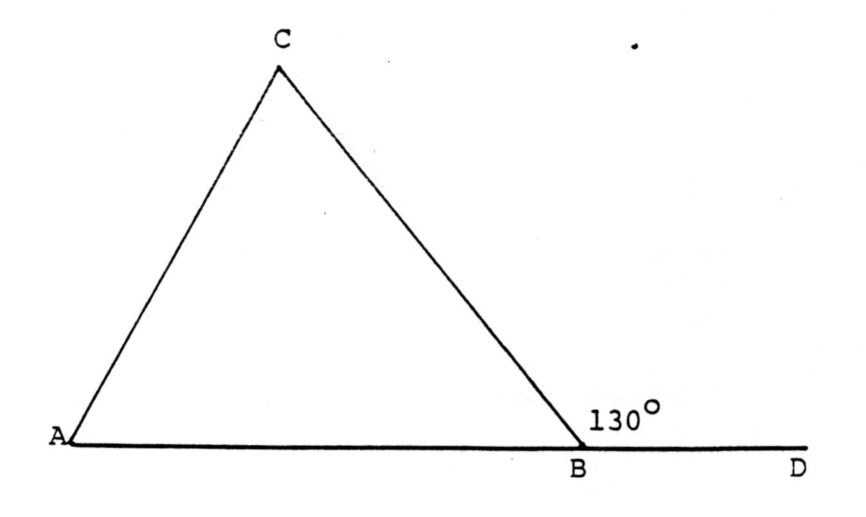

Unit 6 Review

1. Identify the order property illustrated in each of the following:

   a. If $10 > 2$ and $2 > -5$, then $10 > -5$.

   b. If $2 > -3$ and $\frac{1}{4} = .25$, then $2 + \frac{1}{4} > -3 + .25$.

   c. If $10 > 3$ and $8 = \frac{16}{2}$, then $10 - 8 > 3 - \frac{16}{2}$.

   d. If $3 > 2$ and 5 is positive, then $15 > 10$.

   e. If $4 < 10$ and -2 is negative, then $-8 > -20$.

   f. If $75 > 2$ and $5 = \frac{10}{2}$, then $5 - 75 < \frac{10}{2} - 2$.

2. Fill in with the appropriate word, phrase, or algebraic expression.

   a. If $-4 < 0$ and $0 < 8$, then ____________. (Transitive Law)

   b. If c and d are real numbers then ______, __________, or ________. (Trichotomy Law)

   c. If $3 < 8$ and 4 is positive, then 12____32. (Multiplication Property, part 1)

3. Write an open sentence with the same meaning as the following. statements.

   a. Three times x decreased by five is positive.

   b. Y is non-positive.

   c. Z is negative and greater than or equal to -21.

   d. Y is between 3 and 11.

4. Solve the following inequalities. Justify each step.

   a. $3(x + 4) - 2(7 - x) > 0$

   b. $6 - \frac{3y - 4}{7} \geq 0$

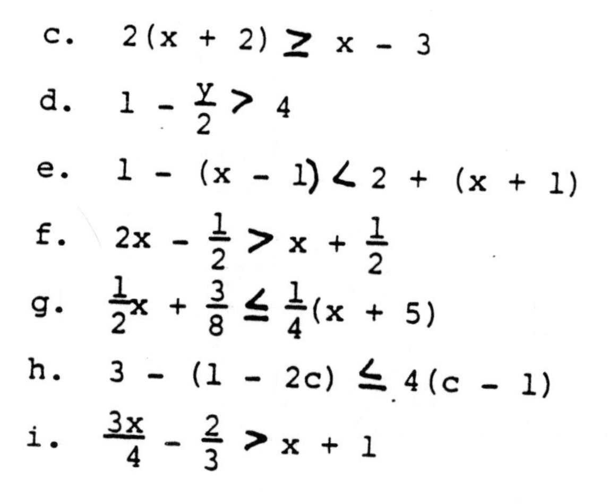

c. $2(x + 2) \geq x - 3$

d. $1 - \frac{y}{2} > 4$

e. $1 - (x - 1) < 2 + (x + 1)$

f. $2x - \frac{1}{2} > x + \frac{1}{2}$

g. $\frac{1}{2}x + \frac{3}{8} \leq \frac{1}{4}(x + 5)$

h. $3 - (1 - 2c) \leq 4(c - 1)$

i. $\frac{3x}{4} - \frac{2}{3} > x + 1$

5. a.

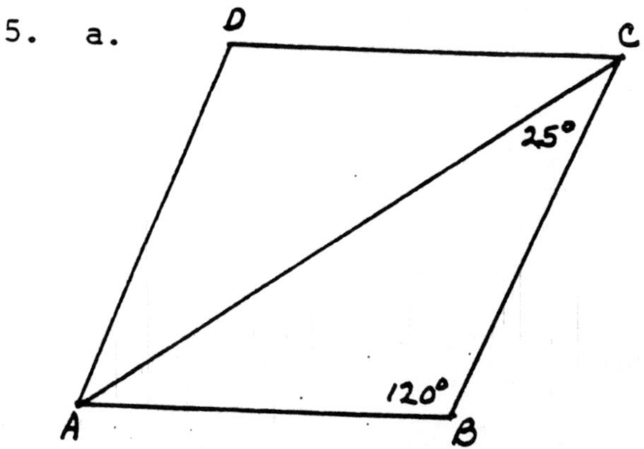

ABCD is a parallelogram. In $\triangle$ ACD which is the longest side and why? Which is the shortest?

b.

C
90°
30°
A
B

In $\triangle$ABC, which side is the longest?

c.

F
12
15
D
17
E

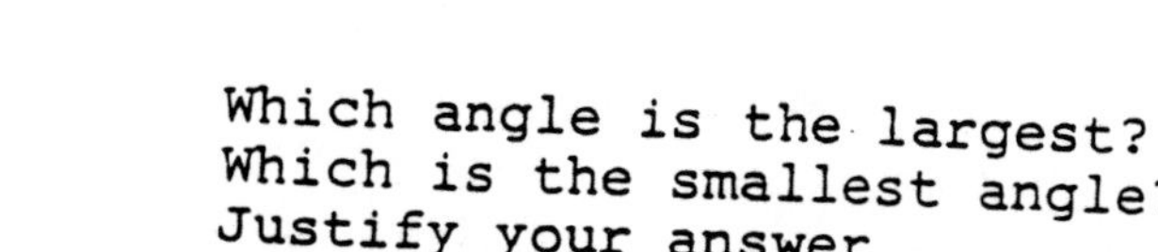

Which angle is the largest? Which is the smallest angle? Justify your answer.

d.

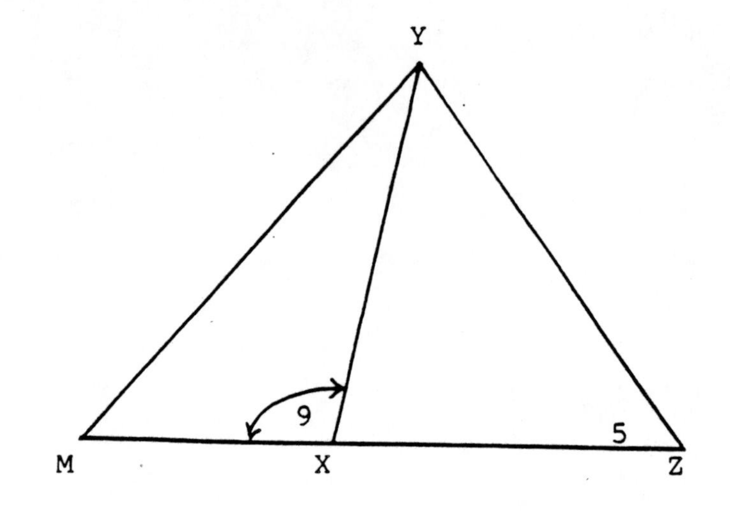

$\angle 9 = 108^\circ$ and $\angle 5 = 54^\circ$. $\angle 9$ is an exterior angle to $\triangle XYZ$.

Explain why $\overline{YZ}$ is longer than $\overline{XZ}$ and $\overline{XY}$.

Unit 7

Area

Learning Objectives:

1. Given real number values the student will compute the area of each of the following: rectangle, square, parallelogram, triangle, trapezoid, and circle.

2. Given real values for all but one of the variables in the formula for a given area, the student will find the remaining variable.

3. Using combinations of the area formulae, the student will find the area of various geometric figures.

4. Using the area formulae, the student will use them to solve real world application problems.

5. The student will demonstrate his mastery of the definitions of circumference, circle, radius, and diameter by writing them and applying them to selected problems.

6. The student will demonstrate his mastery of the following theorem by writing it and applying it to selected problems.

   Theorem: The ratio of the circumference to the diameter is the same for all circles.

AREA

Before discussing the concept of area, some rules concerning radicals and quadratic equations need to be refreshed.

Radicals Review

Rule #1: $\sqrt{ab} = \sqrt{a} \cdot \sqrt{b}$

Two major applications for this rule are as follows:

I. Simplifying Radicals

1. $\sqrt{98}$

First, split the radical into two radicals. In the first radical place all the perfect squares and, in the second, place all other factors. In $\sqrt{98}$, $49 = 7 \cdot 7$ is a perfect square factor.

$$\sqrt{98} = \sqrt{49}\sqrt{2} = \sqrt{7 \cdot 7}\sqrt{2} = 7\sqrt{2}$$

2. $\sqrt{75} = \sqrt{25}\sqrt{3} = \sqrt{5 \cdot 5}\sqrt{3} = 5\sqrt{3}$

3. $\sqrt{27}$ = _____ · _____ = _____ · _____ = _____

4. $\sqrt{81} = \sqrt{9 \cdot 9} = 9$

5. $\sqrt{64}$ = _____ = _____

6. $\sqrt{252}$

The factors of 252 are not readily apparent. Prime factorization is useful here.

```
2 |252
2 |126
3 | 63
3 | 21
     7
```

$$\sqrt{252} = \sqrt{2 \cdot 2 \cdot 3 \cdot 3 \cdot 7}$$

Note 252 has a pair of twos and a pair of threes. The product of these pairs is 36 which is a perfect square. Then $\sqrt{252} = \sqrt{36}\sqrt{7} = \sqrt{6 \cdot 6}\sqrt{7} = 6\sqrt{7}$.

3. $\sqrt{9}\sqrt{3}$
$\sqrt{3 \cdot 3}\sqrt{3}$
$3\sqrt{3}$

5. $\sqrt{8 \cdot 8} = 8$

7. $\sqrt{192}$

Prime factoring:

$$\begin{array}{r|l} 2 & 192 \\ 2 & 96 \\ 2 & 48 \\ 2 & 24 \\ 2 & 12 \\ 2 & 6 \\ & 3 \end{array}$$

$\sqrt{192} = \sqrt{\underline{2 \cdot 2} \cdot \underline{2 \cdot 2} \cdot \underline{2 \cdot 2} \cdot 3}$

Note there are three pairs of twos. Multiplying these three pairs of factors, you can obtain the perfect square part.

$\sqrt{192} = \sqrt{64}\,\sqrt{3} = \sqrt{8 \cdot 8}\,\sqrt{3} = 8\sqrt{3}$

8. $\sqrt{250}$

a) Prime factor first:

$\underline{|250} =$ ____________

b) $\sqrt{250} =$ ____________

9. $\sqrt{147} =$ ____________

$=$ ____________

$=$ ____________

II. Multiplying Radicals

1. $\sqrt{2} \cdot \sqrt{8} = \sqrt{16} = 4$

2. $\sqrt{3} \cdot \sqrt{12} = \sqrt{36} = 6$

3. $\sqrt{5} \cdot \sqrt{20} =$ _____ $=$ _____ $=$ _____

8.

a)
$$\begin{array}{r|l} 5 & 250 \\ 5 & 50 \\ 5 & 10 \\ & 2 \end{array}$$

b) $\sqrt{5 \cdot 5}\,\sqrt{5 \cdot 2}$

$= \sqrt{25}\,\sqrt{10}$

$= 5\sqrt{10}$

9. $\sqrt{7 \cdot 7 \cdot 3}$

$= \sqrt{49}\,\sqrt{3}$

$= 7\sqrt{3}$

3. $\sqrt{100} = \sqrt{10 \cdot 10} = 10$

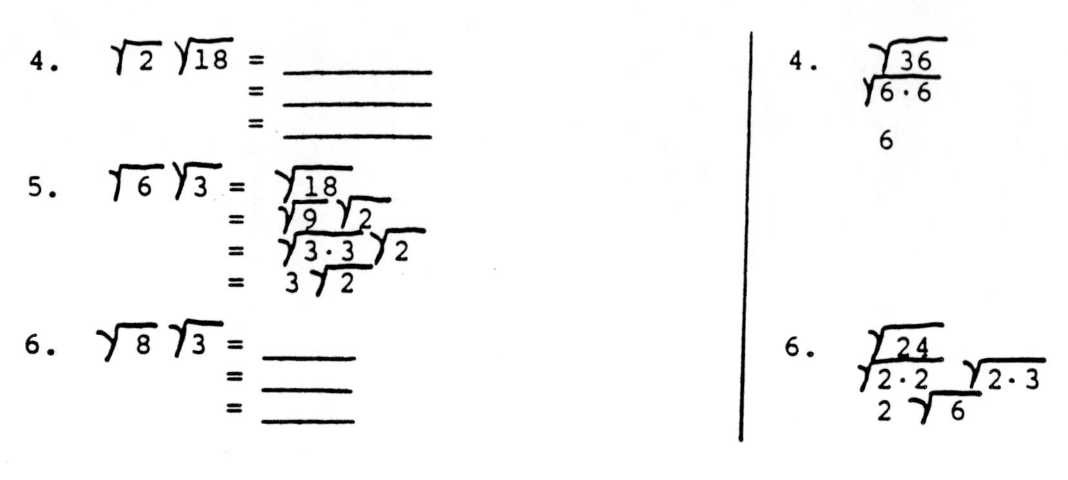

4. $\sqrt{2}\sqrt{18}$ = ________
= ________
= ________

5. $\sqrt{6}\sqrt{3} = \sqrt{18}$
$= \sqrt{9}\sqrt{2}$
$= \sqrt{3 \cdot 3}\sqrt{2}$
$= 3\sqrt{2}$

6. $\sqrt{8}\sqrt{3}$ = ________
= ________
= ________

4. $\sqrt{36}$
$\sqrt{6 \cdot 6}$
6

6. $\sqrt{24}$
$\sqrt{2 \cdot 2}\ \sqrt{2 \cdot 3}$
$2\sqrt{6}$

Rule # 2: $\sqrt{\frac{a}{b}} = \frac{\sqrt{a}}{\sqrt{b}}$

Use Rule # 2 for simplifying quotients of radicals and for simplifying the square roots of fractions under the radical sign.

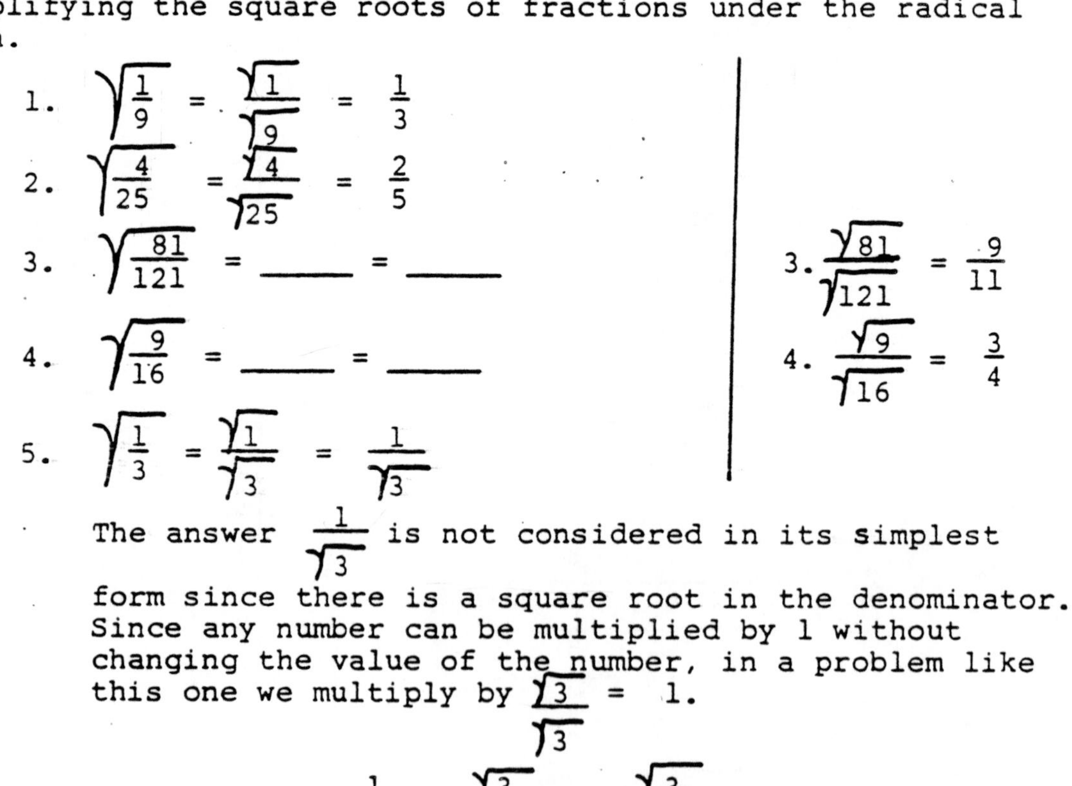

1. $\sqrt{\frac{1}{9}} = \frac{\sqrt{1}}{\sqrt{9}} = \frac{1}{3}$

2. $\sqrt{\frac{4}{25}} = \frac{\sqrt{4}}{\sqrt{25}} = \frac{2}{5}$

3. $\sqrt{\frac{81}{121}}$ = ________ = ________

4. $\sqrt{\frac{9}{16}}$ = ________ = ________

5. $\sqrt{\frac{1}{3}} = \frac{\sqrt{1}}{\sqrt{3}} = \frac{1}{\sqrt{3}}$

3. $\frac{\sqrt{81}}{\sqrt{121}} = \frac{9}{11}$

4. $\frac{\sqrt{9}}{\sqrt{16}} = \frac{3}{4}$

The answer $\frac{1}{\sqrt{3}}$ is not considered in its simplest form since there is a square root in the denominator. Since any number can be multiplied by 1 without changing the value of the number, in a problem like this one we multiply by $\frac{\sqrt{3}}{\sqrt{3}} = 1$.

$$\frac{1}{\sqrt{3}} \cdot \frac{\sqrt{3}}{\sqrt{3}} = \frac{\sqrt{3}}{\sqrt{9}}$$

$\frac{\sqrt{3}}{\sqrt{3}}$ was chosen so the denominator will become the square root of a perfect square.

$$\frac{\sqrt{3}}{\sqrt{9}} = \frac{\sqrt{3}}{\sqrt{3\cdot 3}} = \frac{\sqrt{3}}{3}$$

With the square root eliminated from the denominator, the answer is now in its simplest form.

6. $\sqrt{\frac{2}{5}} = \frac{\sqrt{2}}{\sqrt{5}} \cdot \frac{\sqrt{5}}{\sqrt{5}} = \frac{\sqrt{10}}{\sqrt{25}} = \frac{\sqrt{10}}{\sqrt{5\cdot 5}} = \frac{\sqrt{10}}{5}$

7. $\sqrt{\frac{5}{6}}$ = ____________

= ____________

= ____________

= ____________

8. $\sqrt{\frac{4}{5}} = \frac{\sqrt{4}}{\sqrt{5}} = \frac{2}{\sqrt{5}}$

$= \frac{2}{\sqrt{5}} \cdot \frac{\sqrt{5}}{\sqrt{5}}$

$= \frac{2\sqrt{5}}{\sqrt{25}}$

7. $\frac{\sqrt{5}}{\sqrt{6}} \cdot \frac{\sqrt{6}}{\sqrt{6}}$

$\frac{\sqrt{5\cdot 6}}{\sqrt{6\cdot 6}}$

$\frac{\sqrt{30}}{\sqrt{36}}$

$\frac{\sqrt{30}}{6}$

Note: This answer does not reduce because 30 is under the radical and 6 is not.

Note: The 2 remains outside the radical sign. It is a whole number and not a square root.

$$\frac{2\sqrt{5}}{\sqrt{25}} = \frac{2\sqrt{5}}{\sqrt{5\cdot 5}} = \frac{2\sqrt{5}}{5}$$

9. $\sqrt{\frac{64}{7}} = \frac{\sqrt{}}{\sqrt{}}$

$= \left(\frac{8}{}\right) \cdot \left(\frac{}{}\right)$

= ____________

= ____________

9. $\frac{\sqrt{64}}{\sqrt{7}}$

$\left(\frac{8}{\sqrt{7}}\right) \cdot \left(\frac{\sqrt{7}}{\sqrt{7}}\right)$

$\frac{8\sqrt{7}}{\sqrt{7\cdot 7}}$

$\frac{8\sqrt{7}}{7}$

Some problems take a slightly different twist. The denominator can be partially simplified before completely removing the radical.

10. $\frac{\sqrt{7}}{\sqrt{8}}$ First $\sqrt{8} = \sqrt{4} \cdot \sqrt{2} = 2\sqrt{2}$.

$= \frac{\sqrt{7}}{\sqrt{8}} \quad \frac{\sqrt{7}}{2\sqrt{2}}$ Now multiplying by $\frac{\sqrt{2}}{\sqrt{2}}$ will rationalize the denominator as follows:

$$\frac{\sqrt{7}}{2\sqrt{2}} \cdot \frac{\sqrt{2}}{\sqrt{2}} = \frac{\sqrt{7 \cdot 2}}{2\sqrt{2 \cdot 2}} = \frac{\sqrt{14}}{2 \cdot 2} = \frac{\sqrt{14}}{4}$$

Observe the final answer cannot be further simplified. True 2 divides 4 and also 14. However, observe that 14 is under the radical sign and 4 is not.

11. $\frac{\sqrt{5}}{\sqrt{27}} = \frac{\sqrt{5}}{\sqrt{9}\sqrt{3}} =$ ________

$=$ ________

$=$ ________

$=$ ________

$=$ ________

12. $\sqrt{\frac{4}{6}} = \frac{\sqrt{4}}{\sqrt{6}} = \frac{2}{\sqrt{6}} \cdot \frac{\sqrt{6}}{\sqrt{6}} = \frac{2\sqrt{6}}{\sqrt{6 \cdot 6}}$

$= \frac{2\sqrt{6}}{6}$ In this example

2 in the numerator and 6 in the denominator have a common factor of 2. Both 2 and 6 are not under a radical sign. Consequently $\frac{2\sqrt{6}}{6}$ can be reduced to $\frac{\sqrt{6}}{3}$.

13. $\sqrt{\frac{25}{10}} = \frac{\sqrt{25}}{\sqrt{10}} = \frac{5}{\sqrt{10}} =$ ____ ____

11. $\frac{\sqrt{5}}{3\sqrt{3}}$

$\frac{\sqrt{5}}{3\sqrt{3}} \cdot \frac{\sqrt{3}}{\sqrt{3}}$

$\frac{\sqrt{5 \cdot 3}}{3\sqrt{3 \cdot 3}}$

$\frac{\sqrt{15}}{3 \cdot 3}$

$\frac{\sqrt{15}}{9}$

13. $\frac{5}{\sqrt{10}} \cdot \frac{\sqrt{10}}{\sqrt{10}}$

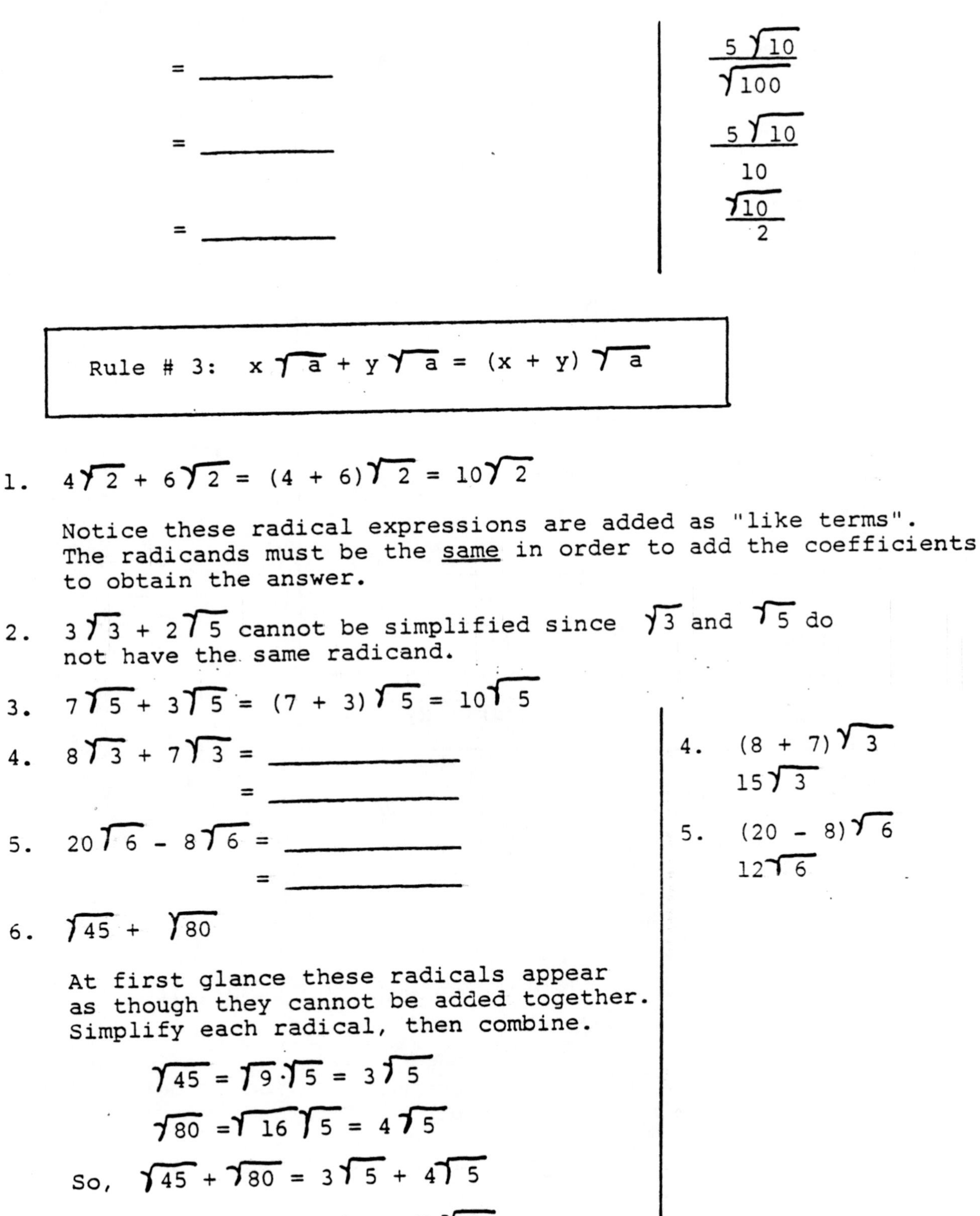

$= \underline{\qquad\qquad}$

$= \underline{\qquad\qquad}$

$= \underline{\qquad\qquad}$

$\frac{5\sqrt{10}}{\sqrt{100}}$

$\frac{5\sqrt{10}}{10}$

$\frac{\sqrt{10}}{2}$

Rule # 3: $x\sqrt{a} + y\sqrt{a} = (x + y)\sqrt{a}$

1. $4\sqrt{2} + 6\sqrt{2} = (4 + 6)\sqrt{2} = 10\sqrt{2}$

   Notice these radical expressions are added as "like terms". The radicands must be the <u>same</u> in order to add the coefficients to obtain the answer.

2. $3\sqrt{3} + 2\sqrt{5}$ cannot be simplified since $\sqrt{3}$ and $\sqrt{5}$ do not have the same radicand.

3. $7\sqrt{5} + 3\sqrt{5} = (7 + 3)\sqrt{5} = 10\sqrt{5}$

4. $8\sqrt{3} + 7\sqrt{3} = \underline{\qquad\qquad}$

   $= \underline{\qquad\qquad}$

5. $20\sqrt{6} - 8\sqrt{6} = \underline{\qquad\qquad}$

   $= \underline{\qquad\qquad}$

6. $\sqrt{45} + \sqrt{80}$

   At first glance these radicals appear as though they cannot be added together. Simplify each radical, then combine.

   $\sqrt{45} = \sqrt{9}\cdot\sqrt{5} = 3\sqrt{5}$

   $\sqrt{80} = \sqrt{16}\sqrt{5} = 4\sqrt{5}$

   So, $\sqrt{45} + \sqrt{80} = 3\sqrt{5} + 4\sqrt{5}$

   $= (3 + 4)\sqrt{5}$

   $= 7\sqrt{5}$

4. $(8 + 7)\sqrt{3}$

   $15\sqrt{3}$

5. $(20 - 8)\sqrt{6}$

   $12\sqrt{6}$

7. $5\sqrt{7} + \sqrt{28} =$ ______

$=$ ______

$=$ ______

$=$ ______

8. $3\sqrt{20} + 4\sqrt{5} = 3\sqrt{4}\sqrt{5} + 4\sqrt{5}$

$= 3 \cdot 2\sqrt{5} + 4\sqrt{5}$

$= 6\sqrt{5} + 4\sqrt{5}$

$= (6 + 4)\sqrt{5}$

$= 10\sqrt{5}$

9. $\sqrt{12} + \sqrt{27} + \sqrt{75} =$ ______

$=$ ______

$=$ ______

$=$ ______

7. $5\sqrt{7} + \sqrt{4}\sqrt{7}$

$5\sqrt{7} + 2\sqrt{7}$

$(5 + 2)\sqrt{7}$

$7\sqrt{7}$

9. $\sqrt{4}\sqrt{3} + \sqrt{9}\sqrt{3} + \sqrt{25}\sqrt{3}$

$2\sqrt{3} + 3\sqrt{3} + 5\sqrt{3}$

$(2 + 3 + 5)\sqrt{3}$

$10\sqrt{3}$

Rule # 4: $(x\sqrt{a}) \cdot (y\sqrt{b}) = xy\sqrt{ab}$

1. $2\sqrt{2} \cdot 3\sqrt{6}$
Multiplying coefficients $2 \cdot 3 = 6$.
Multiplying radicals $\sqrt{2} \cdot \sqrt{6} = \sqrt{12}$.

Hence, $2\sqrt{2} \cdot 3\sqrt{6} = 6\sqrt{12}$

$= 6(2\sqrt{3})$ Rule #1 ($\sqrt{12}$ simplifies to $2\sqrt{3}$)

$= 12\sqrt{3}$

2. $4\sqrt{8} \cdot 5\sqrt{6} = 4 \cdot 5\sqrt{8}\sqrt{6}$ using commutative and associative laws to group coefficients and radicals.

$= 20\sqrt{48}$ multiplying coefficients and radicals.

$= 20\sqrt{16}\sqrt{3}$ simplifying the radical.

$= 20 \cdot 4\sqrt{3}$ taking the square root.

$= 80\sqrt{3}$ multiplying coefficients.

3. $(3\sqrt{3})\left(\frac{-1}{2}\sqrt{12}\right)$

Multiplying coefficients $3 \cdot \frac{-1}{2} = \frac{-3}{2}$.

Multiplying radicands $\sqrt{3} \cdot \sqrt{12} = \sqrt{36} = 6$.

Hence, $(3\sqrt{3})\left(\frac{-1}{2}\sqrt{12}\right) = \frac{-3}{2} \cdot 6 = -9$

4. $3\sqrt{7} \cdot 5\sqrt{14} =$ __________

= __________

= __________

= __________

= __________

5. $(25\sqrt{3})\left(\frac{-1}{5}\sqrt{6}\right) =$ __________

= __________

= __________

= __________

= __________

4. $3 \cdot 5\sqrt{7}\sqrt{14}$

$15\sqrt{98}$

$15\sqrt{49}\sqrt{2}$

$15 \cdot 7\sqrt{2}$

$105\sqrt{2}$

5. $25 \cdot \frac{-1}{5}\sqrt{3}\sqrt{6}$

$-5\sqrt{18}$

$-5\sqrt{9}\sqrt{2}$

$-5 \cdot 3\sqrt{2}$

$-15\sqrt{2}$

6. $(4\sqrt{2})^3$

$= (4\sqrt{2})(4\sqrt{2})(4\sqrt{2})$ using the definition of exponent

$= 64\sqrt{8}$ using Rule # 4

$= 64\sqrt{4}\sqrt{2}$ using Rule #1

$= 64 \cdot 2\sqrt{2}$ simplifying

$= 128\sqrt{2}$ multiplying coefficients.

7. $(2\sqrt{3})^3 =$ __________

= __________

= __________

= __________

= __________

4. $(2\sqrt{3})(2\sqrt{3})(2\sqrt{3})$

$8\sqrt{27}$

$8\sqrt{9}\sqrt{3}$

$8 \cdot 3 \cdot \sqrt{3}$

$24\sqrt{3}$

Rule # 5 Distributive Law $a\sqrt{b}\left(c\sqrt{d} + e\sqrt{f}\right) = ac\sqrt{bd} + ae\sqrt{bf}$

1. $5\sqrt{6}\left(4\sqrt{3} + 2\sqrt{10}\right)$

$= 5\cdot 4\sqrt{6}\sqrt{3} + 5\cdot 2\sqrt{6}\sqrt{10}$ using Rule #5

$= 20\sqrt{18} + 10\sqrt{60}$ using Rule # 4 and Rule # 5

$= 20\sqrt{9}\sqrt{2} + 10\sqrt{4}\sqrt{15}$ using Rule # 1

$= 20\cdot 3\sqrt{2} + 10\cdot 2\sqrt{15}$ taking the square root

$= 60\sqrt{2} + 20\sqrt{15}$ multiplying coefficients

(Note: These radicals cannot be added since the radicands are different.)

2. $3\sqrt{2}\left(4\sqrt{2} + 5\sqrt{6}\right)$ = ______________

= ______________

= ______________

= ______________

2. $12\sqrt{4} + 15\sqrt{12}$

$12\cdot 2 + 15\sqrt{4}\sqrt{3}$

$24 + 15\cdot 2\sqrt{3}$

$24 + 30\sqrt{3}$

Exercises-Radicals Review

I. Simplify each of the following:

1. $\sqrt{32}$
2. $3\sqrt{8}$
3. $\sqrt{\frac{3}{16}}$
4. $\sqrt{128}$
5. $\frac{1}{\sqrt{3}}$
6. $\frac{7}{\sqrt{7}}$
7. $(5\sqrt{3})(4\sqrt{2})$
8. $(2\sqrt{3})(3\sqrt{3})$
9. $\sqrt{125} - \sqrt{20}$
10. $\frac{\sqrt{15}}{\sqrt{5}}$
11. $3\sqrt{5}\cdot\sqrt{15}$
12. $5\sqrt{3}\cdot 8\sqrt{2}$
13. $\sqrt{20} + \sqrt{45} - 4\sqrt{5}$
14. $\sqrt{12} + \sqrt{27} + \sqrt{48}$
15. $(8\sqrt{3})(5\sqrt{27})$
16. $6\sqrt{2}\left(5\sqrt{3} + 4\sqrt{2}\right)$
17. $\frac{1}{2}\sqrt{3}\left(5\frac{1}{2} + 2\sqrt{3}\right)$
18. $\frac{7}{3}\ (6\sqrt{2})(5\sqrt{6})$
19. $\frac{4}{3}\ (4\sqrt{3})^{3}$
20. $\sqrt{448}$

Quadratic Equations Review

The general form of a quadratic equation is $Ax^2 + Bx + C = 0$.

I. To solve a quadratic equation by FACTORING, follow this procedure.

Example: $4x^2 = 13x - 3$

a. Set the equation equal to zero.

$$4x^2 = 13x - 3$$
$$4x^2 - 13x + 3 = 0$$

b. Factor the left member of the equation.

$$(x-3)(4x-1) = 0$$

c. Set each factor equal to zero and solve for x.

$x - 3 = 0$ and $4x - 1 = 0$

$x = 3$ $\qquad 4x = 1$

$x = \frac{1}{4}$

d. Check your answer by substitution.

If $x = 3$, then $4x^2 = 13x - 3$

becomes
$$4(3)^2 = 13(3) - 3$$
$$4(9) = 39 - 3$$
$$36 = 36$$

If $x = \frac{1}{4}$, then $4x^2 = 13x - 3$

becomes
$$4\left(\frac{1}{4}\right)^2 = 13\left(\frac{1}{4}\right) - 3$$
$$4\left(\frac{1}{16}\right) = \frac{13}{4} - \frac{3}{1}$$
$$\frac{4}{16} = \frac{13}{4} - \frac{12}{4}$$
$$\frac{1}{4} = \frac{1}{4}$$

II. You will discover very quickly in step (b) of the factoring method above whether or not a quadratic equation is factorable. If the equation is prime, then use the QUADRATIC FORMULA to obtain the solutions. If the equation is in general form, the formula is given by

$$x = \frac{-B \pm \sqrt{B^2 - 4AC}}{2A}$$

where A is the coefficient of x^2; B is the coeffieient of x; and, C is the constant term.

Example: $3x^2 - 11x = -6$

a. Set the equation equal to zero.

$$3x^2 - 11x + 6 = 0$$

b. Select A = 3, B = -11, and C = 6. Substitute these values directly into the formula.

$$x = \frac{-(-11) \pm \sqrt{(-11)^2 - 4(3)(6)}}{2(3)}$$

c. Solve for x.

$$x = \frac{11 \pm \sqrt{121 - 72}}{6}$$

$$x = \frac{11 \pm \sqrt{49}}{6}$$

$$x = \frac{11 \pm 7}{6}$$

$$x = \frac{11 + 7}{6} = \frac{18}{6} = 3 \quad \text{and}$$

$$x = \frac{11 - 7}{6} = \frac{4}{6} = \frac{2}{3}$$

d. The solution set is $\left\{\frac{2}{3}, 3\right\}$.

Exercises-Quadratic Equations Review

I. Solve each of the following equations by <u>factoring</u>.

1. $x^2 - 8x + 15 = 0$
2. $12x^2 - 19x + 5 = 0$
3. $5x^2 = -2x$
4. $6x^2 + 11x - 30 = 0$
5. $4x^2 - 9 = 0$
6. $2x^2 = x + 6$
7. $4x^2 + 9 = 12x$
8. $10x^2 + 21x - 10 = 0$

II. Solve each of the following equations using the <u>quadratic formula</u>.

1. $x^2 + 6x + 5 = 0$
2. $2x^2 - 7 = 0$
3. $2x^2 - 5x + 3 = 0$
4. $x^2 + 2x - 11 = 0$
5. $3x^2 = 7$
6. $3x^2 - 2x = 6$
7. $x(x - 8) + 2(2x - 5) = 3(x^2 - 3)$
8. $8x^2 - 8x + 2 = 0$

AREA

Area deals with the number of square units in a figure. For example, in a rectangle 3 ft. by 4 ft., there will be 12 sq. ft. or 12 blocks each 1 ft. by 1 ft. This number can be obtained by counting.

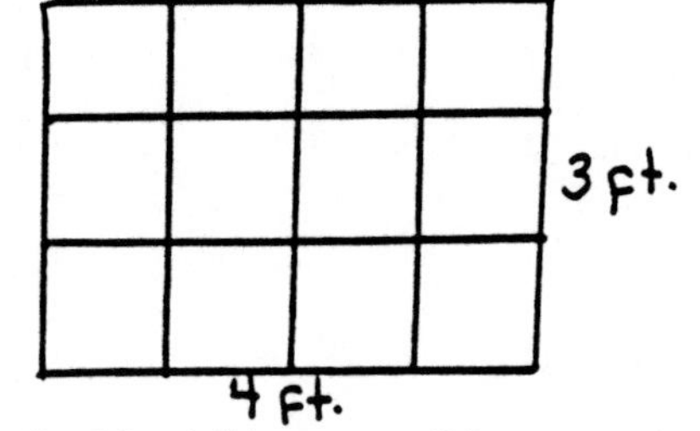

The procedure of dividing a figure into square blocks and counting these units is tedious, not to mention time consuming as well. A procedure needs to be developed to do this for all figures whose area will be of concern. As you progress through these sections you will need to make a list of formulas.

Area of:

Rectangle $A = LW$

Square $A = S^2$

Parallelogram A = Base X HEIGHT

Triangle $A = \frac{1}{2}bh$

Trapezoid $A = \frac{1}{2}h(b_1 + b_2)$

Postulate 1:	The area of a rectangle is the product of the length and the width. (If L = length, W = width, and A = area, then A = LW.)

Example:

If the length of a rectangle is 5 in. and the width is 6 in. Find the area.

A = LW

A = (5 in.)(6 in.)

= 30 sq. in.

Example:

If the length of a rectangle is $4\frac{3}{4}$ ft. and the width is $5\frac{1}{8}$ ft. Find the area.

$A = LW$

$A = \left(4\frac{3}{4} \text{ ft.}\right)\left(5\frac{1}{8} \text{ ft.}\right)$

$A = \left(\frac{19}{4}\right)\left(\frac{41}{8}\right)$

$A = \frac{779}{32}$

$A = 24\frac{11}{32}$ sq. ft.

1. If the length of a rectangle is 12 ft. and the width is 8.7 ft. Find the area.

   A = LW

   A = ____________

   A = ____________ sq. ft.

1. (12)(8.7)

   104.4

Example:

Find the cost of covering a rectangular floor 5 yd. by 4 yd. if the carpeting cost $9 per sq. yd.

Solution:

First determine the area of the rectangle.

(5 yd.)(4 yd.) = 20 sq. yd. = A

Each square yard costs $9 and we have 20 sq. yd.

(20)(9) = $180 = cost of carpeting the room

2. A hallway 20 ft. by $4\frac{1}{2}$ ft. is to be covered with carpeting that costs $4 a

square foot. Determine the cost of carpeting the hallway.

a) Area of hall = ____________

b) Cost of carpeting = Area . Cost per sq. ft.

= ______________

Example:

Mr. Brown owns a farm that is 16 mi. longer than it is wide. If the area of the farm is 80 sq. mi., find the dimensions of his rectangular farm.

$$x = \text{width}$$

$$x + 16 = \text{length}$$

$$\text{Area} = \text{length} \cdot \text{width}$$

$$80 = (x + 16)x$$

$$80 = x^2 + 16x$$

$$-80 \qquad\qquad -80$$

$$0 = x^2 + 16x - 80$$

$$0 = (x + 20)(x - 4)$$

$$x + 20 = 0 \qquad x - 4 = 0$$

$$x = -20 \qquad x = 4$$

$x = -20$ is not a suitable choice for width since we measure figures with positive numbers only.

$x = 4$ is the width.

$x + 16 = 4 + 16 = 20$ is the length.

2.

a) $(20)\left(4\frac{1}{2}\right)$

$(20)\left(\frac{9}{2}\right)$

90 sq. ft.

b)

$4 \cdot 90 = \$360$

3. Mr. Smith owns a farm whose length is five times its width and whose area is 320 sq. mi. Find the dimensions of his rectangular garden.

x = width

5x = length

$5x^2 = 320$ = equation

Solution = ______

The negative root would be discarded since dimensions are expressed in positive numbers only.

width = ______

length = ______

3.

x

5x

$5x^2 = 320$

$x^2 = 64$

$x = \pm\sqrt{64} = \pm 8$

8 mi.

40 mi.

Example: The Wildlife Service has a rectangular park they wish to fence. The rangers know the park is three times longer than it is wide and the area is 75 sq. mi. How much fence will they need to purchase?

Solution:

This problem deals with both area and perimeter. The fencing is placed on the perimeter of the park. Before determining the perimeter, the length and width need to be obtained.

From the problem:

x = width

3x = length

(width) (length) = area

$(x)\quad(3x) = 75$

$3x^2 = 75$

$x^2 = 25$

$x = \pm 5$

$x = 5$ width

$3x = 15$ length

Perimeter of a rectangle is the distance around the rectangle. It takes two lengths and two widths to **equal the perimeter.**

2 widths + 2 lengths = perimeter

$2(5) + 2(15) = 10 + 30$

$= 40$ mi. of fence needed for the park.

4. Tom and Betty wish to fence in their yard so they can get a puppy. Their lot is twice as long as it is wide and is 20,000 sq. ft. in area. How much fence will they need for their rectangular yard?

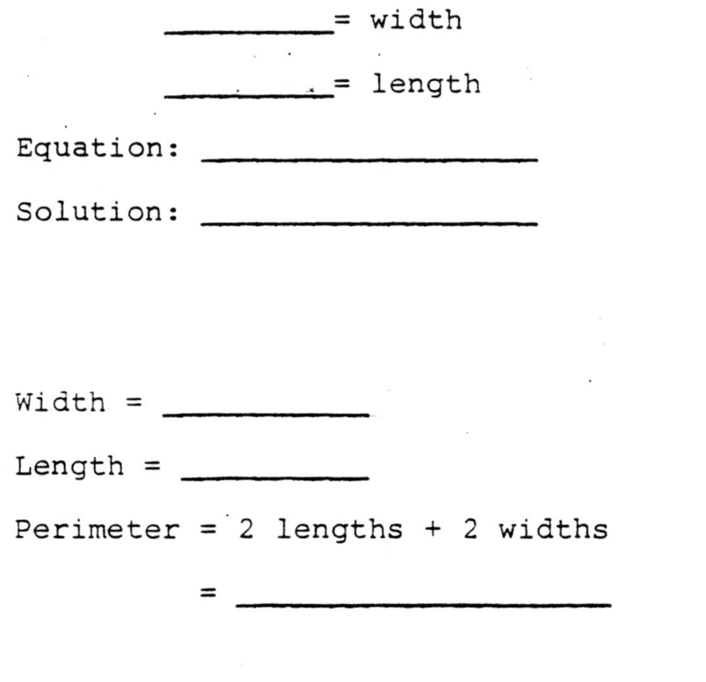

__________ = width

__________ = length

Equation: ____________________

Solution: ____________________

Width = ____________

Length = ___________

Perimeter = 2 lengths + 2 widths

= ______________________

Amount of fencing needed: _______________

4.

x

$2x$

$(x)(2x) = 20,000$

$2x^2 = 20,000$

$x^2 = 10,000$

$x = \pm 100$

100

200

$2(200) + 2(100)$

$400 + 200$

600 ft.

600 ft.

Example:

Find the area of the given region.

Method 1:

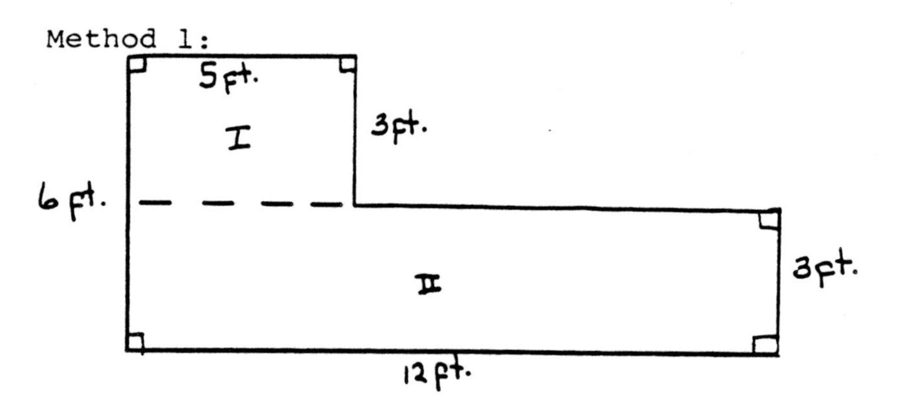

The region can be divided into two rectangles. (See illustration.) The top region (I) is 5 ft. by 3 ft. The lower region (II) is 12 ft. by 3 ft.

Area I: 5 ft. x 3 ft. = 15 sq. ft.

Area II: 12 ft. x 3 ft. = 36 sq. ft.

Total area of entire region------------------ = 51 sq. ft.

Method II:

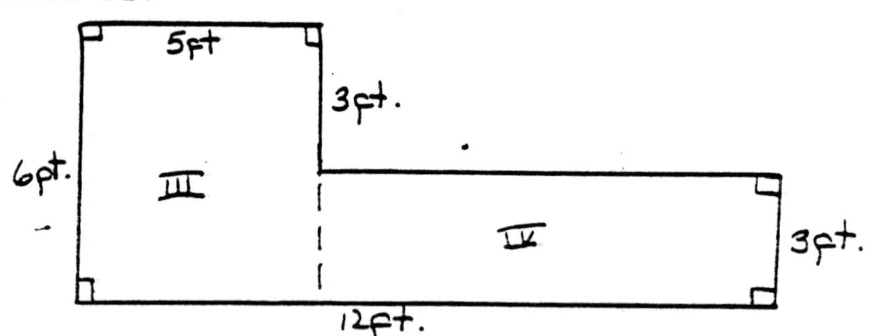

A second way to divide into regions is illustrated. The left region (III) is 6 ft. by 5 ft. The right region (IV) is 3 ft. by (12 - 5) ft.

Area III: 6 ft. x 5 ft. = 30 sq. ft.

Area IV: 3 ft. x 7 ft. = 21 sq. ft.

Area = 51 sq. ft.

Either method is acceptable.

5. Determine the area of this region.

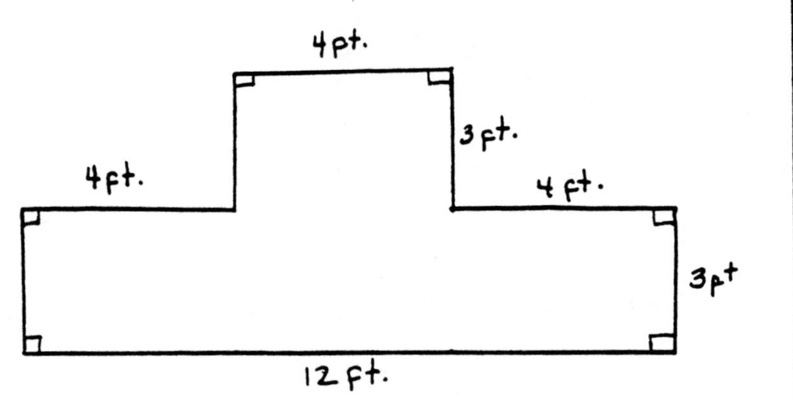

One possible method is:

5.

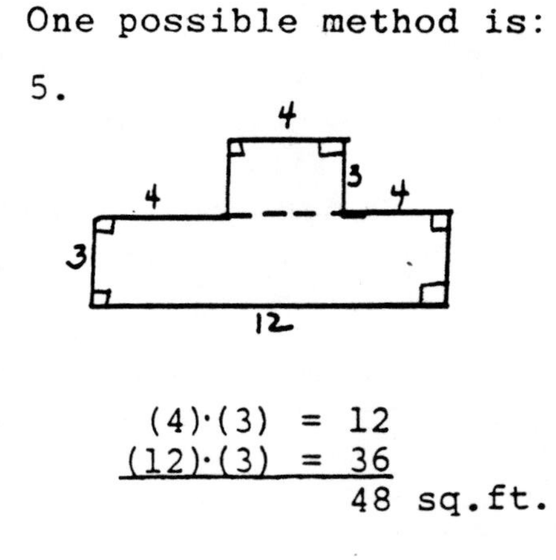

$$\begin{array}{r} (4)\cdot(3) = 12 \\ \underline{(12)\cdot(3) = 36} \\ 48 \text{ sq.ft.} \end{array}$$

> **Theorem 1:** The area of a square with side, s, is given by the formula $A = s^2$.

A square is a special type of rectangle. It is a rectangle with two adjacent sides equal. The area of a rectangle is length times the width. In a square the length equals the width; therefore, we can readily see why the area of a square is side times side or side squared.

Example:

If the side of a square is 7 ft., find its area.

$A = (7\text{ft.})^2 = (7 \text{ ft.})(7 \text{ ft.}) = 49$ sq. ft.

Example:

If the side of a square is $4\sqrt{2}$ in., find the area.

$A = (4\sqrt{2})^2 = (4\sqrt{2})(4\sqrt{2}) = 16\sqrt{4} = 32$ sq. in.

6. If the side of a square is $4\frac{1}{2}$ ft., find its area.

A = ____________________

6. $A = \left(\frac{9}{2}\right)^2$

$= \frac{81}{4} = 20\frac{1}{4}$

sq. ft.

Example:

The perimeter of a square is 48 in. Find its area.

Since the distance all the way around a figure is the perimeter and all four sides of the square are equal, then

$$P = 2x + 2x$$

or

$$P = 4x$$

where x is the length of the side of the square.

$$P = 4x$$

$$48 = 4x$$

$$12 = x$$

If $x = 12$, then $A = x^2$. Area $= 12^2 = 144$ sq. in.

7. If the perimeter of a square is 40 ft. Find its area.
   Let x be the length of a side.

   Equation: ____________________

   Length of a side: __________ ft.

   Area: __________ sq. ft.

7.

$P = 4x$
$40 = 4x$
$10 = x$

$(10)(10) = 100$

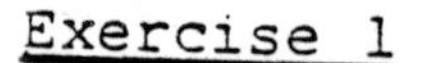

Exercise 1

1. Find the area of the following rectangles:

   a. L = 12 in., W = 10 in.

   b. L = 20 ft., W = 6 ft.

   c. L = $\sqrt{7}$ yd., W = $\sqrt{7}$ yd.

   d. L = $3\frac{1}{2}$m., W = $\frac{77}{5}$ m.

2. In a rectangle, if:

   a. A = 12 sq. in., L = 6 in., find W.

b. $A = \sqrt{14}$ sq. cm., L = 5 cm., find W.

c. $A = 3\frac{1}{3}$ sq. mm., $W = 2\frac{1}{2}$ mm., find L.

3. Find the area of each geometric figure depicted below:

a.

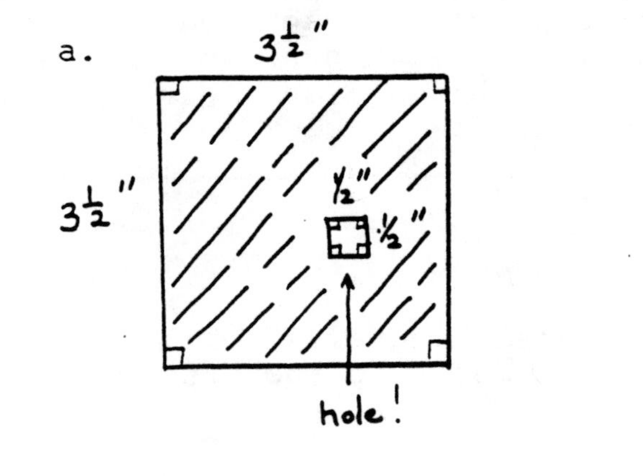

b.

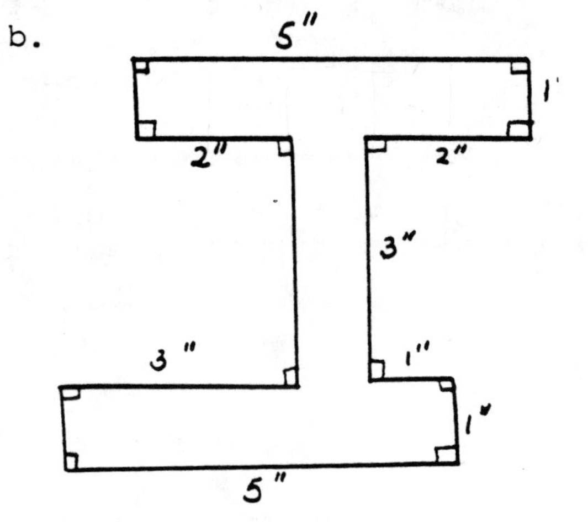

c.

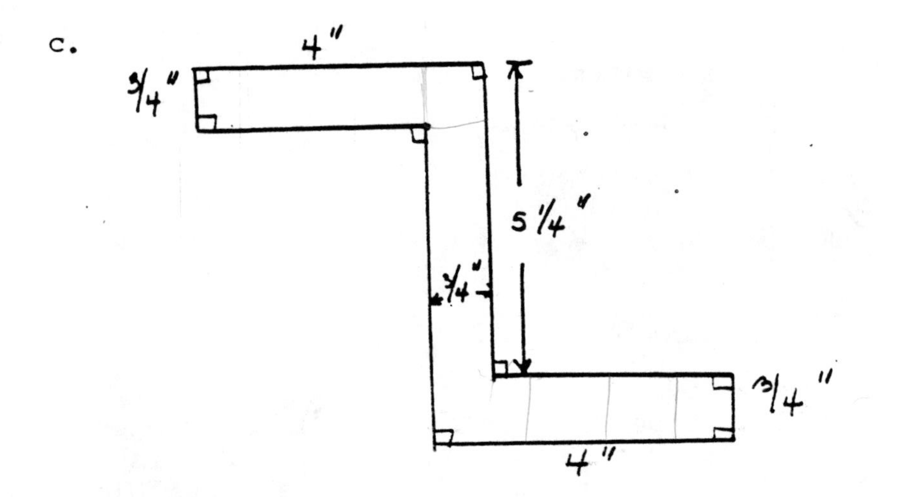

4. Find the measures of the sides of a rectangular room if the length is three feet longer than the width and the area equals 180 sq. ft.

5. To completely border a rectangular lot that measures 54 ft. by 90 ft., the owner must purchase at least how many feet of fencing?

6. If the side of a square if 5 ft., then find its area.

7. If the perimeter of a square is 6 in., find its area.

The next quadrilateral whose area we'll discuss is a parallelogram.

> Theorem 2: The area of a parallelogram is the product of the base and height. In symbols A = bh, where b is the base, h is the height, and A is the area.

The height of a parallelogram is the length of the line drawn from a vertex perpendicular to the other side (base).

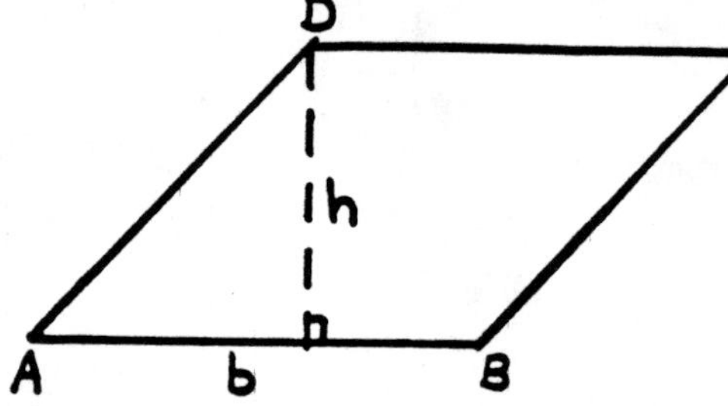

In parallelogram ABCD, h is the height and b is the base.

Let's examine the formula closer.

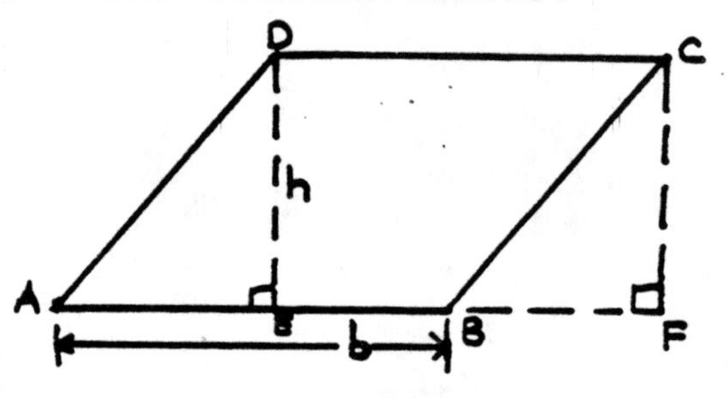

In parallelogram ABCD a perpendicular line ($\overline{CF}$) is dropped to the extended base. By doing this, a rectangle, DCFE, is formed having the same base (b) and altitude (h) as the parallelogram. Since $\triangle ADE \cong \triangle BCF$ and $\triangle BCF$ is added on one side of the parallelogram and $\triangle ADE$ is removed from the other, the area is unchanged. Consequently, the formula for the area of a parallelogram is concluded:

$$A = bh$$

Example:

In parallelogram ABCD, the height is 6 ft. and the base is $10\frac{1}{2}$ ft. Find the area.

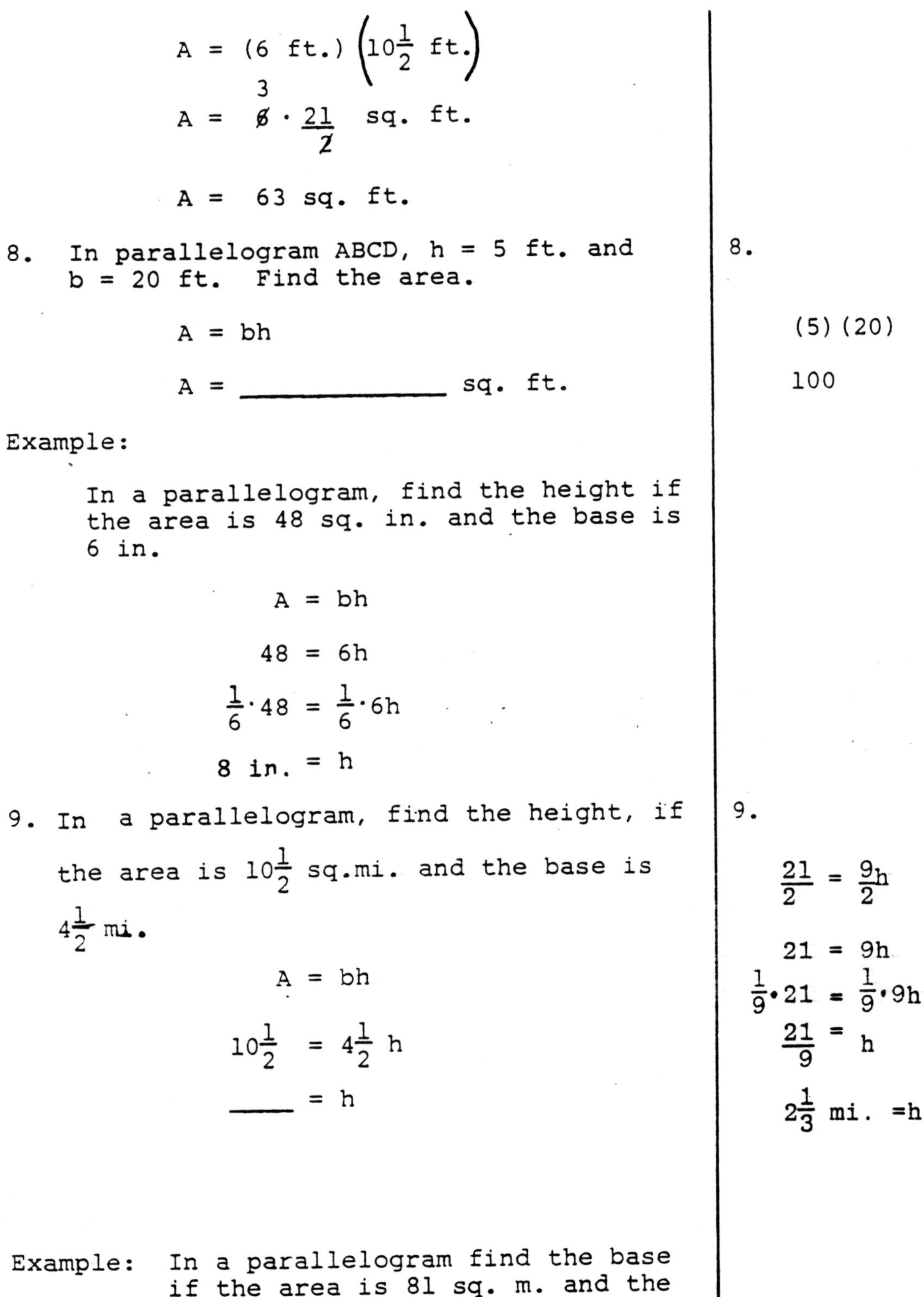

$A = (6 \text{ ft.})\left(10\frac{1}{2} \text{ ft.}\right)$

$A = \overset{3}{\cancel{6}} \cdot \frac{21}{\cancel{2}}$ sq. ft.

$A = 63$ sq. ft.

8. In parallelogram ABCD, h = 5 ft. and b = 20 ft. Find the area.

$A = bh$

$A =$ ______________ sq. ft.

8.

(5)(20)

100

Example:

In a parallelogram, find the height if the area is 48 sq. in. and the base is 6 in.

$A = bh$

$48 = 6h$

$\frac{1}{6} \cdot 48 = \frac{1}{6} \cdot 6h$

$8 \text{ in.} = h$

9. In a parallelogram, find the height, if the area is $10\frac{1}{2}$ sq.mi. and the base is $4\frac{1}{2}$ mi.

$A = bh$

$10\frac{1}{2} = 4\frac{1}{2}\ h$

_____ $= h$

9.

$\frac{21}{2} = \frac{9}{2}h$

$21 = 9h$

$\frac{1}{9} \cdot 21 = \frac{1}{9} \cdot 9h$

$\frac{21}{9} = h$

$2\frac{1}{3} \text{ mi.} = h$

Example: In a parallelogram find the base if the area is 81 sq. m. and the height is three times the base.

x = base

$3x$ = height

$A = bh$

$81 = x \cdot 3x$

$81 = 3x^2$

$27 = x^2$

$3\sqrt{3}$ m. $= x$

10. In a parallelogram find the base if the area is 100 sq. cm. and the height is five times the base.

x = base

____ = height

100 = ________

____ $= x^2$

____ $= x$

10.

$5x$

$100 = x \cdot 5x$

$100 = 5x^2$

$20 = x^2$

$2\sqrt{5} = x$

Exercise 2

1. Find the area of the following parallelograms:

   a. $b = 6$ in., $h = 4$ in.

   b. $b = 13\frac{1}{3}$ ft., $h = 1.7$ ft.

   c. $b = 3\sqrt{5}$ yd., $h = 7\sqrt{72}$ yd.

   d. $b = 7$ cm., $h = 3\frac{2}{3}$ cm.

   e. $b = 8\frac{1}{5}$ dm., $h = 25$ dm.

2. Find the missing dimension in the following:

   a. $A = 40$ sq. m., $h = 5$ m., find b.

   b. $A = 17\sqrt{3}$ sq. cm., $b = \sqrt{3}$ cm., find h.

c. $A = 16\frac{2}{3}$ sq. mi., $b = 5$ mi., find h.

d. $A = 17.1$ sq. m., $h = .03$ m., find b.

From the parallelogram, the formula for the area of a triangle can be obtained.

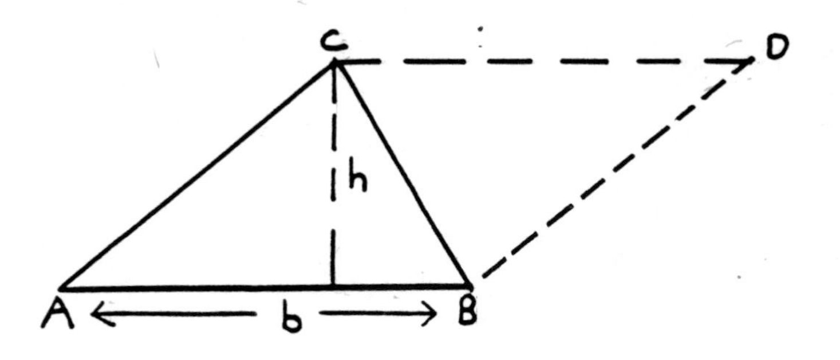

Given a triangle, $\triangle ABC$, a line segment can be constructed through C parallel to $\overline{AB}$ and the same measure as $\overline{AB}$. Likewise, a line segment can be constructed through B parallel to $\overline{AC}$ and the same measure as $\overline{AC}$. We have just formed parallelogram ABDC having the same base and altitude as the triangle. The area of parallelogram ABDC is $A = bh$. The area of $\triangle ABC$ is

$$A = \frac{1}{2} bh$$

Why? $\overline{BC}$ is a diagonal of parallelogram ABDC. The diagonal of a parallelogram divides it into two congruent triangles. Congruent triangles have the same area. Therefore, we can conclude the following:

> Theorem 3: The area, A, of a triangle is given by $A = \frac{1}{2} bh$, where h is the height and b is the base.

Example:

In triangle ABC if $h = 6$ ft. and $b = 7\frac{1}{2}$ ft., find the area.

$$A = \frac{1}{2} b h$$

$A = \frac{1}{2} \cdot 6 \cdot 7\frac{1}{2}$

$A = \frac{1}{\cancel{2}_1} \cdot \overset{3}{\cancel{6}} \cdot \frac{15}{2}$

$A = \frac{45}{2} = 22\frac{1}{2}$ sq. ft.

11. In a triangle if h = 12 ft. and b = $7\frac{1}{5}$ ft., find the area.

$A = \frac{1}{2} \cdot b \cdot h$

A = ____________

11.

$\frac{1}{\cancel{2}_1} \cdot \overset{6}{\cancel{12}} \cdot 7\frac{1}{5}$

$1 \cdot 6 \cdot \frac{36}{5}$

$\frac{216}{5} = 43\frac{1}{5}$ sq.ft.

Example: If the area of a triangle is 40 sq. in. and the base is 8 in., find the height.

$A = \frac{1}{2}$ bh

$40 = \frac{1}{\cancel{2}_1} \cdot \overset{4}{\cancel{8}} \cdot h$

$40 = 4h$

10 in.= h

12. If the area of a triangle is 75 sq.m. and the base is 5 m., find the height.

$A = \frac{1}{2}$ b h

________ = h

12.

$75 = \frac{1}{2} \cdot 5 \cdot h$

$75 = \frac{5}{2}$ h

30 m. = h

Example: The height of a triangle is five more than twice the base. If the area is 12 sq. m., find the base.

x = base

$2x + 5$ = height

$$12 = \frac{1}{2}(x)(2x + 5)$$

$$0 = 2x^2 + 5x - 24$$

$$x = \frac{-5 \pm \sqrt{25 - 4 \cdot 2 \cdot -24}}{4}$$

$$x = \frac{-5 \pm \sqrt{25 + 192}}{4}$$

$$x = \frac{-5 \pm \sqrt{217}}{4}$$

The solution for the base is $\frac{-5 + \sqrt{217}}{4}$ since dimensions are measured in positive numbers.

13. The height of a triangle is five more than four times the base. Find the height and base if the area is $25\frac{1}{2}$ sq.m.

x = base

$4x + 5$ = height

equation: ______________________

solution: ______________________

13.

$\frac{1}{2}x(4x+5) = 25\frac{1}{2}$

$4x^2+5x-51=0$

$(4x+17)(x-3)=0$

$4x+17=0$ $\quad$ $x-3=0$

$4x=-17$ $\quad$ $x=3$

$x=\frac{-17}{4}$

base = 3 m.

height = 4(3)+5=17 m.

Exercise 3

1. Find the area of the following triangles:

   a. $b = 6$ in., $h = 10$ in.

   b. $b = 5$ ft., $h = 3$ ft.

   c. $b = \sqrt{7}$ cm., $h = \sqrt{5}$ cm.

   d. $b = \sqrt{2}$ m., $h = 2\sqrt{2}$m.

   e. $b = \frac{2}{3}$ mi., $h = \frac{1}{2}$ mi.

2. In a triangle, if:

   a. $A = 16$ sq. ft., $b = 4$ ft., find h.

   b. $A = 14$ sq. in., $b = \sqrt{3}$ in., find h.

   c. $A = 2\frac{1}{4}$ sq. yd., $b = \frac{1}{2}$ yd., find h.

3. The base of a triangle is 8.6 m. and the height is 4.4 m., find the area.

4. If the base is 3 more inches than the height of the triangle and the area equals 90 sq. in., find the measure of the base.

5. The area of a triangle is 40 sq. cm. If the base is 5 cm. then find the measure of the height.

6. The floor of a triangular bathroom is illustrated as:

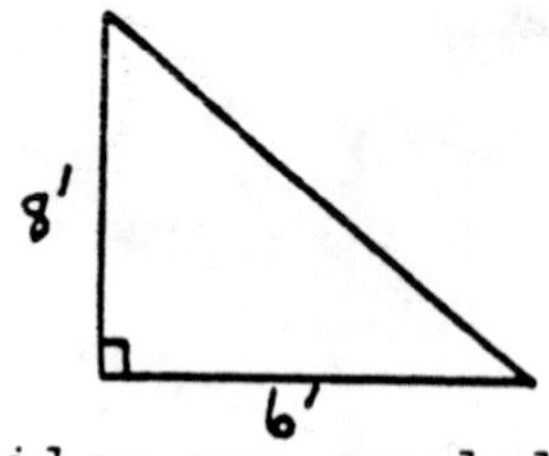

   8'

   6'

   How many tiles are needed to cover the floor if each tile is 1 sq. ft.?

Recall that a trapezoid is a quadrilateral with only one pair of sides parallel. Examining a trapezoid, let us determine a wav to find the area. Begin with trapezoid ABCD.

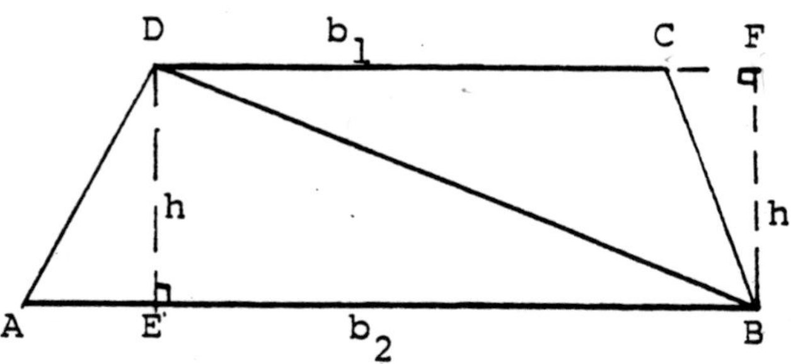

Divide trapezoid ABCD into two triangles by drawing in diagonal $\overline{BD}$. For triangle ABD draw altitude $\overline{DE}$ and call its length h. Area of $\triangle$ABD is $\frac{1}{2}hb_2$. Likewise, for triangle BCD draw altitude $\overline{BF}$. Note $\overline{BF}$ is exterior to the triangle and is perpendicular to the extention of $\overline{DC}$. The length of $\overline{BF}$ is equal to h since $\overline{AB} \parallel DF$ and the perpendicular segments between parallel lines are everywhere equidistant. The area of triangle DBC = $\frac{1}{2}hb_1$. The area of trapezoid ABCD = area of $\triangle$ ADB plus the area of $\triangle$ DBC.

Therefore, $A = \frac{1}{2}hb_2 + \frac{1}{2}hb_1$

$= \frac{1}{2}h(b_2 + b_1)$ (factoring)

> Theorem 4: The area, A, of a trapezoid is given by the formula $A = \frac{1}{2}h(b_1 + b_2)$ where h is the altitude and b_1 and b_2 are the bases.

14. Find the area of the following trapezoids.

a. $b_1 = 10$ in., $b_2 = 18$ in., $h = 9$ in.

$A = \frac{1}{2}h(b_1 + b_2)$

$A = \frac{1}{2} \cdot 9(10 + 18)$

$A = \frac{9}{\not{2}_1}(\not{28}^{14})$

A = 126 sq. in.

b. $b_1 = 15$ in., $b_2 = 20$ in., $h = 10$ in.

$A = \frac{1}{2}h(b_1 + b_2) =$ ____________

14.b.

$\frac{1}{2} \cdot 10(15 + 20)$

$\frac{1}{2} \cdot 10(35)$

$5(35)$

175 sq. in.

Example: Find the height of a trapezoid, if the bases are 12 inches and 14 inches and the area is 60 square inches.

$$A = \frac{1}{2}h(b_1 + b_2)$$
$$60 = \frac{1}{2}h(12 + 14)$$
$$60 = \frac{1}{2}h(26)$$
$$60 = 13h$$
$$4\frac{8}{13} \text{ in.} = h$$

15. Find the height of a trapezoid, if the bases are 8 meters and 10 meters and the area is 72 square meters.

h = ____________meters

15.

$72 = \frac{1}{2}h(8 + 10)$

$72 = \frac{1}{2}h(18)$

$72 = 9h$

$8 = h$

Example: Find each base of a trapezoid in which the larger base is four more than the smaller , the height is 6 inches, and the area is 20 sq. in.

x = smaller base (base 1)

x + 4 = larger base (base 2)

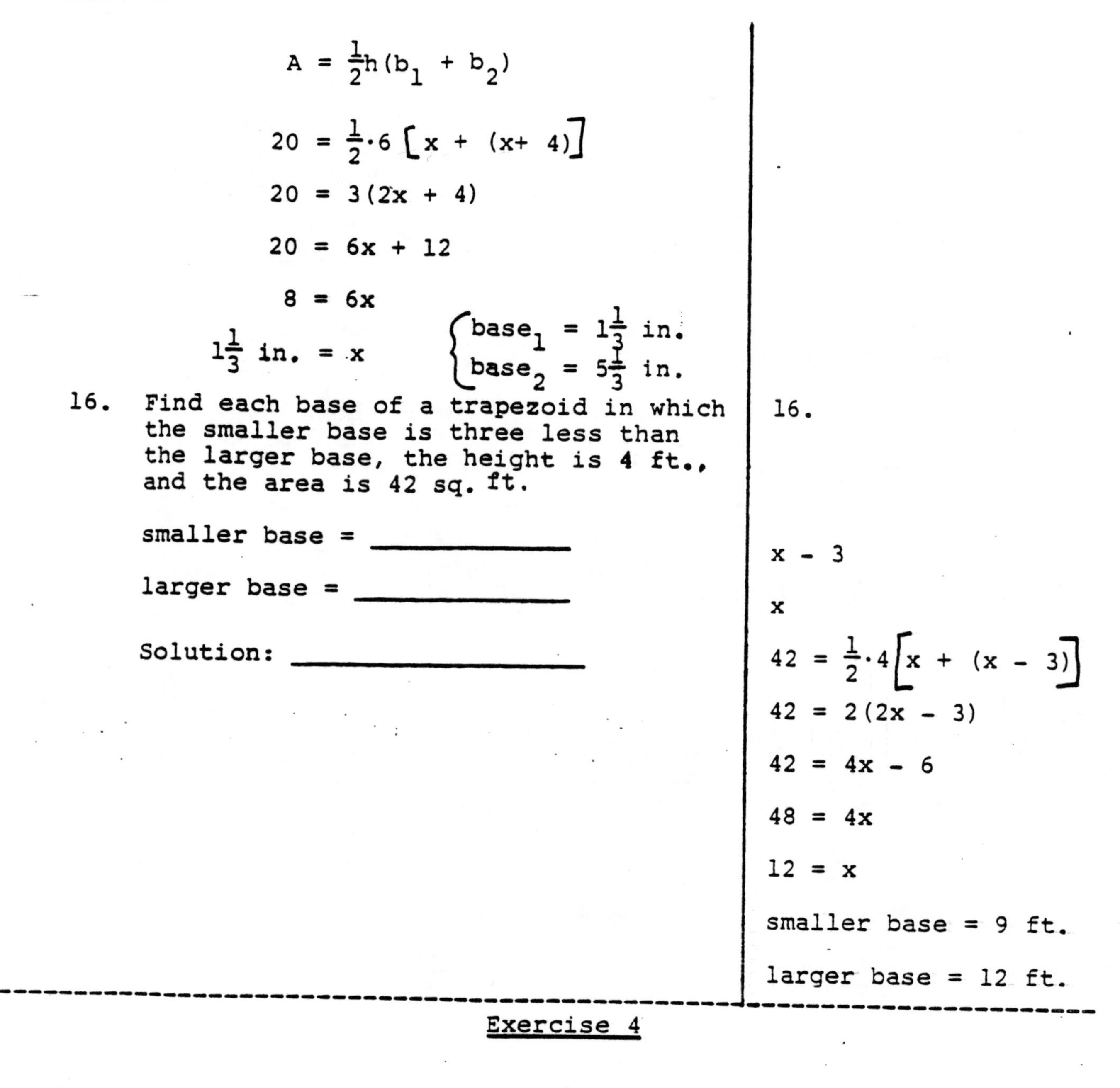

$$A = \frac{1}{2}h(b_1 + b_2)$$

$$20 = \frac{1}{2} \cdot 6 \left[x + (x + 4)\right]$$

$$20 = 3(2x + 4)$$

$$20 = 6x + 12$$

$$8 = 6x$$

$$1\frac{1}{3} \text{ in.} = x \qquad \begin{cases} \text{base}_1 = 1\frac{1}{3} \text{ in.} \\ \text{base}_2 = 5\frac{1}{3} \text{ in.} \end{cases}$$

16. Find each base of a trapezoid in which the smaller base is three less than the larger base, the height is 4 ft., and the area is 42 sq. ft.

smaller base = ______________

larger base = ______________

Solution: ______________

16.

$x - 3$

x

$$42 = \frac{1}{2} \cdot 4 \left[x + (x - 3)\right]$$

$$42 = 2(2x - 3)$$

$$42 = 4x - 6$$

$$48 = 4x$$

$$12 = x$$

smaller base = 9 ft.

larger base = 12 ft.

Exercise 4

1. Find the area of the following trapezoids:

   a. $b_1 = 12$ m., $b_2 = 14$ m., $h = 7$ m.

   b. $b_1 = 30$ cm., $b_2 = 40$ cm., $h = 10$ cm.

   c. $b_1 = 3\sqrt{5}$ mm., $b_2 = 5\sqrt{5}$ mm., $h = 6\sqrt{5}$ mm.

   d. $b_1 = 7\frac{1}{2}$ in., $b_2 = 3\frac{1}{3}$ in., $h = \frac{1}{2}$ in.

2. In a trapezoid, if:

   a. b_1 = 12 mi., b_2 = 10 mi., find h if A = 55 sq. mi.

   b. h = 12 m., b_1 = 14 m., find b_2 if A = 180 sq. m.

3. Find the bases of a trapezoid if the upper base is 2 less than the lower base; the altitude is 7 ft. and the area is 35 sq. ft.

4. Find the sum of the bases of a trapezoid if the area is 63 sq. ft. and the altitude is 3 ft.

5. A base is one less than twice the second base and the area is 40 sq. cm. Find the altitude of the trapezoid if the altitude is the length of the second base.

The topic of area is not complete without including the area of a circle. A brief discussion of the definition and circumference of a circle will precede the topic of area of a circle.

Circles

The top of a can, the end of a pipe, the end of pistons in automobile engines, the tires on cars, and the burner on the gas range are a few examples of circles we all encounter.

One could simply say that a circle is a flat round object. But that discription is certainly not the definition for a circle. For our discussion, let's draw a picture of this object.

Take your ruler and pencil and do the following: From point P below, make dots all a distance of 2 cm. from P.

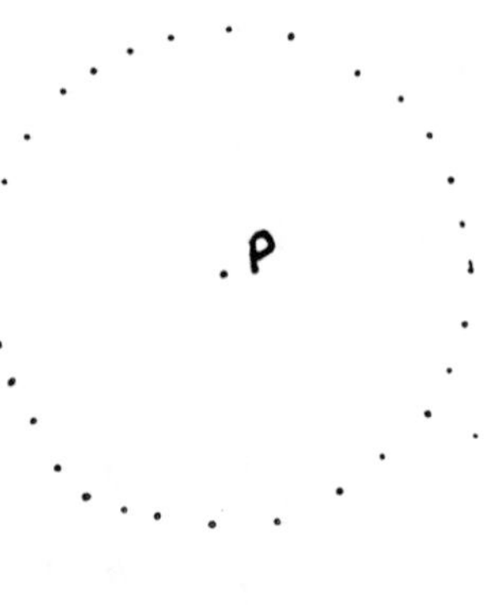

Figure 7.1

Notice these dots, when connected, form a 'round' figure. Note two things about this figure: (1) we started with a fixed point (P) and (2) we used a fixed distance (2 cm.) each time. This exercise leads us to the following definition.

> Definition: A circle is the set of all possible points in a plane a fixed distance from a given point. The fixed distance is called the radius of the circle and the given point is the center.

17. a. In the diagram below what is the radius? ______________________

    b. What is the center? __________

17. a. $\overline{OW}$

    b. O

If two radii are joined to form a straight line segment, this line segment is called a diameter. The measure of one diameter is equivalent to the length of two radii.

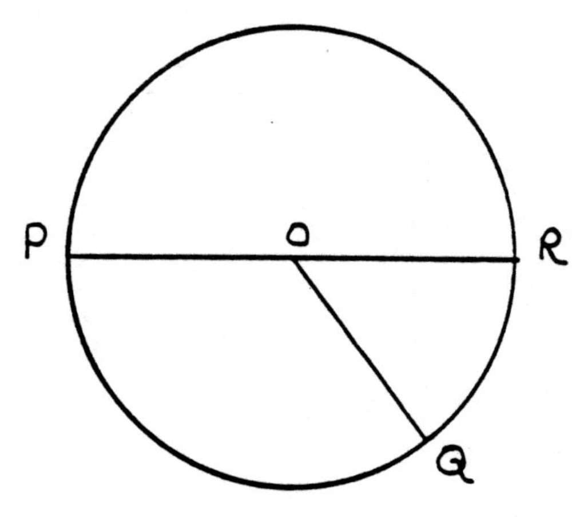

Radii $\overline{OP}$ and $\overline{OR}$ form diameter $\overline{PR}$. $\overline{OR}$ and $\overline{OQ}$ do not form a diameter. A diameter divides a complete circle into two equal parts. Each part is called a semicircle.

18. Does a diameter always pass through the center of a circle? ____________ | 18. Yes

19. Is a diameter equal to the length of two radii? ____________ | 19. Yes

Circumference

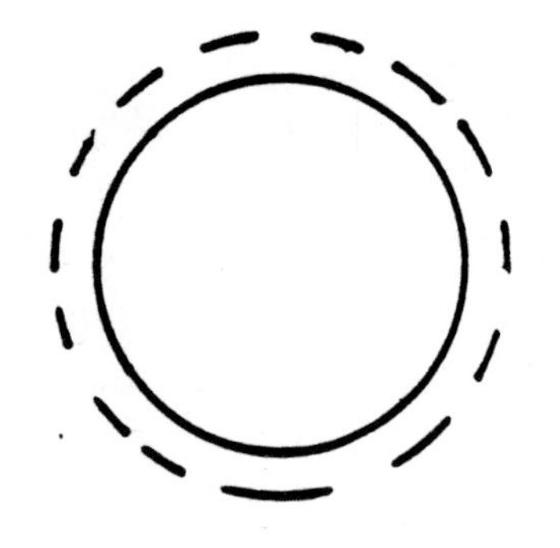

Figure 7.2

The circumference is to a circle as perimeter is to a rectangle. However, circumference can not be defined in the same way as perimeter. Perimeter is defined in terms of adding the lengths of line segments. Since a circle is not composed of line segments, the definition of perimeter is not quite satisfactory. We will, therefore, describe the circumference of a circle simply as the distance around the outer rim of that circle. (See Figure 7.2)

Theorem 1: The ratio of the circumference to the diameter is the same for all circles.

(We will not include the proof since the steps do involve the study of "limits" which is beyond the scope of this text.)

The theorem tells us that for a circle of radius r, circumference C, and diameter d, the ratio of circumference to diameter, $\frac{C}{d}$, equals a constant number. This constant number has been traditionally symbolized by the Greek letter π, pronounced as "pi". Therefore, $\frac{C}{d} = \pi$ or $C = \pi d$. This formula may sometimes be stated in terms of radius (r).

$$\frac{C}{2r} = \pi \text{ or } C = 2\pi r$$

The number, π, which relates the circumference to the diameter of a circle, is an irrational number. π is a constant number that **in its decimal form it never ends nor repeats.** Values such as $3\frac{1}{7}$, 3.14, and $\frac{22}{7}$ are most commonly used for π but π is not exactly equal to any of these numbers. The value of pi has been calculated on modern computers to 500,000 decimal places. Rounded to only the first nine decimal places, π equals approximately 3.141592654. However, in this unit we will express all our answers in terms of π rather than approximate the values.

Example: In a circle if the diameter is 6 cm., find the circumference.

Solution: The formula is $C = \pi d$.
Substituting d = 6, then $C = \pi \cdot 6$
$C = 6\pi$ cm.

20. Find the circumference of a circle if d = 8 ft. The formula is ________.

C = ____________

20.
$C = \pi d$
$C = 8\pi$ ft.

Example: In a circle $r = 2\frac{1}{5}$ m. Find the circumference.

Solution: $C = 2\pi r$ formula

$C = 2\pi \cdot \left(2\frac{1}{5}\right)$ substitution

$C = 2\pi \cdot \left(\frac{11}{5}\right)$ simplify

$C = \frac{22}{5}\pi$ m.

21. If the radius of a circle is $6\sqrt{3}$ in., find the circumference.

The formula is ____________.
The circumference is ________.

21.
$C = 2\pi r$
$C = 12\pi\sqrt{3}$ in.

22. If the radius of a circle is $9\frac{1}{4}$ in., find the circumference.
The formula is ______________.
The circumference is __________.

22.
$C = 2\pi r$
$C = 2 \cdot 9\frac{1}{4}\pi$
$C = \cancel{2}^{1} \cdot \frac{37}{\cancel{4}_{2}}\pi$
$C = \frac{37}{2}\pi$

Example: If the circumference is 28π cm., find the radius.

$$C = 2\pi r$$

$$28\pi = 2\pi r \quad \text{substitution}$$

$$\frac{28\pi}{2\pi} = \frac{2\pi r}{2\pi} \quad \text{solve for r}$$

$$14 \text{ cm.} = r$$

23. If the circumference is 48π cm., find the radius.

The formula is _______________.
Radius is __________________.

24. If the circumference is 10π ft., find the diameter.

23.

$C = 2\pi r$
$48\pi = 2\pi r$
24 cm. $= r$

24. $10\pi = \pi d$
10 ft. $= d$

Area of a Circle

The area of a circle is represented by the shaded area in Figure 7.3.

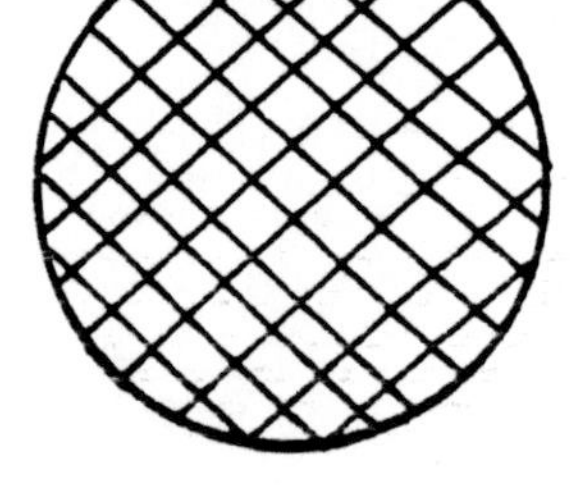

Figure 7.3

The area of a circle can be found by the formula $A = \pi r^2$, where A is the area, π is an irrational constant, and r is the radius of the circle.

Example: Find the area of a circle with radius 5 ft.

$$A = \pi r^2$$

$$A = \pi \cdot 5^2$$

$$A = 25\pi \text{ sq. ft.}$$

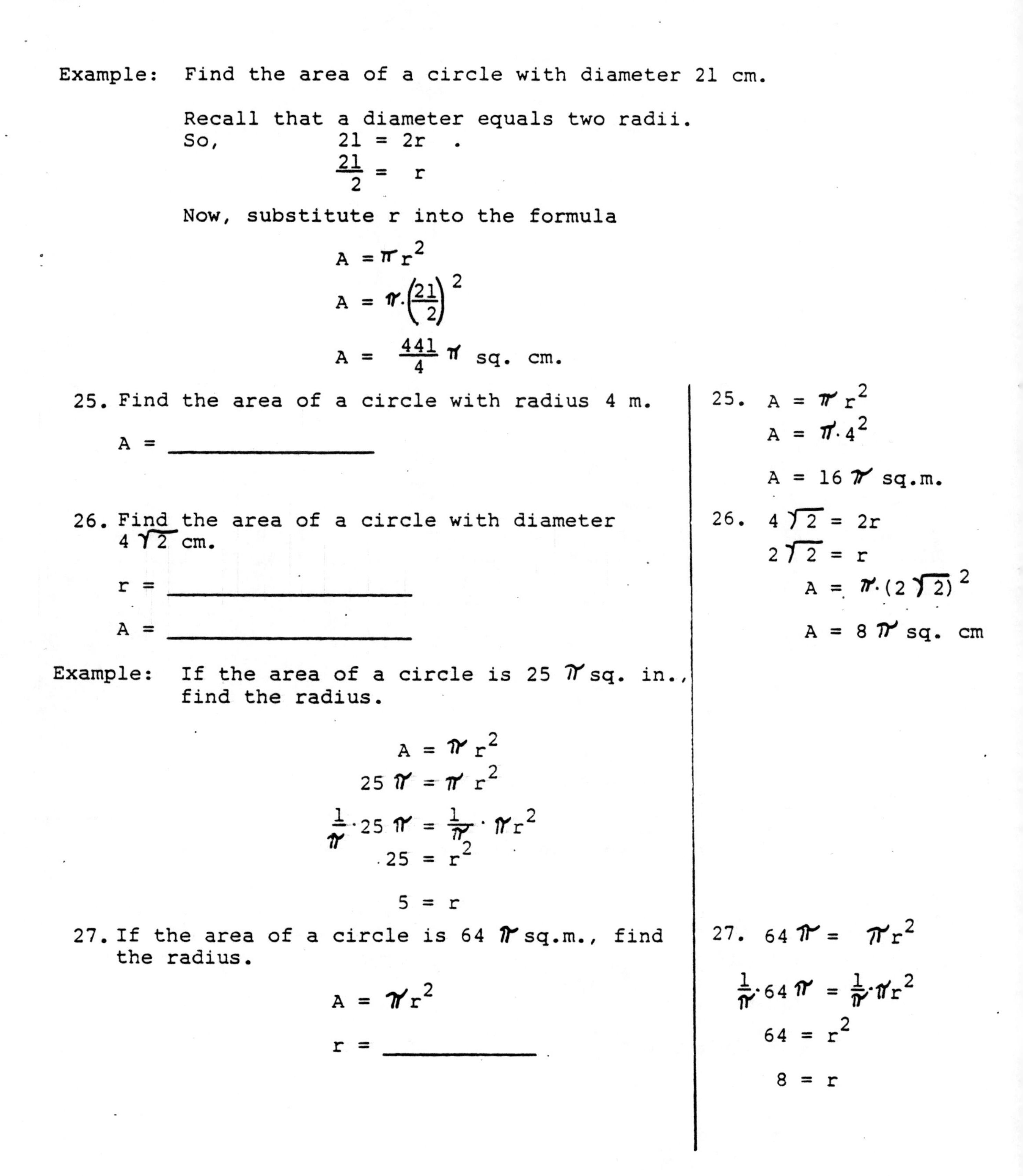

Example: Find the area of a circle with diameter 21 cm.

Recall that a diameter equals two radii.
So, $21 = 2r$.
$\frac{21}{2} = r$

Now, substitute r into the formula

$$A = \pi r^2$$
$$A = \pi \cdot \left(\frac{21}{2}\right)^2$$
$$A = \frac{441}{4}\pi \text{ sq. cm.}$$

25. Find the area of a circle with radius 4 m.

A = ________________

26. Find the area of a circle with diameter $4\sqrt{2}$ cm.

r = ________________

A = ________________

Example: If the area of a circle is 25π sq. in., find the radius.

$$A = \pi r^2$$
$$25\pi = \pi r^2$$
$$\frac{1}{\pi}\cdot 25\pi = \frac{1}{\pi}\cdot \pi r^2$$
$$25 = r^2$$
$$5 = r$$

27. If the area of a circle is 64π sq.m., find the radius.

$A = \pi r^2$

r = ________________

25. $A = \pi r^2$
$A = \pi \cdot 4^2$
$A = 16\pi$ sq.m.

26. $4\sqrt{2} = 2r$
$2\sqrt{2} = r$
$A = \pi \cdot (2\sqrt{2})^2$
$A = 8\pi$ sq. cm

27. $64\pi = \pi r^2$
$\frac{1}{\pi}\cdot 64\pi = \frac{1}{\pi}\cdot \pi r^2$
$64 = r^2$
$8 = r$

Exercise 5

1. Find the area and circumference of the following circles:

   a. $d = 30$ m.

   b. $r = 2\sqrt{3}$ cm.

   c. $r = \frac{3}{2}$ ft.

2. Find r in each of the following:

   a. $C = 24\pi$ m.

   b. $C = 45\pi$ cm.

   c. $A = 25\pi$ sq. m.

   d. $A = 48\pi$ sq. in.

3. A country club built a jogging track by putting a semicircle on each end of a rectangle. Find the area of the jogging track.

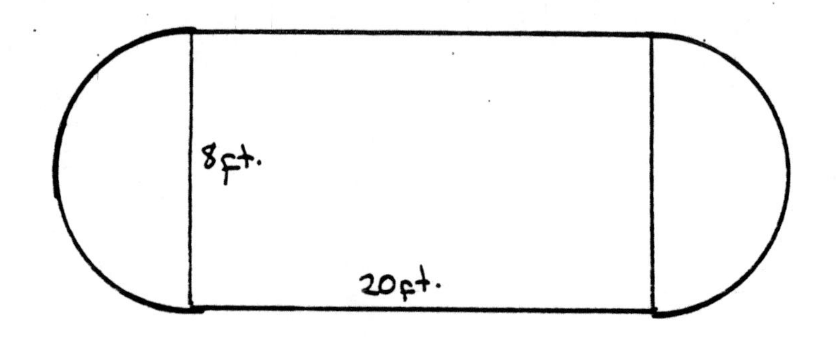

In summary, the formulae you have studied in this unit are as follows:

Figure	Area Formula
Rectangle	$A = LW$
Square	$A = s^2$
Parallelogram	$A = bh$
Triangle	$A = \frac{1}{2}bh$
Trapezoid	$A = \frac{1}{2}h(b_1 + b_2)$
Circle	$A = \pi r^2$

Perimeter of a rectangle--$P = 2L + 2W$

Circumference of a circle-$C = 2\pi r$ (or) $C = \pi d$

Unit 7 Review

1. Find the areas of the figures depicted below:

a. b.

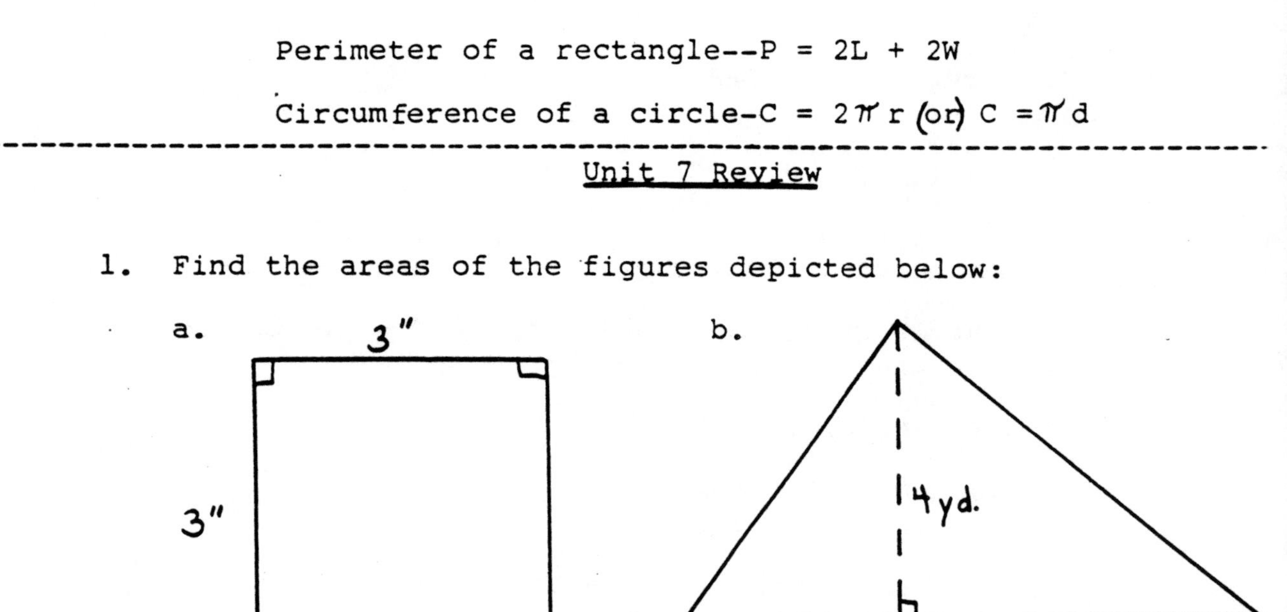

c.

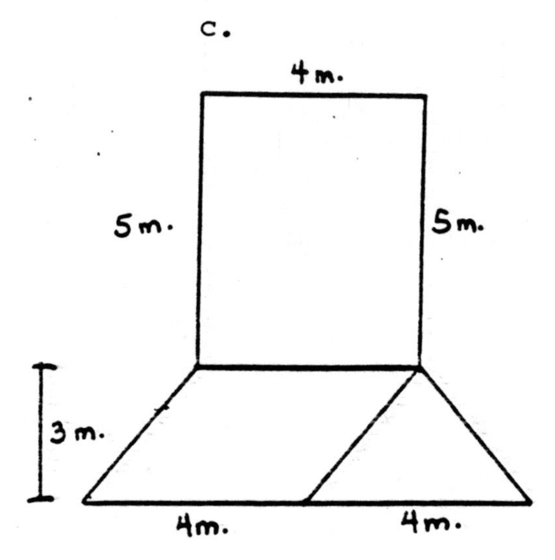

d.

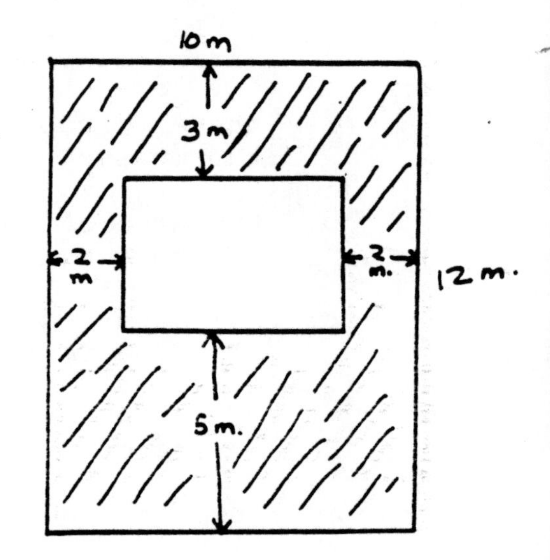

e.

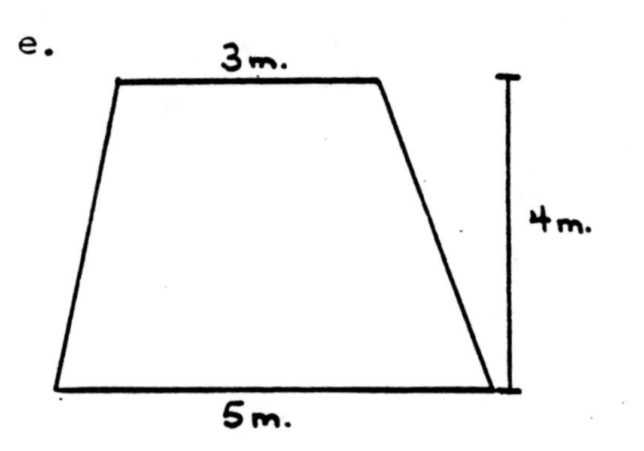

2. In a rectangle, find:

   a. the sides if the rectangle is a square and its area is 12 sq. in.

   b. the perimeter of the rectangle if it is a square and its area is 12 sq. in.

   c. the length of the rectangle if its width is 5 m. and its area is 105 meters squared.

3. In a triangle, find:

   a. the base if the area is 30 centimeters squared and the altitude is 10 cm.

   b. the base if the area is 45 sq. ft. and the altitude is eight more than twice the base.

   c. the area if the base and the altitude are both 9 in.

4. In a parallelogram, find:

   a. the altitude if the base is seven less than one-half the measure of the altitude and the area is 10 sq. cm.

   b. the area of a rhombus whose perimeter is 56 ft. and whose altitude is 3 ft.

   c. the base and altitude if the base is $(4x + 1)$, the altitude is $(3x + 2)$ and the area is represented as $(12x^2 + 35)$.

5. In a trapezoid, find:

   a. area if $b_1 = 25$ inches, $b_2 = 15$ inches and $h = 7$ inches.

   b. the bases if the lower base is twice the upper base, the altitude is 9 inches and the area is 27 sq. in.

   c. the altitude if the bases are 5 ft. and 13 ft. and the area is 11 sq. ft.

6. Find the circumference and area of the following circles.

   a. $r = 2\frac{1}{3}$ cm.

   b. $r = 5\sqrt{2}$ ft.

   c. $d = 1.6$ dm.

   d. $d = 12$ in.

7. Find the radius in each of the following circles:

   a. $C = 6\pi$ m.

   b. $A = 25\pi$ sq. ft.

   c. $d = 9.4$ cm.

   d. $A = 128\pi$ sq. in.

8. A high school built a track field by putting a semicircle on one end of a rectangle. Find the area of the field.

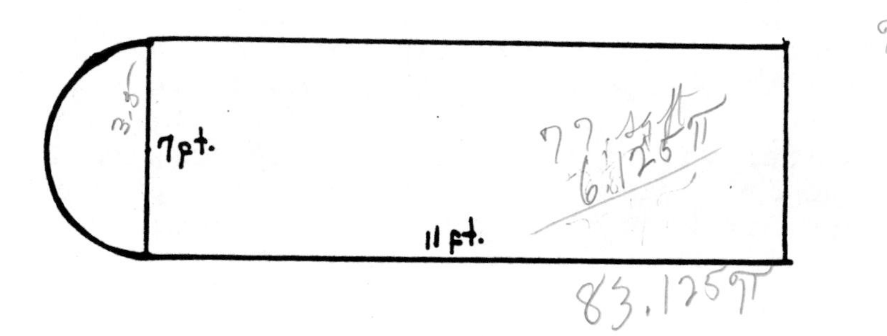

9. Fill in the blanks with an appropriate word or phrase as discussed in this unit.

   a. A ______________ is the set of points a given distance from a fixed point. The given distance is a ___________ of the circle and the fixed point is the ______________.

   b. If two radii are put together to form a straight line segment, this line segment is called a ______________.

   c. The ______________ of a circle is the distance around the outer rim.

Unit 8

Planes in Space and Volume

Learning objectives:

1. The student will list the three methods to determine a plane.
   a. Three non-collinear points will determine a plane.
   b. A line and a point not on that line will determine a plane.
   c. Two intersecting lines will determine a plane.

2. The student will demonstrate his mastery of the following definitions and concepts by writing them and by applying them to the solutions of selected problems.
   <u>Definitions</u>: coplanar, parallel planes, sphere, cylinder, prism, parallelopiped, rectangular solid, cube, pyramid, cone, and volume.
   <u>Concepts</u>: The intersection of two planes will form a straight line.
   Two lines perpendicular to the same plane are parallel.
   Two planes perpendicular to the same line are parallel.
   If a plane is the perpendicular bisector of a line segment, then every point on the plane is equidistant from the endpoints of the line segment.

3. Given real number values the student will compute the volume of each of the following: right rectangular prism, right pyramid, right circular cylinder, right circular cone, and sphere.

4. Given real number values for all but one of the variables in the formula for a given volume the student will find the remaining variable.

Planes in Space and Volume

A plane is a flat surface that extends indefinitely in every direction. A very crude representation of a plane is illustrated in Figure 8.1.

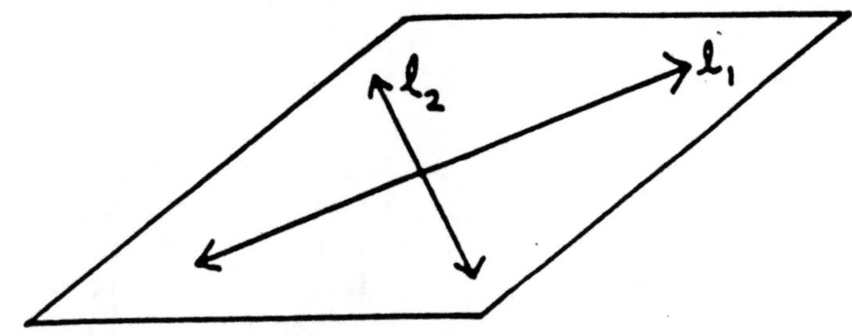

Figure 8.1

A plane may be determined by the following methods:

a) Three non-collinear points will determine a plane.

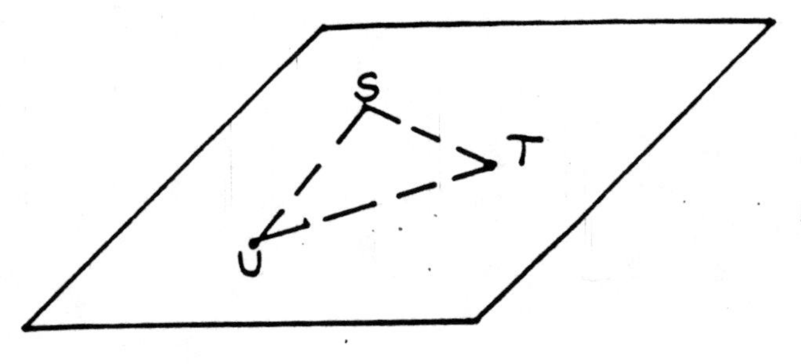

b) A line and a point not on that line will determine a plane.

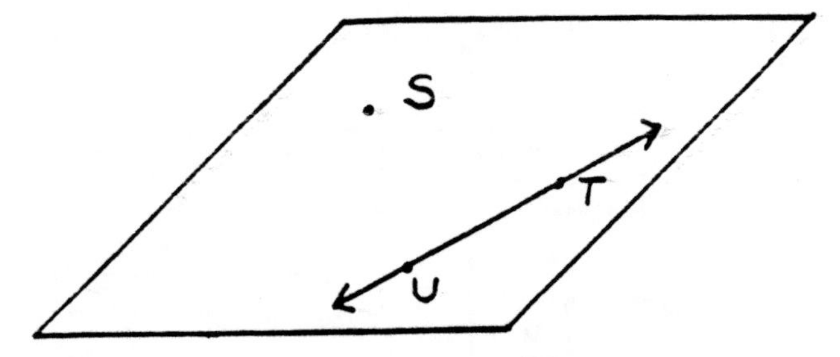

c) Two intersecting lines will determine a plane. (See Figure 8.1.)

A set of points is <u>coplanar</u> if and only if all the points in the set lie in the same plane. For example, the set of points of line l are coplanar to the plane P in Figure 8.2.

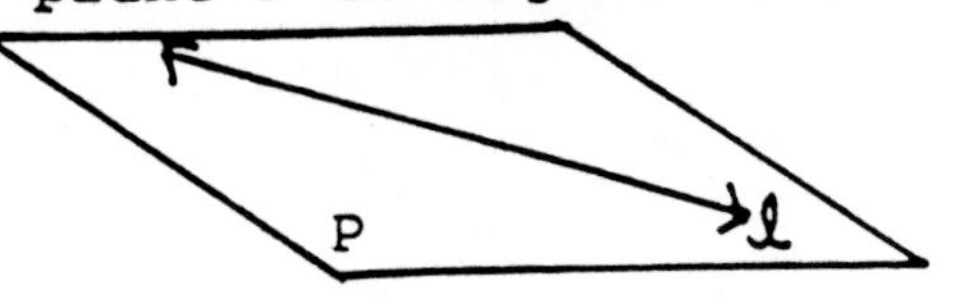

Figure 8.2

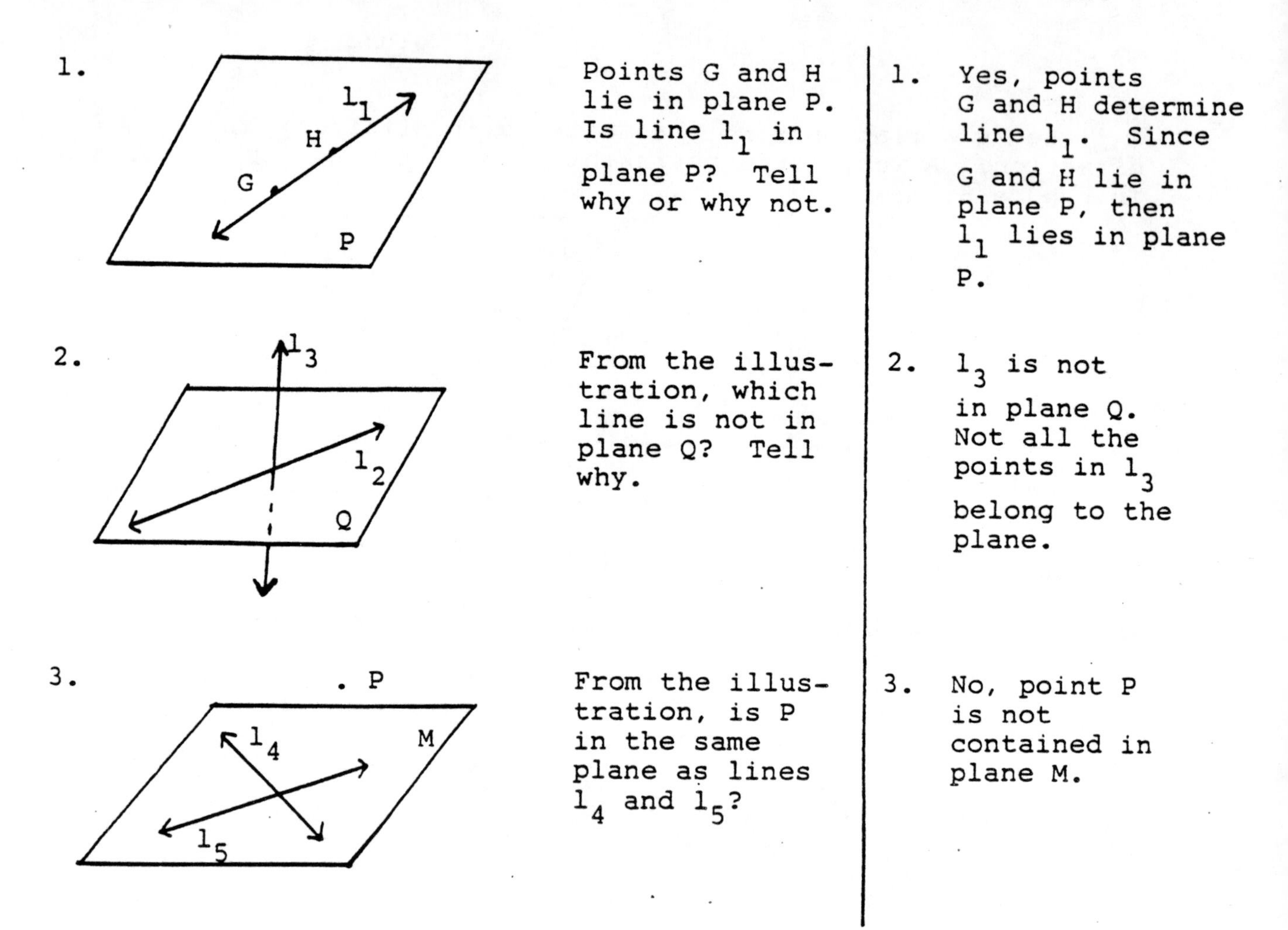

1. Points G and H lie in plane P. Is line l_1 in plane P? Tell why or why not.

   1. Yes, points G and H determine line l_1. Since G and H lie in plane P, then l_1 lies in plane P.

2. From the illustration, which line is not in plane Q? Tell why.

   2. l_3 is not in plane Q. Not all the points in l_3 belong to the plane.

3. From the illustration, is P in the same plane as lines l_4 and l_5?

   3. No, point P is not contained in plane M.

Planes, just as with lines, may intersect or may be parallel.

Definition: Two planes are <u>parallel</u> if they do not intersect.

Planes K and M are parallel to one another.

Figure 8.3

4. If two lines lie in parallel planes, are the lines necessarily themselves parallel?

4. No, the lines may be skewed. See lines l_1 and l_2 in Figure 8.3.

If a second plane intersects plane K, the intersection will form a straight line.

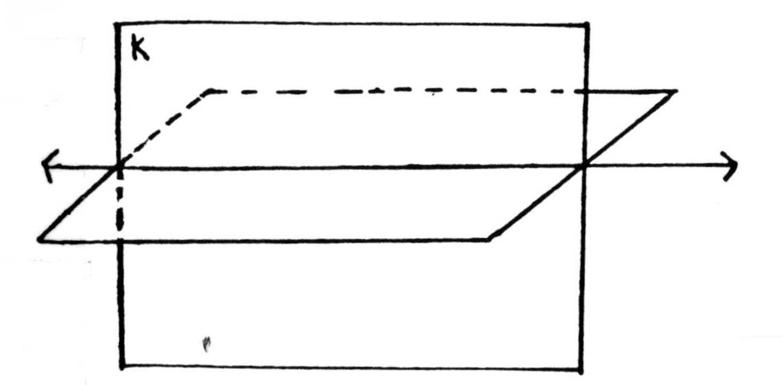

5. If a third plane intersects two other planes that are parallel, will the intersections form two parallel lines? Show why or why not.

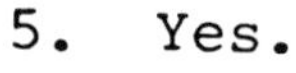

5. Yes.

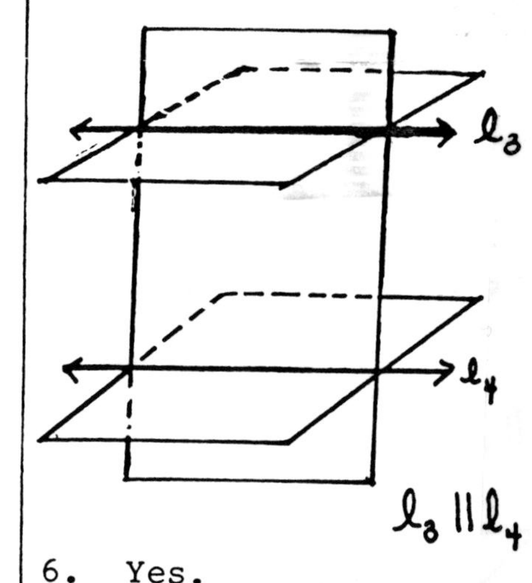

6. If plane S is perpendicular to one of two parallel planes, is plane S perpendicular to the other plane? Show why or why not.

6. Yes.

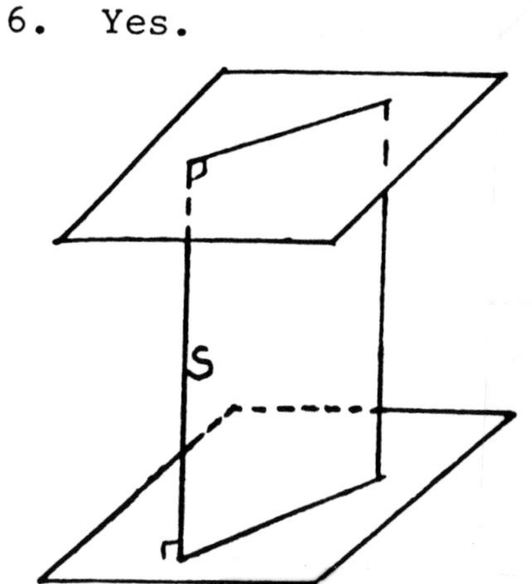

7. In exercise 6, will every line contained in plane S be perpendicular to the parallel planes? Show why or why not.

7. No, not every line.

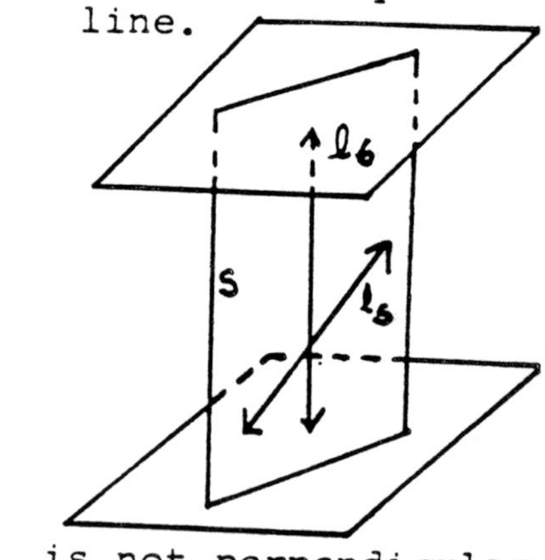

l_5 is <u>not</u> perpendicular to the parallel planes. l_6 <u>is</u> perpendicular to the parallel planes.

Two lines perpendicular to the same plane are parallel.

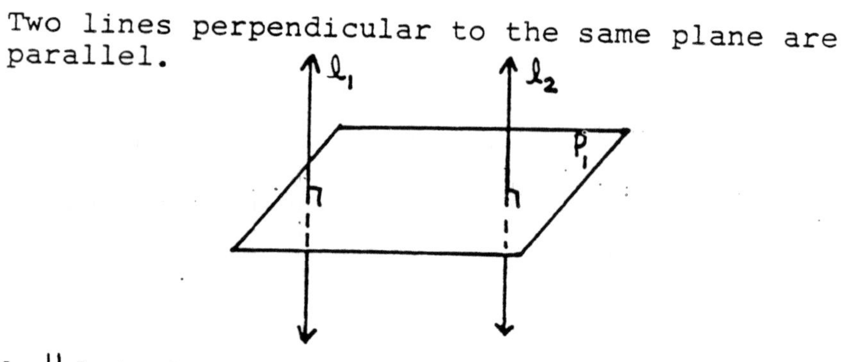

$l_1 \parallel l_2$, l_1 and l_2 are perpendicular to the same plane P_1.

8. What can you say about two planes perpendicular to the same line?

P_2 and P_3 are ____________________.

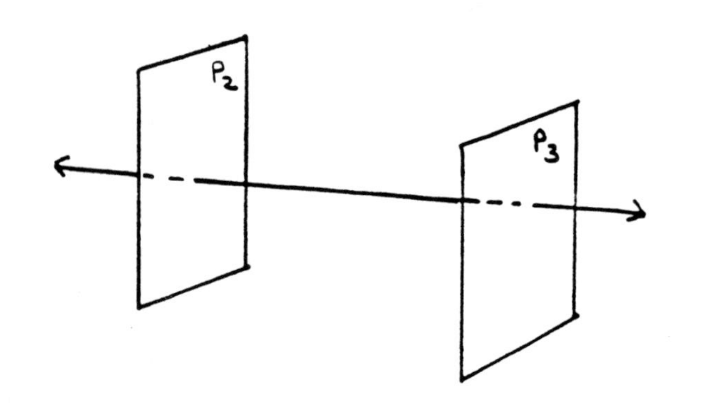

8. Parallel

Any point on the perpendicular bisector of a line segment is equidistant from the endpoints of the line segment.

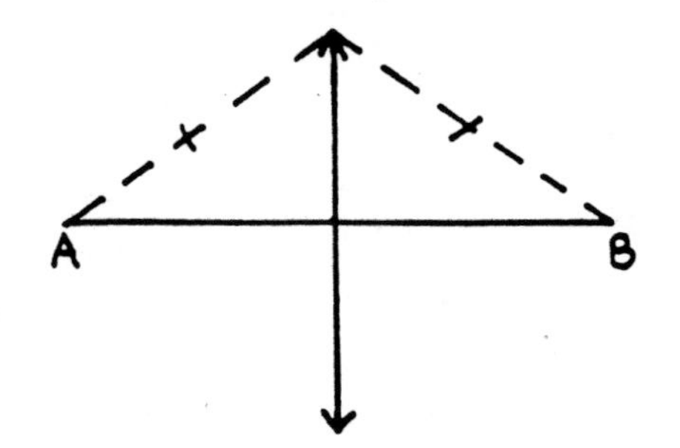

Therefore, the same principle applies to planes.

9. If a plane is the perpendicular bisector of line segment $\overline{AB}$, then every point on the plane is ________________________________.

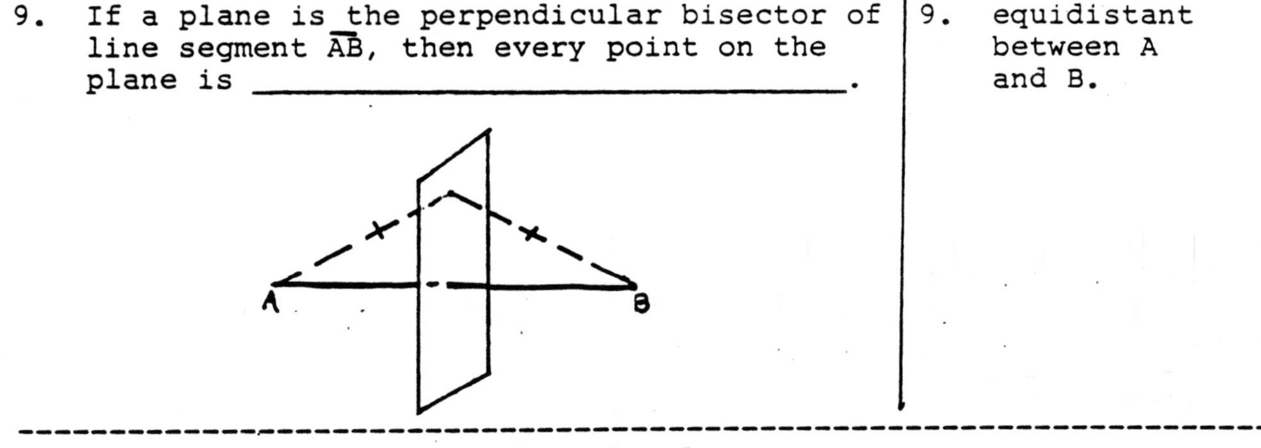

9. equidistant between A and B.

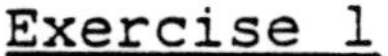

Exercise 1

1. Indicate whether each statement is true or false. Illustrate why.

   a. Two lines parallel to the same plane may be perpendicular to each other.

   b. Two intersecting planes form a straight line.

   c. If a line and a plane are parallel, every line in the plane is parallel to the given line.

2. Illustrate on a sheet of paper each of the following figures:

   a. a plane intersecting a line not in the plane.

b. two intersecting planes

c. two parallel planes

d. a plane that is the perpendicular bisector of a line segment

3. Complete the following phrases.

a. A plane is determined by three ________________points.

b. A plane is determined by a line and ________________.

4. Describe the intersection of each of the following:

a. Two intersecting planes form____________________.

b. A plane intersecting a line not in that plane forms a _______.

SOLIDS

In a plane, the set of all points at a given distance (r) from a given point is called a <u>circle</u> having the given point as its center (O).

In space, the set of all points at a given distance from a given point is called a <u>sphere</u> having the given point as center.

10. List some examples of spheres that you may have come into contact with.

10. soft ball, bowling ball, tennis ball, etc.

The set of all points a given distance from a given line is a pair of lines parallel to the given line. One line is drawn above the given line and one is drawn below the given line.

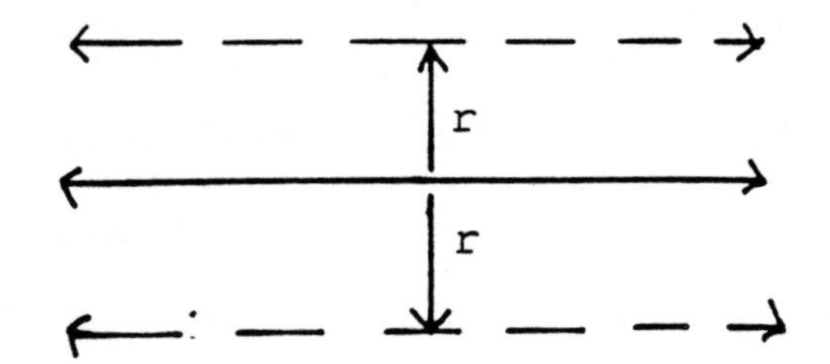

The set of all points a given distance in space from a given line is a circular cylinder having the given line as its axis.

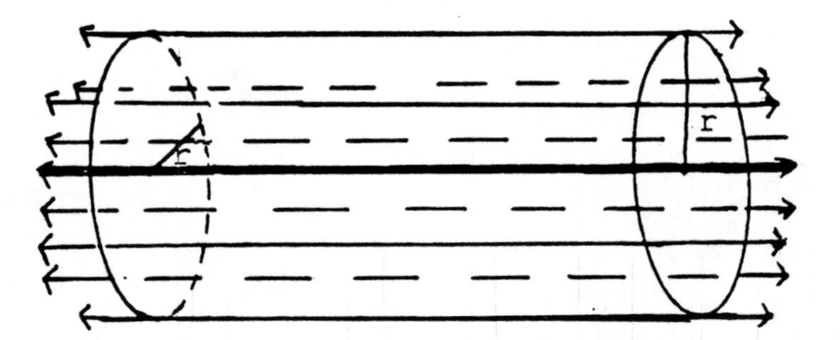

11. Some examples of circular cylinders one might find in daily living are ________.	11. soup can, smoke stack, etc.

A set of planes can be bounded together to form a rectangular solid. (See Figure 8.4.)

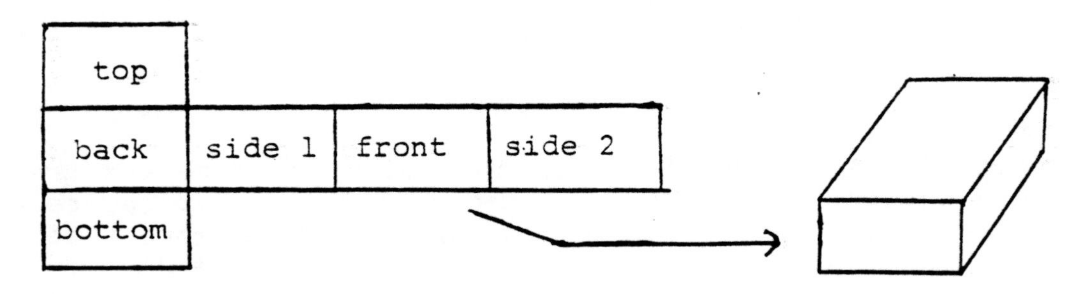

Figure 8.4

Figure 8.4 is called a solid rectangular prism.

A <u>solid</u> is a closed portion of space bounded by planes and/or curved surfaces. The following are examples of solids.

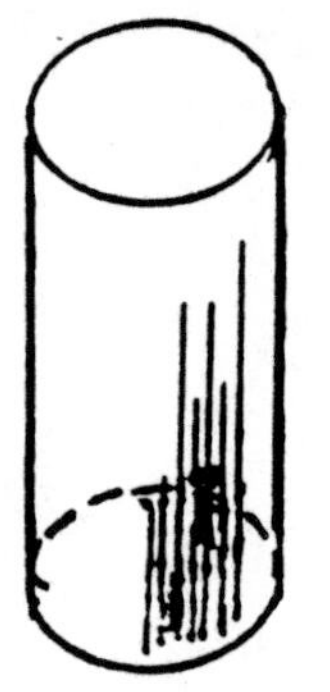

cylinder

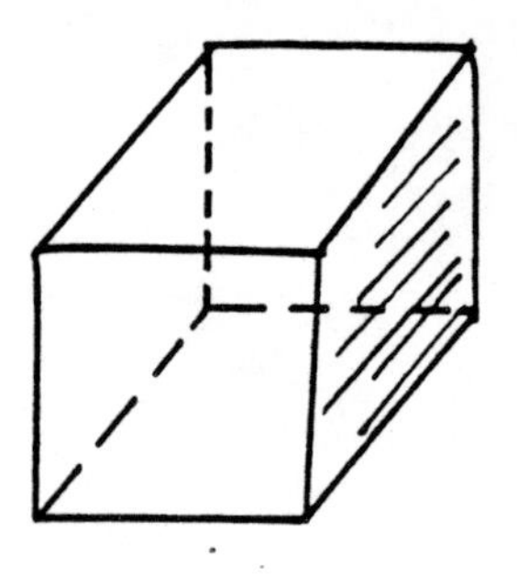

cube

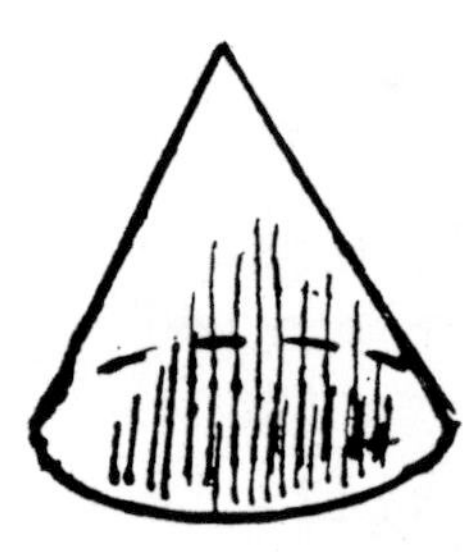

cone

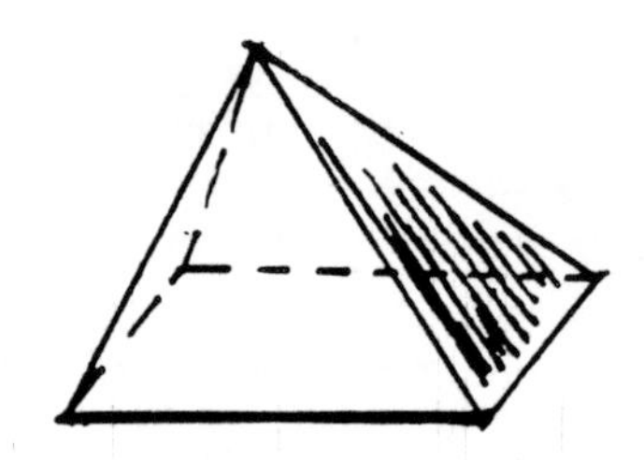

pyramid

sphere

In solid geometry, we disregard the color, surface texture and weight of the solid object. We are more concerned about the relationships of these solid objects to one another and their properties such as size and shape.

Definition: A <u>prism</u> is a solid bounded by planes on all sides.

This figure compares to a quadrilateral in a flat surface.

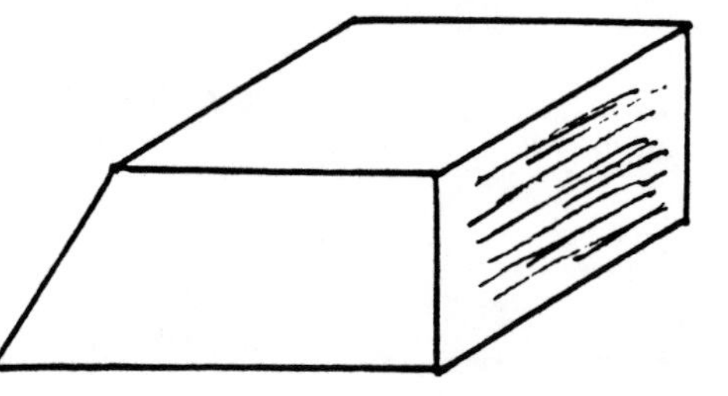

Definition: A <u>parallelopiped</u> is a prism. Each set of opposite sides of this figure are formed by parallel planes.

This figure compares to a parallelogram in a flat surface.

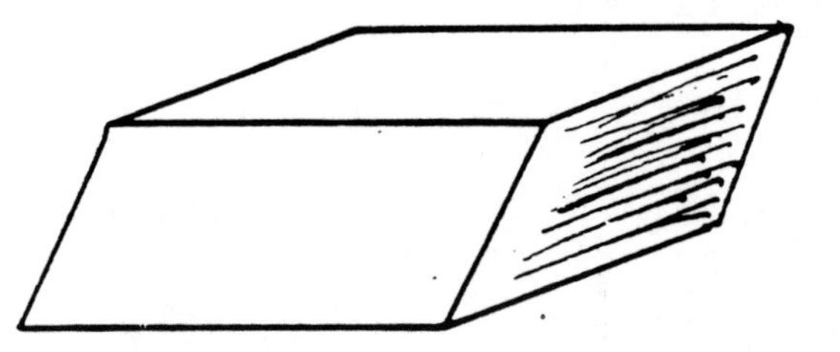

Definition: A <u>rectangular solid</u> is a parallelopiped with each set of consecutive sides formed by perpendicular planes.

This figure compares to a rectangle in a flat surface.

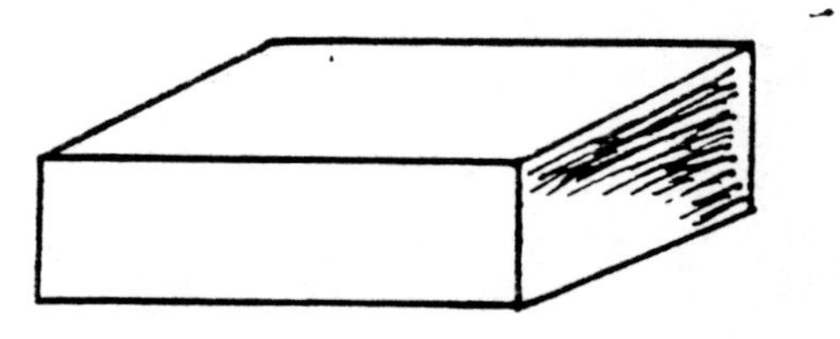

Definition: A <u>cube</u> is a rectangular solid where all of its sides are equal in measure.

This figure compares to a square in a flat surface.

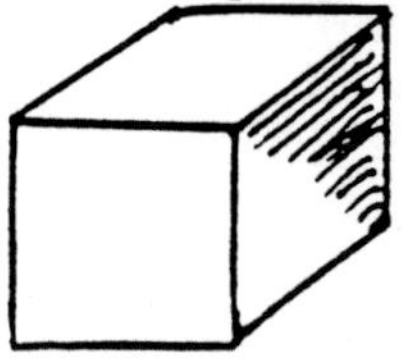

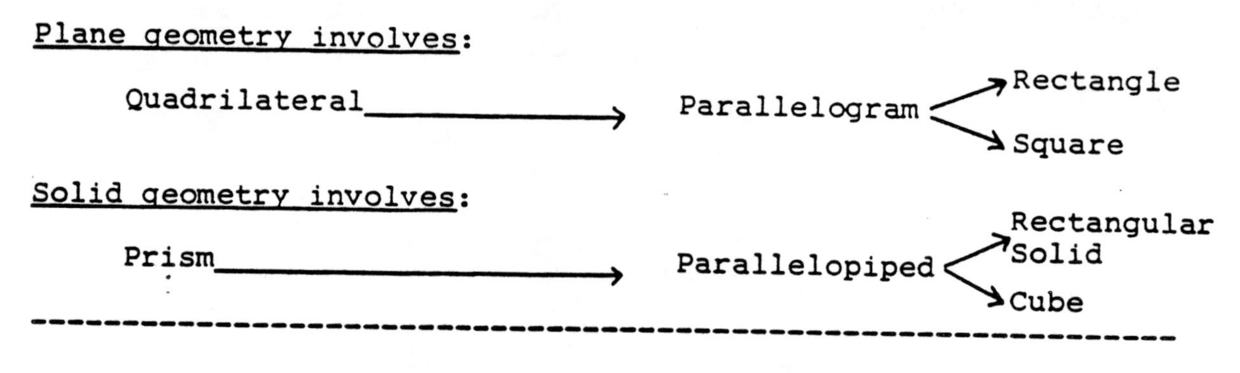

Exercise 2

1. Is a closed cardboard box a prism?
2. a. Is a cube also a rectangular solid?

   b. Describe an example of a cube.
3. Are all parallelopipeds also rectangular solids?
4. Describe the intersection of a plane and a sphere.

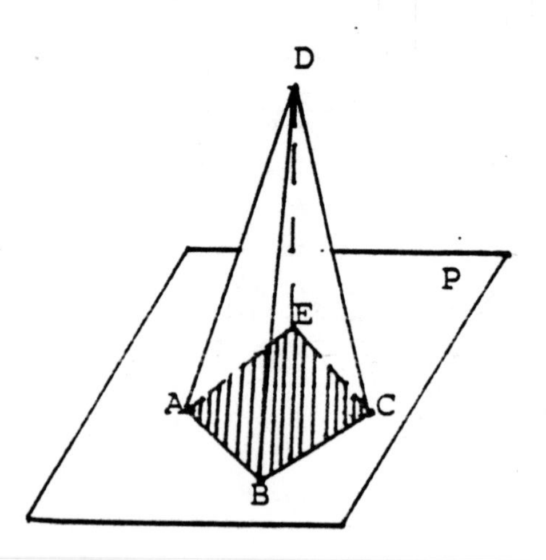

Given plane P with point D not in the plane and quadrilateral ABCE in P. The set of all line segments that join point D to a point of quadrilateral ABCE form a pyramid. Quadrilateral ABCE is called the base of the pyramid. Note: The base of a pyramid may be any closed figure with line segments as sides. Therefore, a pyramid may have a base with three, four, or more sides.

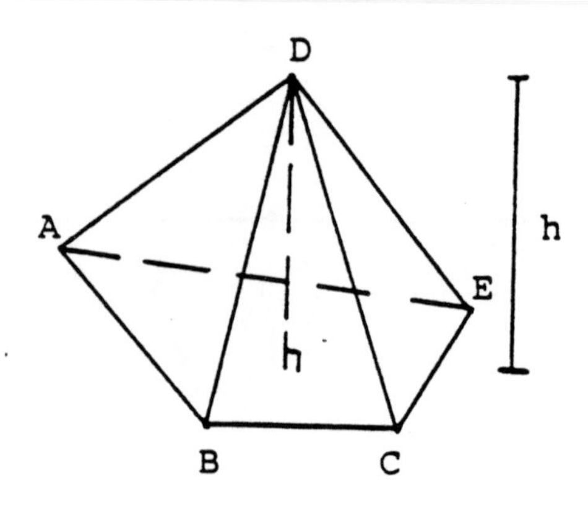

The altitude of the pyramid is the perpendicular line segment joining its vertex D to the plane of the base.

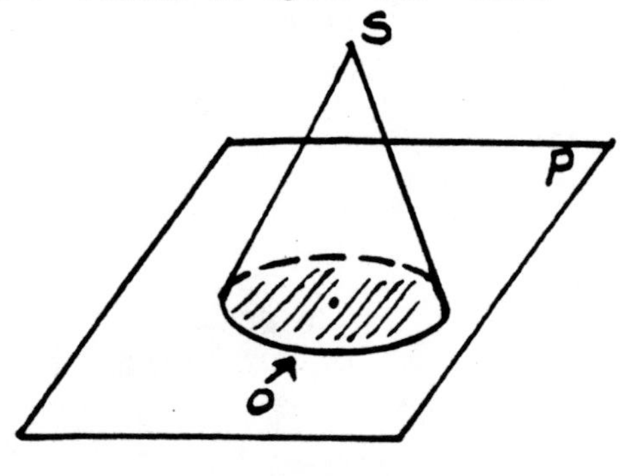

Given plane P with point S not in the plane and circle O in P. The set of all segments that join S to a point of the circle O form a <u>cone</u>. The circle O is called the <u>base</u> of the cone.

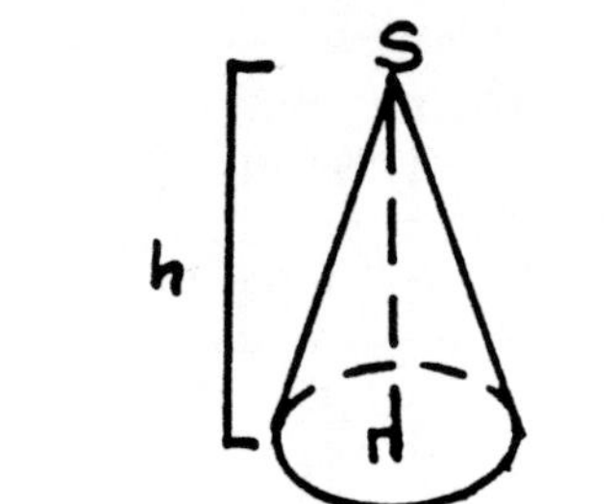

The <u>altitude</u> of the cone is the perpendicular line segment joining its vertex S to the plane of its base.

12. A peaked starchy container holding two scoops of ice cream is called a _________.	12. cone
13. The ancient tombs of the pharoahs in Egypt are called _______________.	13. pyramids
14. A balloon is called a ____________.	14. sphere
15. A 250 gallon oil drum is an example of what type of solid?_______________	15. cylinder

VOLUME OF A SOLID

The volume of a spatial figure is the number of unit cubes that can fit into the figure. We can see that there are 4 layers of 9 cubes.

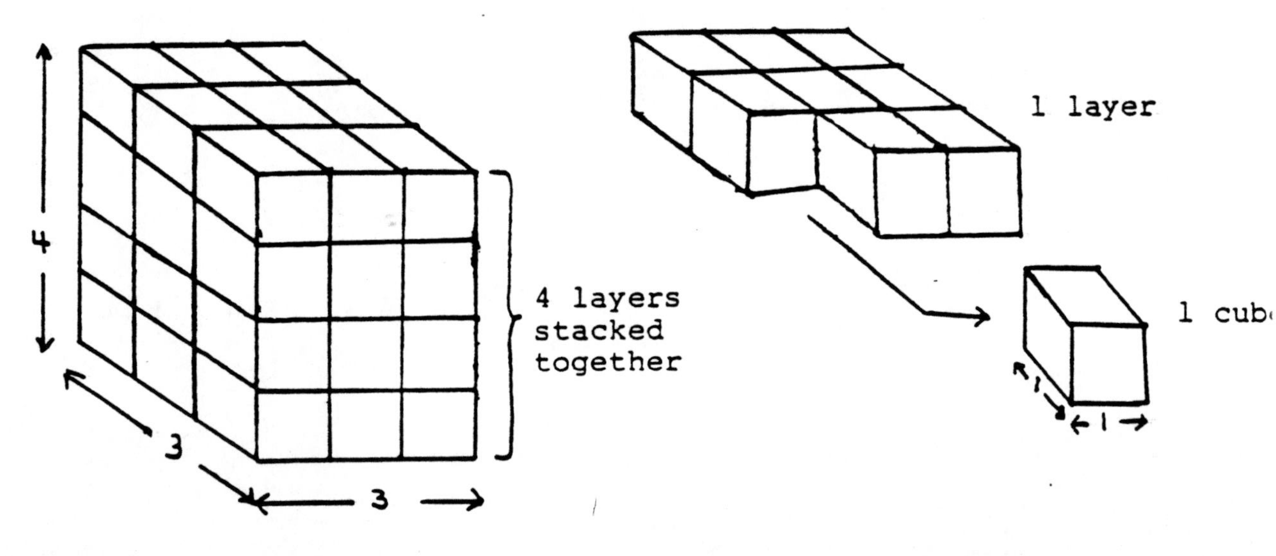

Since each cube is one inch long, one inch wide, and one inch high, the volume of each cube is 1 cubic inch. The total volume, therefore, is 36 cubic inches.

Use the following definition for total volume of a solid. Please note that any solid in this unit will be of uniform shape and size.

Definition:	The volume of a solid is the measure of capacity and represents the space inside a solid. The volume equals the product of the area of the base and the altitude of the solid.

VOLUME FORMULAE

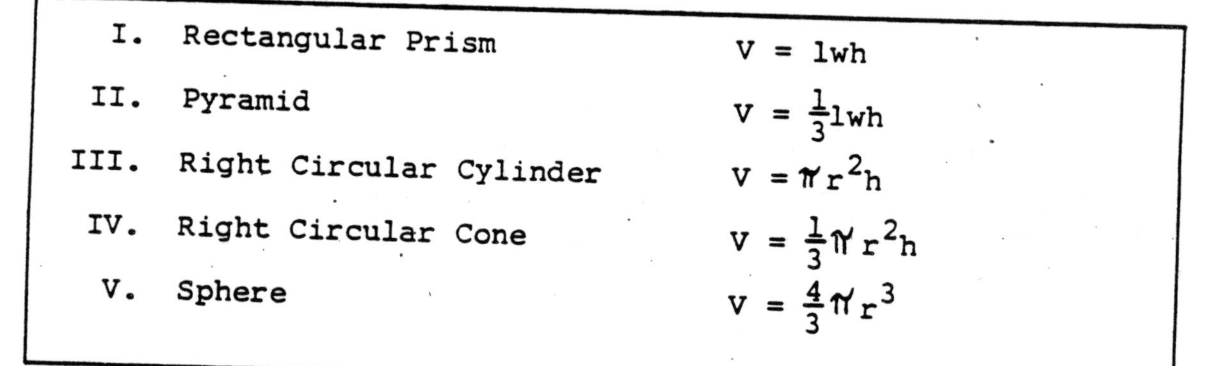

I.	Rectangular Prism	$V = lwh$
II.	Pyramid	$V = \frac{1}{3}lwh$
III.	Right Circular Cylinder	$V = \pi r^2 h$
IV.	Right Circular Cone	$V = \frac{1}{3}\pi r^2 h$
V.	Sphere	$V = \frac{4}{3}\pi r^3$

I. Volume of a rectangular prism is the length times the width times the height.

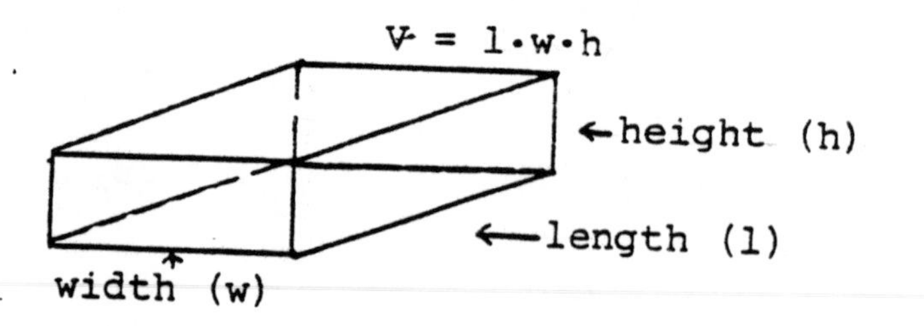

Example: Find the volume of the following rectangular prism.

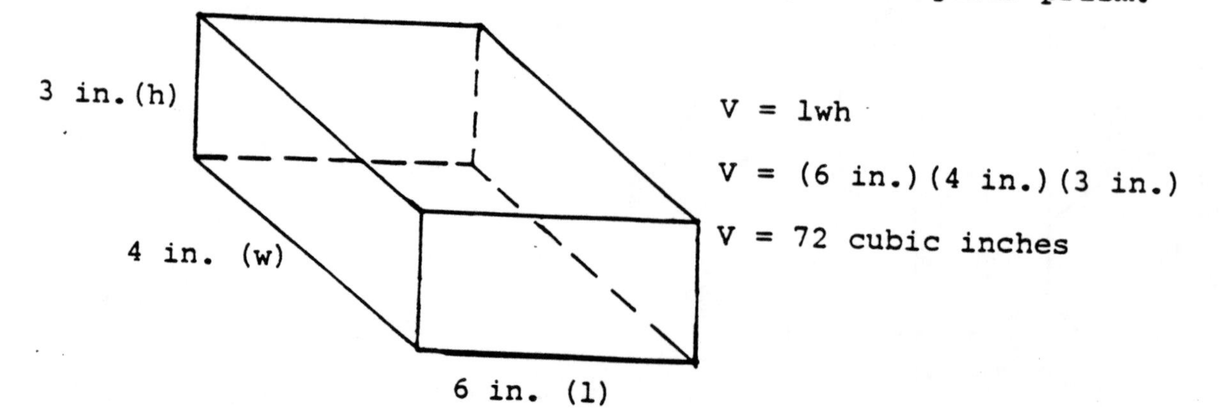

$V = lwh$

$V = (6 \text{ in.})(4 \text{ in.})(3 \text{ in.})$

$V = 72$ cubic inches

16. Find the volume of a rectangular prism with length = 3 cm., width equal 4 cm., and h = 2 cm.

$V = l \cdot w \cdot h$

V = _____

16.

V = (3cm.)(4cm.)(2cm.)

$V = 24 \text{ cm.}^3$ or

V = 24 cubic cm.

17. Find the volume of a rectangular solid with $l = \frac{1}{2}$ ft., w = 5 ft., h = 3 ft.

$V = l \cdot w \cdot h$

V = _____

17.

$V = (\frac{1}{2}\text{ft.})(5\text{ft.})(3\text{ft.})$

$V = \frac{15}{2}(\text{ft.})^3$

$V = 7\frac{1}{2}$ cu. ft.

18. A rectangular glass fish tank, 4 ft. by 2 ft. by 3 ft., is filled with water. What is the volume of the tank?

18.

V = (4ft.)(2ft.)(3ft.)

V = 24 cu. ft.

19. What is the height of a lead bar that is in the shape of a rectangular parallelopiped? Its base has dimensions 2 inches by 6 inches and its volume is $96\sqrt{3}$ cu. in.

19.

$96\sqrt{3} = (2)(6)h$

$96\sqrt{3} = 12h$

$\frac{96\sqrt{3}}{12} = h$

$8\sqrt{3}$ in. = h

Exercise 3

1. Find the volume of the following rectangular prisms:

   a. l = 5 ft., w = 4 ft., h = 7 ft.

   b. $l = 2\sqrt{2}$ m., $w = 5\sqrt{2}$ m., h = 3m.

   c. $l = 1\frac{1}{2}$ cm., w = 3 cm., $h = 4\frac{1}{4}$ cm.

2. Find the length of a rectangular prism if the volume is 56 cu. dm., the width is 7 dm., and the height is 2 dm.

3. Find the width of a rectangular prism if the volume is $16\frac{2}{3}$ cu. ft., the length is $1\frac{2}{3}$ ft., and the height is 5 ft..

4. Find the length of a rectangular prism if the volume is $48\sqrt{6}$ cu. in., the width is $2\sqrt{2}$ in., and the height is $4\sqrt{3}$ in.

5. The Jones family put a swimming pool in their backyard with length 25 ft., width 15 ft. and depth 5 ft. Find the volume of the pool. (The pool is in the shape of a rectangular prism.)

II. The volume of a <u>right pyramid</u> is $\frac{1}{3}(l \cdot w \cdot h)$.

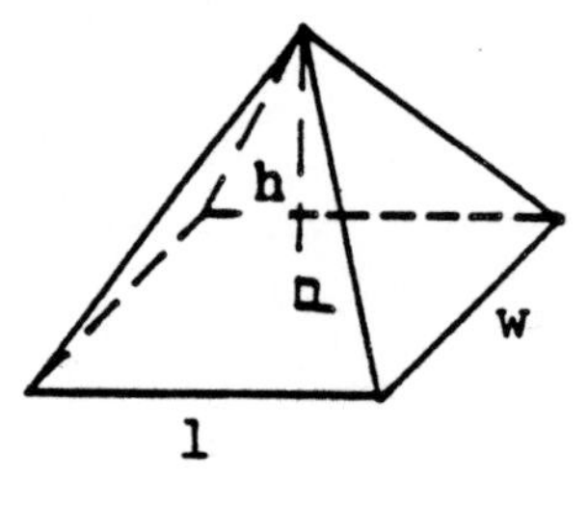

Example: Find the volume of the following right pyramid:

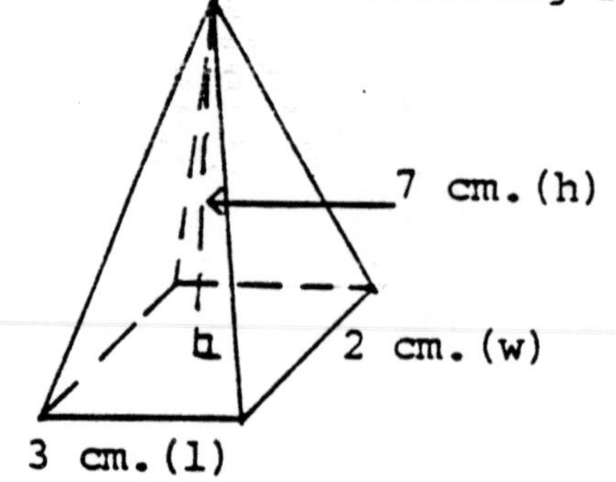

$V = \frac{1}{3}(l \cdot w \cdot h)$

$V = \frac{1}{3}(3 \text{ cm.})(2 \text{ cm.})(7 \text{ cm.})$

$V = 14$ cu. cm.

20.

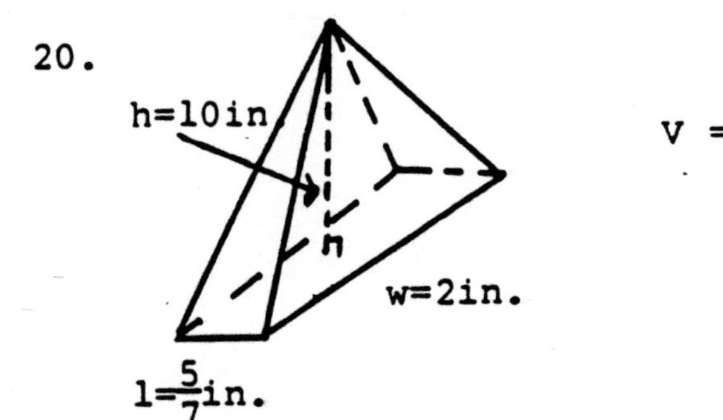

V = ______

21. Find the volume of a right pyramid if l = 15 m., w = 4 m., and h = 20 m.

V = ______

22. Find the height of a right pyramid if the volume is 24 cu. ft., l = 4 ft., and w = 6 ft.

h = ______

23. Find the volume of a pyramid that has a square base with perimeter 16 cm. and its altitude is 7 cm.

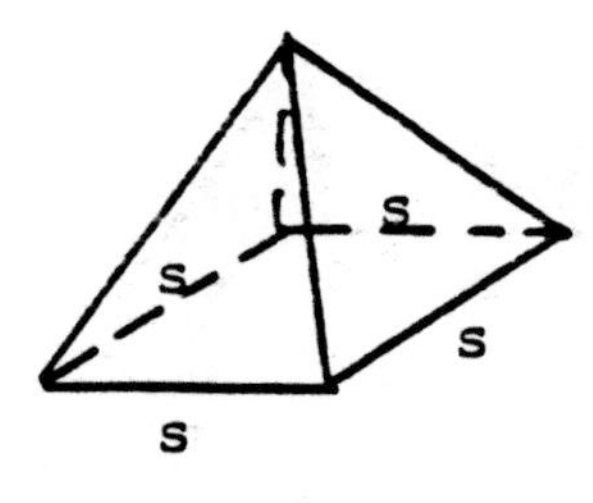

Perimeter = 4s
16 = 4s
____ = s

Volume $= \frac{1}{3}$ (l·w·h)

20.

$V = \frac{1}{3}(10in.)(2in.)(\frac{5}{7}in.$

$V = \frac{100}{21}$ in.^{3}

$V = 4\frac{16}{21}$ cu.in.

21.

$V = \frac{1}{3}(15m.)(4m.)(20m.)$

$V = 400$ m.^{3} or

400 cubic meters

22.

$V = \frac{1}{3}(l \cdot w \cdot h)$

$24ft.^3 = \frac{1}{3}(4ft.)(6ft.)h$

$24ft.^3 = 8ft.^2 h$

3 ft. = h

23.

4 = s

$V = \frac{1}{3}(4cm.)(4cm.)(7cm$

$V = \frac{1}{3}(112)$ cm.^{3}

$V = 37\frac{1}{3}$ cm.^{3}

Exercise 4

1. Find the volume of the following pyramids:

   a. $l = 2\sqrt{3}$ in., $w = \sqrt{3}$ in., $h = 4$ in.

   b. $l = 6\sqrt{7}$ m., $w = 2\sqrt{7}$ m., $h = \sqrt{7}$ m.

2. In a pyramid, if:

   a. $V = \frac{5}{6}$ cu.in., $l = \frac{1}{2}$ in., $w = \frac{3}{4}$ in., find h.

   b. $V = 2\frac{5}{6}$ cu.cm., $w = 1\frac{1}{8}$ cm., $h = 2\frac{1}{3}$ cm., find l.

3. Find the height of a pyramid that has a rectangular base with $l = 6$ in. and $w = 7$ in. The volume of the pyramid is 180 cubic inches.

III. The volume of a <u>right circular cylinder</u> is $\pi r^2 h$.

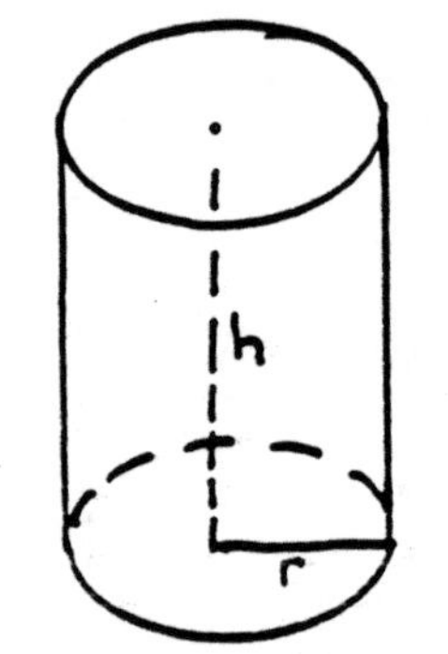

Example: Find the volume of the following right circular cylinder.

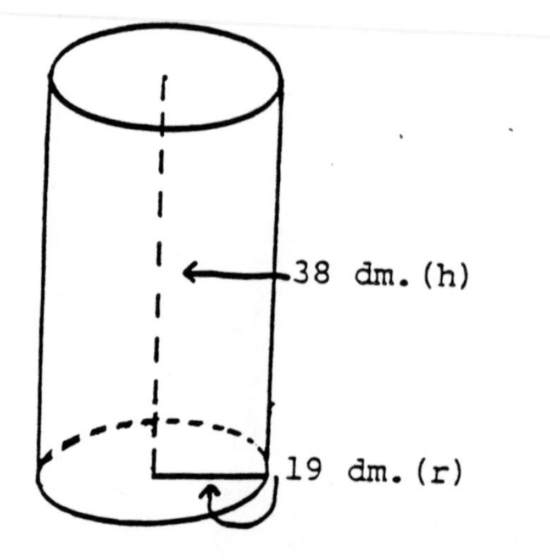

$V = \pi r^2 h$

$V = \pi(19 \text{ dm.})^2(38\text{dm.})$

$V = \pi(361\text{sq.dm.})(38\text{dm.})$

$V = 13{,}718\,\pi \text{ cu.dm.}$

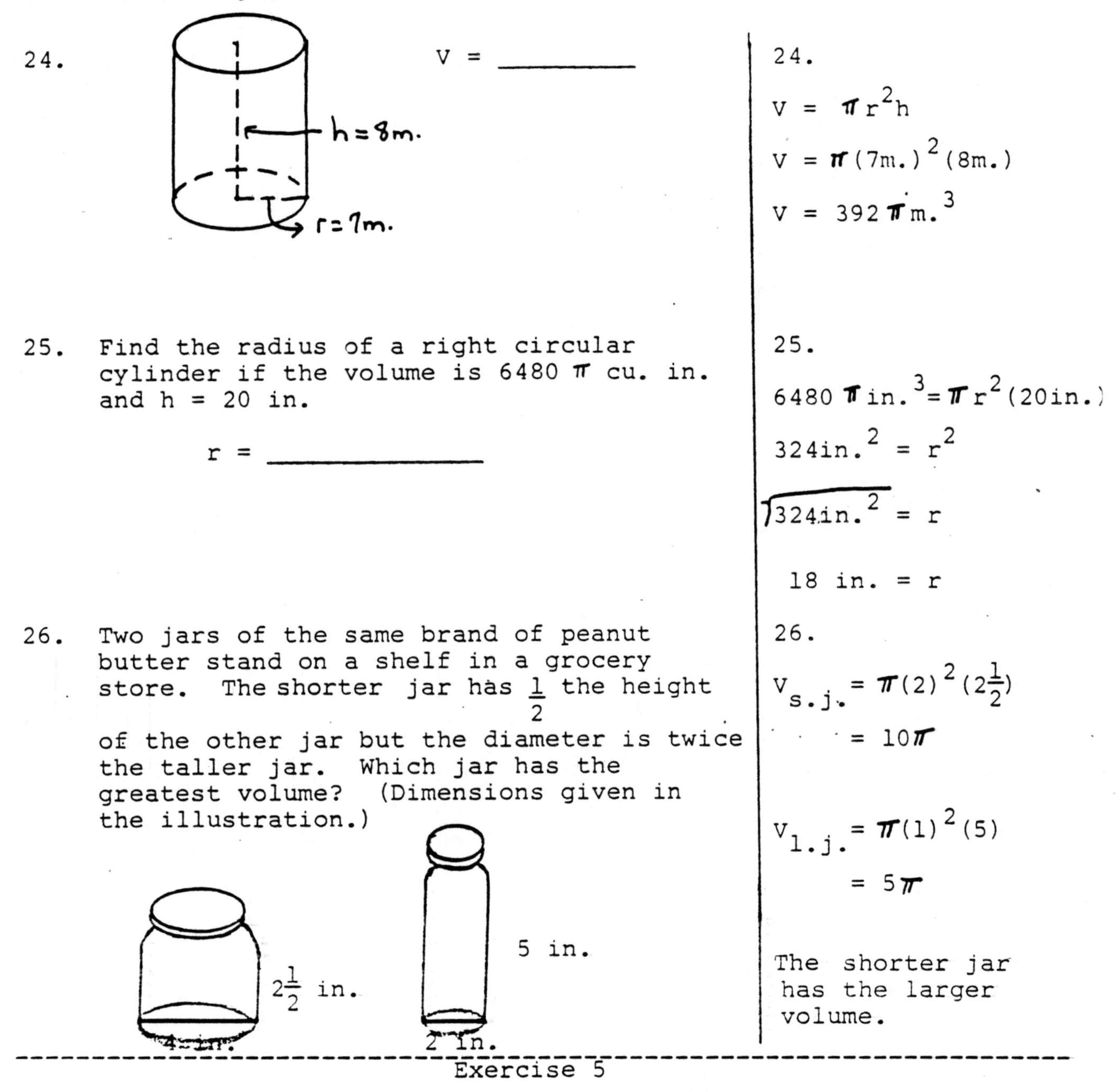

24. $V = $ ________

24.

$V = \pi r^2 h$

$V = \pi (7m.)^2 (8m.)$

$V = 392\pi m.^3$

25. Find the radius of a right circular cylinder if the volume is 6480π cu. in. and h = 20 in.

r = ________

25.

$6480\pi in.^3 = \pi r^2 (20in.)$

$324 in.^2 = r^2$

$\sqrt{324 in.^2} = r$

$18 in. = r$

26. Two jars of the same brand of peanut butter stand on a shelf in a grocery store. The shorter jar has $\frac{1}{2}$ the height of the other jar but the diameter is twice the taller jar. Which jar has the greatest volume? (Dimensions given in the illustration.)

26.

$V_{s.j.} = \pi(2)^2(2\frac{1}{2})$

$= 10\pi$

$V_{l.j.} = \pi(1)^2(5)$

$= 5\pi$

The shorter jar has the larger volume.

Exercise 5

1. Find the volume of the following right circular cylinders:

   a. r = 5 in., h = 12 in.

   b. r = 6 yd., h = 10 yd.

2. In a cylinder, if:

   a. $V = 1000\pi$ cu. in., r = 10 in., find h.

   b. $V = 448\pi$ cu. cm., h = 7 cm., find r.

c. $V = 720\pi$ cu.yd., h = 5 yd., find r.

d. $V = 96\pi$ cu.ft., h = 6 ft., find r.

e. $V = 6\sqrt{2}\,\pi$ cu. m., $h = 2\sqrt{2}$ m., find r.

3. A cylindrical storage tank has a height of 25 ft. and a volume of 3600π cu. ft. Find the radius of the base.

IV. The volume of a <u>right</u> <u>circular</u> <u>cone</u> is $\frac{1}{3}\pi r^2 h$.

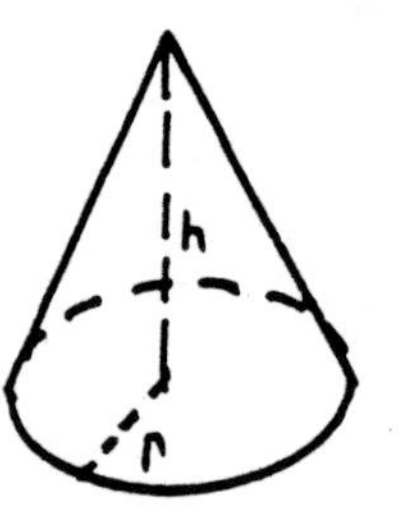

Example: Find the volume of the following right circular cone:

7 in. (r)

15 in. (h)

$V = \frac{1}{3}\pi r^2 h$

$V = \frac{1}{3}\pi (7 \text{ in.})^2 (15 \text{ in.})$

$V = \frac{1}{3}\pi (49 \text{ sq.in.})(15 \text{ in.})$

$V = 245\pi$ cu.in.

27.

a) $r =$ ________

b) $V =$ ________

h = 6 in.

d = 10 in.

28. A conical tank is $10\frac{1}{2}$ ft. deep and its circular top has a radius of 3 ft. How many cubic feet of water will it hold?

$$V = \frac{1}{3}\pi r^2 h$$
$$V = \frac{1}{3}\pi (3)^2 (10\tfrac{1}{2})$$
$$V = \_\_\_\_\_\_\_\_$$

29. Find the height if the volume of a right circular cone is 500π cu. in. and $r = 10$ in.

$h =$ ________

27. a) $d = 2r$
$10 = 2r$
$5 = r$

b) $V = \frac{1}{3}\pi(5 \text{ in.})^2(6 \text{ in.})$
$V = \frac{1}{3}\pi(25 \text{ in.}^2)(6 \text{ in.}$
$V = 50\pi(\text{in.})^3$

or

$V = 50\pi$ cu. in.

28. $31\frac{1}{2}\pi$ cu.ft.

29.
$V = \frac{1}{3}\pi r^2 h$
$500\pi \text{ in.}^3 = \frac{1}{3}\pi(10)^2 h$
$500\pi \text{ in.}^3 = \frac{1}{3}\pi 100 \text{in.}^2 h$
15 in.= h

Exercise 6

1. Find the volume of the following cones:

   a. $r = 6$ m., $h = 9$ m.

   b. $r = 7$ in., $h = 5$ in.

2. In a cone if:

   a. $V = 75\pi$ cu. in., $h = 9$ in., find r.

b. $V = 96\pi$ cu. ft., $h = 8$ ft., find r.

c. $V = 196\pi$ cu. yd., $r = 7$ yd., find h.

d. $V = \frac{200}{3}\pi$ cu. m., $h = 8$ m., find r.

e. $V = 15\sqrt{2}\pi$ cu. cm., $h = 5\sqrt{2}$ cm., find r.

3. A conical ant hill is 4 inches high with a base of 6π in. in circumference. Find the total volume of dirt that the hill contains.

V. The volume of a <u>sphere</u> is $\frac{4}{3}\pi r^3$.

Example: Find the volume of the following sphere:

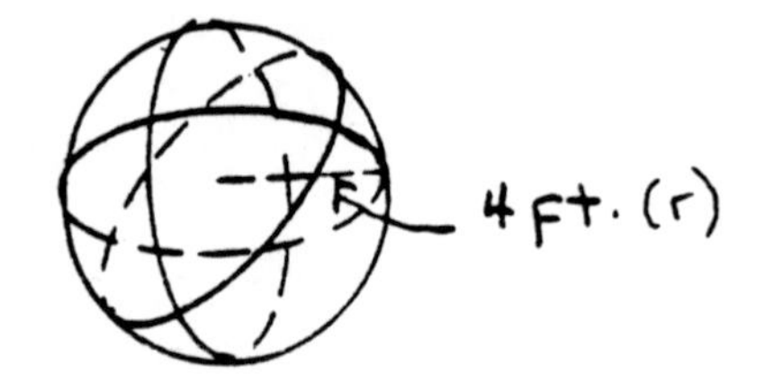

$V = \frac{4}{3}\pi r^3$

$V = \frac{4}{3}\pi(4 \text{ ft.})^3$

$V = \frac{4}{3}\pi(64 \text{ cu. ft.})$

$V = 85\frac{1}{3}\pi$ cu. ft.

30. In a sphere if $r = 5$ m., find V.

30. $V = \frac{4}{3}\pi r^3$

$V = \frac{4}{3}\pi(5 \text{ m.})^3$

$V = 166\frac{2}{3}\pi$ cu. m.

31. In a sphere if d = 3 cm., find V.

31. $2r = d$

$2r = 3$

$r = \frac{3}{2}$

$V = \frac{4}{3}\pi r^3$

$V = \frac{4}{3}\pi\left(\frac{3}{2}\right)^3$

$V = 4\frac{1}{2}\pi$ cu. cm.

32. In a sphere if $V = 1333\frac{1}{3}\pi$ cu ft., find the radius.

32.

$1333\frac{1}{3}\pi = \frac{4}{3}\pi r^3$

$\frac{4000}{3}\pi = \frac{4}{3}\pi r^3$

10 ft. $= r$

33. A spherical storage tank has radius 6 ft.. What is its volume in cubic feet?

33. 288π cu. ft.

Unit 8 Review

1. Find the volumes of each of the following:

   a. right circular cylinder, d = 14 in., h = 5 in.

   b. right circular cone, r = 3 cm., h = 9 cm.

   c. sphere, r = 9 dm.

2. Find the missing values:

   a. right pyramid has volume 54 cu. yd.; its square base has a side measuring 4 yds.; find h.

   b. right circular cylinder has volume of $6\sqrt{2}\pi$ cu.m., and $h = 3\sqrt{2}$ m., find r.

   c. rectangular prism has volume of 728 cu. ft., l = 8 ft., and width = $6\frac{1}{2}$ ft., find h.

3. Define the following terms:

   a. coplanar points

   b. skew lines (Unit 4)

   c. parallel planes

   d. perpendicular planes

   e. plane

4. Illustrate the following:

   a. plane K perpendicular to line m and plane P perpendicular to line m.

   b. two parallel planes

   c. a plane that is the perpendicular bisector of a line segment.

5. Complete the following. (Illustrate each statement.)

   a. Intersecting Plane S and Plane R form ________________.

   b. Intersecting Plane S and a sphere form __________________.

   c. Plane S intersecting line l not in plane S forms a ________.

6. If a cylindrical pipe has the dimensions given in the diagram, find the radius and the volume of the material that is stored in the inner core.

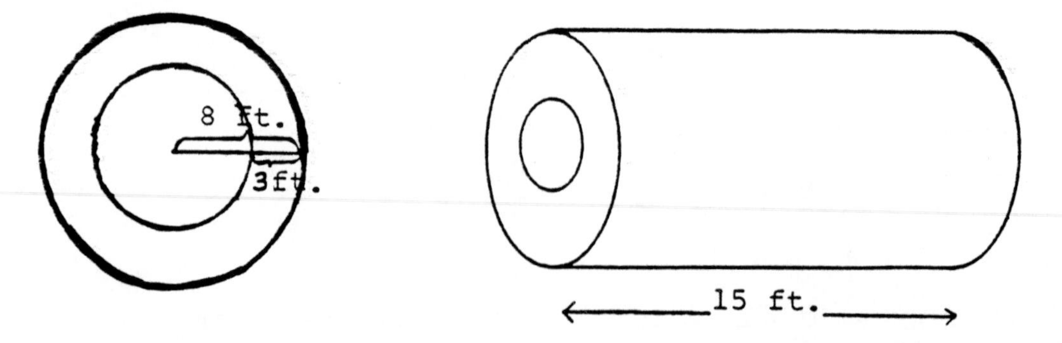

7. An ice cream cone has inside dimensions 5 inches deep and 3 inches across the top. What volume of solid ice cream will the cone hold without overflowing?

8. Find the total volume of the solids shown below:

a. The top part of the figure is half of a cylinder.

3 ft.

21 ft.

10 ft.

b.

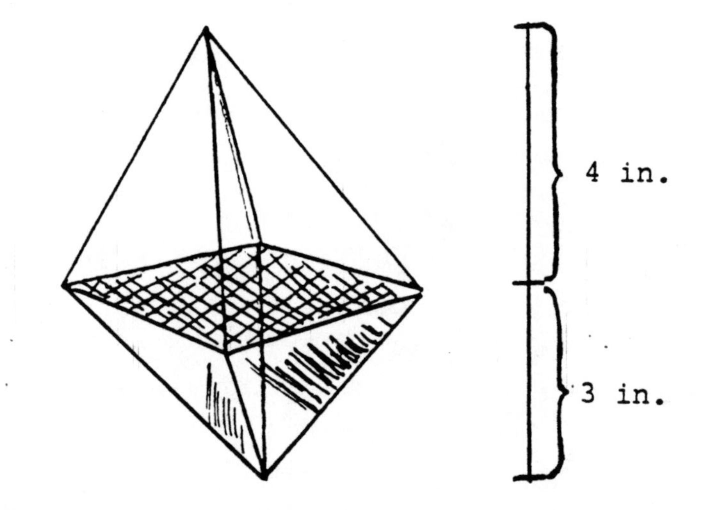

The rectangular base (shaded portion) has l = 7 in. and w = 6 in. This base is common to both pyramids. The heights of each pyramid are shown in the drawings. Find the total volume of the figure.

9. A metal ball has radius of 2 in. and a right cylindrical drinking glass is 6 in. tall with a radius of $2\frac{1}{2}$ in.. Find the volume of each item. If the metal ball is dropped into the tall glass and filled with water, determine what volume of water is needed.

10. A homeowner wishes to pour enough concrete to form four steps leading to his front door. If the dimensions are given below, what volume of concrete would he need?

(Each step has the same measure.)

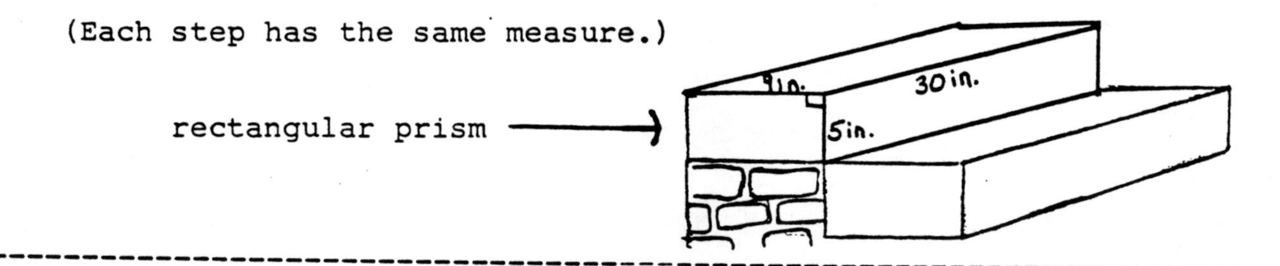

Unit 9

PYTHAGOREAN THEOREM WITH APPLICATIONS

Learning Objectives:

The student will demonstrate his mastery of the following theorems by writing them and by applying them to solutions of selected problems:

Theorem 1: In a right triangle with sides a, b, and c, the hypotenuse squared is equal to the sum of the squares of the sides. ($a^2 + b^2 = c^2$)

Converse: In a triangle with sides a, b, and c, if $a^2 + b^2 = c^2$, then the triangle is a right triangle.

Theorem 2: In a $30^\circ - 60^\circ$ right triangle, the length of the side opposite the 30° angle is x. The length of the hypotenuse is twice x and the length of the side opposite the 60° angle is x multiplied by $\sqrt{3}$.

Theorem 3: In an isosceles 45° right triangle, the hypotenuse is $\sqrt{2}$ times the length of one side.

Theorem 4: An altitude of an equilateral triangle is $\frac{1}{2}\sqrt{3}$ (s) times the length of one side.

PYTHAGOREAN THEOREM WITH APPLICATIONS

One of the best known theorems in mathematics is probably this one, the Pythagorean Theorem. This theorem states that if unit squares are drawn on each of the three sides of a right triangle, the number of squares formed on the hypotenuse will equal in quantity to the sum of the square units formed on the other two sides.

The right triangle below has sides 3, 4, and 5 units. Squares A, B, and C are formed at each side. Notice that the sum of the squares A and B equal in number to square C.

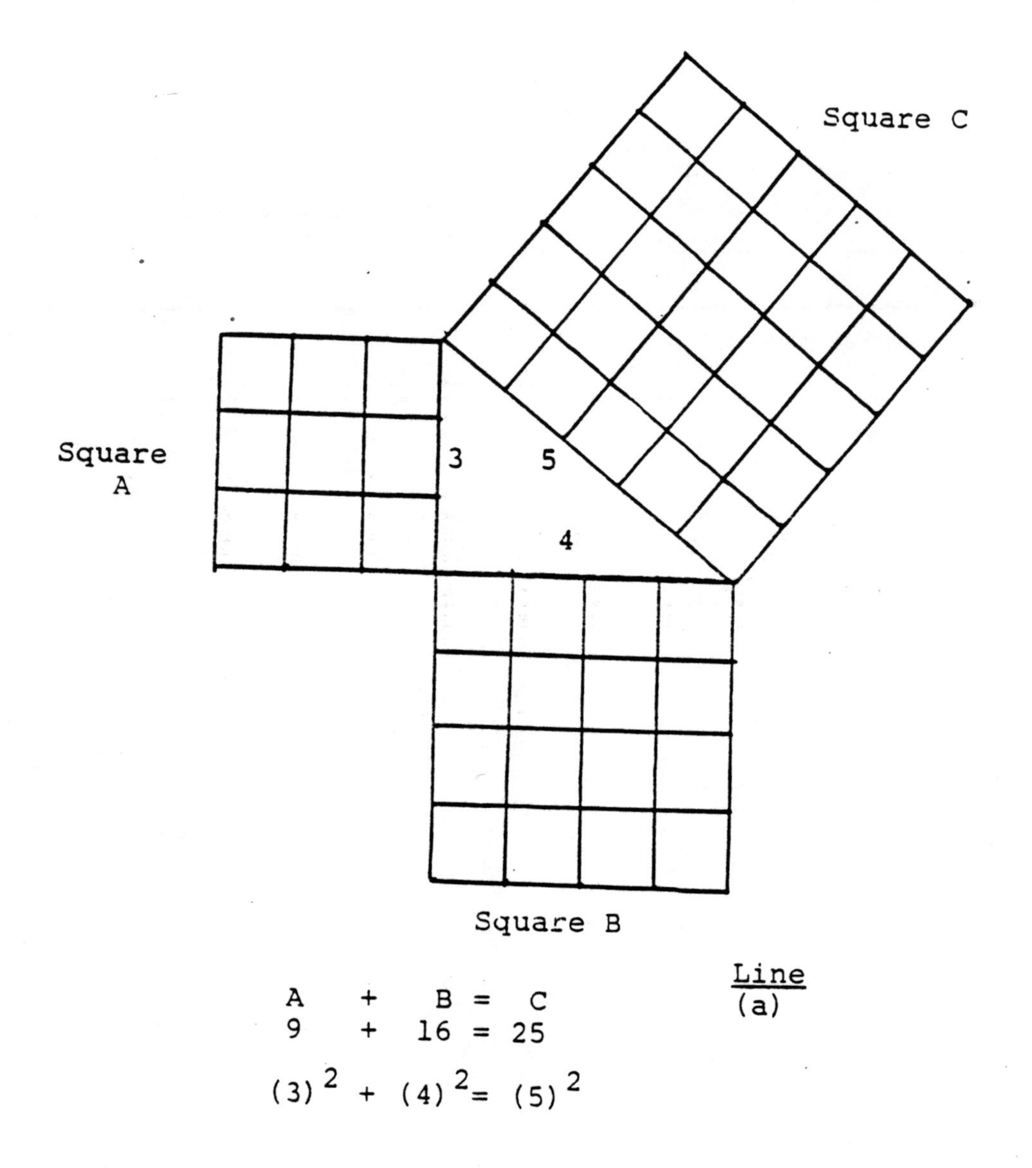

1. Line (a) states that the sum of squares A plus B equals square ____________.

2. Square A has __________blocks.
Square B has __________blocks.
Therefore, the sum of squares A and B is ____________ which has __________blocks.

1. C

2. A = 9
B = 16
C = 25

9 + 16 = 25 can be written as a sum of two squares expressed as:

$$(3)^2 + (4)^2 = (5)^2$$

3, 4, and 5 are the sides of the right triangle.

Theorem 1 : In a right triangle with sides a, b, and c, the hypotenuse squared is equal to the sum of the squares of the sides.

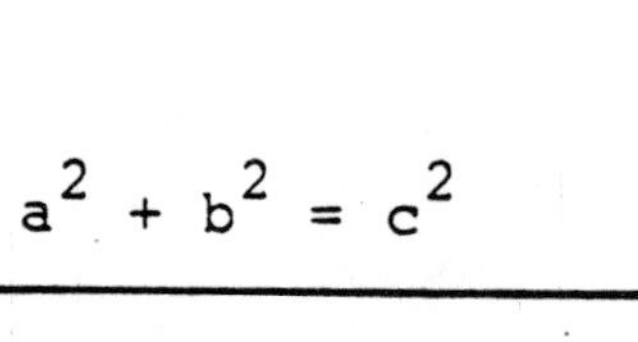

$$a^2 + b^2 = c^2$$

Converse: In a triangle with sides a, b, and c, if $a^2 + b^2 = c^2$, then the triangle is a right triangle.

Example: Show that 2, $1\frac{1}{2}$, $2\frac{1}{2}$ are sides of a figure that form a right triangle.

Solution: The hypotenuse is the longest side-$2\frac{1}{2}$.
Now substitute the values into the formula

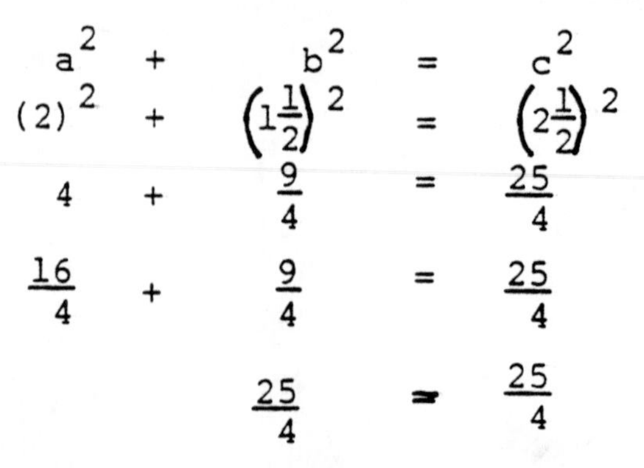

$$a^2 + b^2 = c^2$$
$$(2)^2 + \left(1\frac{1}{2}\right)^2 = \left(2\frac{1}{2}\right)^2$$
$$4 + \frac{9}{4} = \frac{25}{4}$$
$$\frac{16}{4} + \frac{9}{4} = \frac{25}{4}$$
$$\frac{25}{4} = \frac{25}{4}$$

The equation is a true statement; therefore, the figure is a right triangle.

Example: A ladder that is 10 ft. long is placed against a building. The foot of the ladder is 5 feet from the base of the building. How far above the ground does the ladder touch the building?

Solution:

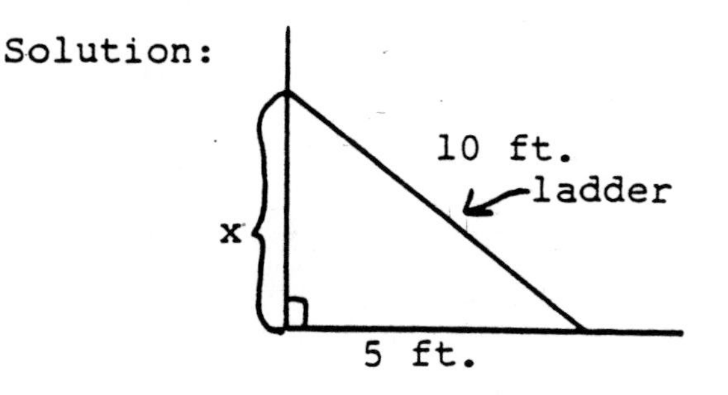

Theoretically, the building is perpendicular to the ground and the ground is a flat level surface. Therefore, the figure formed with the leaning ladder is a right triangle. The height (x) is found by solving:

$$x^2 + (5)^2 = (10)^2$$

$$x^2 + 25 = 100$$

$$x^2 = 75$$

$$x = \sqrt{75}$$

$$x = 5\sqrt{3} \text{ ft.}$$

The ladder will touch the building $5\sqrt{3}$ ft. off the ground.

3. Is triangle with sides 6, 8, 10 a right triangle?
First, choose c = 10 since the longest side is the hypotenuse.
Now, does $(6)^2 + (8)^2 = (10)^2$? ________

4. Is a triangle with sides 4, 4, and $4\sqrt{2}$ a right triangle?

c = ____________________

Does $(4)^2 + (4)^2 = (4\sqrt{2})^2$? ________

3. Yes
$36 + 64 = 100$

4.
$c = 4\sqrt{2}$
yes, $16 + 16 = 32$

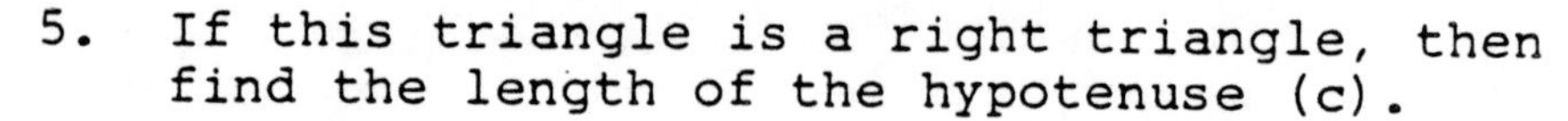

5. If this triangle is a right triangle, then find the length of the hypotenuse (c).

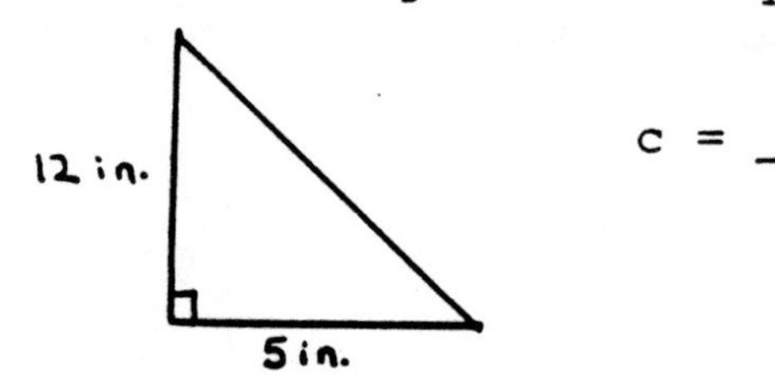

c = ____________

5. $(5)^2 + (12)^2 = c^2$

$25 + 144 = c^2$

$169 = c^2$

$\sqrt{169} = c$

$13 \text{ in.} = c$

6. Show that the triangle below is or is not a right triangle. ____________

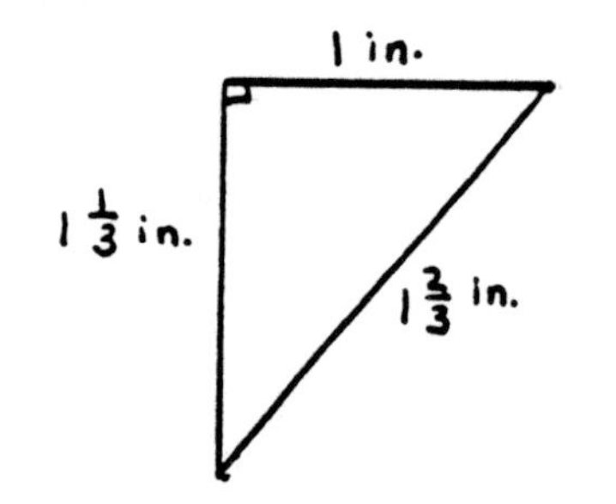

6. It is a right triangle.

$(1)^2 + \left(1\frac{1}{3}\right)^2 = \left(1\frac{2}{3}\right)$

$(1)^2 + \left(\frac{4}{3}\right)^2 = \left(\frac{5}{3}\right)^2$

$1 + \frac{16}{9} = \frac{25}{9}$

$\frac{9}{9} + \frac{16}{9} = \frac{25}{9}$

$\frac{25}{9} = \frac{25}{9}$

7. A ladder that is 13 feet long is placed against a building. The foot of the ladder is 4 feet from the base of the building. How far above the ground does the ladder touch the building?

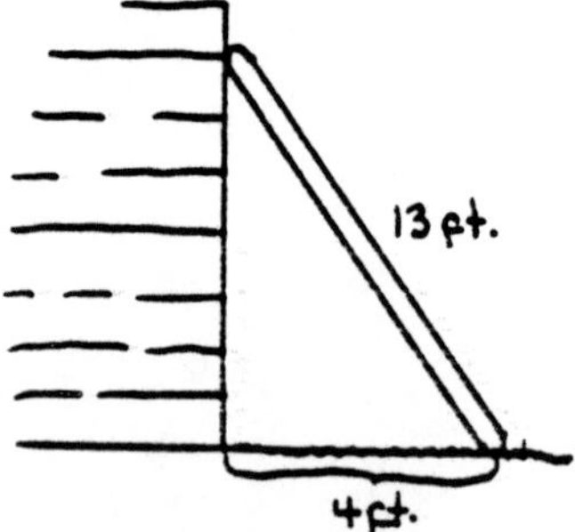

$a^2 + (4)^2 = (13)^2$

Find a. ____________

7. $a^2 + 16 = 169$

$a^2 = 153$

$a = \sqrt{153}$

$a = \sqrt{3 \cdot 3 \cdot 17}$

$a = 3\sqrt{17}$ ft

8. In a rectangle, the diagonal is 10 ft. and one side equals 5 ft. Find the other side and the area of the rectangle.

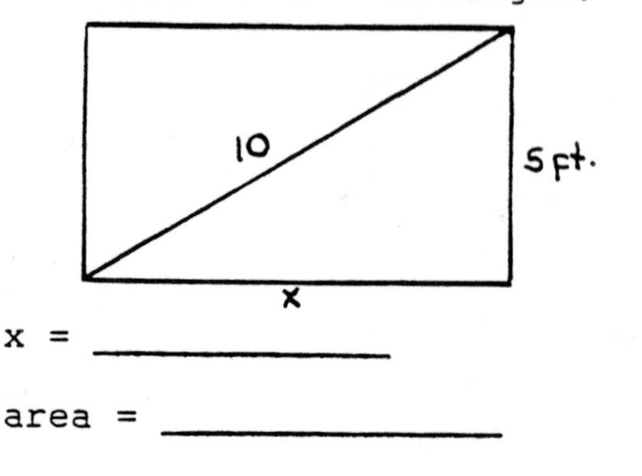

a) x = ____________

b) area = ____________

8.a) $5^2 + x^2 = 10^2$

$25 + x^2 = 100$

$x^2 = 75$

$x = \sqrt{75}$

$x = 5\sqrt{3}$ ft.

b) $A = l \cdot w$

$A = (5)(5\sqrt{3})$

$A = 25\sqrt{3}$ sq. ft.

9. Each side of a square equals 7 cm. Find the length of the diagonal.

7 cm.

x

7 cm.

x = ____________

9. $7^2 + 7^2 = x^2$

$49 + 49 = x^2$

$98 = x^2$

$\sqrt{98} = x$

$7\sqrt{2}$ cm. $= x$

10. In a right triangle, one leg is two more than the second leg and the hypotenuse is 4 more than the second leg. Find all three sides of the triangle.

x

x + 2

x + 4

side 1 = ____________

side 2 = ____8____

hypotenuse = ____________

10. $(x+2)^2+(x)^2=(x+4)^2$

$(x^2+4x+4)+x^2=(x^2+8x+16)$

$2x^2+4x+4 = x^2+8x+16$

$x^2-4x-12 = 0$

$(x-6)(x+2) = 0$

$x = 6$ side 1

$x + 2 = 8$ side 2

$x + 4 = 10$ hypotenuse

Exercise 1

1. Which of the following sets of numbers represent the lengths of the sides of a right triangle? Show why or why not.

   a. 3, 4, 5

   b. 1, $1\frac{1}{3}$, $1\frac{2}{3}$

   c. $2\sqrt{2}$, $3\sqrt{2}$, $4\sqrt{2}$

   d. 8, 15, 17

2. In right triangle ABC, c is the length of the hypotenuse and a and b are the lengths of the legs.

   a. If a = 2 cm. and b = 4 cm., then c = ____________.
   b. If a = 3 m. and c = 6 m., then b = ____________.
   c. If b = 9 in. and c = 10 in., then a = ____________.

3. An isosceles right triangle has hypotenuse 10 ft. Find the length of the sides and the area of the triangle.

4. An 11 ft. ladder is placed against a building. The foot of the ladder is 3 feet from the base of the building. How far above the ground does the ladder touch the building?

SPECIAL RIGHT TRIANGLES

Three more theorems concerning special right triangles should be considered.

Theorem 2: In a 30°-60° right triangle, if the length of the side opposite the 30° angle is x, then the length of the hypotenuse is twice x and the length of the side opposite the 60° angle is x multiplied by the $\sqrt{3}$.

Using the Pythagorean Theorem, we can show that Theorem 2 is a true statement.

$$a^2 + b^2 = c^2$$
$$x^2 + (x\sqrt{3})^2 = (2x)^2$$
$$x^2 + x^2(3) = 4x^2$$
$$4x^2 = 4x^2$$

The sides x, 2x and $x\sqrt{3}$ are legs of a right triangle.

Example: Find the measure of a and b.

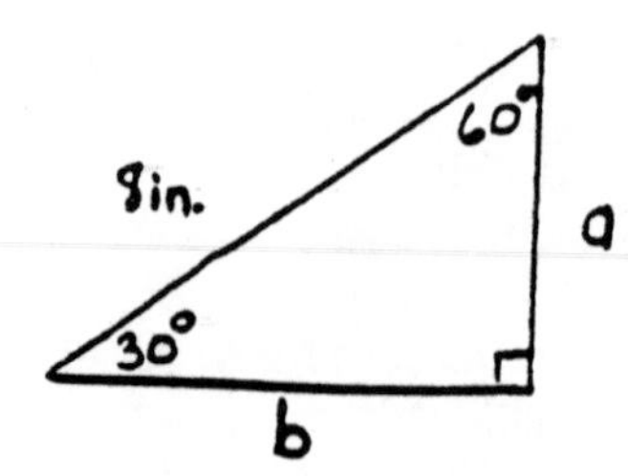

The side opposite the 30° angle is side a. The hypotenuse is c and is twice the measure of a.

$$c = 2a$$
$$8 = 2a$$
$$4 \text{ in.} = a$$

The side opposite the 30° angle measures 4 in. The side opposite the 60° angle is side b. The hypotenuse is twice the measure of b divided by $\sqrt{3}$.

$$c = \frac{2b}{\sqrt{3}}$$

$$8 = \frac{2b}{\sqrt{3}}$$

$$8\sqrt{3} = 2b$$

$$4\sqrt{3} \text{ in.} = b$$

The side opposite the 60° angle measures $4\sqrt{3}$ in.

11.

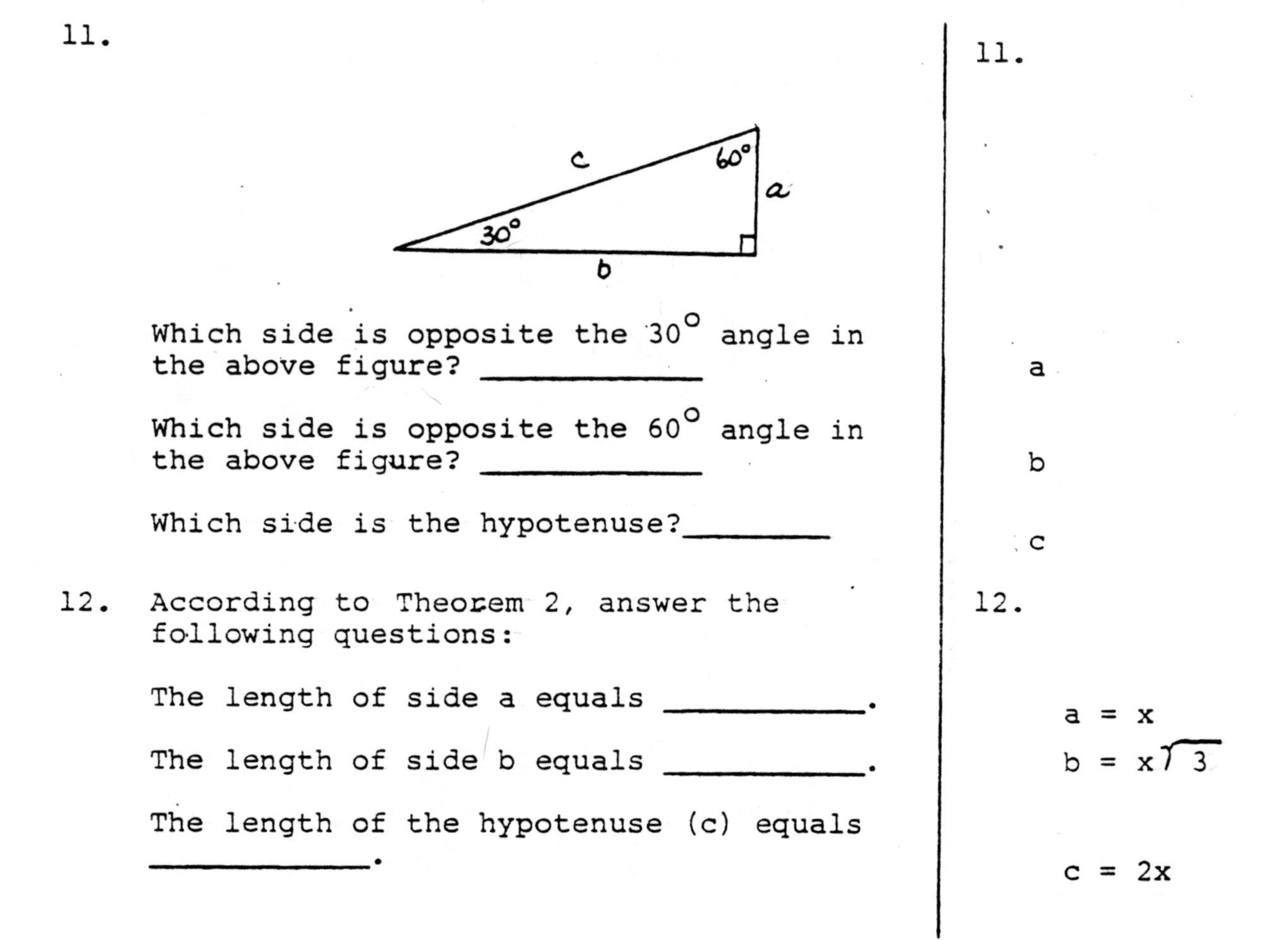

Which side is opposite the 30° angle in the above figure? ____________

Which side is opposite the 60° angle in the above figure? ____________

Which side is the hypotenuse? ________

11.

a

b

c

12. According to Theorem 2, answer the following questions:

The length of side a equals ___________.

The length of side b equals ___________.

The length of the hypotenuse (c) equals ___________.

12.

$a = x$

$b = x\sqrt{3}$

$c = 2x$

13.

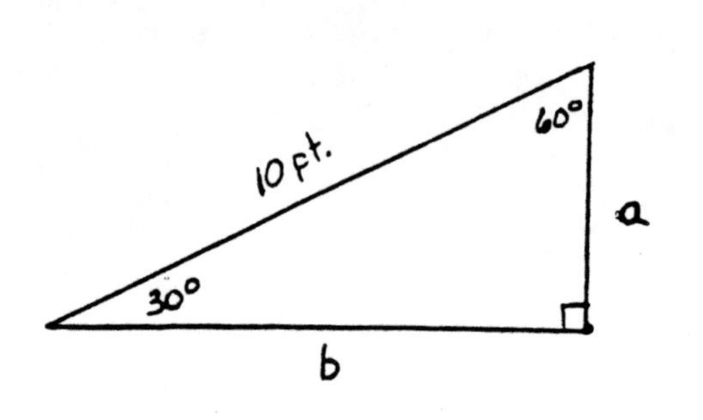

Figure 9.2

In Figure 9.2, since c = 10 ft. and a is the side opposite the 30° angle, then c = 2x. Therefore, a = ______________

In Figure 9.2, if b is the side opposite the 60° angle, then $b = x\sqrt{3}$. Therefore, b = __________.

13.

10 ft. = 2x

5 ft. = x

a is 5 ft.

$b = 5\sqrt{3}$ ft.

14. If c = 21 yd., then the side opposite the 30° angle is __________ and the side opposite the 60° angle is __________.

14. 21 = 2x

$10\frac{1}{2} = x$

$a = 10\frac{1}{2}$ yd.

$b = 10\frac{1}{2}\sqrt{3}$ yd.

15. In Figure 9.3 below, if a = 4 dm., then find c = ______ and b = ________.

15. c = **2(4dm)**

c = 8 dm.

$b = 4\sqrt{3}$ dm.

Figure 9.3

Theorem 2 can be used in determining the altitude (height) in a parallelogram.

Example:

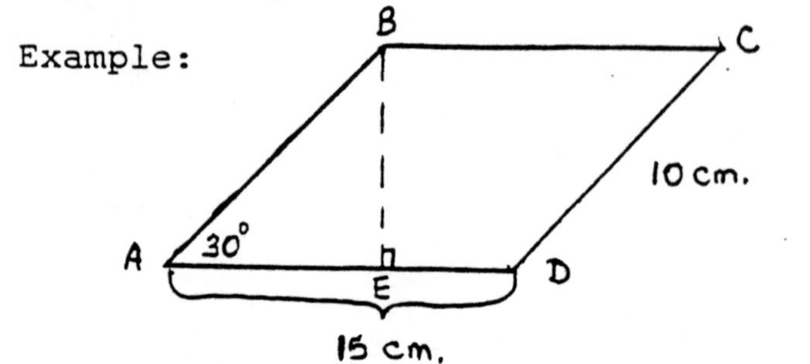

Find the area of ▱ABCD according to the measures given in the illustration.

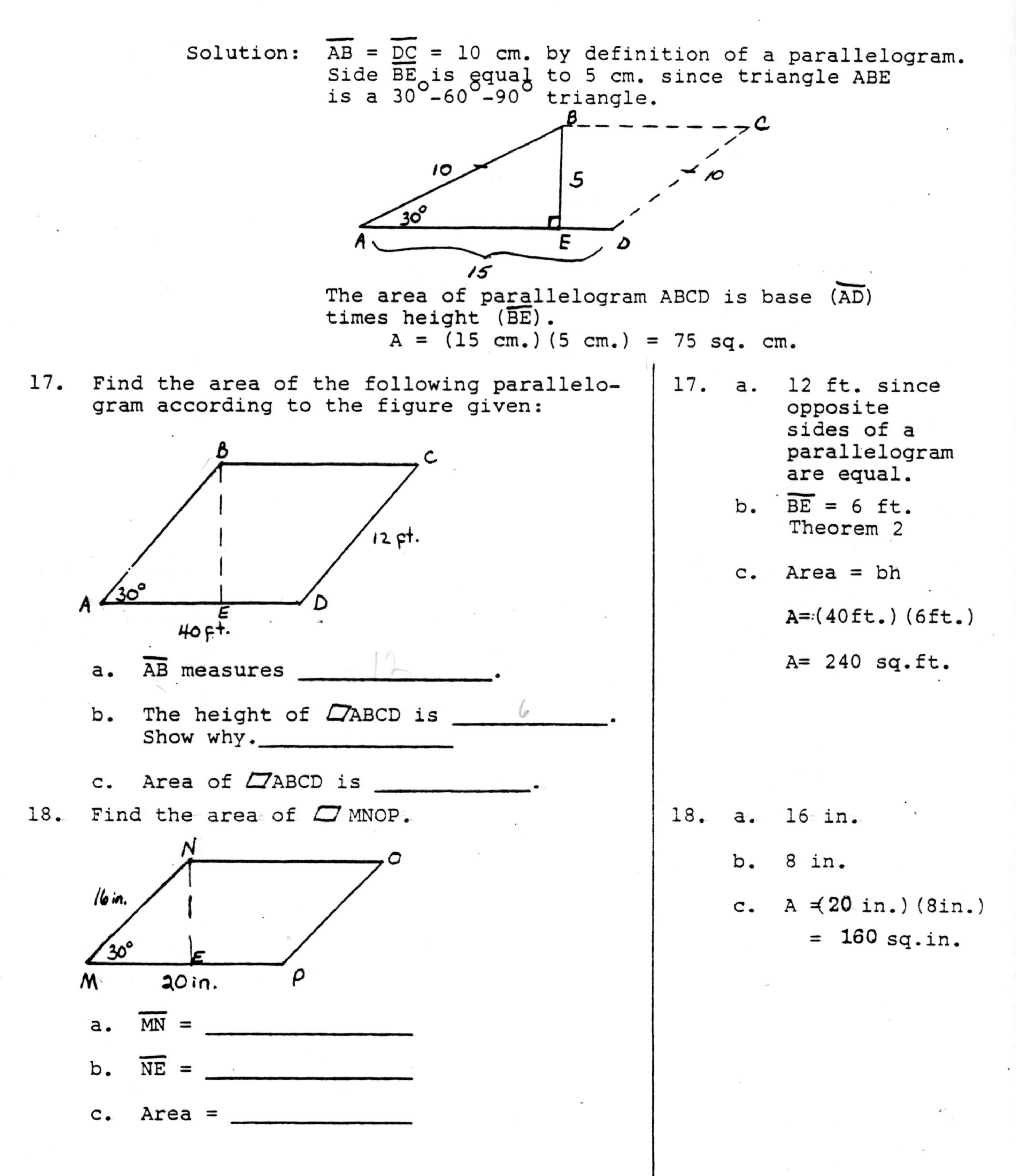

Solution: $\overline{AB} = \overline{DC} = 10$ cm. by definition of a parallelogram. Side $\overline{BE}$ is equal to 5 cm. since triangle ABE is a 30°-60°-90° triangle.

The area of parallelogram ABCD is base $(\overline{AD})$ times height $(\overline{BE})$.

$$A = (15 \text{ cm.})(5 \text{ cm.}) = 75 \text{ sq. cm.}$$

17. Find the area of the following parallelogram according to the figure given:

a. $\overline{AB}$ measures ____12____.

b. The height of ▱ABCD is ____6____. Show why.________

c. Area of ▱ABCD is ________.

18. Find the area of ▱ MNOP.

a. $\overline{MN}$ = ________

b. $\overline{NE}$ = ________

c. Area = ________

17. a. 12 ft. since opposite sides of a parallelogram are equal.

b. $\overline{BE}$ = 6 ft. Theorem 2

c. Area = bh

A = (40ft.)(6ft.)

A = 240 sq.ft.

18. a. 16 in.

b. 8 in.

c. A = (20 in.)(8in.) = 160 sq.in.

19. Find the value of x, then find the area of the parallelogram. $\angle ABC = (4x+30)$

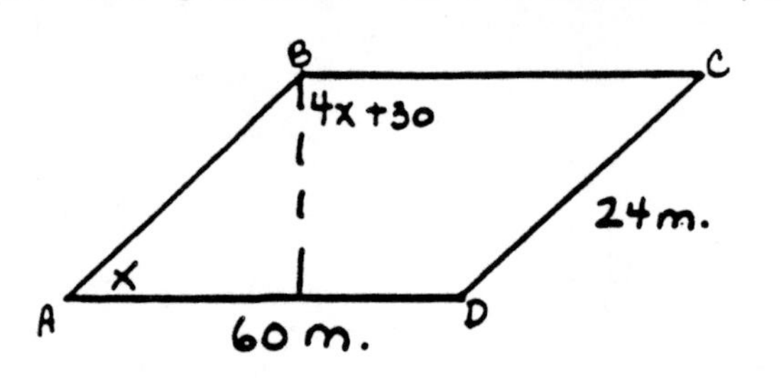

a. $(4x + 30) + (x) = 180^\circ$ Why? ______
Therefore, x = ______

b. height = ______

c. Area = ______

19.a. Consecutive angles of a parallelogram are supplementary.

$$5x + 30^\circ = 180^\circ$$
$$5x = 150^\circ$$
$$x = 30^\circ$$

b. height = 12 m.

c. Area=(60m.) (12 m.)

= 720 sq.m.

20. Find the value of x, then find the area of this equilateral triangle.

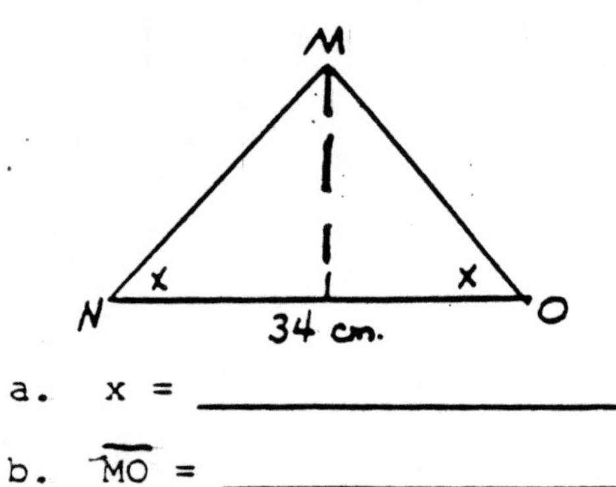

a. x = ______

b. $\overline{MO}$ = ______

c. altitude = ______

d. Area = ______

20.a. By the def. of equilateral triangle, all angles are equal.

$$x+x+x=180^\circ$$
$$3x=180^\circ$$
$$x=60^\circ$$

b. $\overline{MO}$ = 34 cm. by definition of equilateral triangle.

c. Alt.= $\frac{1}{2}(34\sqrt{3})$
$= 17\sqrt{3}$ cm.
by Theorem 2.

d. Area= $\frac{1}{2}$(base) (alt.)

$= \frac{1}{2}(34)(17\sqrt{3})$

$= 289\sqrt{3}$ cm.^{2}

Exercise 2

Solve the following problems:

1. The hypotenuse of a 30°-60° right triangle is 4 inches. Find the length of the legs.

2. The side opposite the 60° angle of a 30°-60° right triangle is 7 inches. What are the lengths of the other sides?

3. Find the area of the triangle in problem 2.

4. One side of an equilateral triangle measures 3 m. Find the length of the altitude and the area of the triangle.

5. Find the value of x and the area of ▱ RUTH.

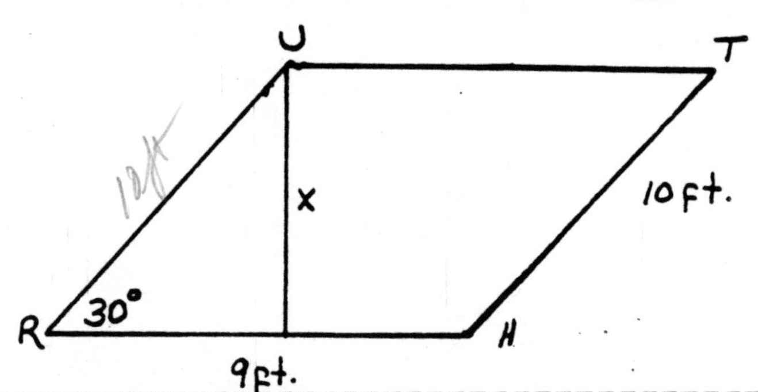

Theorem 3: In an isosceles 45° right triangle, the hypotenuse is $\sqrt{2}$ times the length of one side.

45°, 45°, s, s, $s\sqrt{2}$

Proof of Theorem 3 using the Pythagorean Theorem:

$$a^2 + b^2 = c^2$$

Remember, in an isosceles right triangle sides a and b are equal in length. Use s to represent the equal sides.

$$s^2 + s^2 = c^2$$

$$2s^2 = c^2$$

$$\sqrt{2s^2} = c$$

$s\sqrt{2} = c$ Therefore, the hypotenuse is $\sqrt{2}$ times the length of one side.

Example:

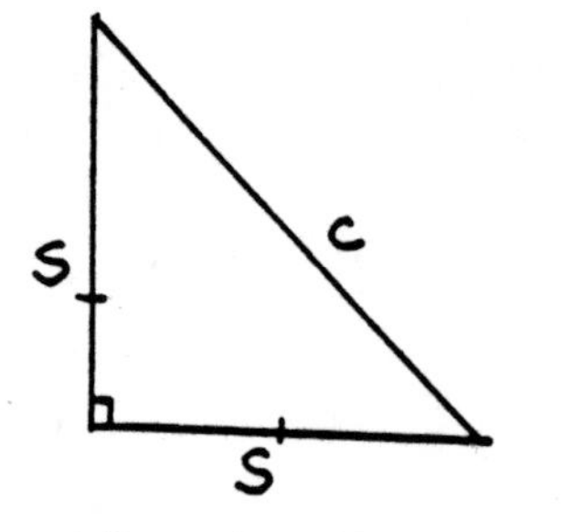

a. If one side of an isosceles right triangle equals 4 in., find the hypotenuse.

Solution: Since $c = s\sqrt{2}$, then $c = 4\sqrt{2}$ in.

b. If the hypotenuse of an isosceles right triangle measured 25 inches, then find the measure of one of the equal legs.

$$c = s\sqrt{2}$$

$$25 = s\sqrt{2}$$

$$\frac{25}{\sqrt{2}} = s$$

$$\frac{25\sqrt{2}}{2} = s$$

$$12\tfrac{1}{2}\sqrt{2} \text{ in.} = s$$

21.

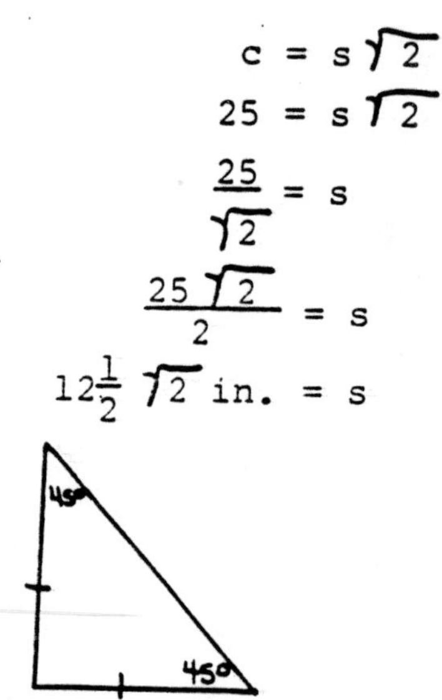

If s = 9 cm. and $c = s\sqrt{2}$ cm., then c must equal ________________.

21. $9\sqrt{2}$ cm.

22.

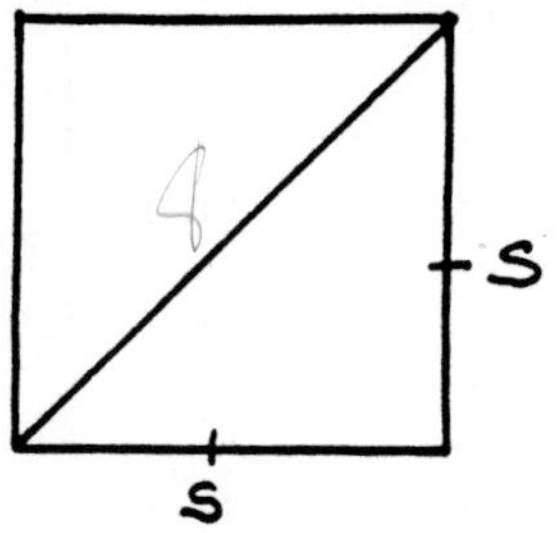

Figure 9.4

If $c = 8$ dm. and $c = s\sqrt{2}$ dm., then s must equal ________________.

23. Using Figure 9.4, if a square has a diagonal of length $5\sqrt{2}$ in., find the length of the sides of the square.

Note: $c = 5\sqrt{2}$ in. The diagonal is also the hypotenuse of an isosceles right triangle.

Remember that c also equals $s\sqrt{2}$. Therefore, $5\sqrt{2}$ in. $= s\sqrt{2}$

________ = s

24. If a square has diagonal 11 ft., find the length of the sides of the square.

$c = 11$ ft. and $c = s\sqrt{2}$

Therefore, 11 ft. $= s\sqrt{2}$

______ = s

25. Find the area of an isosceles right triangle if the hypotenuse is 21 m.

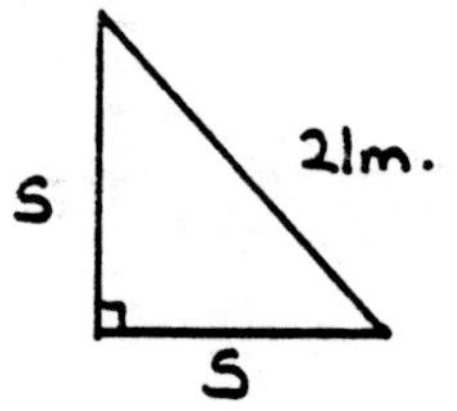

a. s = ______________

b. Area = ______________

---

22. $c = s\sqrt{2}$

$8 = s\sqrt{2}$

$\frac{8}{\sqrt{2}} = s$

$\frac{8\sqrt{2}}{2} = s$

$4\sqrt{2}$ dm. $= s$

23. $5\sqrt{2}$ in. $= s\sqrt{2}$

$\frac{5\sqrt{2}}{\sqrt{2}}$ in. $= s$

5 in. $= s$

24. 11 ft. $= s\sqrt{2}$

$\frac{11}{\sqrt{2}} = s$

$\frac{11\sqrt{2}}{2} = s$

$5\frac{1}{2}\sqrt{2}$ ft. $= s$

25.a. 21 m. $= s\sqrt{2}$

$\frac{21}{\sqrt{2}}$ m. $= s$

$\frac{21\sqrt{2}}{2}$ m. $= s$

$10\frac{1}{2}\sqrt{2}$ m. $= s$

b. A = $\frac{1}{2}$(base)(height)

$A = \frac{1}{2}(10\frac{1}{2}\sqrt{2})(10\frac{1}{2}\sqrt{2})$

$A = \frac{882}{8}$ m^2

$A = 110\frac{1}{4}$ m^2

26. Find the area of an isosceles right triangle if the leg is 5 yds. Also, find the measure of the hypotenuse of the triangle.

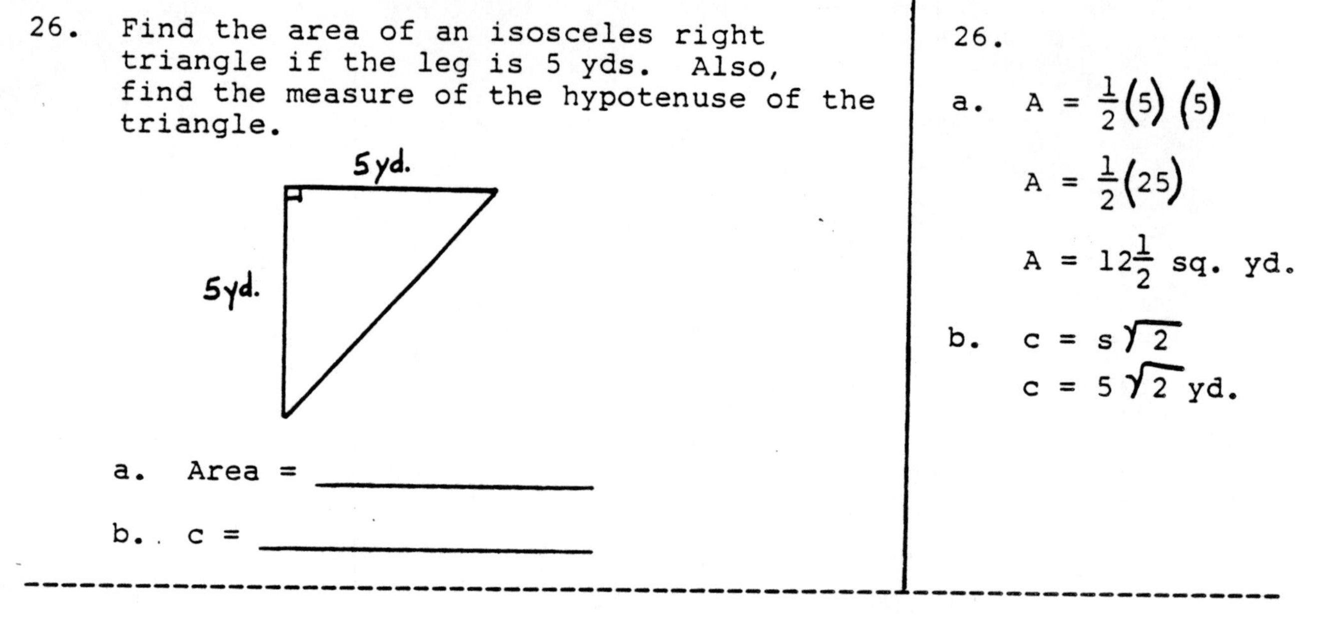

a. Area = ____________

b. c = ____________

26.

a. $A = \frac{1}{2}(5)(5)$

$A = \frac{1}{2}(25)$

$A = 12\frac{1}{2}$ sq. yd.

b. $c = s\sqrt{2}$

$c = 5\sqrt{2}$ yd.

Exercise 3

1. In an isosceles right triangle, if one leg is 13 cm., find the length of the hypotenuse.

2. In a 45°-45°-right triangle, if the hypotenuse is 8 ft., find the length of one of the legs of the triangle.

3. Find the area of the following isosceles-right triangle if the hypotenuse is 6 m.

4. In an isosceles right triangle, if the hypotenuse measures $41\sqrt{2}$ inches, then find the measure of one of the equal sides.

5. If the area of a 45°-45°-right triangle is $24\frac{1}{2}$ sq. ft., then find the measure of one of the equal sides.

6. Find the area of an isosceles right triangle if one of the equal sides is 23 dm.

Theorem 4: An altitude of an equilateral triangle is $\frac{1}{2}\sqrt{3}$ times the length of one side.

s s

altitude equals $\frac{1}{2}\sqrt{3}$ (s)

s

<u>Proof</u> of Theorem 4 using Theorem 2 on page 242.

Divide the equilateral triangle in half forming two 30°-60° right triangles.

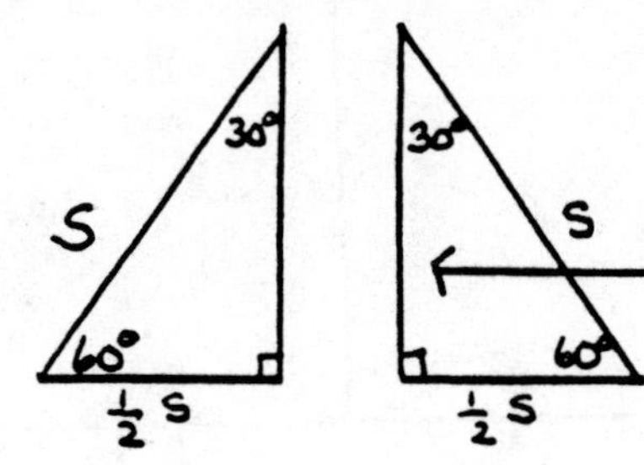

Theorem 2 demonstrates that the altitude is $\frac{1}{2}s\sqrt{3}$ since it is opposite the 60° angle in the figure.

27. Find the altitude of an equilateral triangle if one side is 26 ft.

Altitude = $\frac{1}{2}s\sqrt{3}$

Altitude = ____________

27. $\frac{1}{2}(26)\sqrt{3}$

$13\sqrt{3}$ ft.

28. If the altitude of an equilateral triangle equals $6\sqrt{12}$ in., find the length of one side.

Altitude = $\frac{1}{2}s\sqrt{3}$

$6\sqrt{12}$ in. = $\frac{1}{2}s\sqrt{3}$

________ = s

28. $6\sqrt{12} = \frac{1}{2}s\sqrt{3}$

$12\sqrt{12} = s\sqrt{3}$

$\frac{12\sqrt{36}}{3} = s$

$4\sqrt{36} = s$

$4(6) = s$

24 in. = s

29. If the altitude of an equilateral triangle equals 25 in., find the length of one side.

$25 = \frac{1}{2}s\sqrt{3}$

____ = s

29. $25 = \frac{1}{2}s\sqrt{3}$

$50 = s\sqrt{3}$

$\frac{50}{\sqrt{3}} = s$

$\frac{50\sqrt{3}}{3} = s$

$16\frac{2}{3}\sqrt{3}$ in. = s

30. Find the altitude of an equilateral triangle if one side is 13 dm. Find the area of this triangle.

30.

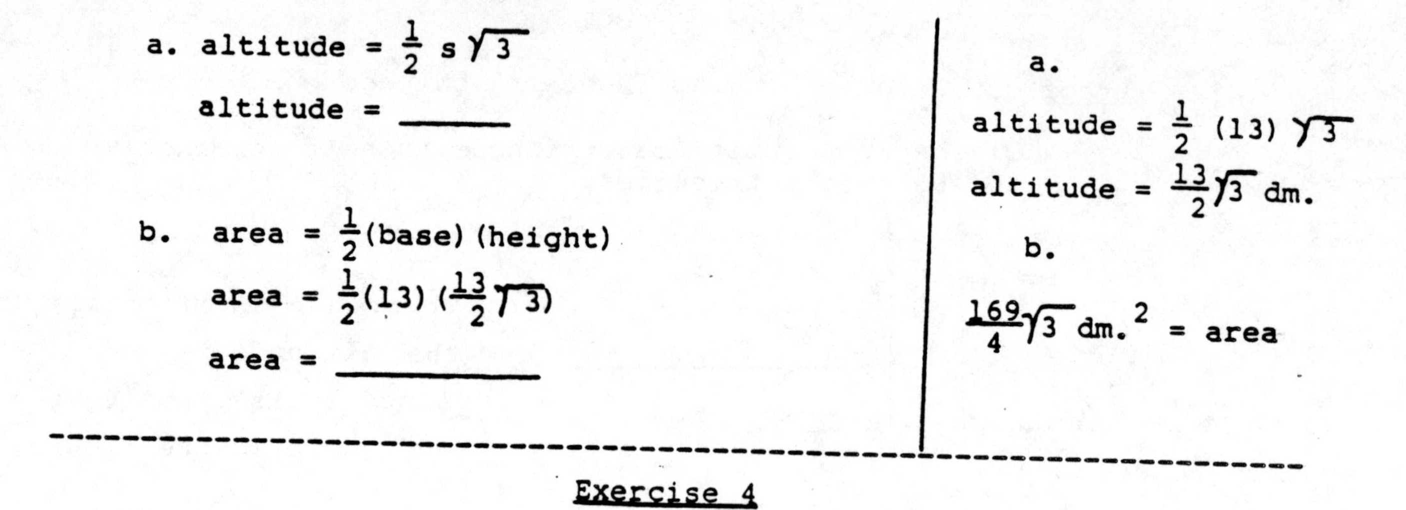

a. altitude = $\frac{1}{2}s\sqrt{3}$

altitude = ______

b. area = $\frac{1}{2}$(base) (height)

area = $\frac{1}{2}(13)\left(\frac{13}{2}\sqrt{3}\right)$

area = ______

a.

altitude = $\frac{1}{2}(13)\sqrt{3}$

altitude = $\frac{13}{2}\sqrt{3}$ dm.

b.

$\frac{169}{4}\sqrt{3}$ dm.^{2} = area

Exercise 4

1. Find the area of an equilateral triangle with one side equal to 12 m.

2. The altitude of an equilateral triangle is 8 inches long. Find the length of a side of the triangle.

3. Find the length of the altitude of an equilateral triangle if one side equals 5 ft.

Unit 9 Review

I. Fill in the requested information.

Theorem	Information
1. Pythagorean Theorem	______ = c^2
2. Theorem 2: 30° - 60° right triangle	Side opposite 30° angle = ______ Side opposite 60° angle = ______
3. Theorem 3: 45° - 45° isosceles right triangle	Hypotenuse = ______ and the two sides are equal to x.
4. Theorem 4: equilateral triangle	Altitude = ______ and all sides are equal to x.

II. Which of the following sets of numbers can be lengths of the sides of a right triangle?

1. 9, 23, 24

2. $\sqrt{2}$, $\sqrt{3}$, $\sqrt{5}$

3. 12, 35, 37

4. 6, 8, 10

5. $1\frac{1}{5}$, $1\frac{3}{5}$, 2

6. 120, 130, 50

III. In each of the problems, find the length of the third side of right triangle ABC.

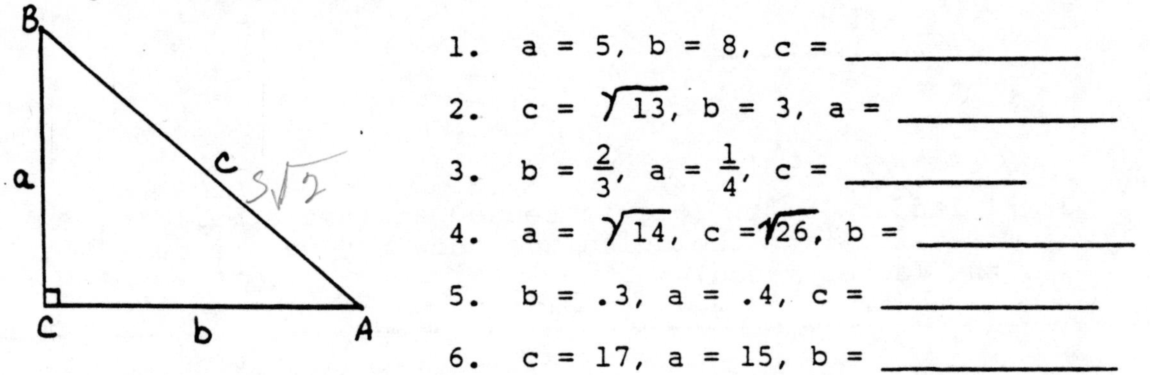

1. a = 5, b = 8, c = __________

2. $c = \sqrt{13}$, b = 3, a = __________

3. $b = \frac{2}{3}$, $a = \frac{1}{4}$, c = __________

4. $a = \sqrt{14}$, $c = \sqrt{26}$, b = __________

5. b = .3, a = .4, c = __________

6. c = 17, a = 15, b = __________

IV. Solve the following:

1. In an isosceles right triangle, find the lengths of the other sides if $c = 5\sqrt{2}$ in.

2. Find the altitude of an equilateral triangle with side of length 7 ft.

3. Find the length of the diagonal of a square with side 10 cm.

4. Find the length of a side of an equilateral triangle if the altitude is $8\sqrt{3}$ m.

5. Find the length of a side of a square whose diagonal is of length 10 ft.

6. Find the area of an equilateral triangle with one side equal to 26 in.

7. Find the measures of the other two sides if the hypotenuse of a 30°-60° right triangle measures $9\sqrt{2}$ cm.

V. 1.

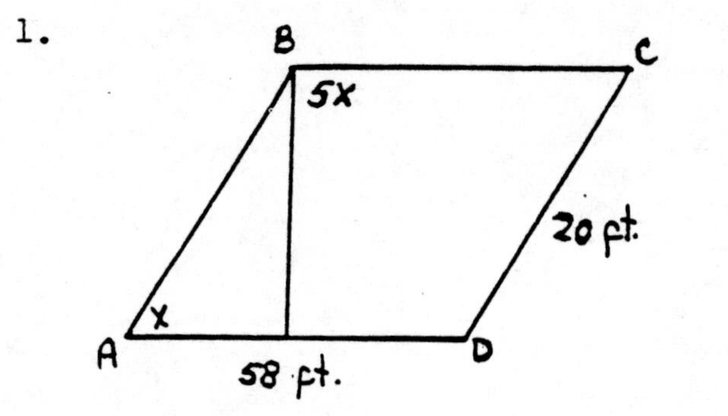

In this parallelogram, find the value of x, the height, and the area. $\angle ABC = 5x$

2. 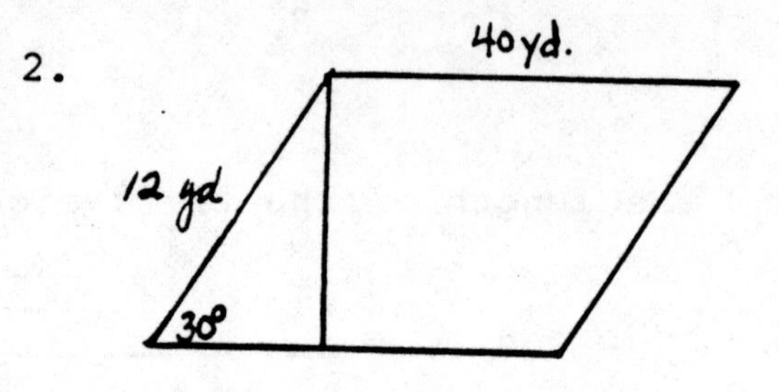

In this parallelogram, find the height and the area.

3. A ladder 25 ft. long is leaned against a building a distance of 6 ft. from the building. How far up the building does the ladder reach?

Unit 10

Ratio, Proportion and Similarity

Learning objectives;

The student will demonstrate his mastery of the following definitions, properties, and theorems by writing them and by applying them to solutions of selected problems:

Definitions: ratio, proportion, means, extremes, constant of proportionality, fourth proportional, mean proportional, similar triangles, and "$\sim$".

Properties:

1. If $\frac{a}{b} = \frac{c}{d}$ and a, b, c, d $\neq 0$, then ad = bc. The product of the means equals the product of the extremes in a proportion.

2. If $\frac{a}{b} = \frac{c}{d}$ and a, b, c, d $\neq 0$, then $\frac{a}{c} = \frac{b}{d}$. The means may be exchanged in any proportion.

3. If $\frac{a}{b} = \frac{c}{d}$ and a, b, c, d $\neq 0$, then $\frac{d}{b} = \frac{c}{a}$. The extremes may be exchanged in any proportion.

4. If $\frac{a}{b} = \frac{c}{d}$ and a, b, c, d $\neq 0$, then $\frac{b}{a} = \frac{d}{c}$.

   Property of inversion.

Theorems:

1. Two triangles are similar if and only if two angles of one triangle are equal respectively to two angles of a second triangle.

2. If corresponding sides of two triangles are proportional, the triangles are similar.

3. A line parallel to one side of a triangle divides the other two sides proportionately.

As a part of our daily routine, frequently we make comparisons between any number of items. For example, we compare wins to losses in athletic events, weight gains vs. losses, the price per pound of meats, EPA ratings on automobiles in miles per gallon, the daily inflation index or even the P/E ratio of common stocks on the stock exchanges.

The comparisons of any two real numbers are termed ratios.

Definition: A ratio is a comparison of two numbers, a and b indicated by their quotient, $\frac{a}{b}$, where $b \neq 0$.

Quite simply, a ratio is a fraction.

Example 1:

a) The ratio of 34 boys to 21 girls in a math class is expressed as $\frac{34}{21}$.

b) The ratio of 34 boys to the total number of students in the math class in part (a) is expressed as $\frac{34}{55}$. (Number in the entire class equals the number of males plus the number of females.)

Example 2: The ratio of 3 days to one week is initially expressed as 3 days to 7 days, then written as $\frac{3}{7}$. (Note: All terms must have the same exact units of measure whenever you are making a comparison between them.)

1. Write the ratio of 4 ft. to 4 yds.

   Recall that 4 yds. is equivalent to 12 ft. since there are three ft. per yd. Overall, 4 ft. to 4 yds. is expressed as 4 ft. to 12 ft. or what ratio?____________

2. There are 15 females and 9 males in a biology class.

   a) Find the ratio of females to males. ____________

   b) Find the ratio of females to the entire class. ____________

   c) Find the ratio of males to females. ____________

Answers:

1. $\frac{4}{12}$ reduced to $\frac{1}{3}$

2.
   a) $\frac{15}{9} = \frac{5}{3}$
   b) $\frac{15}{24} = \frac{5}{8}$
   c) $\frac{9}{15} = \frac{3}{5}$

d) Is the ratio of males to females equivalent to the ratio of females to males? Show why or why not.

d) No. $\frac{3}{5} \neq \frac{5}{3}$

Example 3: A recipe for pie crust uses 2 cups of flour and $\frac{2}{3}$ cup shortening. Find the ratio of flour to shortening. The ratio of 2 to $\frac{2}{3}$ is written as $\frac{(2)}{\left(\frac{2}{3}\right)}$ and worked out as follows:

$$\frac{(2)}{\left(\frac{2}{3}\right)} = \left(\frac{2}{1} \div \frac{2}{3}\right) = \left(\frac{2}{1} \times \frac{3}{2}\right) = \frac{6}{2} = \frac{3}{1}$$

3. Find the ratio of 4 teaspoons of cinnamon to $2\frac{1}{3}$ teaspoons of sugar.

$$\frac{4}{2\frac{1}{3}} = \frac{4}{1} \div 2\frac{1}{3} = \_\_\_\_\_\_\_\_$$

3. $\frac{4}{1} \div \frac{7}{3}$

$\frac{4}{1} \times \frac{3}{7} = \frac{12}{7}$

4. Find the ratio of 8 sec. to 1 hour. 1 hour = ________ seconds; the ratio is ________.

4. 3600 seconds; $\frac{8}{3600} = \frac{1}{450}$

Definition: A <u>proportion</u> is a statement that two ratios are equal.
If $\frac{a}{b}$ and $\frac{c}{d}$ are two ratios, then $\frac{a}{b} = \frac{c}{d}$ is called a proportion where a, b, c, d $\neq$ 0.
(i) b and c are called <u>means</u>.
(ii) a and d are called <u>extremes</u>.

Example:

$\frac{2}{3} = \frac{14}{21}$ is a true proportion since each fraction simplifies to $\frac{2}{3}$.

5. Is $\frac{4}{20} = \frac{3}{15}$ a true proportion?

5. Yes, each fraction simplifies to $\frac{1}{5}$.

6. What two numbers are called the means in exercise 5?

6. 20 and 3

7. Is $\frac{2a}{6b} = \frac{12a}{16b}$ a true proportion?

7. False, $\frac{2a}{6b}$ and $\frac{12a}{16b}$ do not reduce to the same fraction.

8. What two numbers are called the extremes in exercise 7?

8. 2a and 16b

Definition: Two sequences of positive numbers a, b, c,... and p, q, r, ... are <u>proportional</u> if and only $\frac{a}{p} = \frac{b}{q} = \frac{c}{r} = \ldots = k$ (where k is the constant to which each ratio of the sequence is equal.) K is called the <u>constant of proportionality</u>.

Example: Show that the following pair of sequences are proportional by finding the constant of proportionality.

4, 2, 3 and 12, 6, 9

Examine the ratios $\frac{4}{12}$, $\frac{2}{6}$, $\frac{3}{9}$. Since each simplifies to $\frac{1}{3}$ we may write $\frac{4}{12} = \frac{2}{6} = \frac{3}{9} \Rightarrow$ k and k = $\frac{1}{3}$. Therefore, the sequences are proportional with constant of proportionality $\frac{1}{3}$.

9. Show that the following pair of sequences are proportional.

2, 5, 15, 7 and 8, 20, 60, 28

a) $\frac{2}{8} = \frac{5}{20} = \frac{15}{60} = \frac{7}{28}$ = k; k=__________

b) Are the sequences proportional?_____

9.

a) $k = \frac{1}{4}$

b) yes

10. Show that the following pair of sequences are <u>not</u> proportional.

10 , 4, 6, 12, 14 and 5, 2, 4, 6, 8

a) $\frac{10}{5} = \frac{4}{2} = \frac{6}{4} = \frac{12}{6} = \frac{14}{8}$ = k; k = ________

b) Are the sequences proportional?______

10.

a) k = no possible answer

b) No, because there is no constant of proportionality.

Exercise 1

1. Fill in the blank with the appropriate word or phrase.

   a. A ____________ is the comparison of two numbers by their indicated quotient.

   b. A _______ is a statement that two ratios are equal.

   c. In comparing pairs of sequences, each ratio must be equal to the ______________ before the pair of sequences are considered to be proportional.

2. Find the ratio of each of the following:

   a. In a basketball season, one team won 8 games and lost 6. Find the ratio of wins to losses.

   b. Find the ratio of losses to the total number of games played in part a.

   c. Find the ratio of 3 minutes to 4 hours.

   d. Find the ratio of $\frac{5}{4}$ cup of flour to $\frac{1}{8}$ cup of walnuts.

3. Determine if each of the following proportions is <u>true</u> or <u>false</u>.

   a. $\frac{3x}{6y} = \frac{3x}{9y}$

   b. $\frac{10}{5} = \frac{\frac{1}{2}}{\frac{1}{4}}$

   c. $\frac{12y}{4x} = \frac{57y}{19x}$

   d. $\frac{7}{8} = \frac{21}{25}$

4. Show that the following pairs of sequences are or are not proportional by finding the constant of proportionality.

   a. 1, 7, 11 and 3, 21, 33 $k =$ ___________

   b. 18, 15, 39 and 8, 5, 13 $k =$ ___________

   c. $\frac{1}{2}, \frac{1}{3}, \frac{1}{4}$ and $\frac{1}{3}, \frac{2}{9}, \frac{1}{6}$ $k =$ ___________

Algebraic Properties of Proportions

Algebraically, a proportion is an equation and the usual rules of algebra apply. These rules allow us to change the form of any proportion in the following ways:

Property 1: If $\frac{a}{b} = \frac{c}{d}$ and a, b, c, d $\neq$ 0, then ad = bc.

The product of the means equals the product of the extremes in a proportion.

Example: If $\frac{2}{5} = \frac{4}{10}$, then (2)(10) = (5)(4).

20 = 20

Property 2: If $\frac{a}{b} = \frac{c}{d}$ and a, b, c, d $\neq$ 0, then $\frac{a}{c} = \frac{b}{d}$.

The <u>means</u> may be exchanged in any proportion.

Example: If $\frac{2}{5} = \frac{4}{10}$, then $\frac{2}{4} = \frac{5}{10}$ is a true proportion.

Property 3: If $\frac{a}{b} = \frac{c}{d}$ and a, b, c, d $\neq$ 0, then $\frac{d}{b} = \frac{c}{a}$.

The <u>extremes</u> may be exchanged in any proportion.

Example: If $\frac{2}{5} = \frac{4}{10}$, then $\frac{10}{5} = \frac{4}{2}$ is a true proportion.

Property 4: If $\frac{a}{b} = \frac{c}{d}$ and a, b, c, d $\neq$ 0, then $\frac{b}{a} = \frac{d}{c}$.

Property by <u>inversion</u>.

Example: If $\frac{2}{5} = \frac{4}{10}$, then $\frac{5}{2} = \frac{10}{4}$ is a true proportion.

11. If the following pair of sequences are proportional, then demonstrate the truth of property 4.

Given: 2a, 14a, 28a and 3b, 21b, 42b

The proportion is $\frac{2a}{3b} = \frac{14a}{21b} = \frac{28a}{42b}$

a. Proportion by inversion is ________.

b. Is the proportion true or false?

____

12. If $\frac{7}{4} = \frac{91}{52}$ is a true proportion, then demonstrate the truth of property 2.

a. The means are exchanged to be ________.

b. Is the proportion true or false?

____

13. If $\frac{7}{4} = \frac{91}{52}$ is a true proportion then demonstrate the truth of property 3.

a. The extremes are exchanged to be ________.

b. Is the proportion true or false?

____

14. If $\frac{7}{4} = \frac{91}{52}$ is a true proportion then demonstrate the truth of property 1.

a. The means and extremes property is ________.

b. Are the products equal?________

11.

a. $\frac{3b}{2a} = \frac{21b}{14a} = \frac{42b}{28a}$

b. True since $k = \frac{3b}{2a}$

12.

a. $\frac{7}{91} = \frac{4}{52}$

b. True, both ratios reduce to $\frac{1}{13}$.

13.

a. $\frac{52}{4} = \frac{91}{7}$

b. True, both ratios reduce to $\frac{13}{1}$.

14.

a. (7)(52) = (91)(4)

b. Yes, 364 = 364

Find the Fourth Proportional

Definition: In a proportion $\frac{a}{b} = \frac{c}{d}$, a, b, c, d $\neq$ 0, d is known as the <u>fourth proportional</u>.

Example: In a proportion, find the fourth proportional to 2, 5, and 7. (d is the unknown)

$$\frac{2}{5} = \frac{7}{d}$$

Use the means and extremes property (1) to find the value of d.

$$\frac{2}{5} = \frac{7}{d}$$
$$(2)d = (5)(7)$$
$$2d = 35$$
$$d = \frac{35}{2}$$

15. In a proportion, find the fourth proportional to 3, 7, and 15 using property 1.
 a. The proportion is ________________.
 b. The fourth proportional is ________.

16. In a proportion, find the fourth proportional to $\frac{1}{2}$, 3, and 4 using property 1.
 a. The proportion is________________.
 b. The fourth proportional is ________.

15.
 a. $\frac{3}{7} = \frac{15}{d}$
 b. $3d = 105$
 $d = 35$

16.
 a. $\frac{\left(\frac{1}{2}\right)}{3} = \frac{4}{d}$
 b. $\left(\frac{1}{2}\right)d = (4)(3)$
 $\left(\frac{1}{2}\right)d = 12$
 $d = (12)\left(\frac{2}{1}\right)$
 $d = 24$

Finding the Mean Proportion

Definition: In a proportion $\frac{a}{b} = \frac{c}{d}$, if b = c, then b and c are the mean proportional between a and d.

Example: Find the mean proportional between 7 and 14.
Let x be the mean proportional and use property 1 to solve for x.

$$\frac{7}{x} = \frac{x}{14}$$

$$x^2 = 98$$

$$x = \pm\sqrt{98}$$

$$x = \pm 7\sqrt{2}$$

17. Find the mean proportional between 5 and 15.

    a. The proportion is ________.

    b. The mean proportional equals ________.

17.

a. $\frac{5}{x} = \frac{x}{15}$

b. $x^2 = 75$

$x = \pm\sqrt{75}$

$x = \pm 5\sqrt{3}$

18. Find the mean proportional between 4 and 16.

    a. The proportion is ________.

    b. The mean proportional equals ________.

18.

a. $\frac{4}{x} = \frac{x}{16}$

b. $x^2 = (4)(16)$

$x^2 = 64$

$x = \pm\sqrt{64}$

$x = \pm 8$

Given any three terms of a proportion, the value of the unknown term can be obtained by using the Means and Extremes Property.

Example:

$$\frac{x}{\sqrt{7}} = \frac{6\sqrt{8}}{\sqrt{14}}$$

$$(\sqrt{14})x = (6\sqrt{8})(\sqrt{7})$$

$$\sqrt{14}\,x = 6\sqrt{56}$$

$$\frac{1}{\sqrt{14}}(\sqrt{14})x = \frac{1}{\sqrt{14}} \cdot 6\sqrt{56}$$

$$x = \frac{6\sqrt{56}}{\sqrt{14}}$$

$$x = 6\sqrt{4}$$

$$x = (6)\cdot(2)$$

$$x = 12$$

Find the value of x in each of the following:

19. $\frac{x}{3} = \frac{4}{5}$

$5x = (3)(4)$

$5x =$ ________

$x =$ ________

20. $\frac{x}{\sqrt{5}} = \frac{3\sqrt{2}}{\sqrt{10}}$

$\sqrt{10}\,x = (3\sqrt{2})(\sqrt{5})$

$\sqrt{10}\,x =$ ________

$x =$ ________

21. $\frac{x}{x-3} = \frac{5}{3}$

$3x = 5(x - 3)$

$3x =$ ________

$x =$ ________

22. $\frac{x+3}{4} = \frac{x-3}{6}$

$6(x + 3) = 4(x - 3)$

$x =$ ________

23. $\frac{3}{x-2} = \frac{x}{3}$

$9 = x(x - 2)$

________ $= x$

19.

$5x = 12$

$x = \frac{12}{5}$

20.

$\sqrt{10}\,x = 3\sqrt{10}$

$x = 3$

21.

$3x = 5x - 15$

$-2x = -15$

$x = \frac{15}{2}$

22. $6x + 18 = 4x - 12$

$2x = -30$

$x = -15$

23.

$9 = x^2 - 2x$

$0 = x^2 - 2x - 9$

$x = \frac{2 \pm \sqrt{4 - 4(1)(-9)}}{2}$

$x = \frac{2 \pm \sqrt{4 + 36}}{2}$

24. $\dfrac{x + 3}{4} = \dfrac{3}{x - 4}$

$x =$ ____________

$$x = \frac{2 \pm \sqrt{40}}{2}$$

$$x = \frac{2 \pm 2\sqrt{10}}{2}$$

$$x = 1 \pm \sqrt{10}$$

24.

$$(x + 3)(x - 4) = (3)(4)$$

$$x^2 - x - 12 = 12$$

$$x^2 - x - 24 = 0$$

$$x = \frac{1 \pm \sqrt{1 - 4(1)(-24)}}{2}$$

$$x = \frac{1 \pm \sqrt{1 + 96}}{2}$$

$$x = \frac{1 \pm \sqrt{97}}{2}$$

Exercise 2

1. State which algebraic property of proportion is illustrated.

   a. If $\frac{a}{3} = \frac{4}{5}$, then $\frac{3}{a} = \frac{5}{4}$. ____________

   b. If $\frac{x}{7} = \frac{2}{y}$, then $\frac{y}{7} = \frac{2}{x}$. ____________

   c. If $\frac{m}{10} = \frac{x}{y}$, then $my = 10x$. ____________

   d. If $\frac{m}{10} = \frac{x}{y}$, then $\frac{m}{x} = \frac{10}{y}$. ____________

2. In a proportion, find the fourth proportional to:

   a. 3, 5, 12 ________

   b. 8, 9, 4 ________

   c. $\frac{2}{5}$, $\frac{3}{10}$, $\frac{1}{4}$ ________

3. In a proportion, find the mean proportional between:

   a. 2 and 8 ________

   b. 14 and 6 ________

4. Solve for x.

a. $\frac{11}{x} = \frac{5}{9}$

b. $\frac{x}{x - 5} = \frac{4}{5}$

c. $\frac{x + 5}{3} = \frac{x - 13}{6}$

d. $\frac{7}{x + 1} = \frac{x}{6}$

e. $\frac{x + 3}{5} = \frac{x + 7}{x + 10}$

SIMILARITY

Congruent figures may informally be described as "having the same size and shape". Similar figures may informally be described as "having the same shape, but not necessarily the same size" as illustrated in figure 10.1.

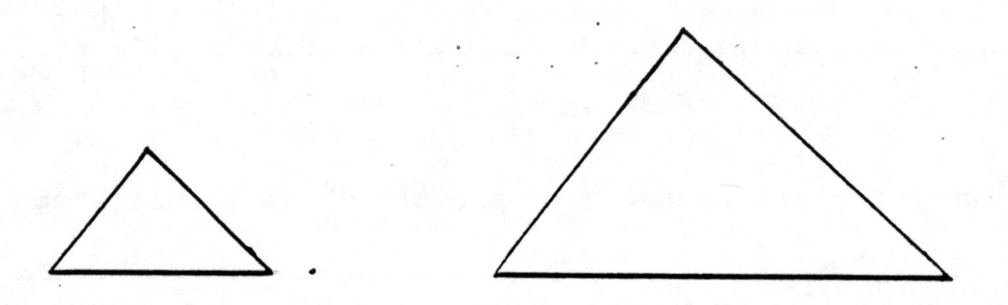

Figure 10.1

Similarity is a concept that photographers generally employ in the darkroom in printing pictures. By using a single negative under magnification, a photographer could enlarge or reduce the negative image as he desires. The overall shape of the image does not appear changed which suggests that although the line or curved segments may become enlarged, the measures of the corresponding angles or arcs remain constant.

Surveyors, housing contractors, and architects all deal with similarity in the use of designs, site plans, and blue prints.

In this unit, we will primarily be concerned with similarity involving triangles.

> Definition: Two triangles are <u>similar</u> if three angles of one triangle equal three angles of a second triangle and all corresponding sides are in proportion. The symbol used to denote similarity is "$\sim$".

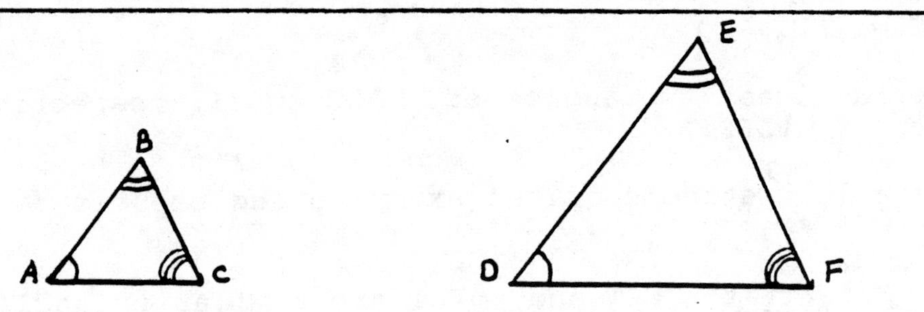

Given that $\triangle ABC \sim \triangle DEF$, then $\angle A = \angle D$, $\angle B = \angle E$, and $\angle C = \angle F$. The definition also states that corresponding sides are proportional. Use precautions in writing out the proportions. Be sure only corresponding sides are compared.

In the diagram above, $\overline{AB}$ corresponds to $\overline{DE}$, $\overline{BC}$ to $\overline{EF}$, and $\overline{AC}$ to $\overline{DF}$. Hence,

$$\frac{\overline{AB}}{\overline{DE}} = \frac{\overline{BC}}{\overline{EF}} = \frac{\overline{AC}}{\overline{DF}}$$

> Theorem 1: Two triangles are similar if and only if two angles of one triangle are equal respectively to two angles of a second triangle.

This theorem is actually the statement and its converse combined.

<u>Statement</u>: If two triangles are similar, then two angles of one triangle equal two angles of the second triangle.

<u>Converse</u>: If two angles of one triangle equal respectively to two angles of a second triangle, the triangles are similar.

The proof of the statement is as follows:

Let $\triangle ABC$ and $\triangle DEF$ be similar triangles.

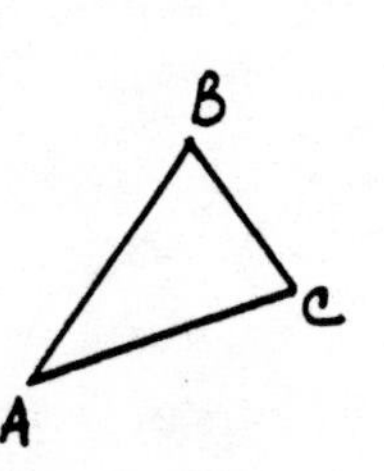

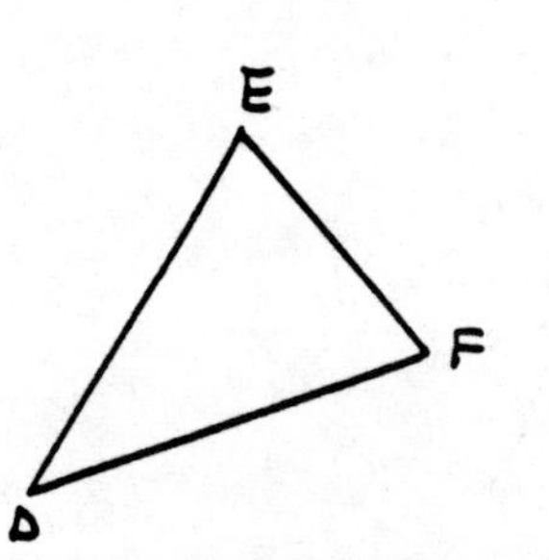

Statements	Reasons
1. $\Delta ABC \sim \Delta DEF$	1. Given
2. $\angle A = \angle D$ $\angle B = \angle E$ $\angle C = \angle F$	2. Definition of similarity

Therefore, at least two angles of ΔABC equal, respectively, two angles of ΔDEF.

Now, take a look at some of the examples and exercises using this theorem.

Example: Show that ΔSRT and ΔVUT are similar triangles if $\overline{SR} \parallel \overline{UV}$.

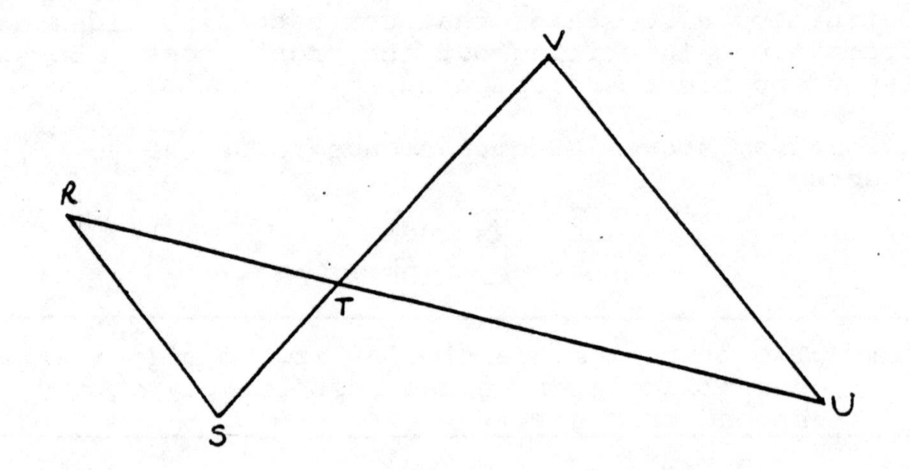

Solution: $\angle R = \angle U$ since alternate interior angles of parallel lines are equal.

$\angle RTS = \angle VTU$ since vertical angles are equal.

$\Delta SRT \sim \Delta VUT$ by theorem 1: If two angles of one triangle equal respectively two angles of a second triangle, then the triangles are similar.

25. Given: $\overline{ZY} \parallel \overline{WX}$. Show that $\Delta UZY \sim \Delta UWX$

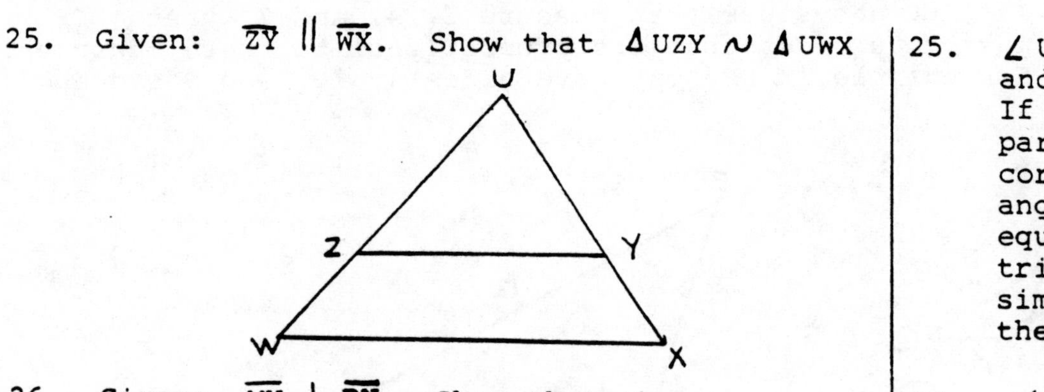

25. $\angle UZY = \angle W$ and $\angle UYZ = \angle X$. If lines are parallel, then corresponding angles are equal. The triangles are similar by theorem 1.

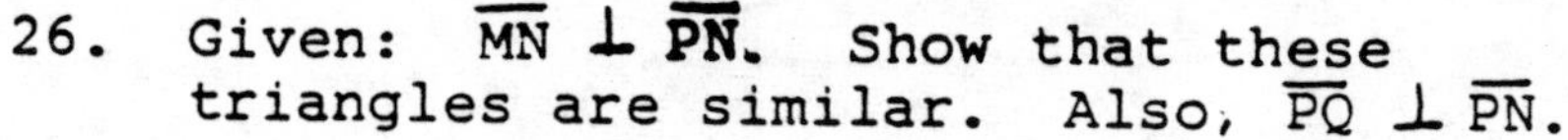

26. Given: $\overline{MN} \perp \overline{PN}$. Show that these triangles are similar. Also, $\overline{PQ} \perp \overline{PN}$.

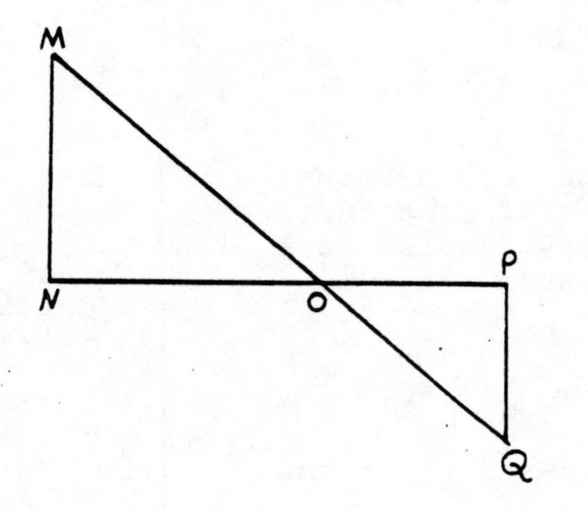

26. $\angle N = \angle P$ Right angles are equal. $\angle MON = \angle QOP$ Vertical angles are equal. Therefore, the triangles are similar by theorem 1.

27. Given: $\overline{AB} \parallel \overleftrightarrow{DC}$ and $\overline{AD} \parallel \overline{BC}$. Show that these triangles are similar.

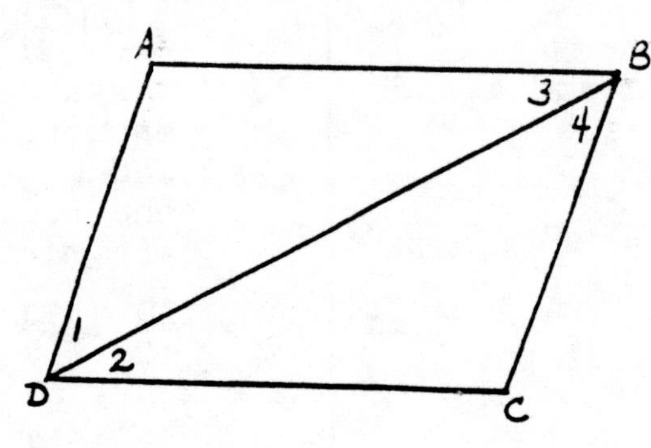

27. $\angle 2 = \angle 3$
$\angle 1 = \angle 4$
Alternate interior angles are equal if lines are parallel.

The triangles are congruent by ASA = ASA and similar by theorem 1.

> Theorem 2: If corresponding sides of two triangles are proportional, the triangles are similar.

Example: ΔABC and ΔDEF in Figure 10.2 are similar triangles. If ΔABC has sides that measure 2, 4, and 3, then ΔDEF must have sides that are proportional to these measures. For example, ΔDEF may have sides 6, 12, and 9 since

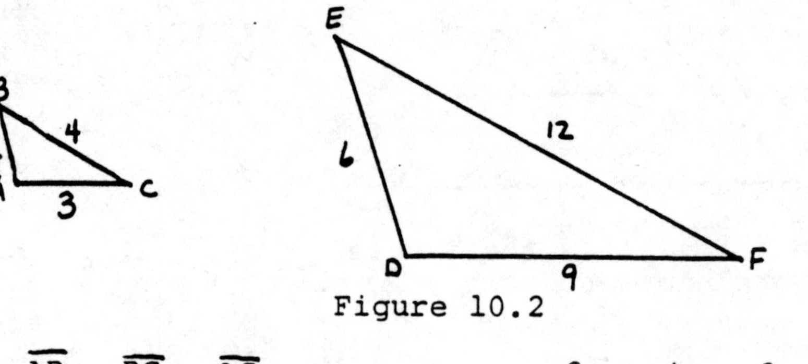

Figure 10.2

$$\frac{\overline{AB}}{\overline{DE}} = \frac{\overline{BC}}{\overline{EF}} = \frac{\overline{AC}}{\overline{DF}} \quad \text{(or)} \quad \frac{2}{6} = \frac{4}{12} = \frac{3}{9}$$

28. If the sides of one triangle measure 10, 7, 5 and the sides of a second triangle measure 20, 14, and 10, are these triangles similar?

28. Yes, their sides are proportional.

$$\frac{10}{20} = \frac{7}{14} = \frac{5}{10} = k$$

$$k = \frac{1}{2}$$

29. Are triangles with sides 2, 5, 6 and 3, $\frac{17}{2}$, 12 similar?

29. No, their sides are not proportional.

$$\frac{2}{3} \neq \frac{5}{\left(\frac{17}{2}\right)} \neq \frac{6}{12}$$

30. Examine the following figures and decide whether or not these triangles are similar. State why.

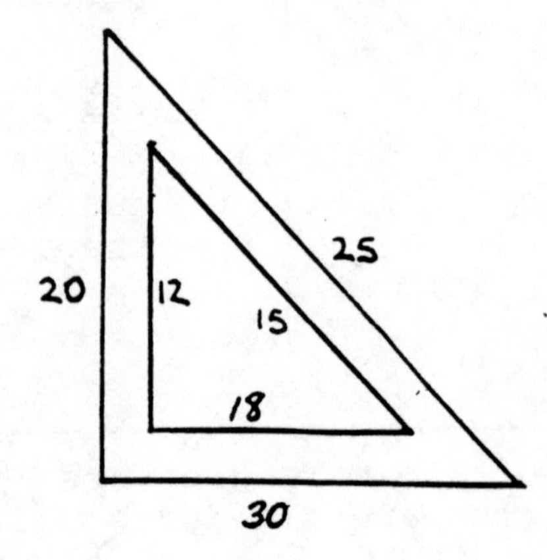

30. The sides are proportional and by theorem 2, the triangles are similar.

$$\frac{20}{12} = \frac{30}{18} = \frac{25}{15} = k$$

$$k = \frac{5}{3}$$

Example: Given $\overline{AB} \parallel \overline{DE}$, find $\overline{BC}$ and $\overline{DE}$.

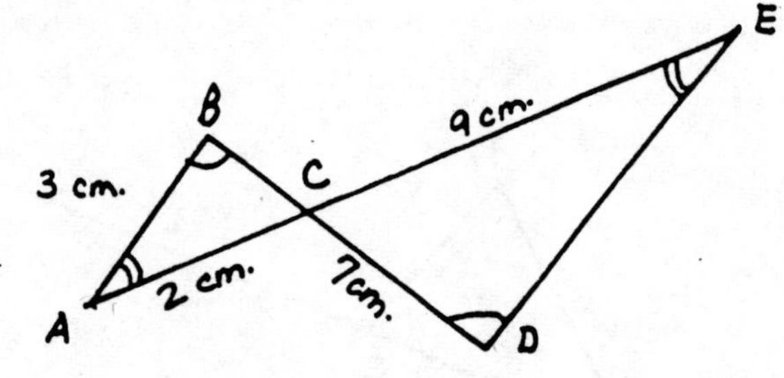

Since $\overline{AB} \parallel \overline{DE}$, $\angle B = \angle D$ (alternate interior angles, $\overline{BD}$ is the transversal) and $\angle A = \angle E$ (alternate interior angles, $\overline{AE}$ is the transversal). From theorem 1, $\triangle ABC \sim \triangle EDC$. Therefore, the sides are proportional by the definition of similarity.

The proportions must be set up to compare corresponding sides.

$$\frac{\overline{AB}}{\overline{ED}} = \frac{\overline{AC}}{\overline{EC}} = \frac{\overline{BC}}{\overline{DC}}$$

Next, substitute the values that were given into the proportion. $\overline{AB}$ = 3 cm., $\overline{AC}$ = 2 cm., $\overline{CE}$ = 9 cm., and $\overline{CD}$ = 7 cm.

Therefore, $$\frac{3}{\overline{ED}} = \frac{2}{9} = \frac{\overline{BC}}{7}$$

Not all of the values of the proportion were given; however, by using the Means and Extremes Property you will be able to solve for the unknowns.

Select a proportion that will give you three known values such as:

$$\left[\frac{3}{\overline{ED}} = \frac{2}{9}\right] \quad \text{and} \quad \left[\frac{2}{9} = \frac{\overline{BC}}{7}\right]$$

Now, solve each proportion for the unknown.

$$\frac{3}{\overline{ED}} = \frac{2}{9}$$
$$(3)(9) = (2)\overline{ED}$$
$$27 = 2\overline{ED}$$
$$\frac{27}{2}\text{ cm.} = \overline{ED}$$

$$\frac{2}{9} = \frac{\overline{BC}}{7}$$
$$(2)(7) = (9)\overline{BC}$$
$$14 = 9\overline{BC}$$
$$\frac{14}{9}\text{ cm.} = \overline{BC}$$

31. Given: $\overline{AB} \parallel \overline{DE}$, find $\overline{BC}$ and $\overline{DE}$.

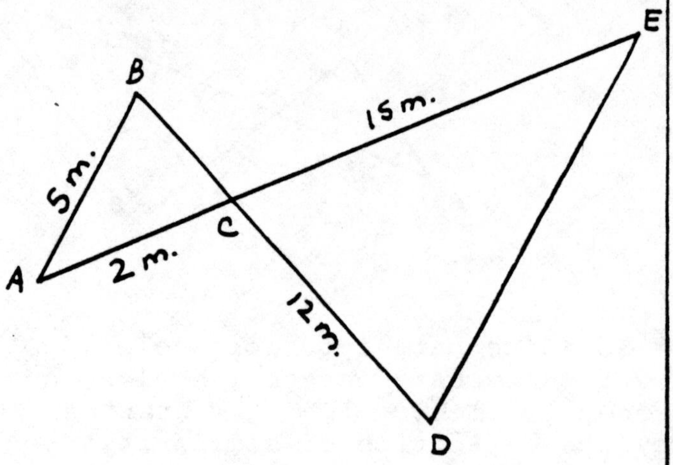

a. Why is $\triangle ABC \sim \triangle EDC$? ____________

b. The sides are proportional in what way and what values are substituted?

____

c. What two proportions did you select in order to solve for the unknown terms? ____________ ____________

d. $\overline{ED}$ = ____________

$\overline{BC}$ = ____________

31.

a. Theorem 1 and $\angle A = \angle E$ and $\angle B = \angle D$ by alternate interior angles.

b. $\frac{\overline{AB}}{\overline{ED}} = \frac{\overline{AC}}{\overline{EC}} = \frac{\overline{BC}}{\overline{DC}}$

$\frac{5}{\overline{ED}} = \frac{2}{15} = \frac{\overline{BC}}{12}$

c. $\left[\frac{5}{\overline{ED}} = \frac{2}{15}\right]$ and $\left[\frac{2}{15} = \frac{\overline{BC}}{12}\right]$

d. $\frac{5}{\overline{ED}} = \frac{2}{15}$

$75 = 2\overline{ED}$

$\frac{75}{2}\text{m.} = \overline{ED}$

$\frac{2}{15} = \frac{\overline{BC}}{12}$

$24 = 15 \cdot \overline{BC}$

$\frac{24}{15} = \overline{BC}$

$\frac{8}{5}\text{ m.} = \overline{BC}$

32. Given similar triangles $\triangle DEF$ and $\triangle LMN$. Find the values of $\overline{FE}$ and $\overline{DF}$.

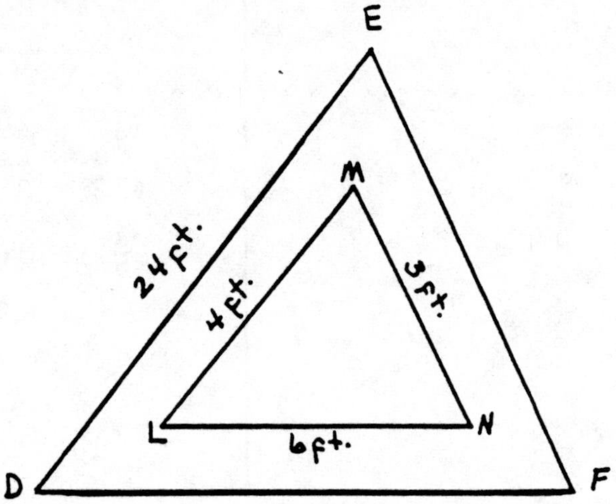

D, E, F, L, M, N; 24ft., 4ft., 3ft., 6ft.

33. A, B, C, D, E; a, b, c, d

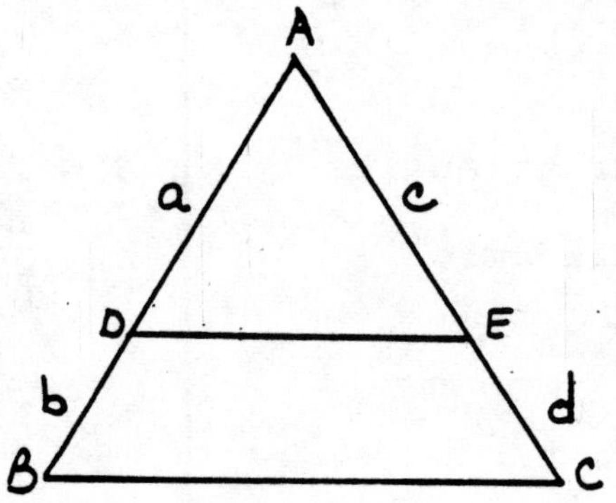

In $\triangle ABC$, $\overline{DE} \parallel \overline{BC}$.

a. If a = 8 m., $\overline{AB}$ = 10 m., d = 6 m., find c.

b. If b = 5 m., a = 9 m., $\overline{AC}$ = 15 m., find d.

32.
$$\frac{\overline{LM}}{\overline{DE}} = \frac{\overline{MN}}{\overline{FE}} = \frac{\overline{NL}}{\overline{DF}}$$
$$\frac{4}{24} = \frac{3}{\overline{FE}} = \frac{6}{\overline{DF}}$$
$$\frac{4}{24} = \frac{3}{\overline{FE}}$$
$$72 = 4\overline{FE}$$
$$18\text{ft.} = \overline{FE}$$

$$\frac{4}{24} = \frac{6}{\overline{DF}}$$
$$144 = 4\overline{DF}$$
$$36\text{ft.} = \overline{DF}$$

33.

a. 10, 8, c, 6

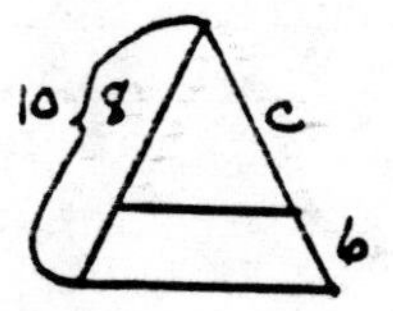

$$\frac{8}{10} = \frac{c}{c + 6}$$
$$8(c + 6) = 10c$$
$$8c + 48 = 10c$$
$$24 \text{ m.} = c$$

b. 9, 15, 5, d

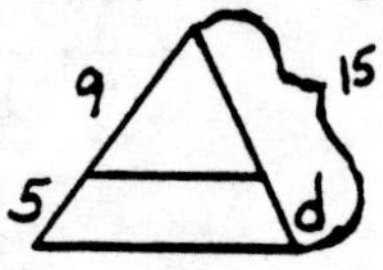

$$\frac{9}{14} = \frac{15 - d}{15}$$
$$135 = 210 - 14d$$
$$-75 = -14d$$
$$5\frac{5}{14} \text{ m.} = d$$

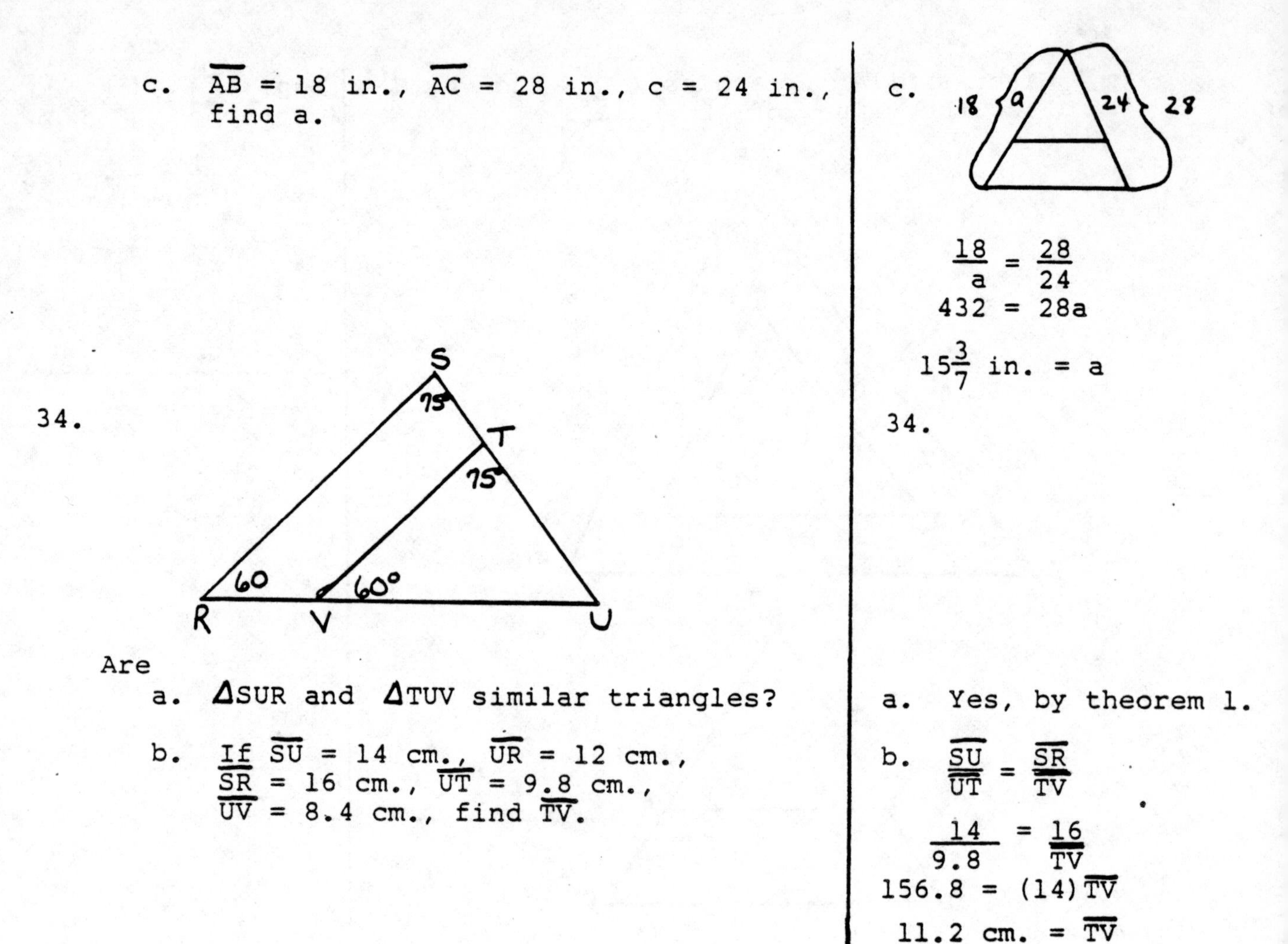

c. $\overline{AB}$ = 18 in., $\overline{AC}$ = 28 in., c = 24 in., find a.

34.

Are

a. ΔSUR and ΔTUV similar triangles?

b. If $\overline{SU}$ = 14 cm., $\overline{UR}$ = 12 cm., $\overline{SR}$ = 16 cm., $\overline{UT}$ = 9.8 cm., $\overline{UV}$ = 8.4 cm., find $\overline{TV}$.

c.

$$\frac{18}{a} = \frac{28}{24}$$

$$432 = 28a$$

$$15\tfrac{3}{7} \text{ in.} = a$$

34.

a. Yes, by theorem 1.

b. $$\frac{\overline{SU}}{\overline{UT}} = \frac{\overline{SR}}{\overline{TV}}$$

$$\frac{14}{9.8} = \frac{16}{\overline{TV}}$$

$$156.8 = (14)\overline{TV}$$

$$11.2 \text{ cm.} = \overline{TV}$$

Example: A man 6 feet tall casts a 10 foot shadow. In the same location and at the same time, a flagpole casts a 40 foot shadow. How high is the flagpole?

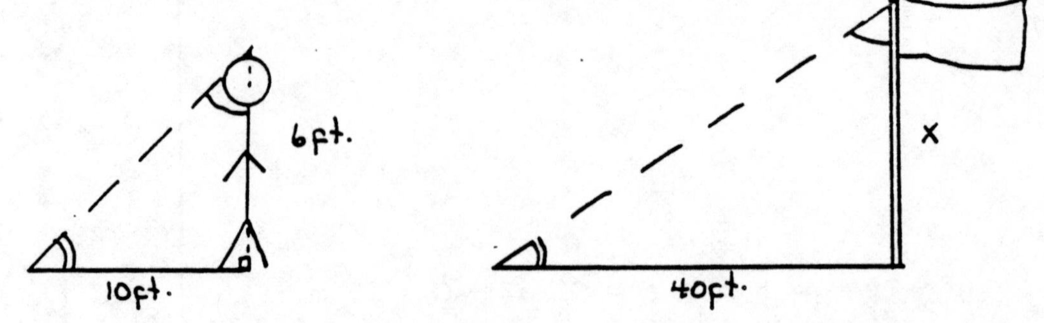

We must assume the man and the flagpole are perpendicular to the ground and the ground is level surface. The sun's ray makes the same angle with the man and flagpole at the same time of day. Since two angles of one triangle equal two angles of the other, the triangles are similar.

Because the triangles are similar, the sides are proportional. The following is an example of one possible proportion that you could set up in order to solve for x:

$$\frac{6}{x} = \frac{10}{40}$$

$$10x = 240$$

$$x = 24 \text{ ft., height of the flagpole}$$

35. At 4 p.m. two trees side by side cast 10 ft. and 25 ft. shadows, respectively. If the first tree is 5 ft. tall, find the height of the second tree. (First, illustrate the exercise on a sheet of paper, then set up your proportion.)

35. $\frac{5}{x} = \frac{10}{25}$

$10x = 125$

$x = 12.5$ ft.

36. If a 6 foot man casts an 8 ft. shadow, at the same time, how tall is a tree that casts a 20 ft. shadow in the same location?

36. $\frac{6}{x} = \frac{8}{20}$

$120 = 8x$

15ft. $= x$

37. A small tree, 3 feet tall, casts a shadow of 2 feet, and the large oak tree casts a shadow of 90 feet. Use your knowledge of similar triangles to compute the height of the large tree.

37. $\frac{3}{x} = \frac{2}{90}$

$270 = 2x$

135ft.$= x$

Example: In a scale drawing, a triangular region is an given below:

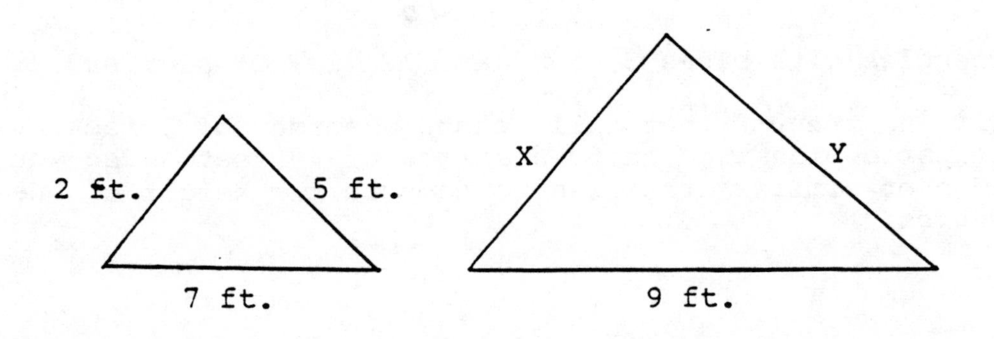

The actual figure has one side 9 ft. Find the other sides. Let one side be x and the other side y.

$$\frac{2}{x} = \frac{7}{9} \qquad \frac{5}{y} = \frac{7}{9}$$

$$7x = 18 \qquad 7y = 45$$

$$x = 2\frac{4}{7} \text{ ft.} \qquad y = 6\frac{3}{7} \text{ ft.}$$

38. In a scale drawing, a triangular region is as given below:

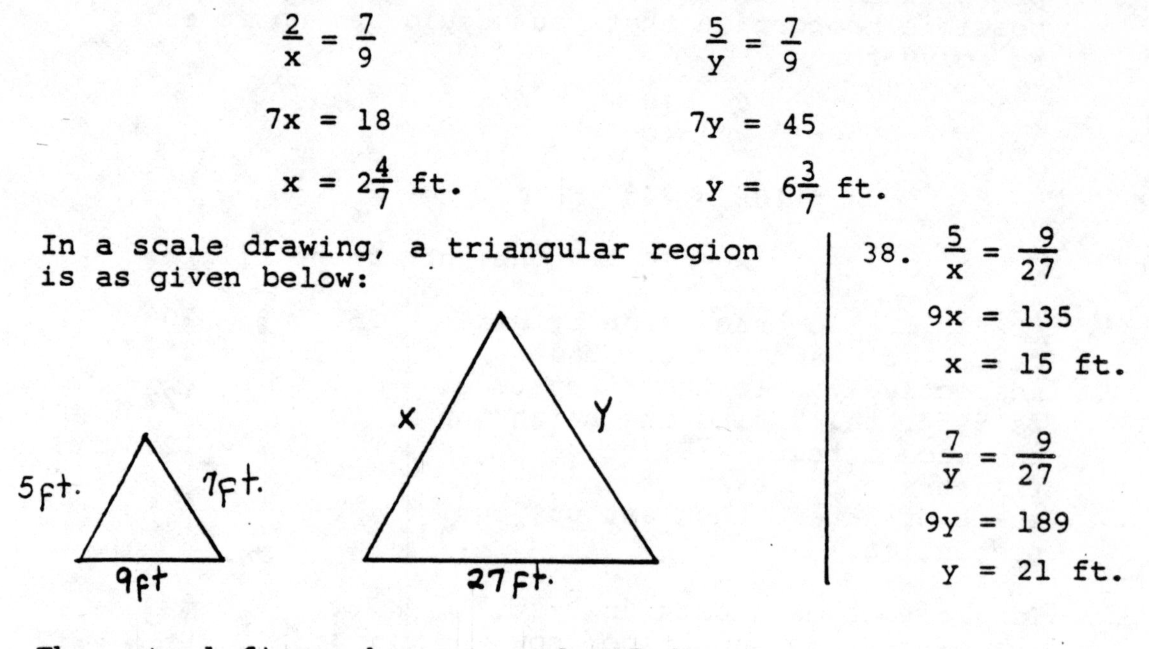

The actual figure has one side 27 ft. Find the other sides.

38. $\frac{5}{x} = \frac{9}{27}$

$9x = 135$

$x = 15$ ft.

$\frac{7}{y} = \frac{9}{27}$

$9y = 189$

$y = 21$ ft.

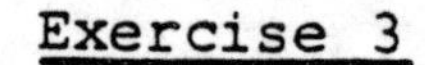

Exercise 3

1. Define similarity and give an example of this concept from your life experiences.

2. Given $\overline{AB} \parallel \overline{DE}$, $\overline{AB} = 3$ in., $\overline{BC} = 5$ in., $\overline{CE} = 7\frac{1}{2}$ in., $\overline{AC} = 3$ in., find $\overline{CD}$ and $\overline{DE}$.

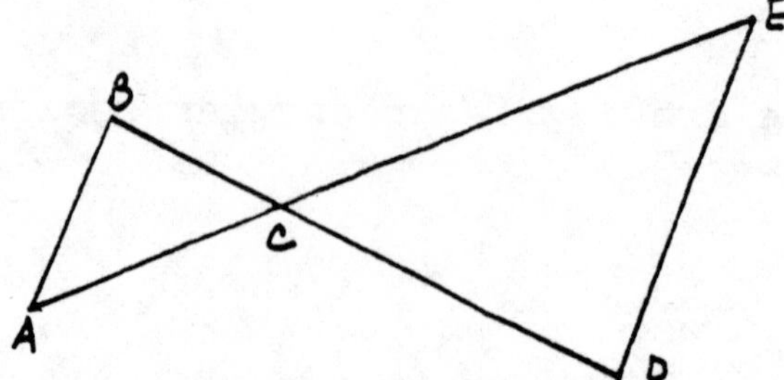

3. Are triangles with sides 3, 5, 7 and 12, 20, 28 similar?

4. A small pine tree, 8 feet tall, casts a shadow of 3 feet, and the large pine tree casts a shadow of 48 feet. Use your knowledge of similar triangles to compute the height of the larger tree.

5. A 5 foot woman casts a 6 foot shadow. In the same location and at the same time a flagpole casts a 30 foot shadow. How high is the flagpole?

6.

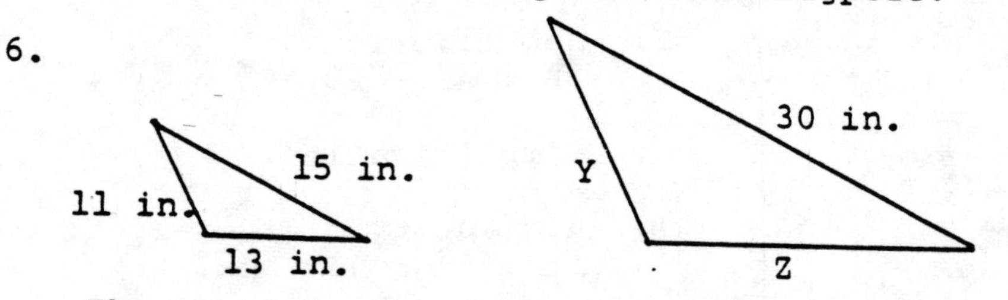

The two triangles illustrated are similar triangles. Find the other sides of the larger figure.

7. Rays of light from the sun form shadows of upright objects on the ground. $\overline{PT}$ represents a large monument; $\overline{TR}$ is a 126 ft. shadow of the monument; $\overline{QS}$ represents a 10 ft. tree which casts an 18 ft. shadow, $\overline{SR}$. Find the height of the monument $\overline{PT}$.

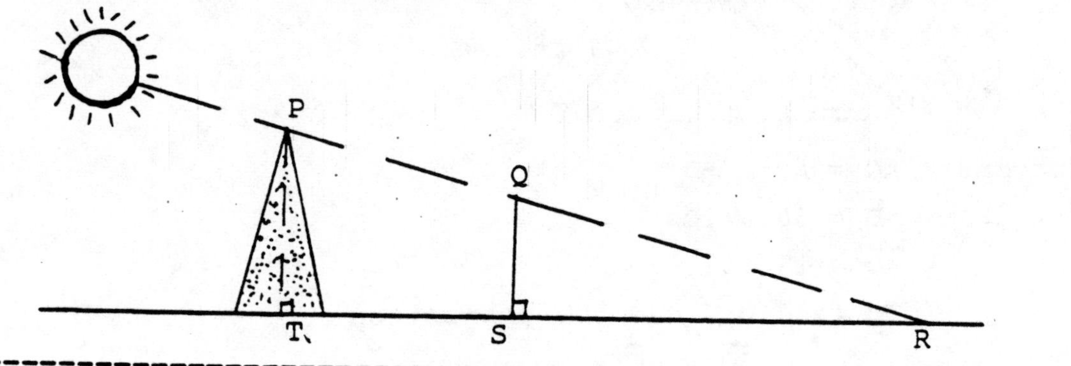

> Theorem 3: A line intersecting two sides of a triangle and parallel to the third side divides the two intersected sides proportionately.

Given: Triangle ABC with $\overline{DE} \parallel \overline{AC}$.

Prove: $\frac{\overline{AD}}{\overline{BD}} = \frac{\overline{EC}}{\overline{BE}}$

Proof:

Statements	Reasons
1. $\overline{DE} \parallel \overline{AC}$	1. Given
2. $\angle A = \angle D$ and $\angle E = \angle C$	2. If two lines are parallel, then their corresponding angles are equal.

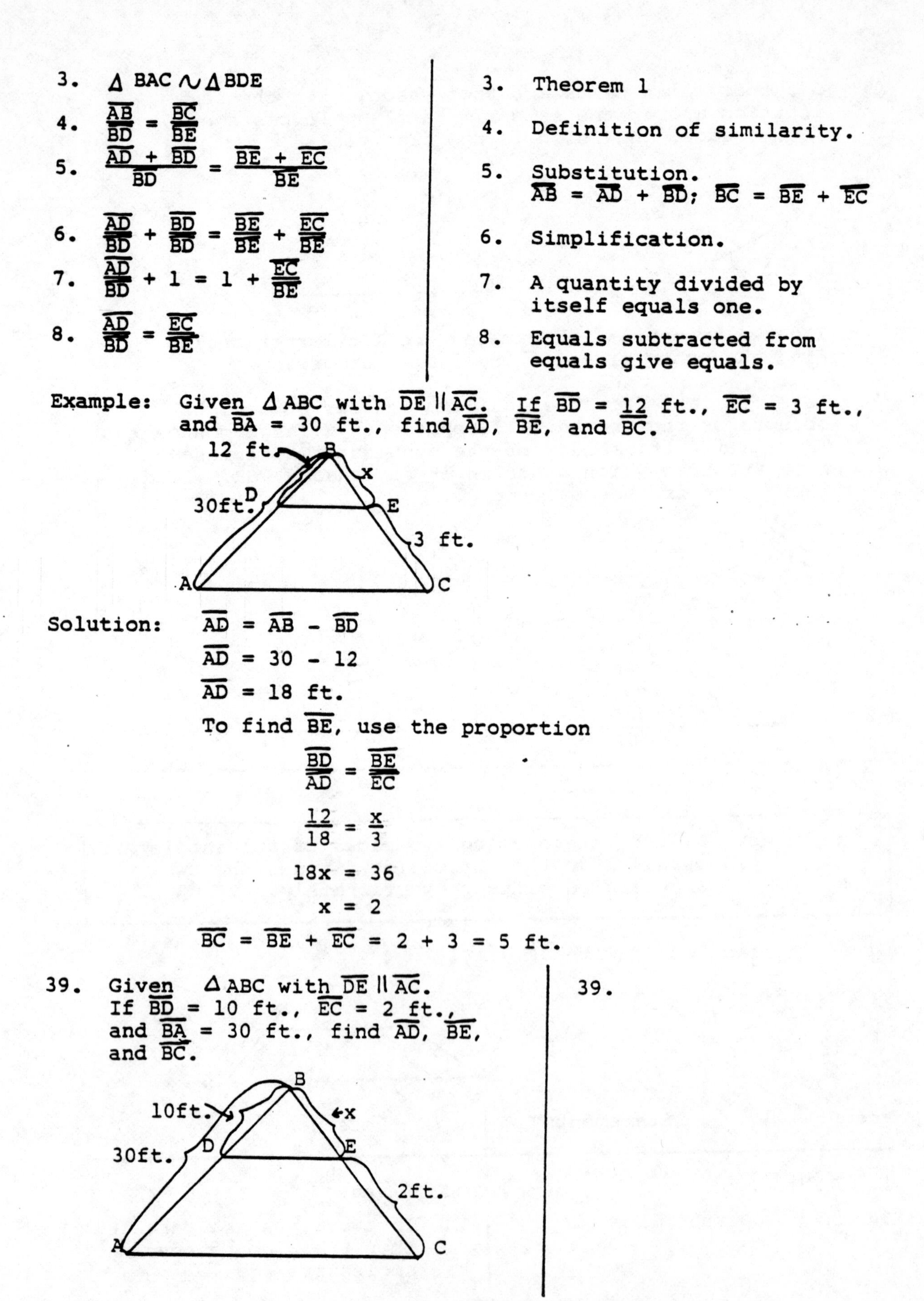

3. $\triangle BAC \sim \triangle BDE$	3. Theorem 1
4. $\dfrac{\overline{AB}}{\overline{BD}} = \dfrac{\overline{BC}}{\overline{BE}}$	4. Definition of similarity.
5. $\dfrac{\overline{AD} + \overline{BD}}{\overline{BD}} = \dfrac{\overline{BE} + \overline{EC}}{\overline{BE}}$	5. Substitution. $\overline{AB} = \overline{AD} + \overline{BD}$; $\overline{BC} = \overline{BE} + \overline{EC}$
6. $\dfrac{\overline{AD}}{\overline{BD}} + \dfrac{\overline{BD}}{\overline{BD}} = \dfrac{\overline{BE}}{\overline{BE}} + \dfrac{\overline{EC}}{\overline{BE}}$	6. Simplification.
7. $\dfrac{\overline{AD}}{\overline{BD}} + 1 = 1 + \dfrac{\overline{EC}}{\overline{BE}}$	7. A quantity divided by itself equals one.
8. $\dfrac{\overline{AD}}{\overline{BD}} = \dfrac{\overline{EC}}{\overline{BE}}$	8. Equals subtracted from equals give equals.

Example: Given $\triangle ABC$ with $\overline{DE} \parallel \overline{AC}$. If $\overline{BD} = 12$ ft., $\overline{EC} = 3$ ft., and $\overline{BA} = 30$ ft., find $\overline{AD}$, $\overline{BE}$, and $\overline{BC}$.

Solution:

$\overline{AD} = \overline{AB} - \overline{BD}$

$\overline{AD} = 30 - 12$

$\overline{AD} = 18$ ft.

To find $\overline{BE}$, use the proportion

$$\frac{\overline{BD}}{\overline{AD}} = \frac{\overline{BE}}{\overline{EC}}$$

$$\frac{12}{18} = \frac{x}{3}$$

$$18x = 36$$

$$x = 2$$

$\overline{BC} = \overline{BE} + \overline{EC} = 2 + 3 = 5$ ft.

39. Given $\triangle ABC$ with $\overline{DE} \parallel \overline{AC}$. If $\overline{BD} = 10$ ft., $\overline{EC} = 2$ ft., and $\overline{BA} = 30$ ft., find $\overline{AD}$, $\overline{BE}$, and $\overline{BC}$.

39.

a. To find $\overline{BE}$, what proportion would you use? ____________

$\overline{BE}$ = ____________

a. Let $\overline{BE} = x$.

$$\frac{\overline{BD}}{\overline{AD}} = \frac{\overline{BE}}{\overline{EC}}$$

$$\frac{10}{20} = \frac{x}{2}$$

$$20x = 20$$

$$x = 1$$

$\overline{BE} = 1$ ft.

b. To find $\overline{AD}$, what equation would you use? ____________

$\overline{AD}$ = ____________

b. $\overline{AD} = \overline{AB} - \overline{BD}$

$\overline{AD} = 30 - 10$

$\overline{AD} = 20$ ft.

c. $\overline{BC}$ = ____________

c. $\overline{BC} = \overline{BE} + \overline{EC}$

$\overline{BC} = 1 + 2$

$\overline{BC} = 3$ ft.

40. Given $\triangle RST$ with $\overline{UV} \parallel \overline{TS}$. If $\overline{RV}$ = 9 cm., $\overline{VS}$ = 3 cm., and $\overline{RU}$ = 7 cm., find $\overline{UT}$, $\overline{RS}$, and $\overline{RT}$.

T
x
U
7cm.
R
9cm.
V
3cm.
S

a. To find $\overline{UT}$, what proportion would you use? ____________

$\overline{UT}$ = ____________

40.

a. Let $\overline{UT} = x$.

$$\frac{\overline{RV}}{\overline{VS}} = \frac{\overline{RU}}{x}$$

$$\frac{9}{3} = \frac{7}{x}$$

$$9x = 21$$

$$x = \frac{21}{9} = 2\frac{1}{3} \text{ cm.}$$

b. To find $\overline{RS}$, what equation would you use? ____________

$\overline{RS}$ = ____________

b. $\overline{RS} = \overline{RV} + \overline{VS}$

$\overline{RS} = 9 + 3$

$\overline{RS} = 12$ cm.

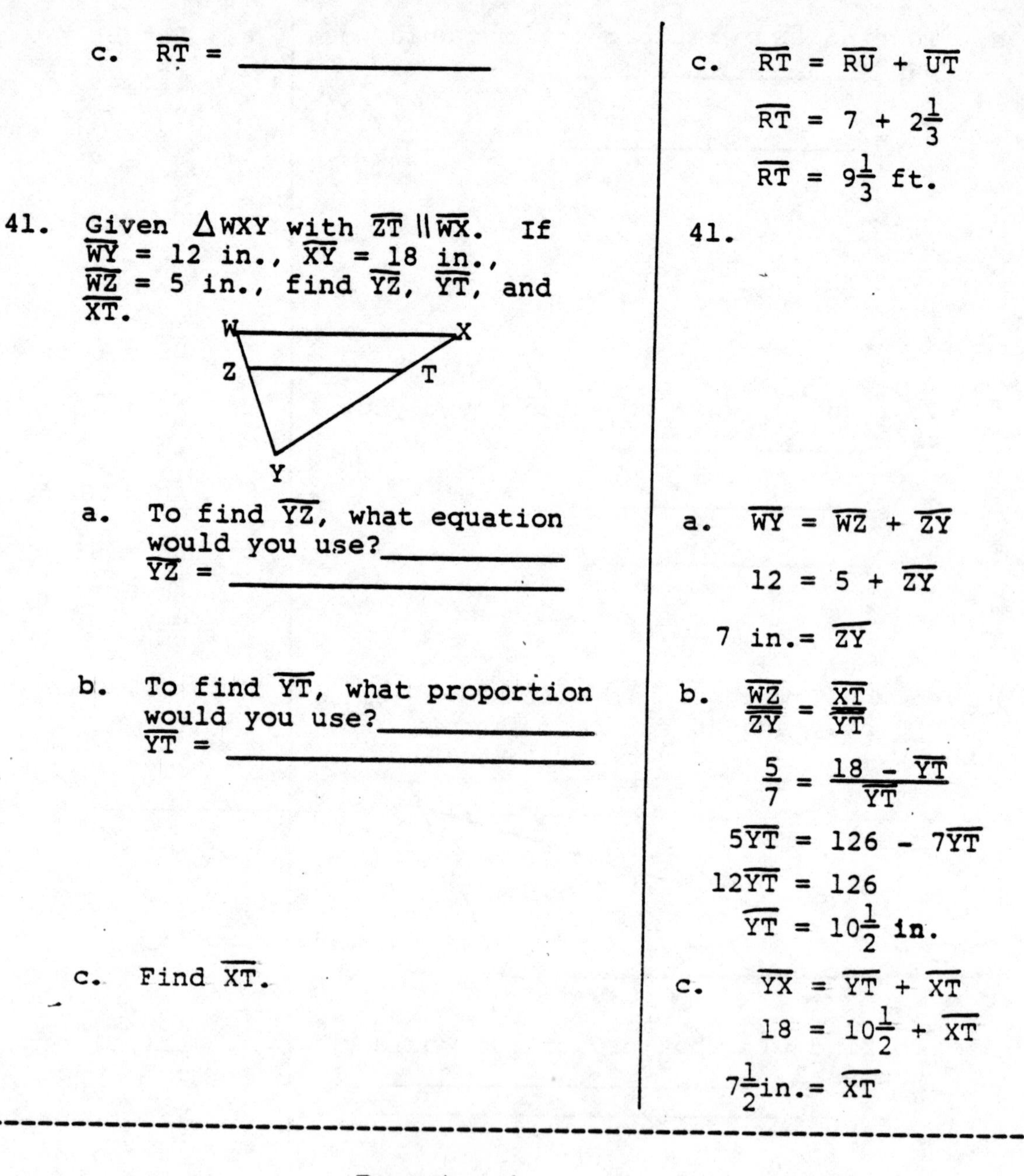

c. $\overline{RT}$ = ________________

c. $\overline{RT} = \overline{RU} + \overline{UT}$

$\overline{RT} = 7 + 2\frac{1}{3}$

$\overline{RT} = 9\frac{1}{3}$ ft.

41. Given $\triangle WXY$ with $\overline{ZT} \parallel \overline{WX}$. If $\overline{WY}$ = 12 in., $\overline{XY}$ = 18 in., $\overline{WZ}$ = 5 in., find $\overline{YZ}$, $\overline{YT}$, and $\overline{XT}$.

a. To find $\overline{YZ}$, what equation would you use? ________________
$\overline{YZ}$ = ________________

b. To find $\overline{YT}$, what proportion would you use? ________________
$\overline{YT}$ = ________________

c. Find $\overline{XT}$.

41.

a. $\overline{WY} = \overline{WZ} + \overline{ZY}$

$12 = 5 + \overline{ZY}$

7 in.= $\overline{ZY}$

b. $\frac{\overline{WZ}}{\overline{ZY}} = \frac{\overline{XT}}{\overline{YT}}$

$\frac{5}{7} = \frac{18 - \overline{YT}}{\overline{YT}}$

$5\overline{YT} = 126 - 7\overline{YT}$

$12\overline{YT} = 126$

$\overline{YT} = 10\frac{1}{2}$ in.

c. $\overline{YX} = \overline{YT} + \overline{XT}$

$18 = 10\frac{1}{2} + \overline{XT}$

$7\frac{1}{2}$in.= $\overline{XT}$

Exercise 4

1. Given: $\overline{WV} \parallel \overline{TU}$. Is $\frac{\overline{WT}}{8} = \frac{5}{6}$ a correct proportion?

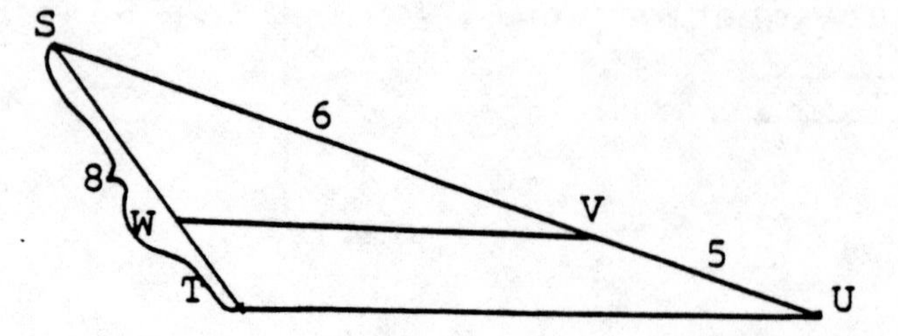

2. $\overline{AC} \parallel \overline{DE}$, $\overline{AD}$ = 4 in., $\overline{DB}$ = 10 in., $\overline{BC}$ = 15 in. and $\angle BDE = 25^{\circ}$. Find $\overline{BE}$, $\overline{EC}$, and $\angle A$.

3. Find x and $\angle BED$.

4. In Δ ABC, $\overline{DE} \parallel \overline{BC}$.

   a. a = 9 m., $\overline{AB}$ = 12 m., d = 7 m., find c.

   b. b = 4 m., a = 8 m., $\overline{AC}$ = 14 m., find d.

   c. $\overline{AB}$ = 20 m., $\overline{AC}$ = 26m., c = 18 m., find a.

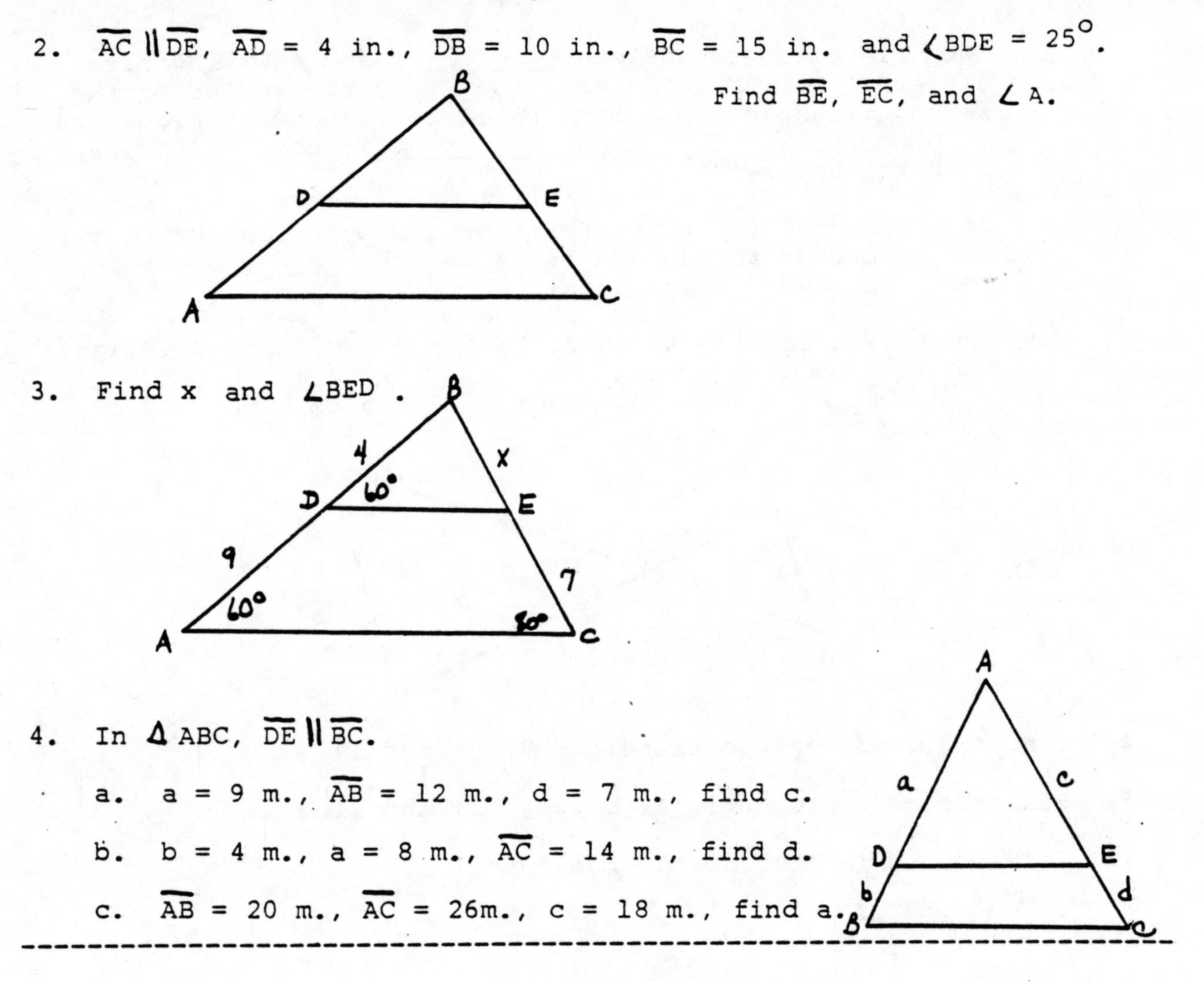

Unit 10 Review

1. Fill in the blank with the appropriate word or phrase.

   a. A __________ is the comparison of two numbers by their indicated quotient.

   b. A ________is a statement that two ratios are equal.

   c. In the proportion $\frac{a}{b} = \frac{c}{d}$ the numbers a and d are called the ___________ of the proportion, and the numbers b and c are called the _________ of the proportion. The single term, d, is called the ____________.

d. Two triangles are called __________ if and only if three angles of one triangle are equal to three angles of the second triangle and all pairs of corresponding sides are in proportion. The symbol __________ is used to indicate such triangles.

e. If the second and third terms of a proportion are equal, the second or third term is called the ________ between the first and fourth terms.

2. Are the triangles with sides 3, 5, 7 and 12, 20, 28 similar?

3. Given $\overline{AB} \parallel \overline{DE}$, $\overline{AB}$ = 3cm., $\overline{BC}$ = 2cm., $\overline{CE}$ = 5 cm., $\overline{CD}$ = 10 cm., find $\overline{AC}$ and $\overline{DE}$.

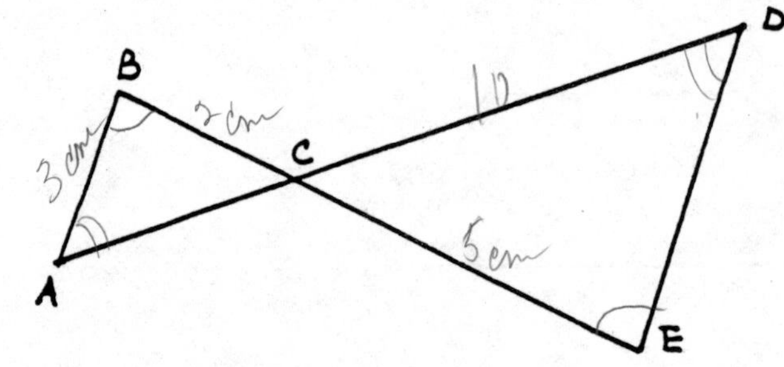

4. Find the fourth proportional to 3, 9, and 17.

5. Find the mean proportional between 11 and 21.

6. Solve for x in each of the following:

a. $\frac{4}{3} = \frac{x + 3}{5}$

d. $\frac{x}{\sqrt{7}} = \frac{\sqrt{8}}{\sqrt{14}}$

b. $\frac{2}{x + 7} = \frac{x + 5}{3}$

e. $\frac{x + 6}{7} = \frac{-4}{x - 5}$

c. $\frac{x}{\sqrt{5}} = \frac{6\sqrt{3}}{\sqrt{2}}$

f. $\frac{x}{x - 6} = \frac{5}{6}$

7. Which of the following proportions are true?

a. $\frac{18}{9} = \frac{\left(\frac{1}{3}\right)}{\left(\frac{1}{2}\right)}$

b. $\frac{7}{8} = \frac{21}{24}$

8. Find the ratio of 7 minutes to 2 hours.

9. A 7 foot upright pole near a vertical tree casts a 5 foot shadow . At the same time, find the height of the tree if its shadow is 35 feet.

10. The two triangles given below are similar. Find x and y.
$\Delta ABC \sim \Delta DEF$

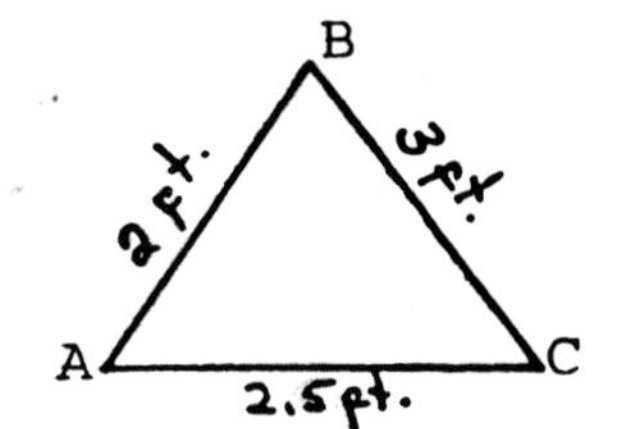

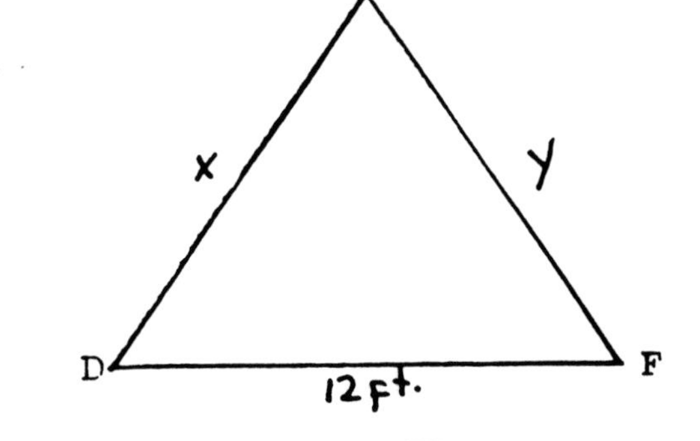

11. $\overline{BD} \parallel \overline{AE}$, $\overline{AB}$ = 6 ft., $\overline{BC}$ = 10 ft., $\overline{DE}$ = 9 ft., find $\overline{DC}$, $\overline{AC}$, $\overline{EC}$, $\angle CBD$, $\angle E$, and $\angle C$.

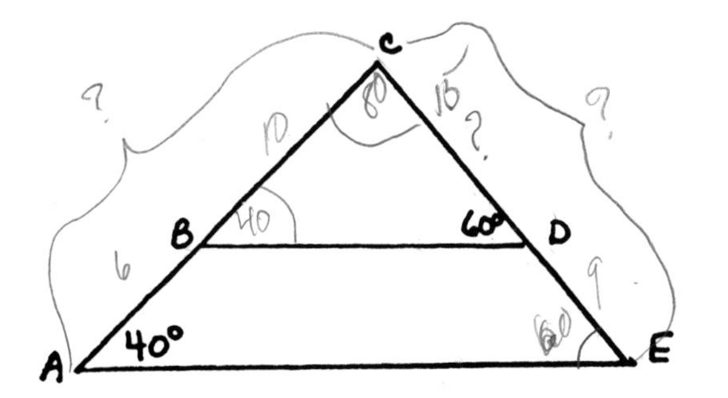

12. $\overline{XY} \parallel \overline{ON}$. $\overline{MO}$ = 21 in., $\overline{MX}$ = 16 in., $\overline{XN}$ = 8 in., find $\overline{MY}$, $\overline{YO}$, and $\overline{MN}$.

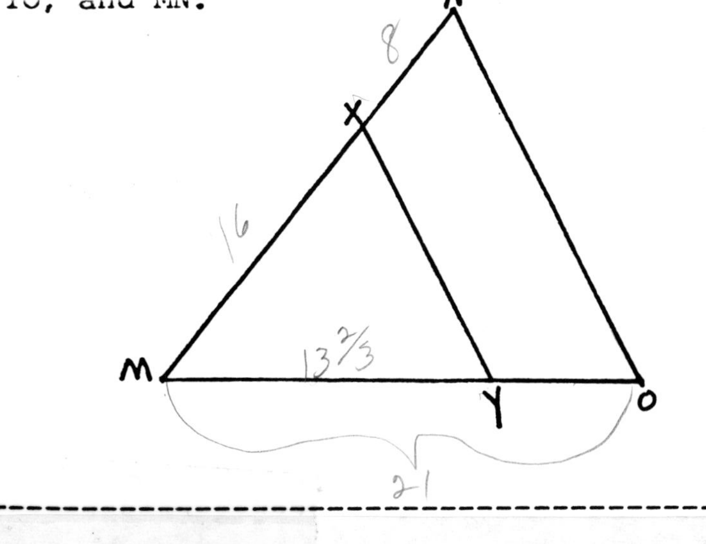

Unit 11

BASIC RIGHT TRIANGLE TRIGONOMETRY

Learning Objectives:

1. The student will demonstrate his mastery of the following definitions by writing them and by applying them to solutions of selected problems: tangent, sine, cosine

2. The student will demonstrate his mastery of using the Square Root Table and Trigonometric Ratios Table by applying them correctly in problem solving.

Unit 11

Basic Right Triangle Trigonometry

The Greek word "trigonometry" means the measurement of triangles. In this unit, however, we will center our attention on the measurement of the acute angles of right triangles.

Study these two right triangles in Figure 11.1 and recall our discussion on similarity from Unit 10. If two angles of one triangle are equal to two corresponding angles of a second triangle, then the triangles are similar. $\Delta A_1B_1C_1 \sim \Delta A_2B_2C_2$ with $\angle B_1 = \angle B_2$ and $\angle C_1 = \angle C_2$.

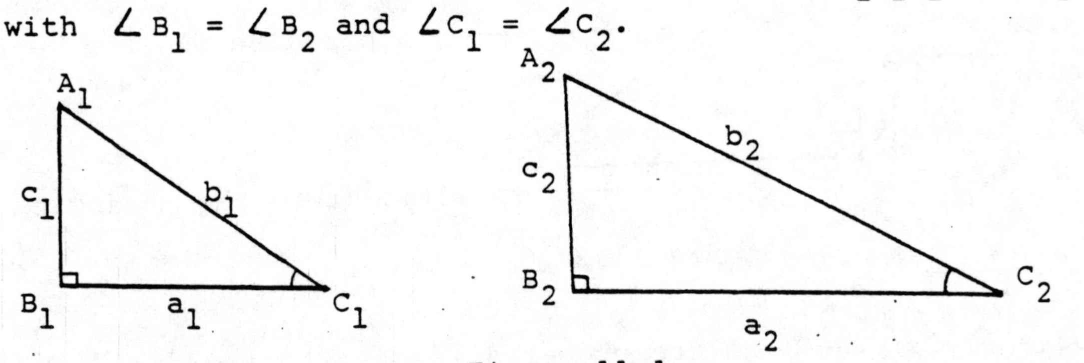

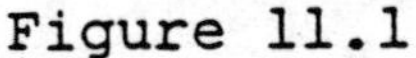

Figure 11.1

By the difinition of similarity, the corresponding sides of the two triangles are proportional.

$$\frac{a_1}{a_2} = \frac{b_1}{b_2} = \frac{c_1}{c_2}$$

By interchanging the means in these equations, it is not difficult to obtain the following relationships. The corresponding ratios of the sides would equal:

$$\frac{a_1}{b_1} = \frac{a_2}{b_2} \text{ and } \frac{b_1}{c_1} = \frac{b_2}{c_2} \text{ and } \frac{a_1}{c_1} = \frac{a_2}{c_2}$$

As illustrated from the new sets of equations, we discovered that there are three basic ratios common to both triangles. These ratios are $\frac{a}{b}$, $\frac{b}{c}$, and $\frac{a}{c}$. These are called trigonmetric ratios. The actual values for these three ratios do not depend on the size of the triangles but upon the measure of the acute angle involved. (The measure of ∠C in Figure 11.1 is the same in both triangles.)

Definition:	The tangent (abbreviated as tan) of an acute angle of a right triangle is the ratio of the length of the opposite leg to the length of the adjacent leg.

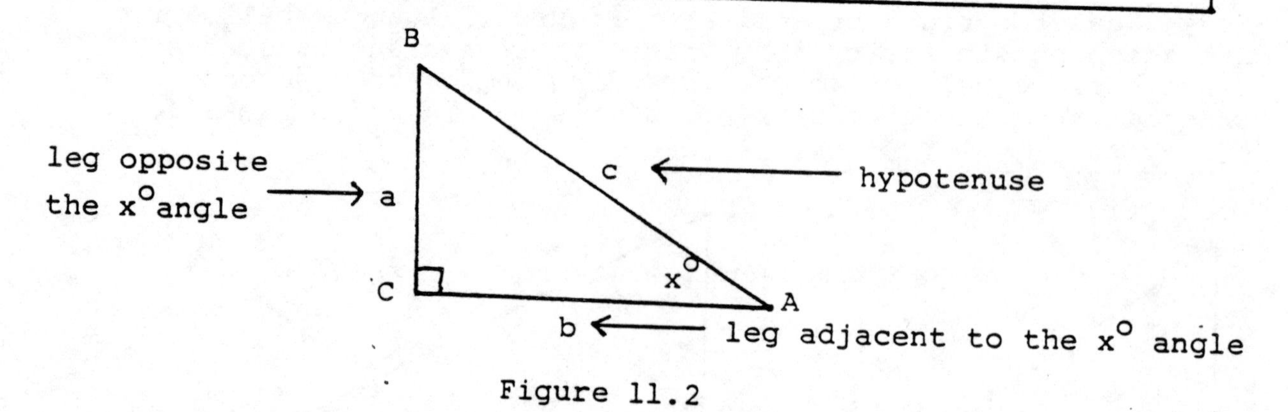

Figure 11.2

Study Figure 11.2. The legs are properly identified and labeled. Therefore, in reference to ∠A, a is the opposite leg and b is the adjacent leg.

$$\tan \angle A = \tan x^\circ = \frac{\text{opposite leg}}{\text{adjacent leg}} = \frac{a}{b}$$

Example: Find the tangent of angle Z in Figure 11.3

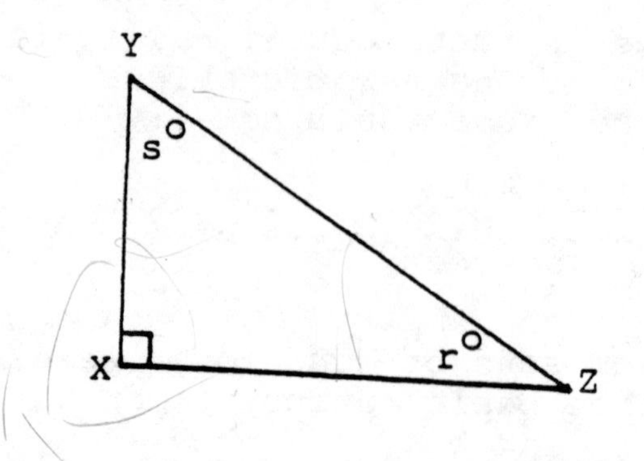

Figure 11.3

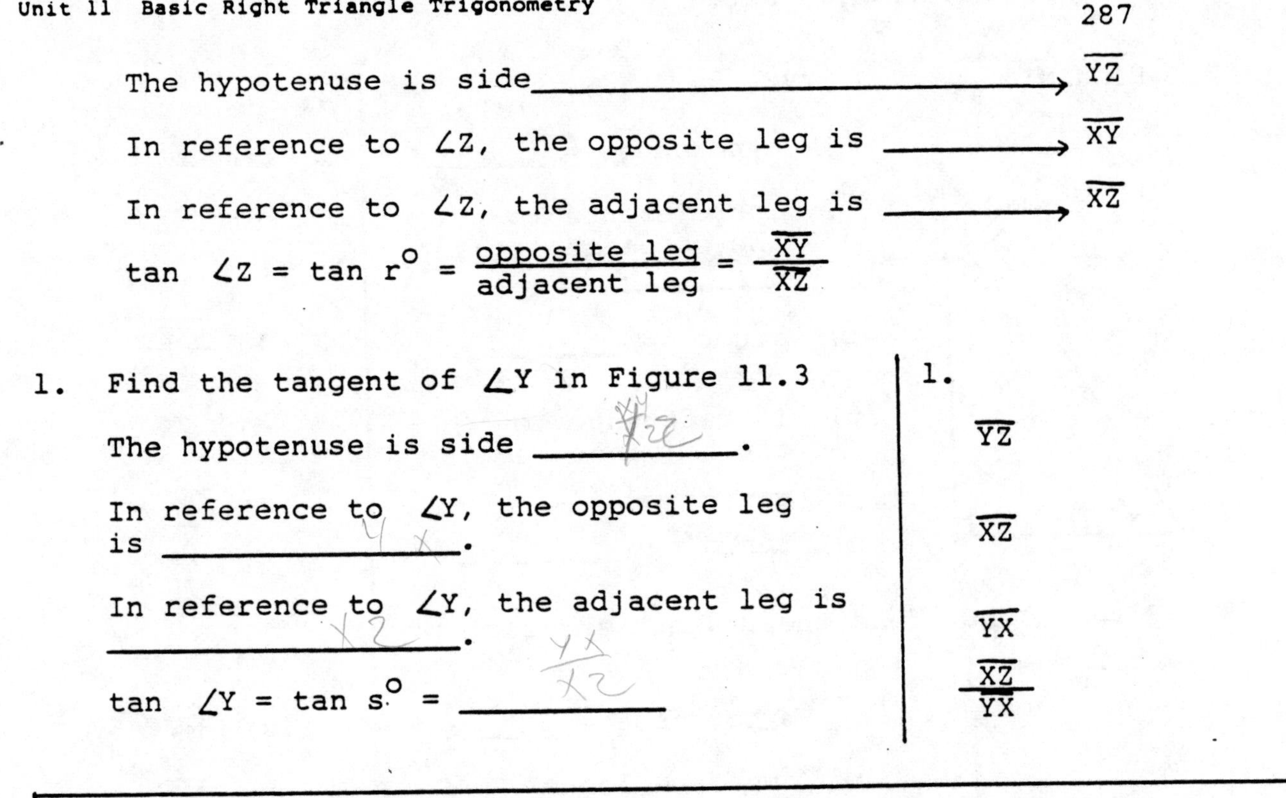

The hypotenuse is side ⟶ $\overline{YZ}$

In reference to $\angle Z$, the opposite leg is ⟶ $\overline{XY}$

In reference to $\angle Z$, the adjacent leg is ⟶ $\overline{XZ}$

$$\tan \angle Z = \tan r^\circ = \frac{\text{opposite leg}}{\text{adjacent leg}} = \frac{\overline{XY}}{\overline{XZ}}$$

1. Find the tangent of $\angle Y$ in Figure 11.3

   The hypotenuse is side ________.

   In reference to $\angle Y$, the opposite leg is ________.

   In reference to $\angle Y$, the adjacent leg is ________.

   $\tan \angle Y = \tan s^\circ =$ ________

Answers:

1. $\overline{YZ}$

   $\overline{XZ}$

   $\overline{YX}$

   $\dfrac{\overline{XZ}}{\overline{YX}}$

> **Definition:** The <u>sine</u> (abbreviated sin) of an acute angle of a right triangle is the ratio of the length of the opposite leg to the length of the hypotenuse.

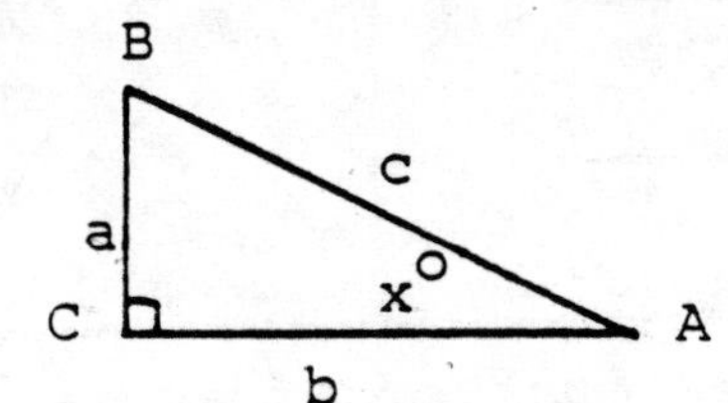

$$\sin \angle A = \sin x^\circ = \frac{\text{opposite leg}}{\text{hypotenuse}} = \frac{a}{c}$$

Example: In ΔXYZ find the tan r° and sin r°.

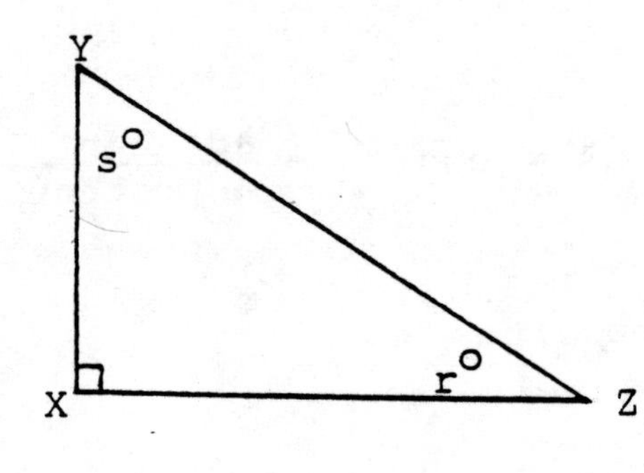

a) hypotenuse = $\overline{YZ}$

b) In reference to $\angle Z$, the opposite leg = $\overline{XY}$.

c) In reference to $\angle Z$, the adjacent leg = $\overline{XZ}$

d) $\tan r^\circ = \dfrac{\text{opposite leg}}{\text{adjacent leg}} = \dfrac{\overline{XY}}{\overline{XZ}}$

e) $\sin r^\circ = \dfrac{\text{opposite leg}}{\text{hypotenuse}} = \dfrac{\overline{XY}}{\overline{YZ}}$

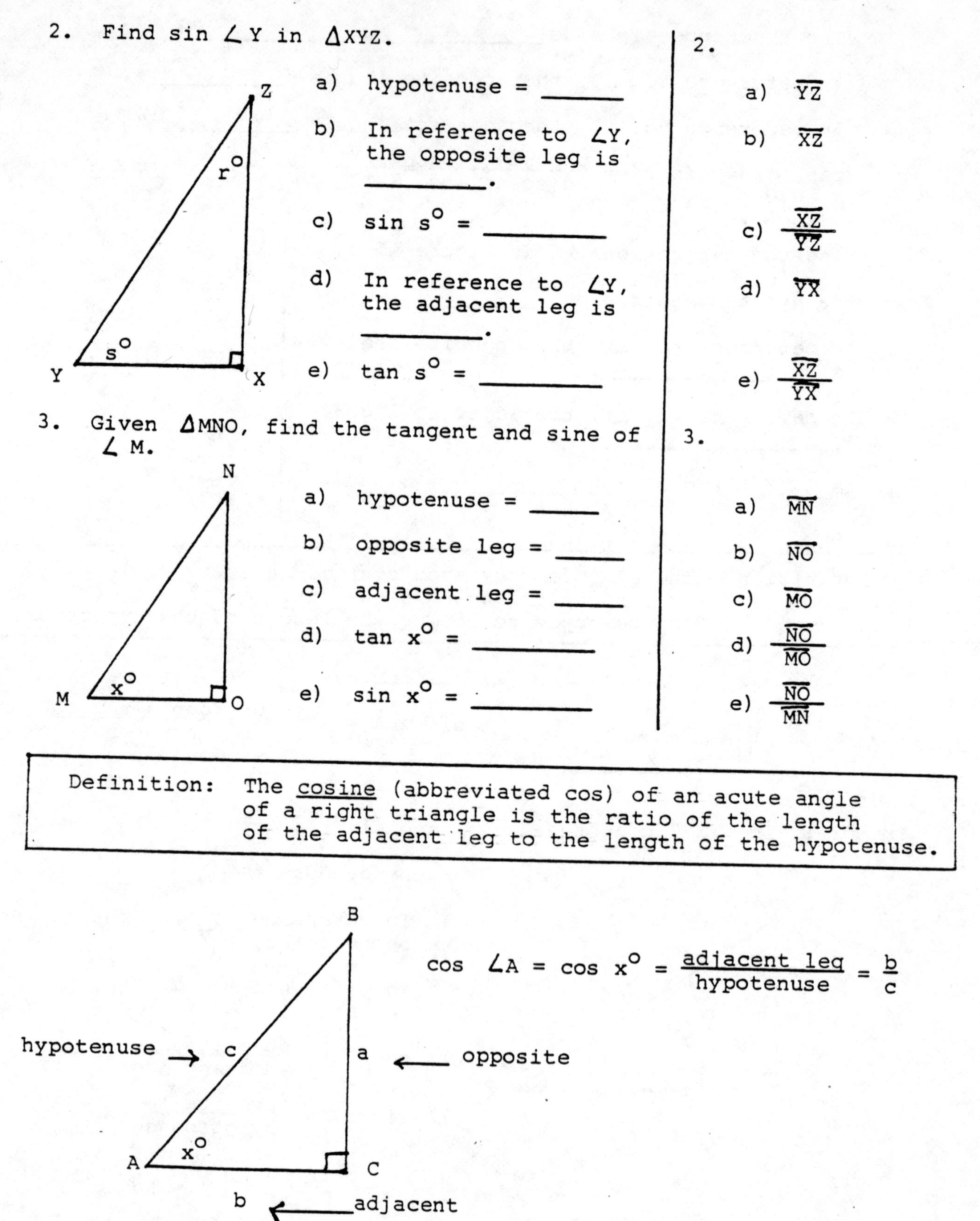

2. Find sin $\angle$Y in ΔXYZ.

   a) hypotenuse = ______

   b) In reference to $\angle$Y, the opposite leg is ______.

   c) $\sin s^o$ = ______

   d) In reference to $\angle$Y, the adjacent leg is ______.

   e) $\tan s^o$ = ______

2.

   a) $\overline{YZ}$

   b) $\overline{XZ}$

   c) $\frac{\overline{XZ}}{\overline{YZ}}$

   d) $\overline{YX}$

   e) $\frac{\overline{XZ}}{\overline{YX}}$

3. Given ΔMNO, find the tangent and sine of $\angle$ M.

   a) hypotenuse = ______

   b) opposite leg = ______

   c) adjacent leg = ______

   d) $\tan x^o$ = ______

   e) $\sin x^o$ = ______

3.

   a) $\overline{MN}$

   b) $\overline{NO}$

   c) $\overline{MO}$

   d) $\frac{\overline{NO}}{\overline{MO}}$

   e) $\frac{\overline{NO}}{\overline{MN}}$

Definition:	The <u>cosine</u> (abbreviated cos) of an acute angle of a right triangle is the ratio of the length of the adjacent leg to the length of the hypotenuse.

$$\cos \angle A = \cos x^o = \frac{\text{adjacent leg}}{\text{hypotenuse}} = \frac{b}{c}$$

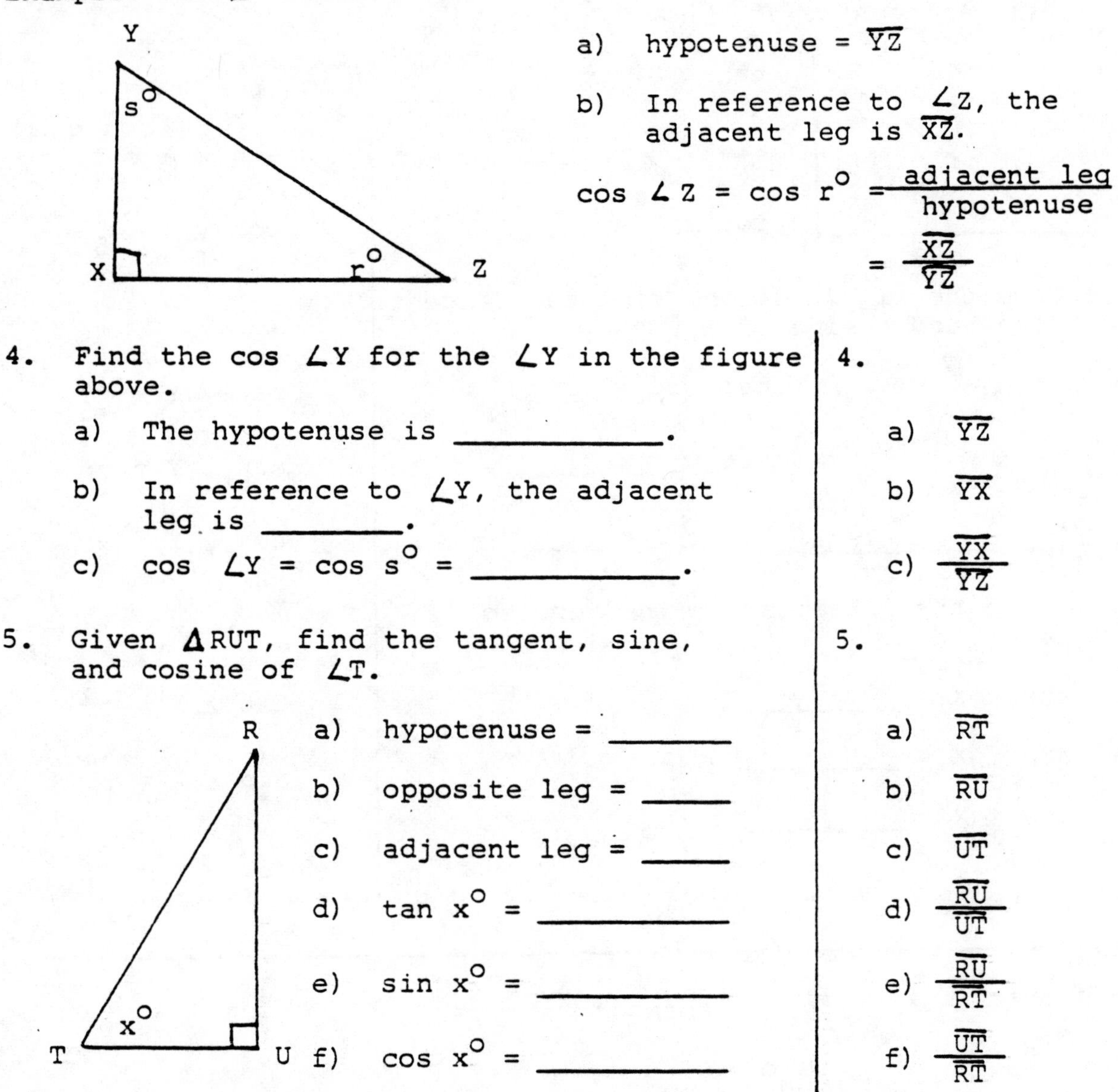

Example: In $\triangle$ XYZ find the cos $\angle$Z.

a) hypotenuse = $\overline{YZ}$

b) In reference to $\angle$Z, the adjacent leg is $\overline{XZ}$.

$$\cos \angle Z = \cos r^{\circ} = \frac{\text{adjacent leg}}{\text{hypotenuse}} = \frac{\overline{XZ}}{\overline{YZ}}$$

4. Find the cos $\angle$Y for the $\angle$Y in the figure above.

a) The hypotenuse is ____________.

b) In reference to $\angle$Y, the adjacent leg is ________.

c) cos $\angle$Y = cos s° = ____________.

4.

a) $\overline{YZ}$

b) $\overline{YX}$

c) $\frac{\overline{YX}}{\overline{YZ}}$

5. Given $\triangle$RUT, find the tangent, sine, and cosine of $\angle$T.

a) hypotenuse = _______

b) opposite leg = _____

c) adjacent leg = _____

d) tan x° = __________

e) sin x° = __________

f) cos x° = __________

5.

a) $\overline{RT}$

b) $\overline{RU}$

c) $\overline{UT}$

d) $\frac{\overline{RU}}{\overline{UT}}$

e) $\frac{\overline{RU}}{\overline{RT}}$

f) $\frac{\overline{UT}}{\overline{RT}}$

When the lengths of the legs and hypotenuse are known, these values may be substituted in the ratios and a numerical value found for the trignometric ratios.

Example: Find the tangent, sine, and cosine of $\angle$M.

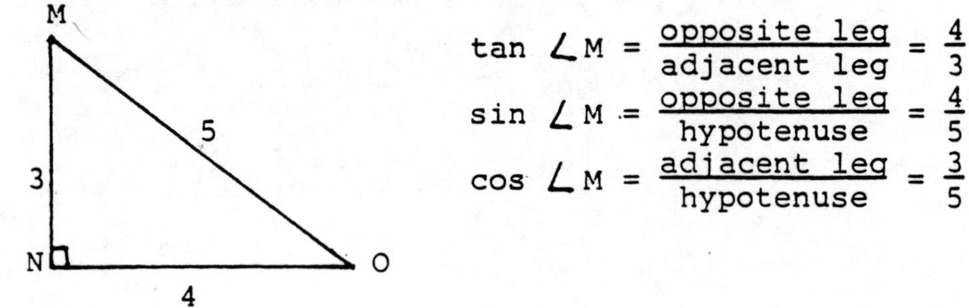

$$\tan \angle M = \frac{\text{opposite leg}}{\text{adjacent leg}} = \frac{4}{3}$$

$$\sin \angle M = \frac{\text{opposite leg}}{\text{hypotenuse}} = \frac{4}{5}$$

$$\cos \angle M = \frac{\text{adjacent leg}}{\text{hypotenuse}} = \frac{3}{5}$$

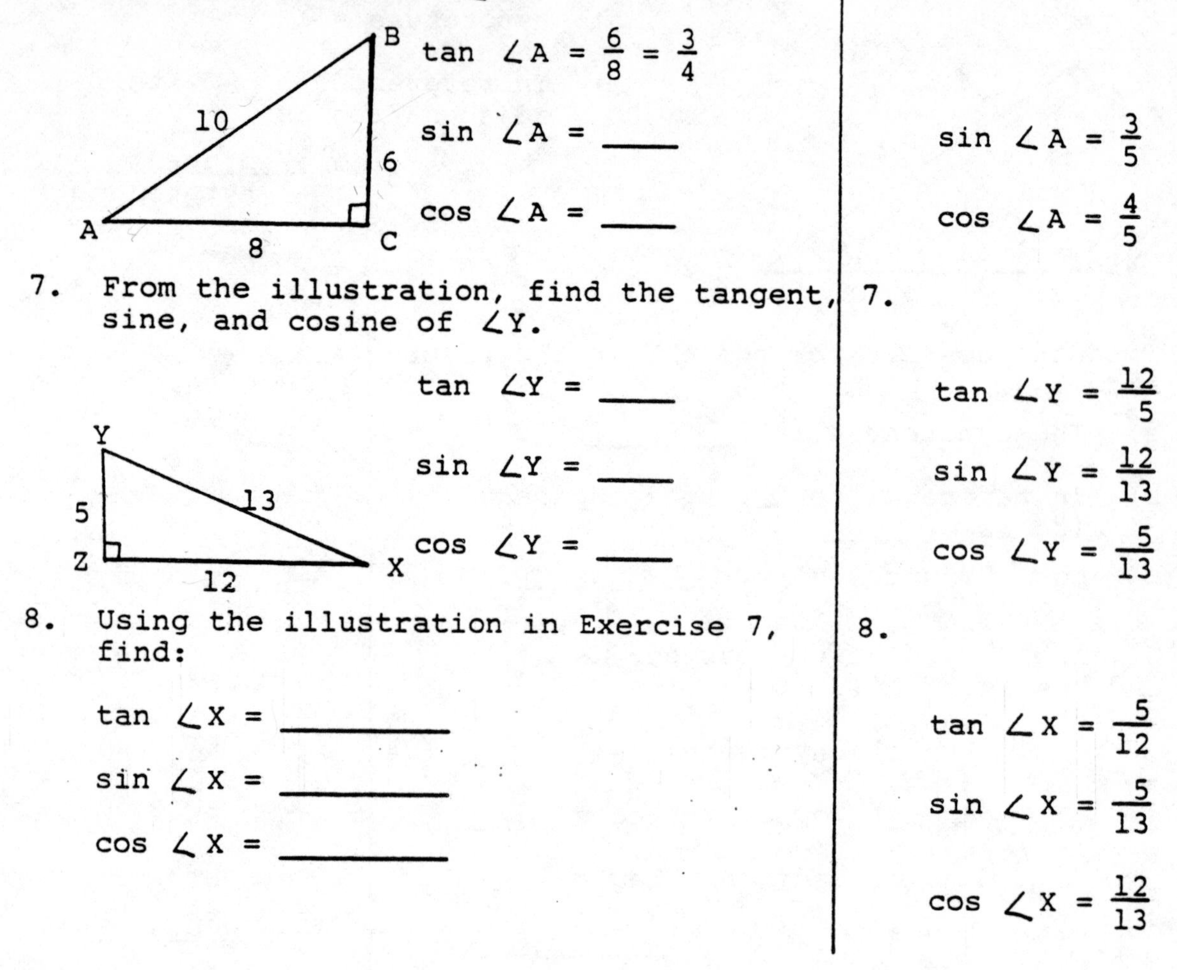

6. From the illustration, find the tangent, sine, and cosine of $\angle A$.

$\tan \angle A = \frac{6}{8} = \frac{3}{4}$

$\sin \angle A =$ ____

$\cos \angle A =$ ____

6. $\sin \angle A = \frac{3}{5}$

$\cos \angle A = \frac{4}{5}$

7. From the illustration, find the tangent, sine, and cosine of $\angle Y$.

$\tan \angle Y =$ ____

$\sin \angle Y =$ ____

$\cos \angle Y =$ ____

7. $\tan \angle Y = \frac{12}{5}$

$\sin \angle Y = \frac{12}{13}$

$\cos \angle Y = \frac{5}{13}$

8. Using the illustration in Exercise 7, find:

$\tan \angle X =$ ________

$\sin \angle X =$ ________

$\cos \angle X =$ ________

8. $\tan \angle X = \frac{5}{12}$

$\sin \angle X = \frac{5}{13}$

$\cos \angle X = \frac{12}{13}$

In summary, given right triangle XYZ,

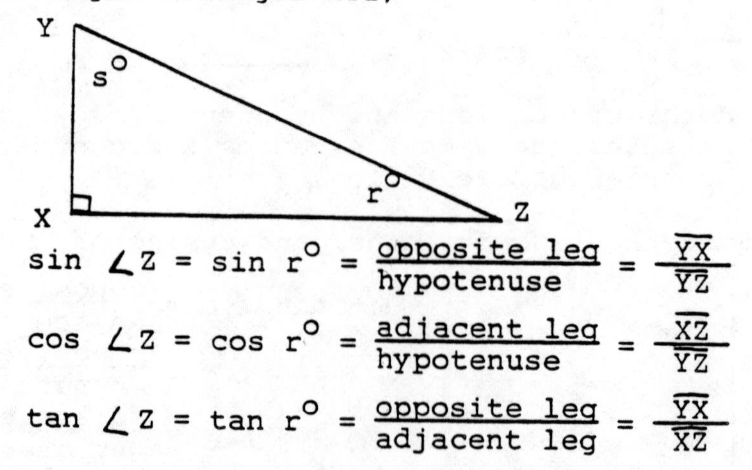

$$\sin \angle Z = \sin r^\circ = \frac{\text{opposite leg}}{\text{hypotenuse}} = \frac{\overline{YX}}{\overline{YZ}}$$

$$\cos \angle Z = \cos r^\circ = \frac{\text{adjacent leg}}{\text{hypotenuse}} = \frac{\overline{XZ}}{\overline{YZ}}$$

$$\tan \angle Z = \tan r^\circ = \frac{\text{opposite leg}}{\text{adjacent leg}} = \frac{\overline{YX}}{\overline{XZ}}$$

$$\sin \angle Y = \sin s^\circ = \frac{\text{opposite leg}}{\text{hypotenuse}} = \frac{\overline{XZ}}{\overline{YZ}}$$

$$\cos \angle Y = \cos s^\circ = \frac{\text{adjacent leg}}{\text{hypotenuse}} = \frac{\overline{YX}}{\overline{YZ}}$$

$$\tan \angle Y = \tan s^\circ = \frac{\text{opposite leg}}{\text{adjacent leg}} = \frac{\overline{XZ}}{\overline{YX}}$$

Exercise 1

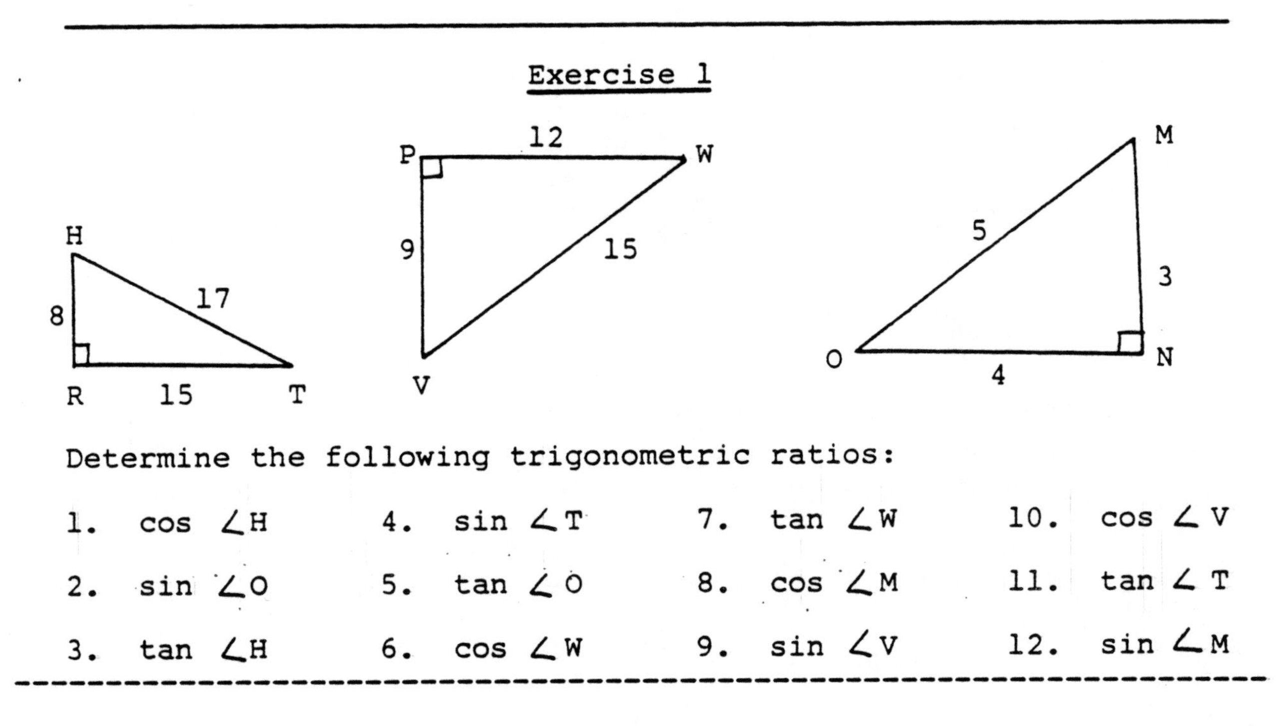

Determine the following trigonometric ratios:

1. cos ∠H	4. sin ∠T	7. tan ∠W	10. cos ∠V
2. sin ∠O	5. tan ∠O	8. cos ∠M	11. tan ∠T
3. tan ∠H	6. cos ∠W	9. sin ∠V	12. sin ∠M

Reading the Trigonometric Table

For some angles, the trigonometric ratios are not difficult to calculate. Specifically, let's examine the isosceles right triangle and the 30°-60° right triangle.

Example: △ABC is an isosceles right triangle. Find sin ∠C, tan ∠C and cos ∠C.

A, B, C; 45°, 45°; 3, 3, $3\sqrt{2}$

Theorem 3 in Unit 9 tells us that the hypotenuse measures $\sqrt{2}$ times the length of one of the sides of the triangle.

$$\sin \angle C = \sin 45^\circ = \frac{3}{3\sqrt{2}} = \frac{1}{\sqrt{2}} \text{ or } \frac{\sqrt{2}}{2}$$

$$\tan \angle C = \tan 45^\circ = \frac{3}{3} = 1$$

$$\cos \angle C = \cos 45^\circ = \frac{3}{3\sqrt{2}} = \frac{1}{\sqrt{2}} \text{ or } \frac{\sqrt{2}}{2}$$

Example: $\triangle ABC$ is a 30°-60° right triangle. Find the sine, cosine, and tangent of $\angle C$ and $\angle A$.

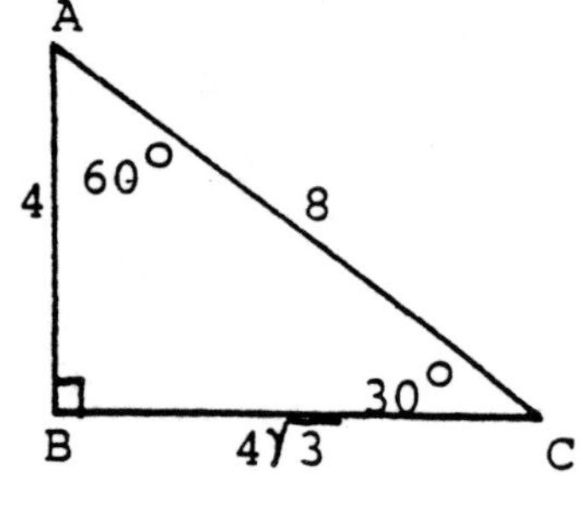

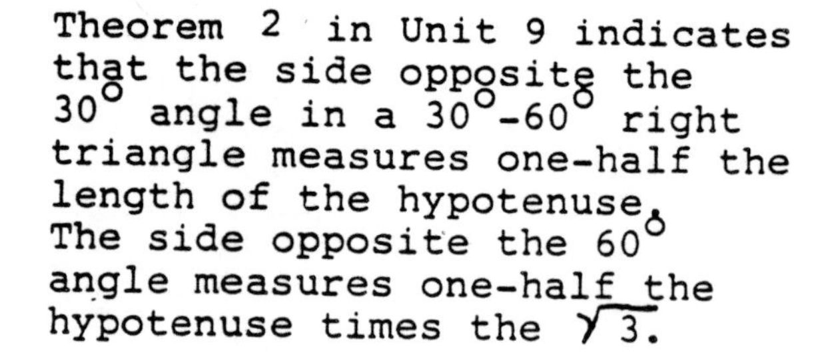

Theorem 2 in Unit 9 indicates that the side opposite the 30° angle in a 30°-60° right triangle measures one-half the length of the hypotenuse. The side opposite the 60° angle measures one-half the hypotenuse times the $\sqrt{3}$.

$$\sin \angle C = \sin 30^\circ = \frac{4}{8} = \frac{1}{2}$$

$$\tan \angle C = \tan 30^\circ = \frac{4}{4\sqrt{3}} = \frac{1}{\sqrt{3}} \text{ or } \frac{\sqrt{3}}{3}$$

$$\cos \angle C = \cos 30^\circ = \frac{4\sqrt{3}}{8} = \frac{\sqrt{3}}{2}$$

$$\sin \angle A = \sin 60^\circ = \frac{4\sqrt{3}}{8} = \frac{\sqrt{3}}{2}$$

$$\tan \angle A = \tan 60^\circ = \frac{4\sqrt{3}}{4} = \sqrt{3}$$

$$\cos \angle A = \cos 60^\circ = \frac{4}{8} = \frac{1}{2}$$

Using the square root tables on page 379, find the decimal approximations for $\sqrt{2}$ and $\sqrt{3}$. Now, substitute these values into the trigonometric ratios listed.

Since $\sqrt{2} = 1.414$ and $\sqrt{3} = 1.732$ then:

$$\begin{cases} \sin 30^\circ = \frac{1}{2} = .500 \\ \tan 30^\circ = \frac{\sqrt{3}}{3} = \frac{1.732}{3} = .577 \\ \cos 30^\circ = \frac{\sqrt{3}}{2} = \frac{1.732}{2} = .866 \end{cases}$$

$$\begin{cases} \sin 45^\circ = \frac{\sqrt{2}}{2} = \frac{1.414}{2} = .707 \\ \tan 45^\circ = 1.000 \\ \cos 45^\circ = \frac{\sqrt{2}}{2} = \frac{1.414}{2} = .707 \end{cases}$$

$$\begin{cases} \sin 60^\circ = \frac{\sqrt{3}}{2} = .866 \\ \tan 60^\circ = \sqrt{3} = 1.732 \\ \cos 60^\circ = \frac{1}{2} = .500 \end{cases}$$

These trigonometric values are arranged in a table similar to the one shown below in Figure 11.4.

Degrees	$\sin x^\circ$	$\cos x^\circ$	$\tan x^\circ$
30°	.500	.866	.577
45°	.707	.707	1.000
60°	.866	.500	1.732

Figure 11.4

Example: Using the table in Figure 11.4, find the cosine of an angle that measures 30°.

Solution: First find the 30° angle under the column headed <u>degrees</u>. From this point move horizontally to the right and stop under the column for cosine. The value you find will be .866.

$\cos 30^\circ = .866$

A complete table of trignometric values can be found in the appendix.

Example: Using the Trigonometric Ratios Table on page 380, find the tangent, cosine, and sine of an 89° angle.

Solution: Locate the 89° measure in the fifth column under the degree heading. From this point move horizontally to the right stopping at each new column. The value you find for sine will be 1.000; .017 for cosine; 57.290 for tangent.

Therefore, $\sin 89^{\circ} = 1.000$

$\cos 89^{\circ} = .017$

$\tan 89^{\circ} = 57.290$

9. Using the Trigonometric Ratios Table, find the value of each of the following.

a) $\tan 37^{\circ}$ = ________

b) $\sin 45^{\circ}$ = ________

c) $\tan 60^{\circ}$ = ________

d) $\cos 37^{\circ}$ = ________

e) $\sin 45^{\circ}$ = ________

f) $\sin 30^{\circ}$ = ________

9.

a) .754

b) .707

c) 1.732

d) .799

e) .707

f) .500

Example: Using the Trigonometric Ratios Table, find the measure of an angle A whose tangent equals .601 .

Solution: First, locate the value .601 in the tangent column. Move horizontally to the left to the degree column. The measure you find will be 31°.

Therefore, if $\tan \angle A = .601$, then $\angle A = 31^{\circ}$.

Example: If $\tan \angle A = 1.000$, find $\angle A$ in the Trigonometric Ratios Table.

$\tan \angle A = 1.000$

Therefore, $\angle A = 45^{\circ}$

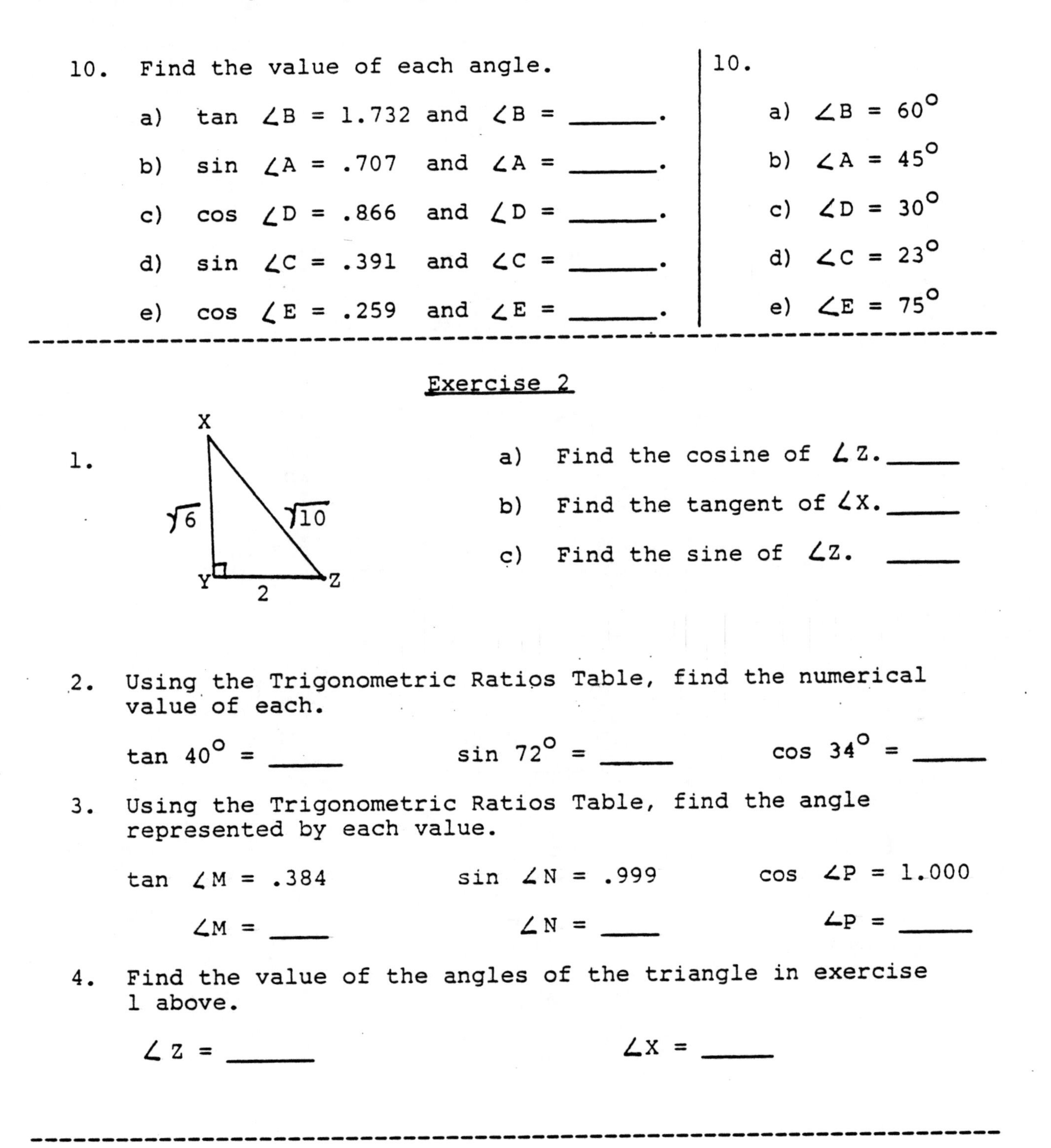

10. Find the value of each angle.

a) tan $\angle B = 1.732$ and $\angle B =$ ______.

b) sin $\angle A = .707$ and $\angle A =$ ______.

c) cos $\angle D = .866$ and $\angle D =$ ______.

d) sin $\angle C = .391$ and $\angle C =$ ______.

e) cos $\angle E = .259$ and $\angle E =$ ______.

10.

a) $\angle B = 60^\circ$

b) $\angle A = 45^\circ$

c) $\angle D = 30^\circ$

d) $\angle C = 23^\circ$

e) $\angle E = 75^\circ$

Exercise 2

1.

a) Find the cosine of $\angle Z$. ______

b) Find the tangent of $\angle X$. ______

c) Find the sine of $\angle Z$. ______

2. Using the Trigonometric Ratios Table, find the numerical value of each.

tan 40° = ______ sin 72° = ______ cos 34° = ______

3. Using the Trigonometric Ratios Table, find the angle represented by each value.

tan $\angle M = .384$ sin $\angle N = .999$ cos $\angle P = 1.000$

$\angle M =$ ____ $\angle N =$ ____ $\angle P =$ ______

4. Find the value of the angles of the triangle in exercise 1 above.

$\angle Z =$ ______ $\angle X =$ ______

Applications of Trigonometric Ratios

Example: An isosceles right triangle has one leg of length 7 in. Find the sine of $\angle B$.

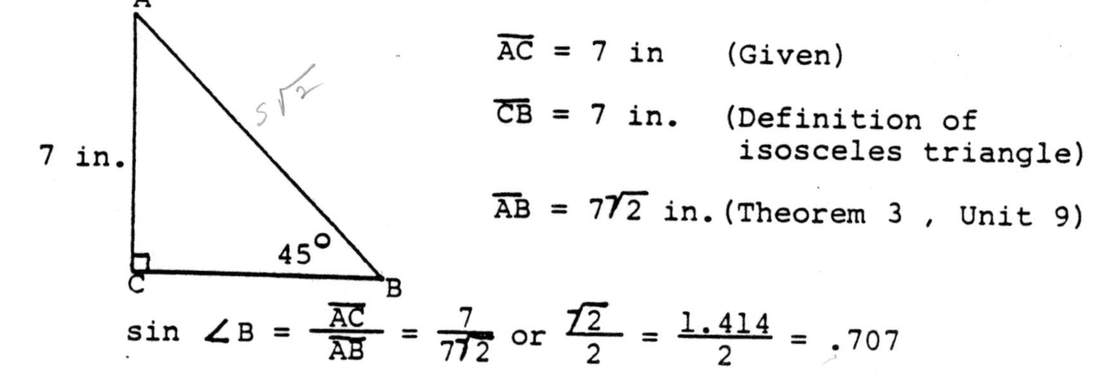

$\overline{AC} = 7$ in (Given)

$\overline{CB} = 7$ in. (Definition of isosceles triangle)

$\overline{AB} = 7\sqrt{2}$ in. (Theorem 3, Unit 9)

$$\sin \angle B = \frac{\overline{AC}}{\overline{AB}} = \frac{7}{7\sqrt{2}} \text{ or } \frac{\sqrt{2}}{2} = \frac{1.414}{2} = .707$$

Using the Trigonometric Ratios Table, $\sin \angle B = \sin 45^\circ = .707$.

Example: Given trapezoid ABCD with $\angle A = 37^\circ$, $\overline{AB} = 10$ in., $\overline{CD} = 8$ in., and $\overline{AC} = 5$ in., find the area of ABCD.

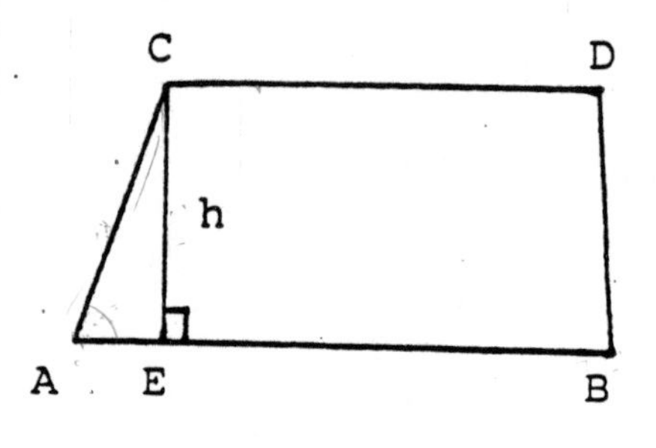

Solution:

$$\sin 37^\circ = .602$$

$$\sin 37^\circ = \frac{h}{\overline{AC}}$$

$$(\sin 37^\circ)(\overline{AC}) = h$$

$$(.602)(5 \text{ in.}) = h$$

$$3.010 \text{ in.} = h$$

The area of trapezoid ABCD $= \frac{1}{2}h(b_1 + b_2)$

$$A = \frac{1}{2}(3.01)(8 + 10)$$

$$A = \frac{1}{2}(3.01)(18)$$

$$A = (3.01)(9)$$

$$A = 27.09 \text{ sq. in.}$$

11. In trapezoid ABCD, find the altitude and the area using the measures that are given.

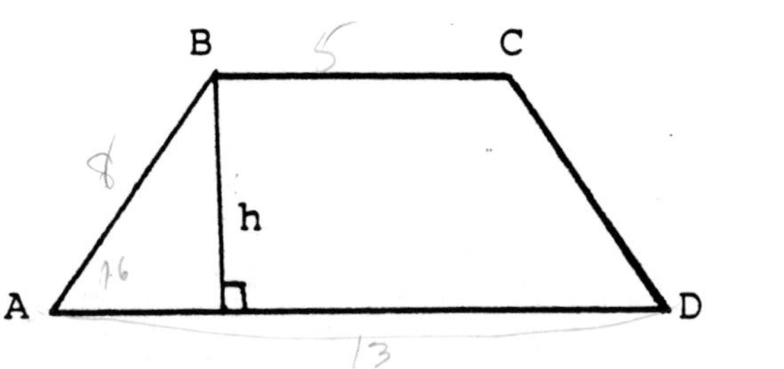

$\overline{AD}$ = 13 in., $\overline{BC}$ = 5 in., $\overline{AB}$ = 8 in., and $\angle A = 16^\circ$.

a) $\sin 16^\circ$ = ______

b) $\sin 16^\circ = \frac{h}{\overline{AB}}$

$(\sin 16^\circ)(\overline{AB}) = h$

____________ = h

c) Area of trapezoid = $\frac{1}{2}(\overline{AD} + \overline{BC})h$

A = ____________

11.

$\frac{1}{2}h(b_1 + b_2)$

a) $\sin 16^\circ = .276$

b)

h = 2.208 in.

c)

A = 19.872 sq.in.

12. In trapezoid ABCD, find the altitude and area using the measures that are given.

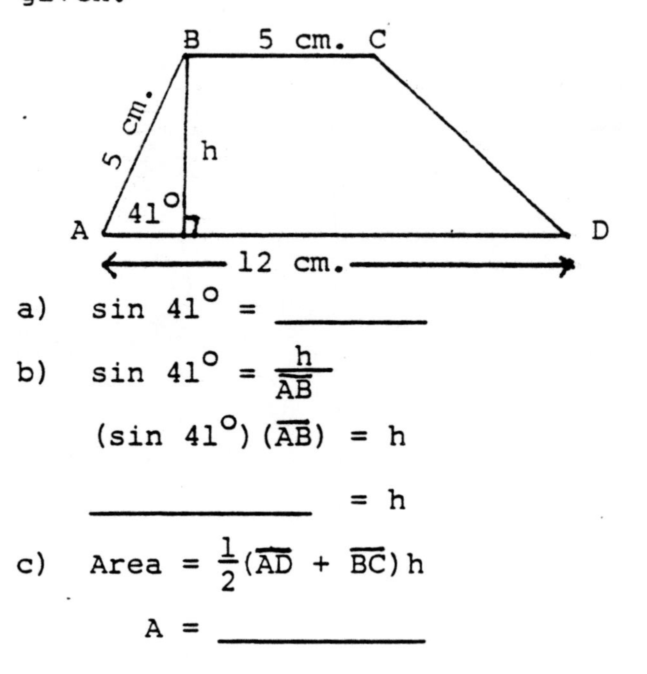

a) $\sin 41^\circ$ = ________

b) $\sin 41^\circ = \frac{h}{\overline{AB}}$

$(\sin 41^\circ)(\overline{AB}) = h$

____________ = h

c) Area = $\frac{1}{2}(\overline{AD} + \overline{BC})h$

A = ____________

12.

a) $\sin 41^\circ = .656$

b)

h = 3.28 cm.

c)

A = 27.88 sq.cm.

13. Find the altitude and the area of parallelogram RUTH using the measures given.

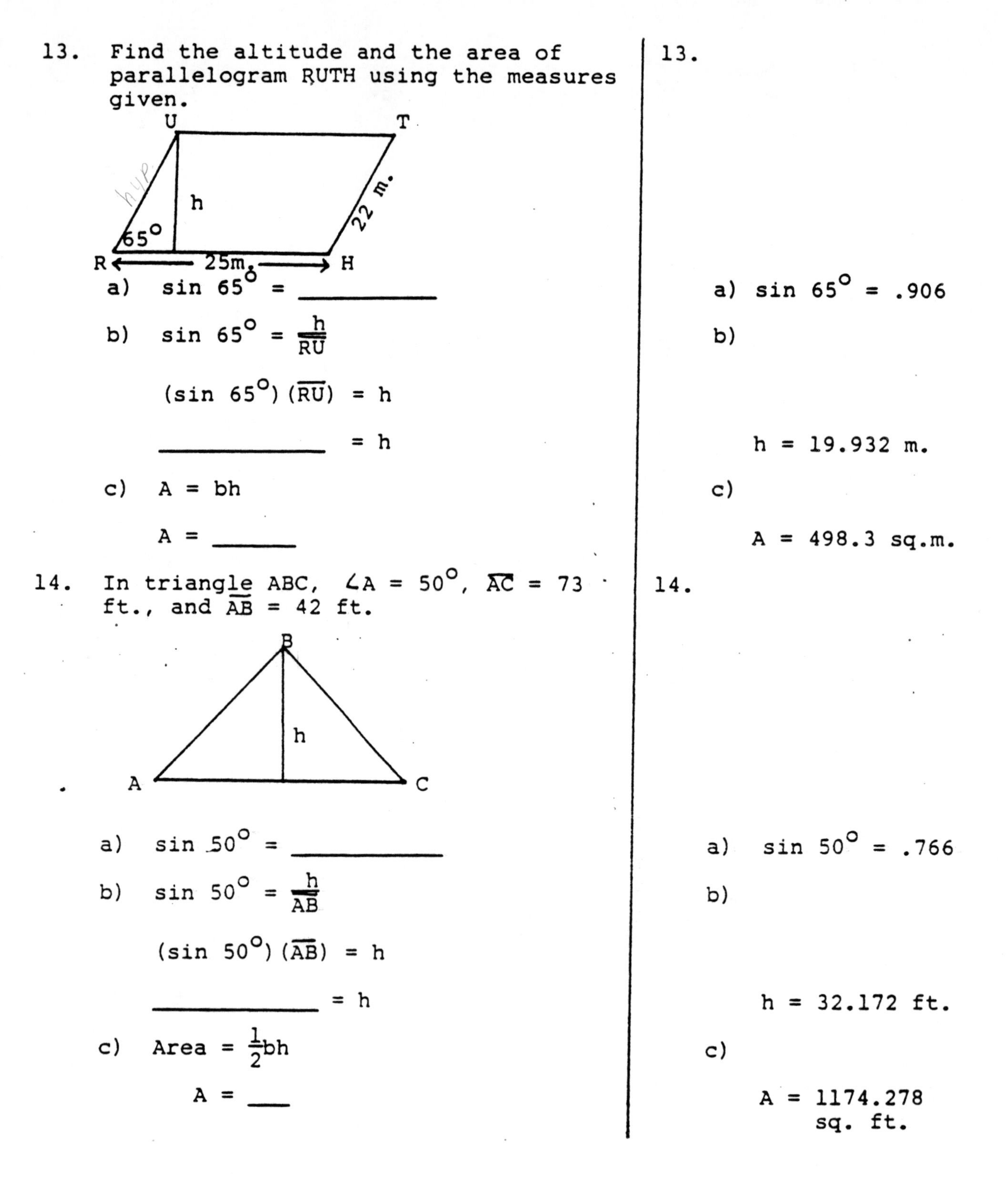

a) $\sin 65^\circ =$ ______

b) $\sin 65^\circ = \frac{h}{\overline{RU}}$

$(\sin 65^\circ)(\overline{RU}) = h$

______ $= h$

c) $A = bh$

$A =$ ______

14. In triangle ABC, $\angle A = 50^\circ$, $\overline{AC} = 73$ ft., and $\overline{AB} = 42$ ft.

B

h

A C

a) $\sin 50^\circ =$ ______

b) $\sin 50^\circ = \frac{h}{\overline{AB}}$

$(\sin 50^\circ)(\overline{AB}) = h$

______ $= h$

c) Area $= \frac{1}{2}bh$

$A =$ ___

13.

a) $\sin 65^\circ = .906$

b)

$h = 19.932$ m.

c)

$A = 498.3$ sq.m.

14.

a) $\sin 50^\circ = .766$

b)

$h = 32.172$ ft.

c)

$A = 1174.278$ sq. ft.

15. In surveying for a new bridge, an engineer marked two points. One marking was placed on each side of the river. A third point, T, was marked such that $\overline{NT} \perp NO$ and angle T measures 44°. What is the distance across the river from point N to O if $\overline{OT}$ = 80ft.?

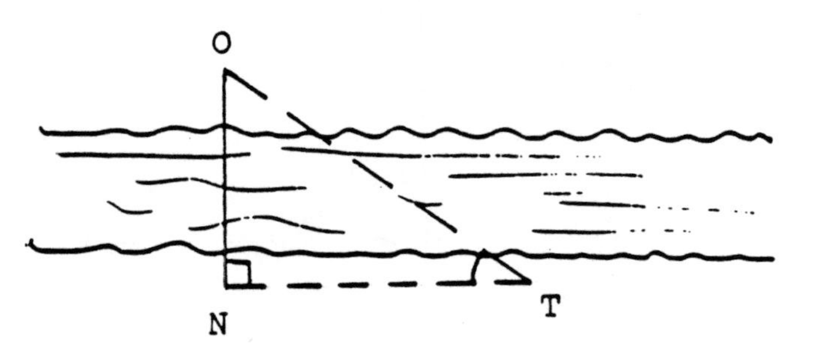

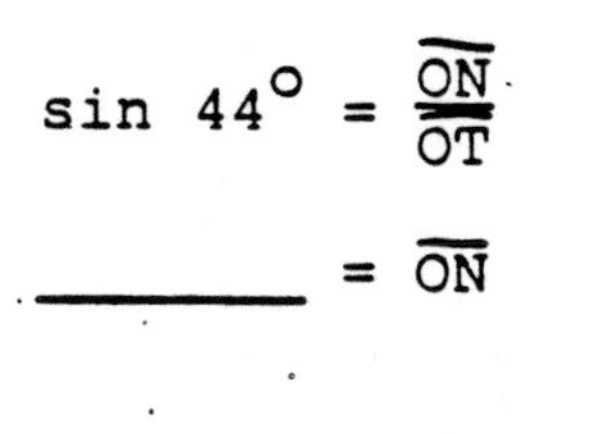

$\sin 44^\circ = \frac{\overline{ON}}{\overline{OT}}$

________ = $\overline{ON}$

15.

$.695 = \frac{\overline{ON}}{80}$

$(.695)(80) = \overline{ON}$

$55.6 \text{ ft.} = \overline{ON}$

16. A six foot man observes the height of a tree that stood 100 ft. away from him. If the man sights the top of the tree at an angle of 10° from the horizon, approximately how tall is the tree?

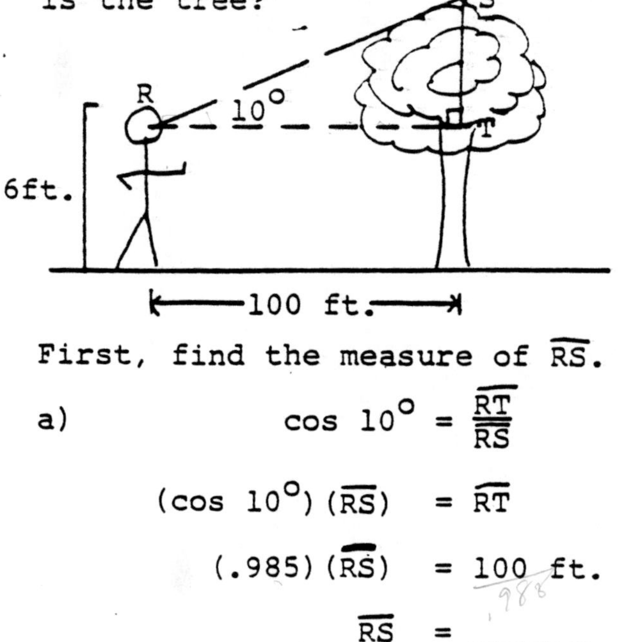

First, find the measure of $\overline{RS}$.

a) $\cos 10^\circ = \frac{\overline{RT}}{\overline{RS}}$

$(\cos 10^\circ)(\overline{RS}) = \overline{RT}$

$(.985)(\overline{RS}) = 100 \text{ ft.}$

$\overline{RS}$ = ________

16.

a)

$\overline{RS} = 101.52 \text{ ft.}$

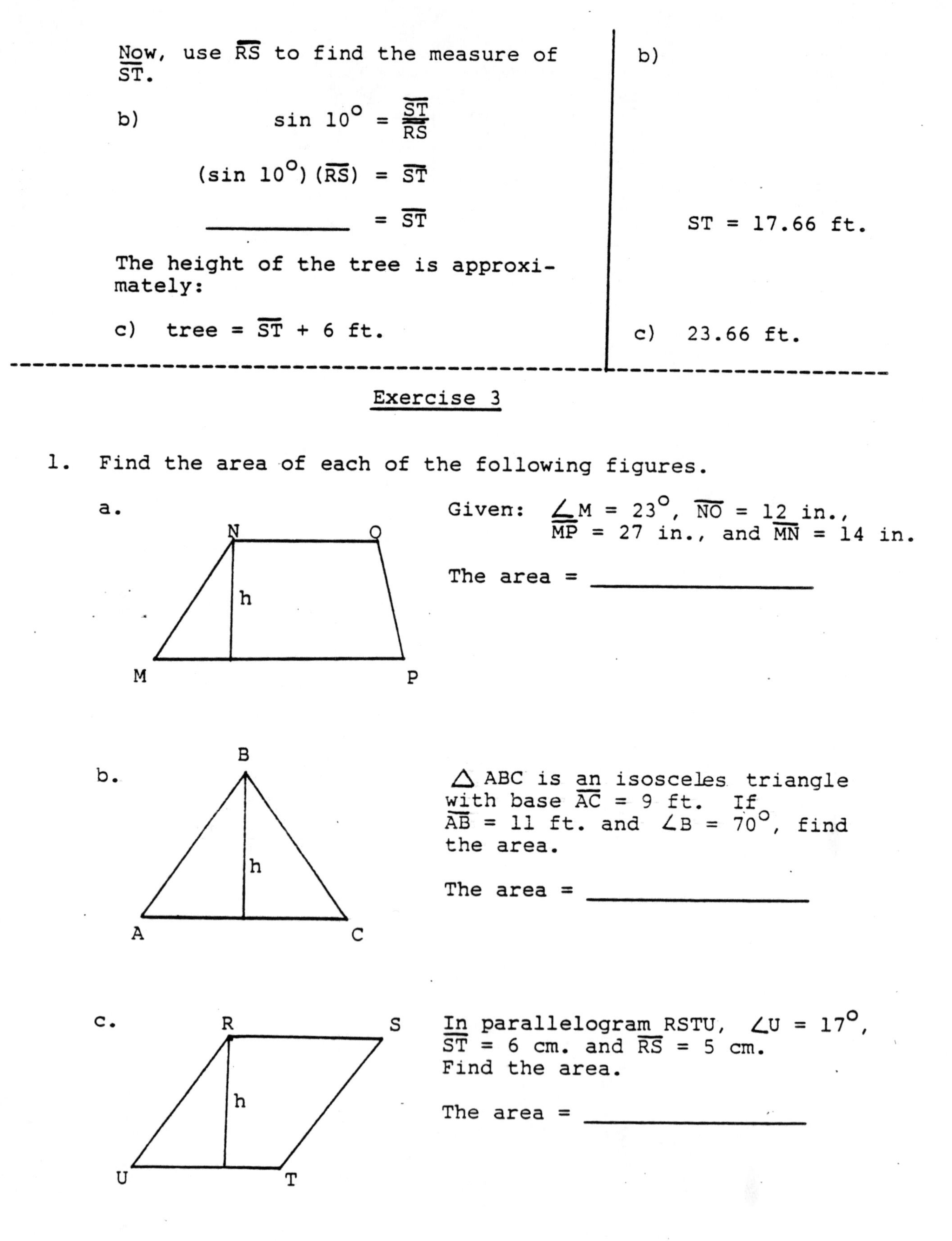

Now, use $\overline{RS}$ to find the measure of $\overline{ST}$.

b)
$$\sin 10^\circ = \frac{\overline{ST}}{\overline{RS}}$$
$$(\sin 10^\circ)(\overline{RS}) = \overline{ST}$$
$$\_\_\_\_\_\_\_\_\_\_ = \overline{ST}$$

The height of the tree is approximately:

c) tree = $\overline{ST}$ + 6 ft.

b)

ST = 17.66 ft.

c) 23.66 ft.

Exercise 3

1. Find the area of each of the following figures.

a.

Given: $\angle M = 23^\circ$, $\overline{NO}$ = 12 in., $\overline{MP}$ = 27 in., and $\overline{MN}$ = 14 in.

The area = ________________

b.

$\triangle$ ABC is an isosceles triangle with base $\overline{AC}$ = 9 ft. If $\overline{AB}$ = 11 ft. and $\angle B = 70^\circ$, find the area.

The area = ________________

c.

In parallelogram RSTU, $\angle U = 17^\circ$, $\overline{ST}$ = 6 cm. and $\overline{RS}$ = 5 cm. Find the area.

The area = ________________

2. A tall radio tower is anchored by guy wires. One such wire, $\overline{XZ}$, was 175 ft. from the base of the tower. If $\angle ZXY = 48^\circ$, find the length of $\overline{XZ}$.

Z

X Y

Unit 11 Review

1.

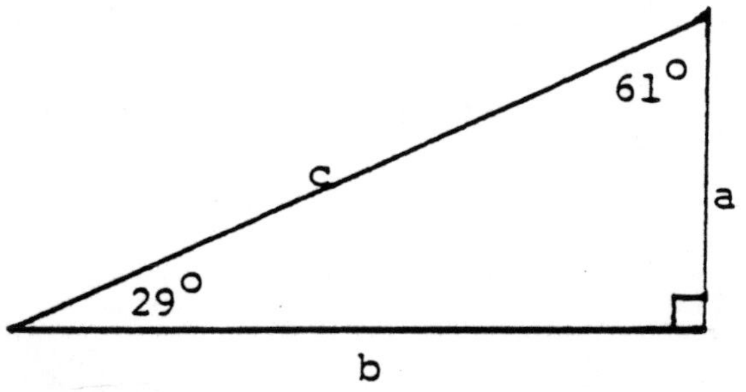

61°

c

a

29°

b

a) The leg opposite the 29° angle is ________________.

b) The leg adjacent to the 61° angle is ________________.

c) The hypotenuse is ______________.

d) The leg opposite the 61° angle is ________________.

e) The leg adjacent to the 29° angle is ________________.

2. Using the triangle in exercise 1 above, write the three trigonometric ratios for the 29° angle in terms of a, b, and c. Do the same for the 61° angle. Leave your answers in terms of a, b, and c.

sin 29° = ______ cos 29° = ______ tan 29° = ______

sin 61° = ______ cos 61° = ______ tan 61° = ______

3. Using the Trigonometric Ratios Table, write the numeric values for the ratios found in exercise 2 above.

sin 29° = _____ cos 29° = _____ tan 29° = _____

sin 61° = _____ cos 61° = _____ tan 61° = _____

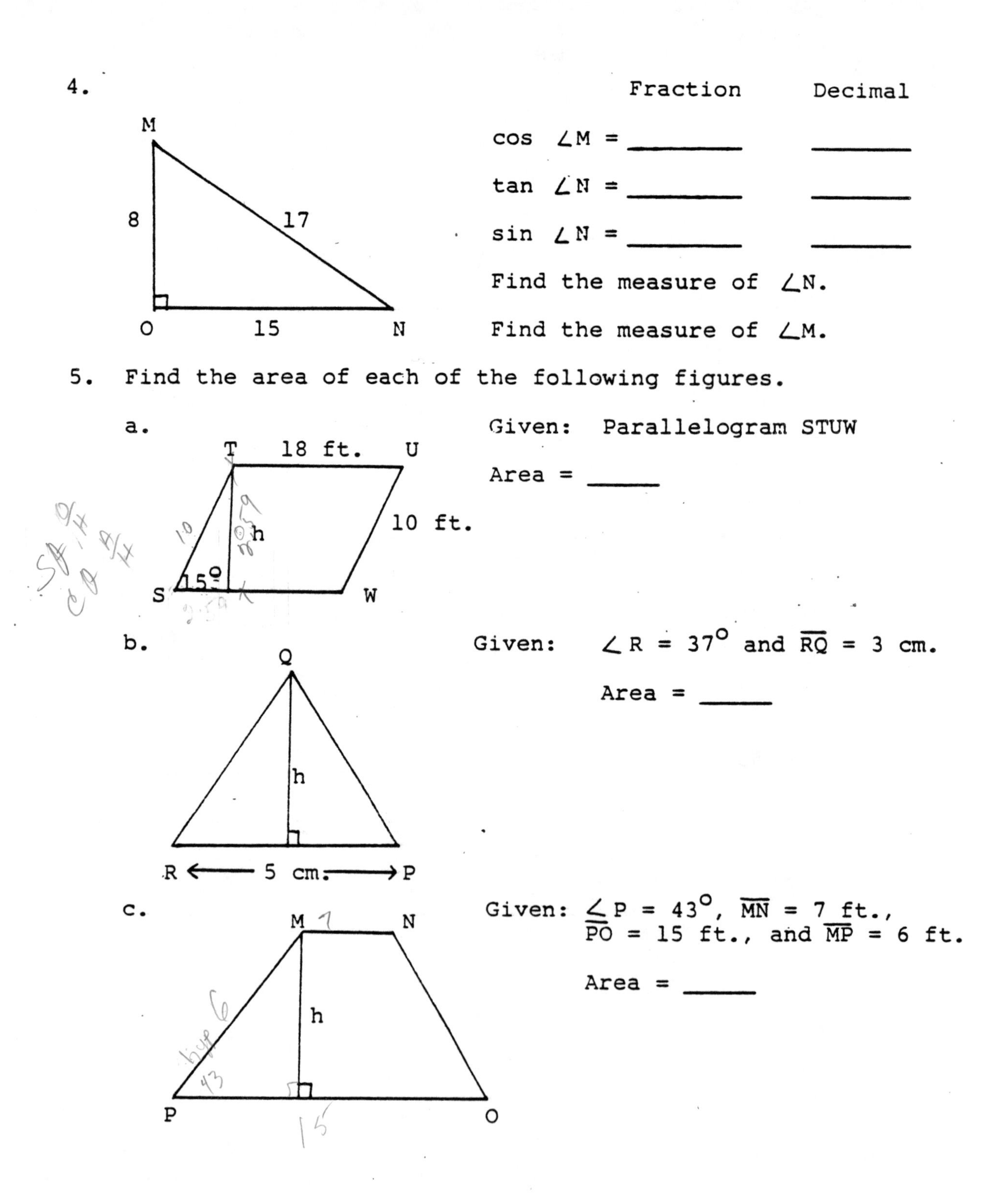

4.

	Fraction	Decimal
cos ∠M =	______	______
tan ∠N =	______	______
sin ∠N =	______	______

Find the measure of ∠N.

Find the measure of ∠M.

5. Find the area of each of the following figures.

a. Given: Parallelogram STUW

Area = ____

b. Given: ∠R = 37° and $\overline{RQ}$ = 3 cm.

Area = ____

c. Given: ∠P = 43°, $\overline{MN}$ = 7 ft., $\overline{PO}$ = 15 ft., and $\overline{MP}$ = 6 ft.

Area = ____

6. A surveyor wants to determine the distance between two points on opposite sides of a ditch. He measures $\overline{AB}$ and angle B. If $\overline{AB}$ = 200 ft. and $\angle B = 25^{\circ}$ then how far across does the ditch measure?

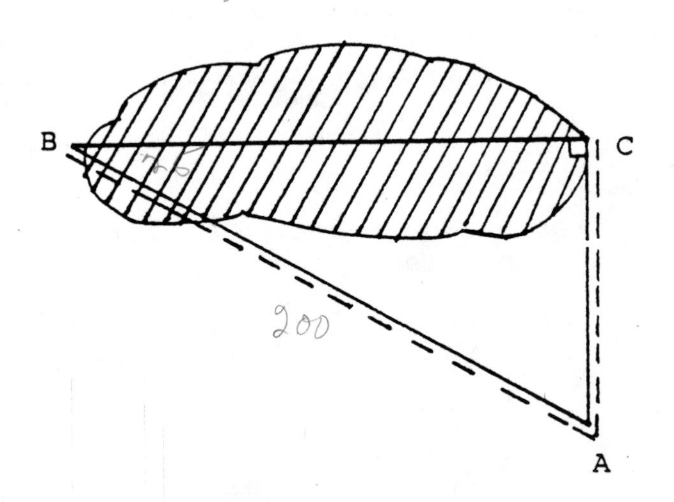

Unit 12

Circles and Sectors

Learning Objectives:

1. The student will demonstrate his mastery of the following definitions, postulate, and theorems by writing them and by applying them to solutions of selected problems:

   Definitions: circle, radius, center, chord, diameter, semicircle, tangent, point of tangency, secant, central angle, inscribed angle, circumference, length of an arc, area of a circle, sector, area of a sector, circumcenter, and incenter.

   Postulate: Two circles are congruent if and only if they have the same radius.

   Theorems:
   1. The measure of a central angle equals the measure of its intercepted arc.
   2. The measure of an inscribed angle is equal to one-half the measure of its intercepted arc.
   3. The ratio of the length of an arc of a circle to the circumference of the circle is the same as the ratio of the measure of the central angle of the arc to 360°.
   4. The ratio of the area of a sector of a circle to the area of the circle is the same as the ratio of the measure of the central angle of the arc to 360°.
   5. A line drawn from the center of a circle perpendicular to a chord bisects the chord and its intercepted arc.
   6. In the same circle or in congruent circles, equal arcs have equal chords.
   7. In a circle, two chords are the same distance from the center of the circle if, and only if, they are equal.
   8. The perpendicular bisectors of the sides of a triangle meet at a point which is equidistant from the verticies of the triangle.

9. The angle bisectors of a triangle meet in a point which is equidistant from the sides of the triangle.

2. The following constructions are required:

   a. To circumscribe a circle about a triangle.

   b. To inscribe a circle within a triangle.

CIRCLES AND SECTORS

Circles are in the environment everywhere: the top of the glass you had your juice from for breakfast; the top of your coffee cup; the plate you eat from; etc. The path a horse on a merry-go-round makes as the merry-go-round turns is a circle. The radius is the distance the horse is from the center of the merry-go round and the center of the circle is the axle on which the merry-go-round turns.

To be more formal, recall the definition of a circle from Unit 7.

> Definition: A circle is the set of all possible points a fixed distance from a given point. The fixed distance is called the radius of the circle and the given point is the center.

1. a. In the diagram below what is the radius? ____________

   b. What is the center? ________

O W

1. a. $\overline{OW}$

   b. O

> Definition: A line segment is a chord if and only if its endpoints are points on the circle.

A specific type of chord that passes directly through the center of the circle is called a diameter as shown in Figure 12.1 on page 307. Because a diameter passes through the center of the circle, it can be defined in terms of the radius of a circle.

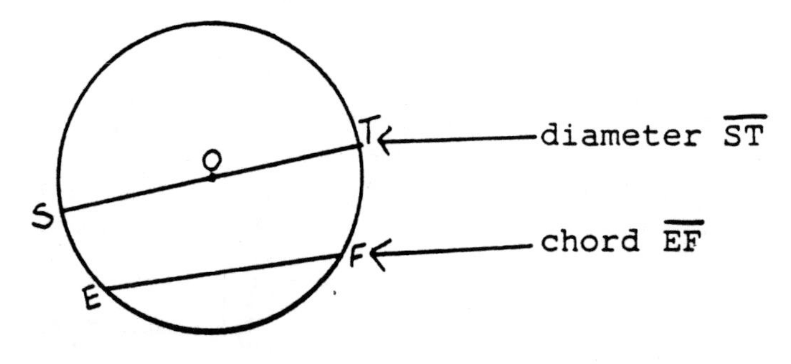

Figure 12.1

If two radii are joined to form a straight line segment, this line segment is called a diameter. The measure of one diameter is equivalent to the length of two radii. ($d = 2r$, where d represents the diameter and r represents the radius.)

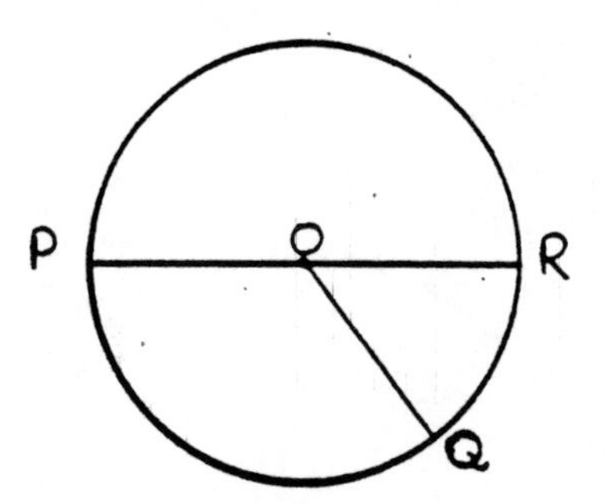

Figure 12.2

Radii $\overline{OP}$ and $\overline{OR}$ form diameter $\overline{PR}$. $\overline{OR}$ and $\overline{OQ}$ do not form a diameter. A diameter divides a complete circle into two equal parts. Each part is called a semicircle.

2. Does a diameter always pass through the center of a circle? ________	2. Yes
3. Is a diameter a chord of a circle? _____	3. Yes

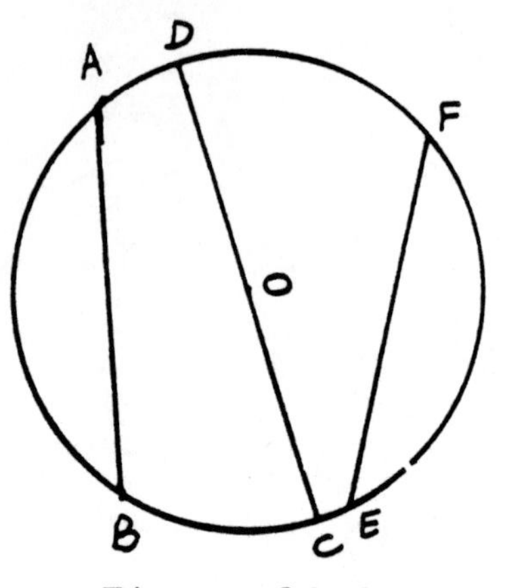

Figure 12.3

$\overline{AB}$, $\overline{CD}$, and $\overline{EF}$ are all chords of this circle.

4. Diameter $\overline{CD}$ is formed by what two radii? ______________

4. $\overline{OD}$ and $\overline{CO}$.

A line which intersects a circle at only one point is called a <u>tangent</u>. The point of intersection of the circle and tangent is called the <u>point of tangency</u>.

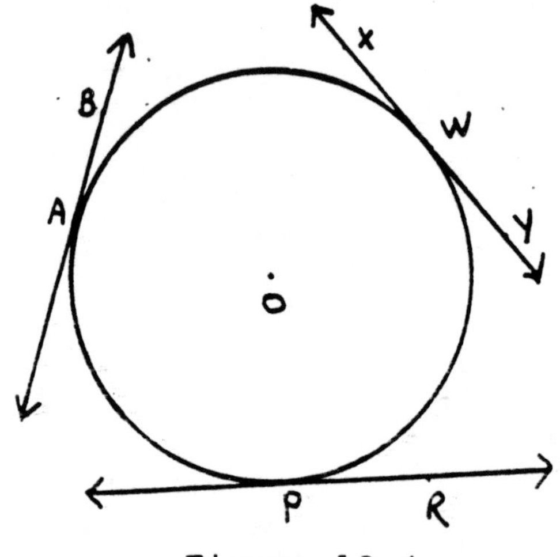

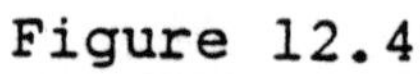

Figure 12.4

$\overleftrightarrow{AB}$ is a tangent with point of tangency, A. $\overleftrightarrow{XY}$ is a tangent with point of tangency, W.

5. Name another tangent and point of tangency in Figure 12.4.
Tangent ______________
Point of tangency ______________

5.
$\overleftrightarrow{PR}$
P

A secant is a line which intersects a circle in two distinct points.

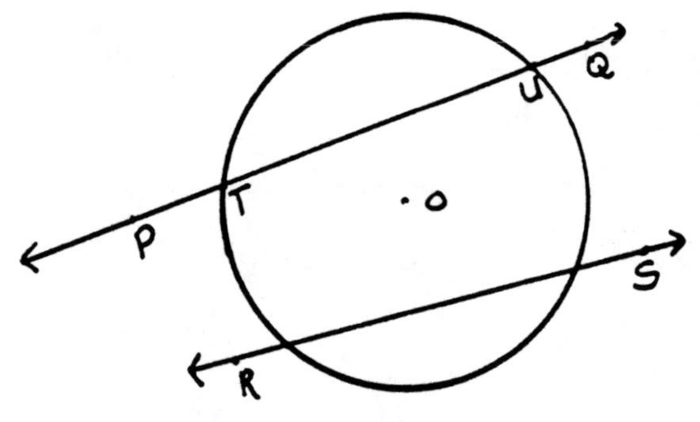

Figure 12.5

$\overleftrightarrow{PQ}$ and $\overleftrightarrow{RS}$ are secant lines. A line is a secant if and only if it contains a chord. In Figure 12.5 secant $\overleftrightarrow{PQ}$ contains chord $\overline{TU}$. Note: Secants and chords both intersect the circle in two distinct points. However, secants are lines and extend indefinitely beyond the circle. Chords are line segments and end on the circle.

6.

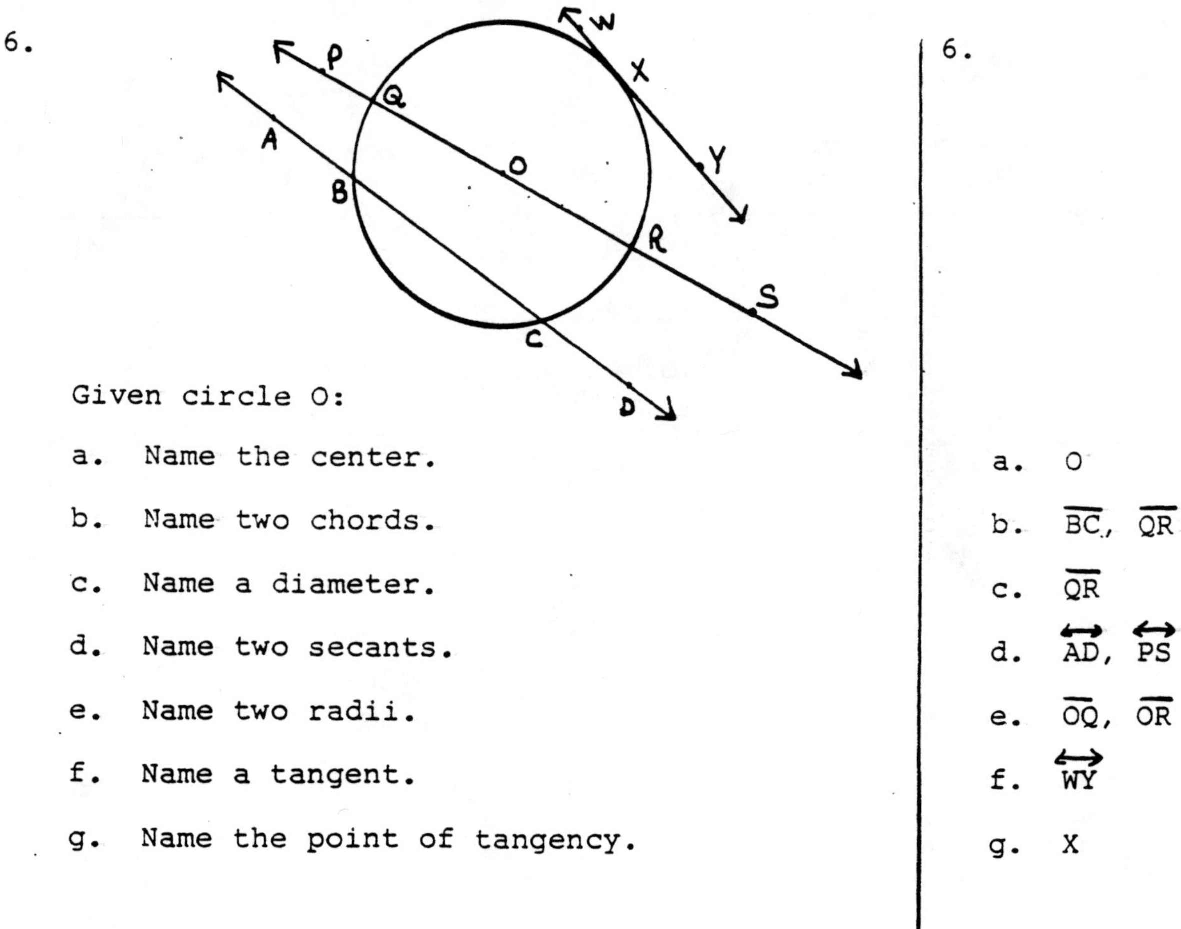

Given circle O:

a. Name the center.

b. Name two chords.

c. Name a diameter.

d. Name two secants.

e. Name two radii.

f. Name a tangent.

g. Name the point of tangency.

6.

a. O

b. $\overline{BC}$, $\overline{QR}$

c. $\overline{QR}$

d. $\overleftrightarrow{AD}$, $\overleftrightarrow{PS}$

e. $\overline{OQ}$, $\overline{OR}$

f. $\overleftrightarrow{WY}$

g. X

Exercise 1

1. Explain why a diameter is a special type of chord.

2. Explain the relationship between the radius of a circle and the diameter of that same circle.

3. Explain the relationship between chord and secant.

4. Distinguish between secant and tangent.

5. In the circle below:

    a. Name the center.

    b. Name three radii.

    c. Name a diameter.

    d. Name three chords.

    e. Name a secant.

    f. Name a tangent.

    g. Name a point of tangency.

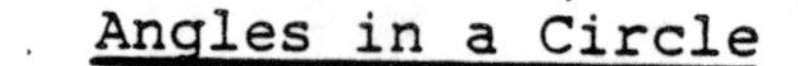

Angles in a Circle

Definition:	A <u>central angle</u> of a circle is an angle with vertex at the center of the circle and with sides as radii of the circle.

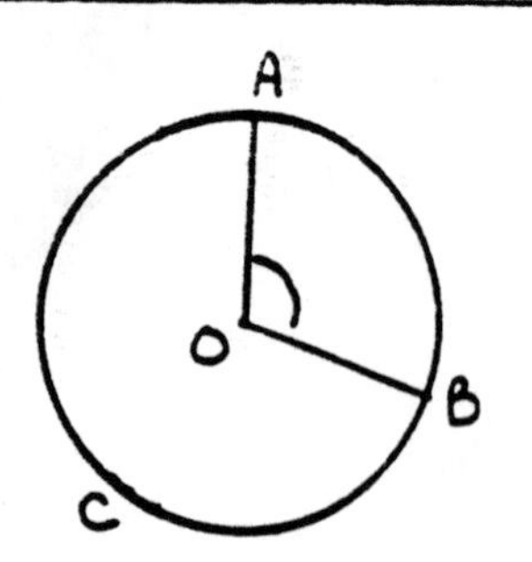

$\angle AOB$ is a central angle.

The set of points between A and B divides the circle into two parts called arcs. Arcs are denoted by letters with the symbol $\frown$. <u>Minor arcs</u> are less than half of a circle and are denoted by two letters. $\overset{\frown}{AB}$ is a minor arc. <u>Major arcs</u> are more than a semicircle and are denoted by three letters. $\overset{\frown}{ACB}$ is a major arc. Since the entire circle measures 360°, a <u>semicircle</u> is an arc of a circle which

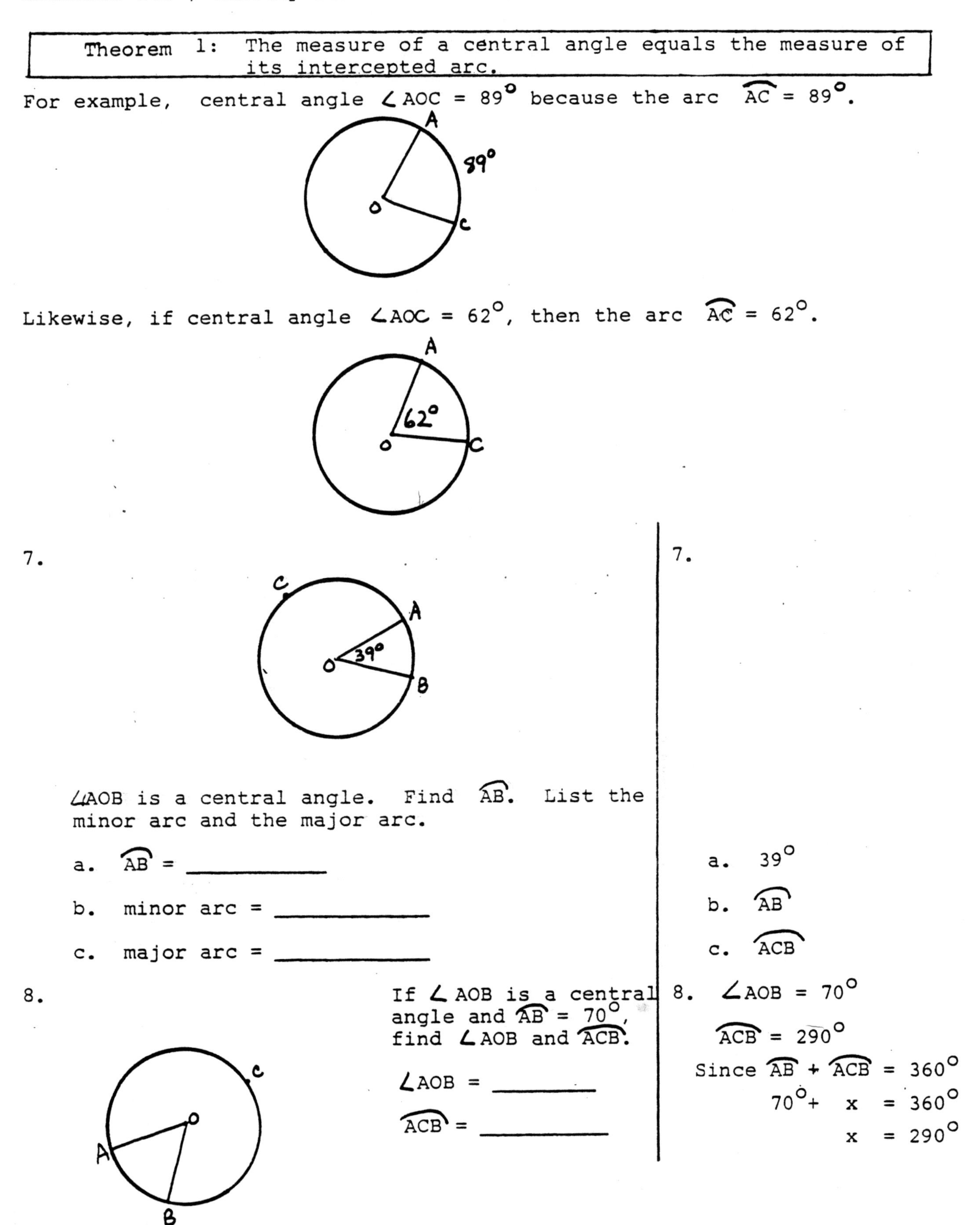

measures 180°, exactly half the measure of a circle.

Theorem 1: The measure of a central angle equals the measure of its intercepted arc.

For example, central angle $\angle AOC = 89^\circ$ because the arc $\widehat{AC} = 89^\circ$.

Likewise, if central angle $\angle AOC = 62^\circ$, then the arc $\widehat{AC} = 62^\circ$.

7.

$\angle AOB$ is a central angle. Find $\widehat{AB}$. List the minor arc and the major arc.

a. $\widehat{AB}$ = ____________

b. minor arc = ____________

c. major arc = ____________

7.

a. 39°

b. $\widehat{AB}$

c. $\widehat{ACB}$

8.

If $\angle AOB$ is a central angle and $\widehat{AB} = 70^\circ$, find $\angle AOB$ and $\widehat{ACB}$.

$\angle AOB$ = __________

$\widehat{ACB}$ = __________

8. $\angle AOB = 70^\circ$

$\widehat{ACB} = 290^\circ$

Since $\widehat{AB} + \widehat{ACB} = 360^\circ$

$70^\circ + x = 360^\circ$

$x = 290^\circ$

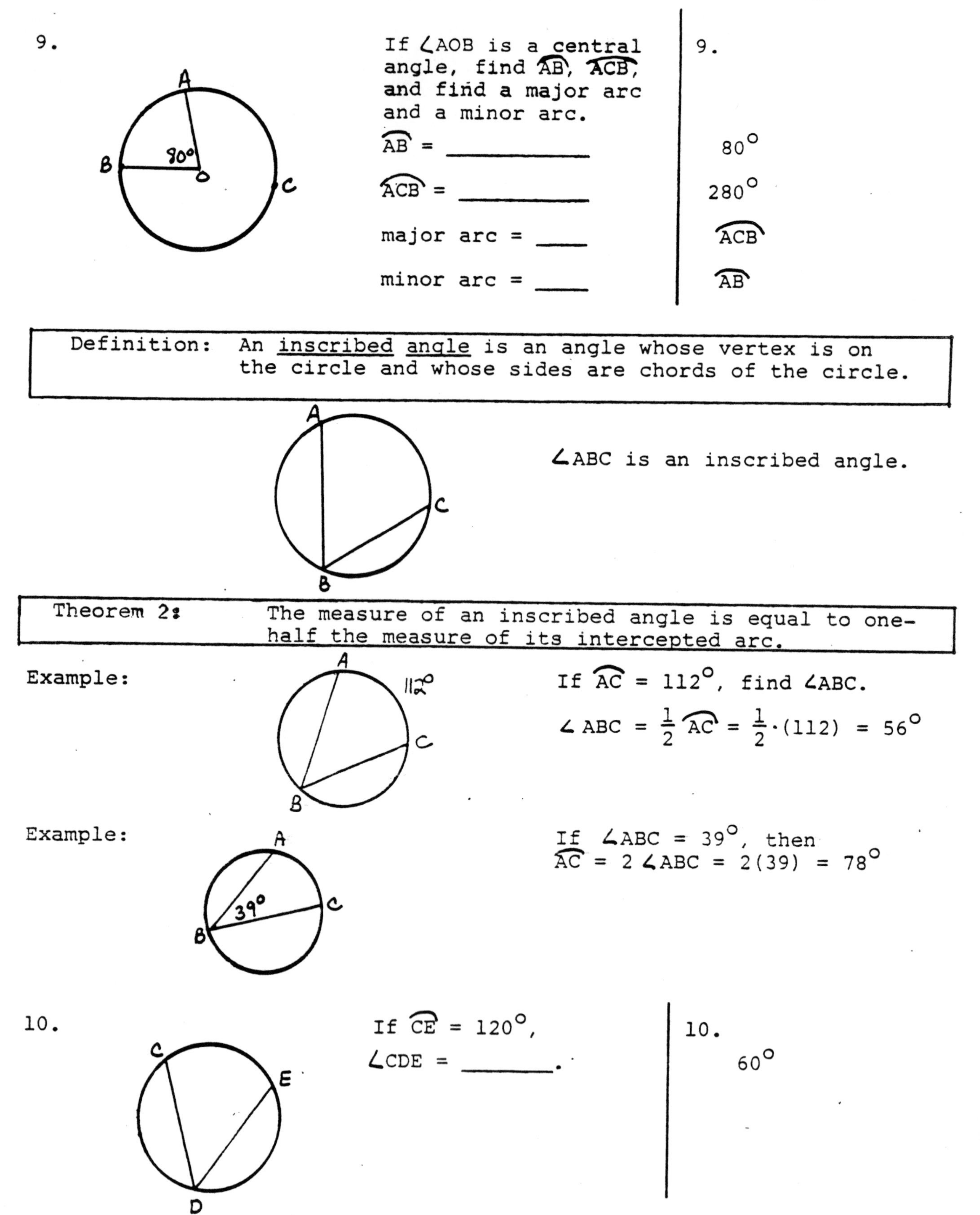

9. If $\angle AOB$ is a central angle, find $\overset{\frown}{AB}$, $\overset{\frown}{ACB}$, and find a major arc and a minor arc.

$\overset{\frown}{AB}$ = ________

$\overset{\frown}{ACB}$ = ________

major arc = ____

minor arc = ____

9.

80°

280°

$\overset{\frown}{ACB}$

$\overset{\frown}{AB}$

Definition: An <u>inscribed angle</u> is an angle whose vertex is on the circle and whose sides are chords of the circle.

$\angle ABC$ is an inscribed angle.

Theorem 2: The measure of an inscribed angle is equal to one-half the measure of its intercepted arc.

Example:

If $\overset{\frown}{AC} = 112^\circ$, find $\angle ABC$.

$\angle ABC = \frac{1}{2}\overset{\frown}{AC} = \frac{1}{2}\cdot(112) = 56^\circ$

Example:

If $\angle ABC = 39^\circ$, then $\overset{\frown}{AC} = 2\angle ABC = 2(39) = 78^\circ$

10. If $\overset{\frown}{CE} = 120^\circ$, $\angle CDE$ = ________.

10. 60°

11.

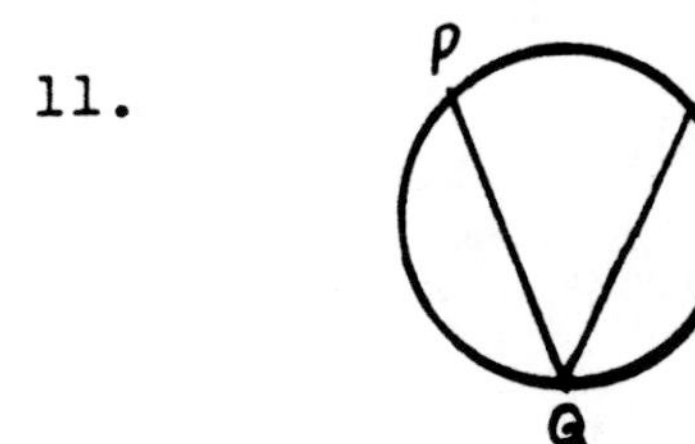

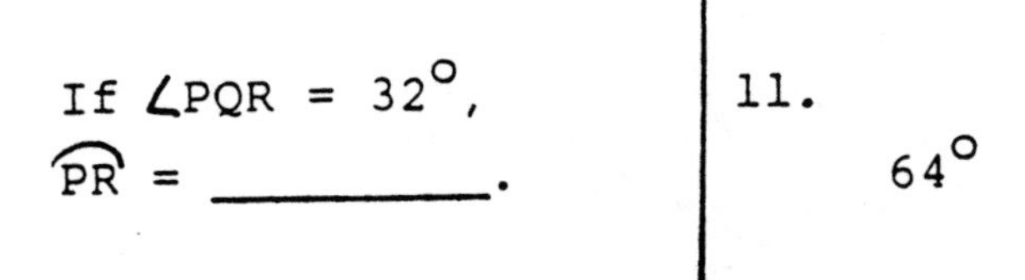

If $\angle PQR = 32^\circ$, $\overset{\frown}{PR}$ = __________.

11. 64°

Exercise 2

1. Find the measures of central angle $\angle AOB$ and inscribed angle $\angle ACB$. Identify the major arcs and the minor arcs.

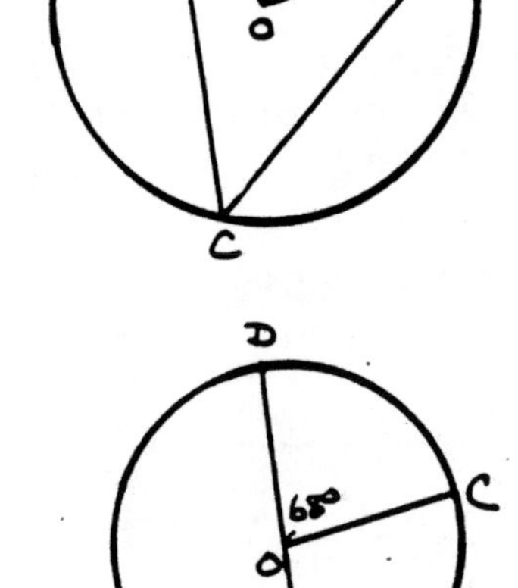

2. Point O is the center of the circle. Find $\overset{\frown}{DC}$, $\angle DOE$ if $\overline{DE}$ is the diameter, $\overset{\frown}{CE}$, and $\angle COE$.

3. $\angle ABC$ is an inscribed angle. If $\angle ABC = 51^\circ$, find $\overset{\frown}{AC}$.

4. Δ ABC is an equilateral triangle. Find $\overset{\frown}{AC}$, $\overset{\frown}{AB}$, and $\overset{\frown}{BC}$.

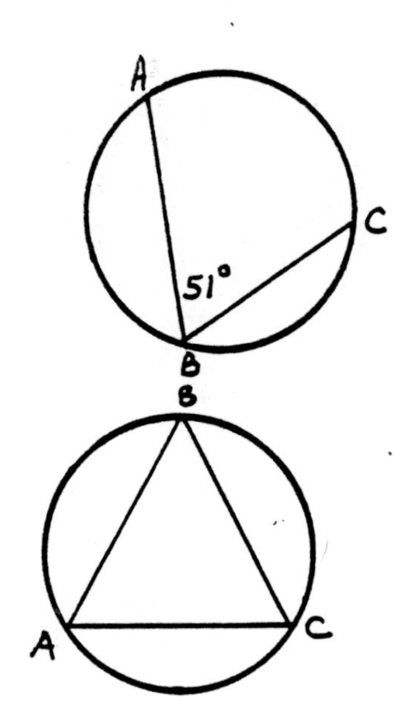

5.

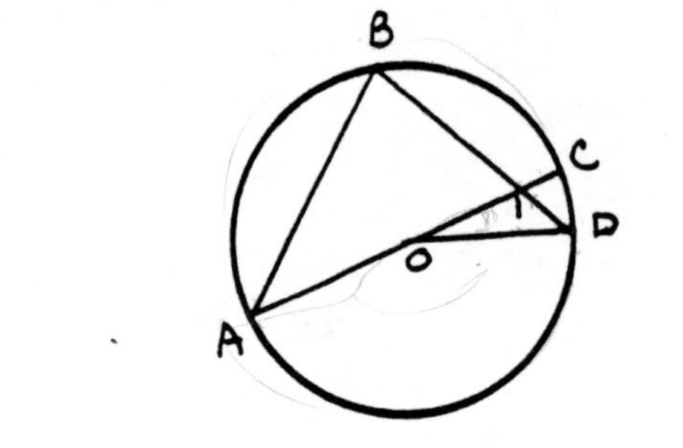

O is the center of the circle.

$\angle BAC = 30^\circ$, $\angle COD = 30^\circ$, $\angle 1 = 105^\circ$.

Find $\angle ABD$, $\angle BDO$, $\overset{\frown}{BC}$, $\overset{\frown}{CD}$, $\overset{\frown}{AB}$, and $\overset{\frown}{AD}$.

Arc Length

Recall the measure of an arc is the measure of its central angle. The length of an arc is the linear measure of the arc, i. e. how long it is.

> Theorem 3: The ratio of the length of an arc of a circle to the circumference of the circle is the same as the ratio of the measure of the central angle of the arc to 360°. The length of an arc is expressed by the following equation:
>
> $$\frac{\text{Length of arc (L)}}{\text{Circumference (C)}} = \frac{\text{Central Angle}}{360^\circ}$$
>
> $$\frac{L}{C} = \frac{\text{Central Angle}}{360^\circ}$$

Example: Find the length of an arc whose central angle is 40° and whose circle has radius 6 in.

In finding the solution, find circumference first. The formula for circumference is $C = 2\pi r$.

$$C = 2 \cdot 6 \cdot \pi$$
$$= 12\pi$$

Sutstituting C and the other values into the above formula to obtain:

$$\frac{L}{12\pi} = \frac{40^\circ}{360^\circ}$$

$$360L = 480\pi$$

$$L = \frac{480\pi}{360} = \frac{4}{3}\pi$$

12. Find the length of an arc whose central angle is 45° and whose circle has radius 4 ft.

Circumference ______________

$\frac{L}{C} = \frac{\text{Central Angle}}{360^\circ}$ ________

12. $C = 2\pi r$

$C = 2\pi \cdot 4$

$C = 8\pi$

$\frac{L}{8\pi} = \frac{45}{360}$

$360L = 360\pi$

$L = 1\pi$

13. Find the length of an arc whose central angle is 60° and whose circle has diameter 10 in.

13. $C = \pi d = 10\pi$ in.

$\frac{L}{10\pi} = \frac{60}{360}$

$L = \frac{5\pi}{3}$ in.

Area of a Sector

Definition: A sector of a circle is the set of all possible points bounded by two radii and their intercepted arc.

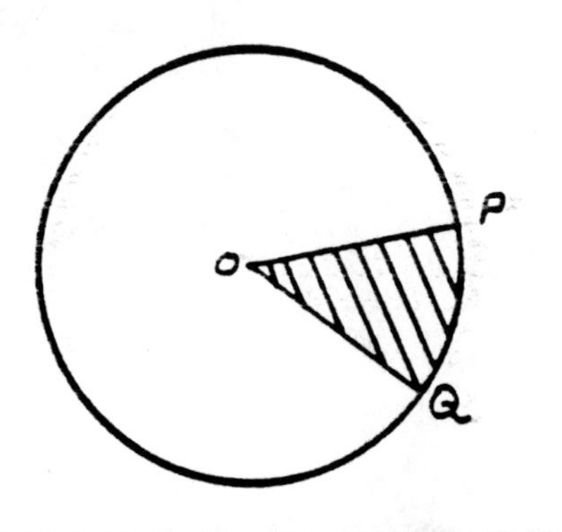

The shaded part represents a sector of a circle that can be described as a slice of pie.

Theorem 4: The ratio of the area of a sector of a circle to the area of the circle is the same as the ratio of the measure of the central angle of the arc to 360°. The area of a sector of a circle is expressed by this equation:

$$\frac{\text{Area of Sector}}{\text{Area of Circle}} = \frac{\text{Central Angle}}{360^\circ}$$

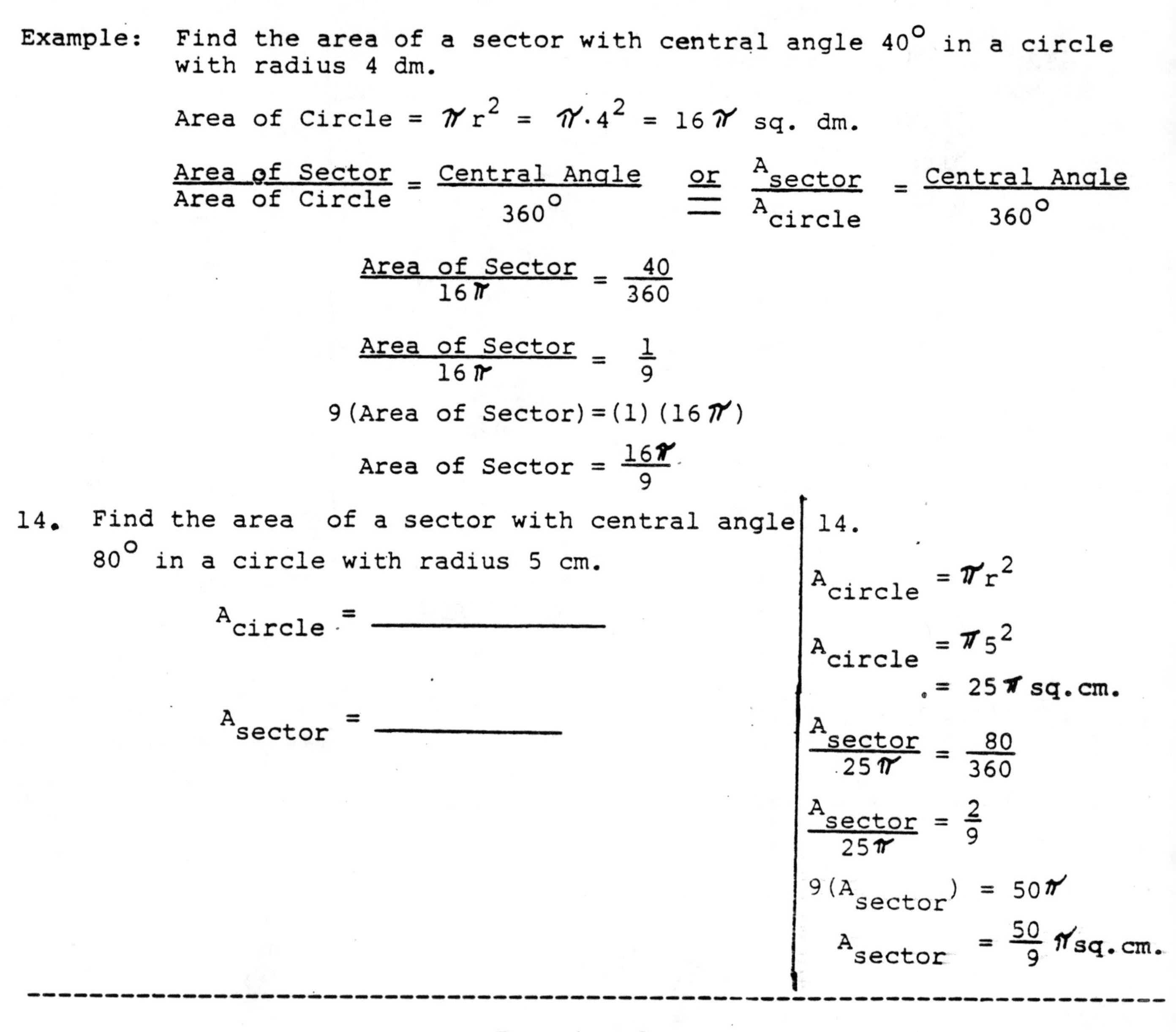

Example: Find the area of a sector with central angle 40° in a circle with radius 4 dm.

Area of Circle $= \pi r^2 = \pi \cdot 4^2 = 16\pi$ sq. dm.

$$\frac{\text{Area of Sector}}{\text{Area of Circle}} = \frac{\text{Central Angle}}{360^\circ} \quad \text{or} \quad \frac{A_{sector}}{A_{circle}} = \frac{\text{Central Angle}}{360^\circ}$$

$$\frac{\text{Area of Sector}}{16\pi} = \frac{40}{360}$$

$$\frac{\text{Area of Sector}}{16\pi} = \frac{1}{9}$$

$$9(\text{Area of Sector}) = (1)(16\pi)$$

$$\text{Area of Sector} = \frac{16\pi}{9}$$

14. Find the area of a sector with central angle 80° in a circle with radius 5 cm.

$A_{circle} =$ ______________

$A_{sector} =$ ______________

14.

$A_{circle} = \pi r^2$

$A_{circle} = \pi 5^2$

$= 25\pi$ sq. cm.

$\frac{A_{sector}}{25\pi} = \frac{80}{360}$

$\frac{A_{sector}}{25\pi} = \frac{2}{9}$

$9(A_{sector}) = 50\pi$

$A_{sector} = \frac{50}{9}\pi$ sq. cm.

Exercise 3

1. In each of the following find the length of the arc and the area of the sector.

   a. central angle 90°, $r = 10$ ft.

   b. central angle 50°, $r = 4\sqrt{2}$ cm.

   c. central angle 12°, $d = 9$ m.

2. If the length of an arc is 20π m. and the circumference of the circle is 120π m., find the central angle.

3. If the area of a sector is 3π sq. m. and the circle has radius 3 m., find the central angle.

4. Find the area of the sector and length of the arc of the circle whose diameter is $4\frac{1}{2}$ in. and whose central angle is 50°.

5.

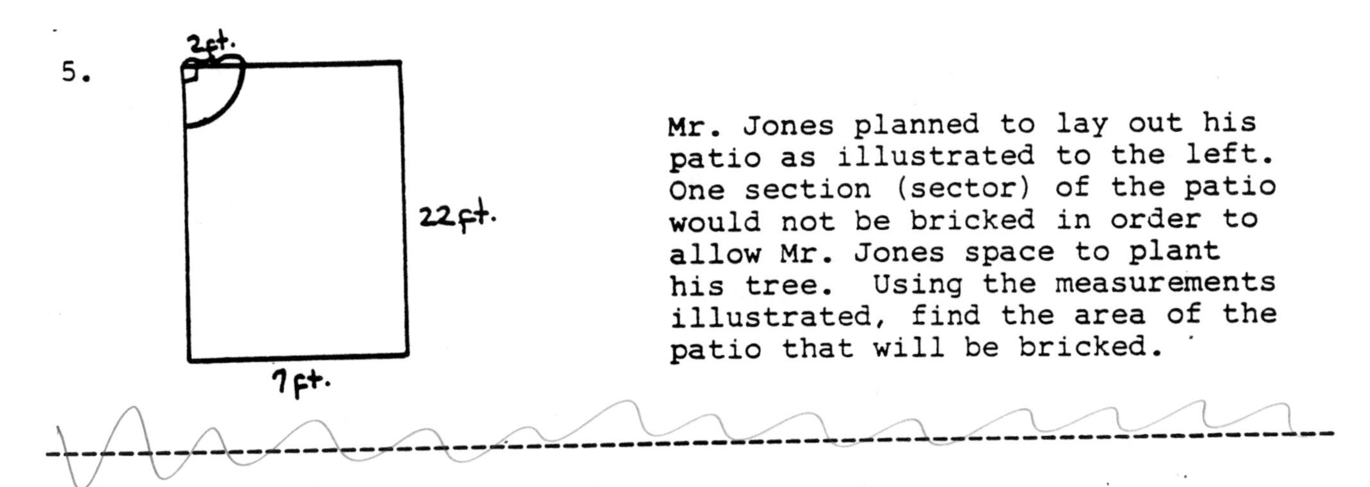

Mr. Jones planned to lay out his patio as illustrated to the left. One section (sector) of the patio would not be bricked in order to allow Mr. Jones space to plant his tree. Using the measurements illustrated, find the area of the patio that will be bricked.

There are several theorems dealing with circles.

Theorem 5:	A line drawn from the center of a circle perpendicular to a chord bisects the chord and its intercepted arc.

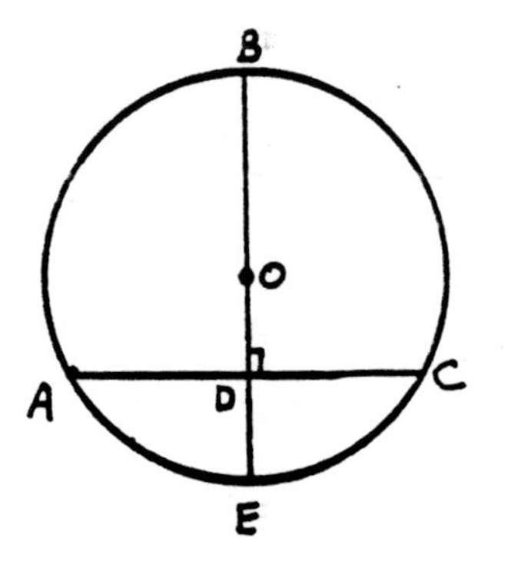

Given: O is the center of the circle. $\overline{OE} \perp \overline{AC}$.

Conclusion: The theorem says that $\overline{OE}$ bisects $\overline{AC}$ which means $\overline{AD} = \overline{DC}$. Also, $\overline{OE}$ bisects $\widehat{AC}$; therefore, $\widehat{AE} = \widehat{EC}$.

Example:

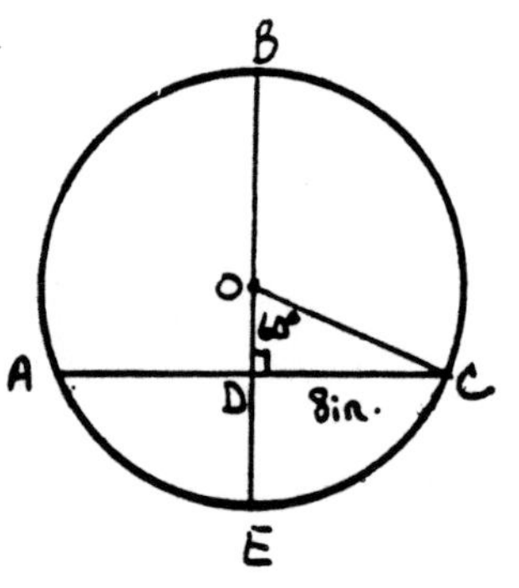

Given: $\overline{OE} \perp \overline{AC}$ and O is the center of the circle.
Find: $\overline{AC}$, $\widehat{CE}$, and $\widehat{AC}$.

$\overline{OE}$ bisects $\overline{AC}$. Since $\overline{DC} = \overline{AD} = 8$ in., $\overline{AC} = 16$ in.
$\angle$ EOC is a central angle. $\widehat{EC}$ has the same measure as its central angle, so $\widehat{EC} = 60^\circ$. Since $\overline{OE}$ bisects $\widehat{AC}$, then $\widehat{AC} = 120^\circ$.

15.

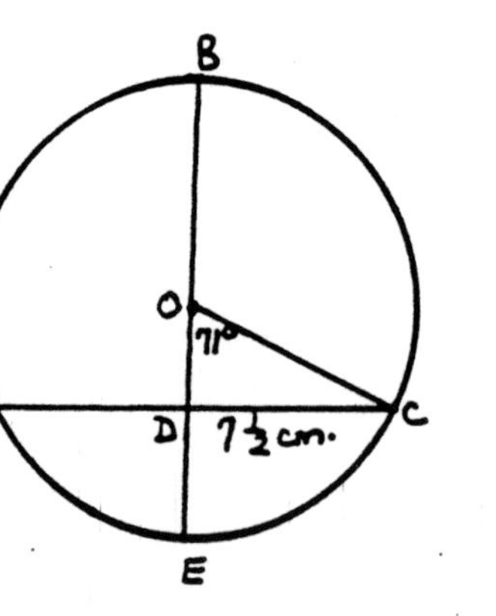

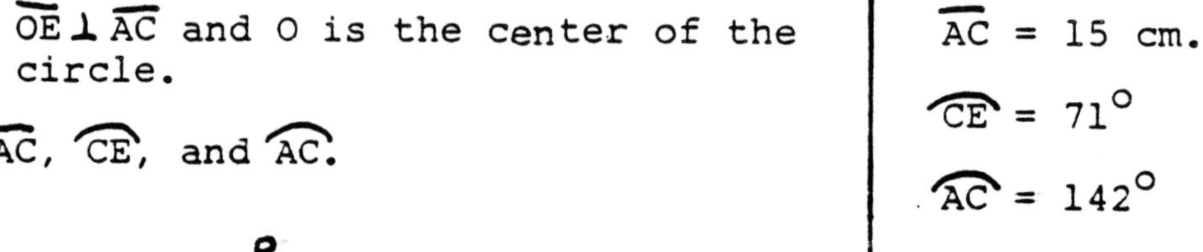

Given: $\overline{OE} \perp \overline{AC}$ and O is the center of the circle.

Find: $\overline{AC}$, $\widehat{CE}$, and $\widehat{AC}$.

15.

$\overline{AC} = 15$ cm.

$\widehat{CE} = 71^\circ$

$\widehat{AC} = 142^\circ$

16.

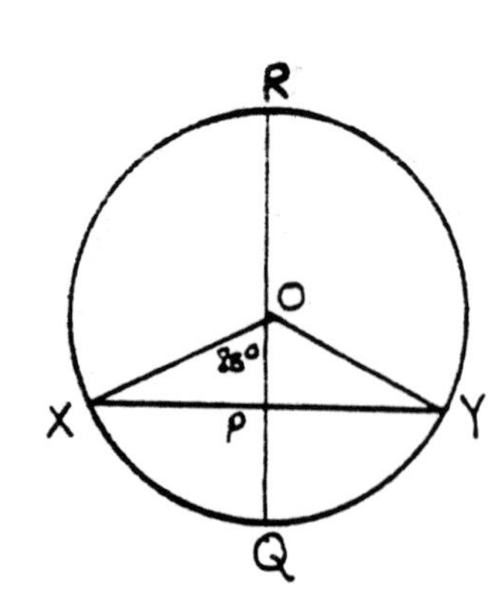

Given: $\overline{OQ} \perp \overline{XY}$ and O is the center of the circle. $\angle XOQ = 85^\circ$ and $\overline{XY} = 27$ in.

Find $\overline{XP}$, $\overline{PY}$, $\angle QOY$ and $\widehat{QY}$.

16.

$\overline{XP} = \frac{1}{2}(27) = 13\frac{1}{2}$ in.

$\overline{PY} = \frac{1}{2}(27) = 13\frac{1}{2}$ in.

$\angle QOY = 85^\circ$

$\widehat{QY} = 85^\circ$

Example: If the radius of a circle is 8 inches and chord $\overline{AB}$ is 5 inches from the center of the circle, what is the measure of chord $\overline{AB}$?

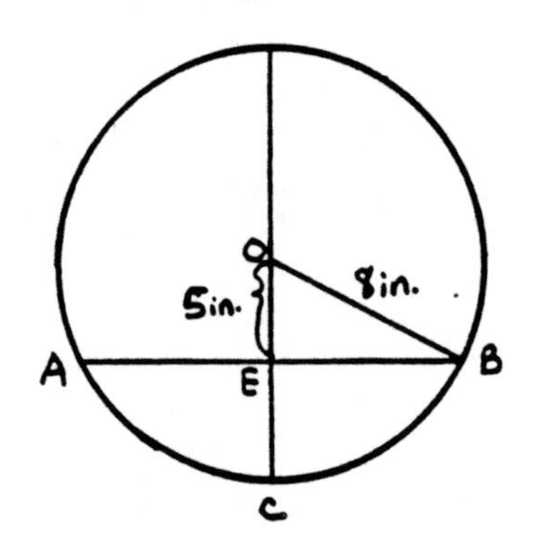

Solution: Draw in radius $\overline{OB}$.
$\triangle$ OEB is a right triangle
Using the Pythagorean theorem,

$$5^2 + \overline{BE}^2 = 8^2$$
$$25 + \overline{BE}^2 = 64$$
$$\overline{BE}^2 = 39$$
$$\overline{BE} = \sqrt{39} \text{ in.}$$
$$\overline{AB} = 2\overline{BE} = 2\sqrt{39} \text{ in.}$$

17.

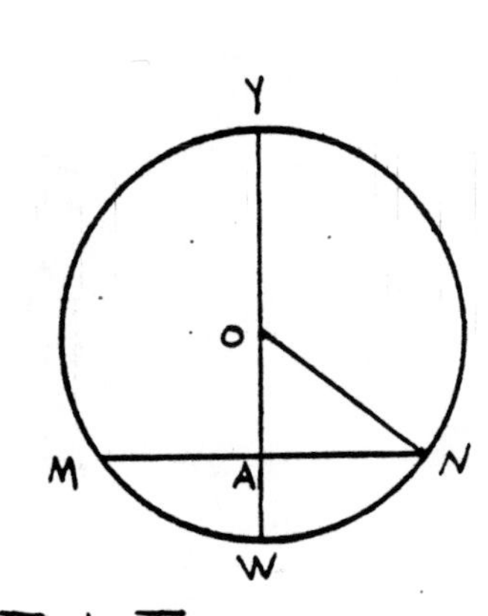

Given: $\overline{OW} \perp \overline{MN}$
If the radius of a circle is ten inches and chord $\overline{MN}$ = 16 in., find $\overline{OA}$.

17.

If $\overline{MN}$ = 16 in.,
$\overline{AN} = \frac{1}{2}(16) = 8$ in
$\triangle$OAN is a right triangle.
Using the Pythagorean theorem,

$$\overline{OA}^2 + \overline{AN}^2 = \overline{ON}^2$$
$$\overline{OA}^2 + 8^2 = 10^2$$
$$\overline{OA}^2 + 64 = 100$$
$$\overline{OA}^2 = 36$$
$$\overline{OA} = 6$$

Postulate: Two circles are congruent if and only if they have the same radius.

> **Theorem 6:** In the same circle or in congruent circles, equal arcs have equal chords.

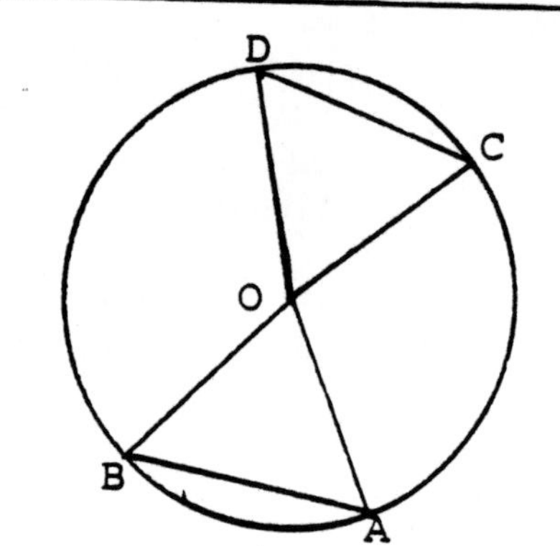

Given: $\overset{\frown}{AB} = \overset{\frown}{CD}$

Prove: $\overline{AB} = \overline{CD}$

Draw in radii $\overline{OB}$, $\overline{OA}$, $\overline{OD}$, and $\overline{OC}$

Statements	Reasons
1. $\overset{\frown}{AB} = \overset{\frown}{CD}$	1. Given
2. $\angle BOA = \angle DOC$	2. The measure of a central angle equals the measure of its intercepted arc. Here the arcs are equal.
3. $\overline{OB} = \overline{OD}$, $\overline{OA} = \overline{OC}$	3. Radii of a circle are equal.
4. $\triangle BOA \cong \triangle DOC$	4. SAS = SAS
5. $\overline{AB} = \overline{CD}$	5. CPCTE

> **Theorem 7:** In a circle, two chords are the same distance from the center of the circle if, and only if, they are equal.

This theorem is a combination of two other theorems. The two are the original statement and its converse.

Statement (1): If two chords are the same distance from the center of a circle, these chords are equal.

Converse (2): If two chords of a circle are equal, then they are the same distance from the center of the circle.

An example of the statement is as follows:

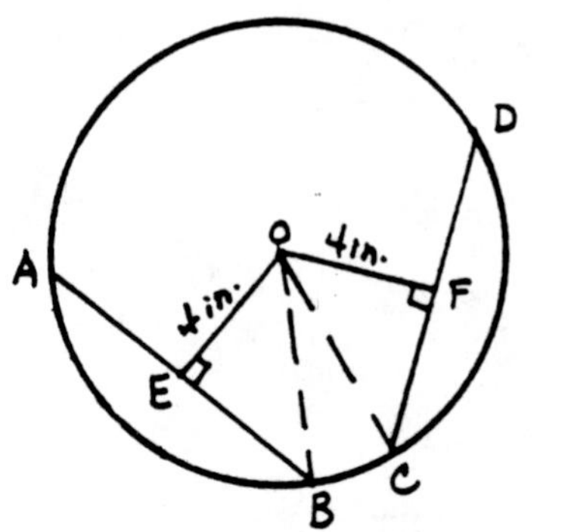

The two chords are drawn equal distance from the center of the circle. In this diagram $\overline{OE}$ and $\overline{OF}$ are both 4 in. Since distance from a point to a line is measured on a perpendicular, $\overline{OE} \perp \overline{AB}$ and $\overline{OF} \perp \overline{CD}$. According to Theorem 7, part (1), $\overline{AB}$ would equal $\overline{CD}$. If $\overline{AB}$ = 6 in., then $\overline{CD}$ = 6 in. also.

The following is an example of the converse:

(2) If two chords of a circle are equal, then they are the same distance from the center of the circle.

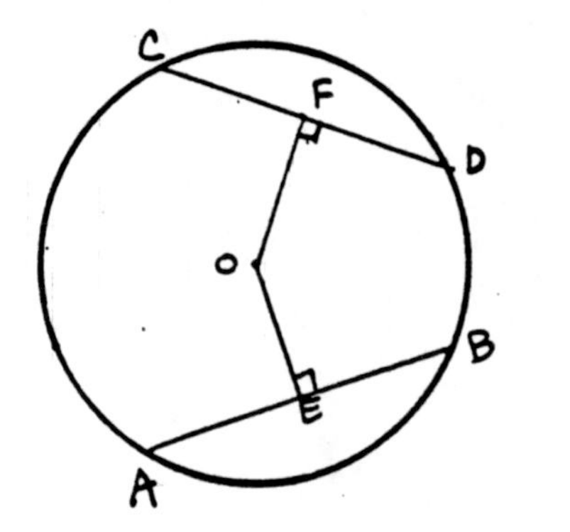

In Theorem 7, Part (2), $\overline{AB} = \overline{CD}$ and from the given it can be concluded that $\overline{OE} = \overline{OF}$.

For example, if $\overline{AB} = \overline{CD}$ and $\overline{OF}$ = 5 ft., then $\overline{OE}$ = 5 ft. by this theorem.

18.

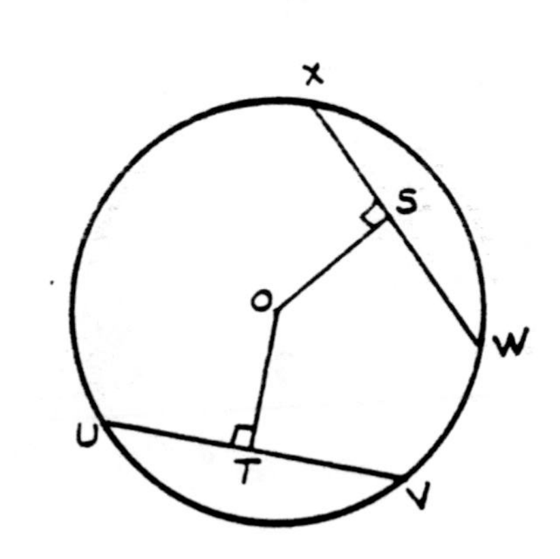

If $\overline{XW} = \overline{UV}$ and $\overline{OS}$ = 2 in., then find $\overline{OT}$. Tell why.

19. In exercise 18, if $\overline{OT} = \overline{OS}$ and $\overline{UV} = 3\frac{1}{2}$ cm., then find $\overline{XW}$. Tell why.

18. Theorem 7, part (2)

$\overline{OT}$ = 2 in.

19. Theorem 7, part (1)

$\overline{XW} = 3\frac{1}{2}$ cm.

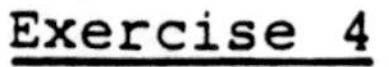

Exercise 4

1.

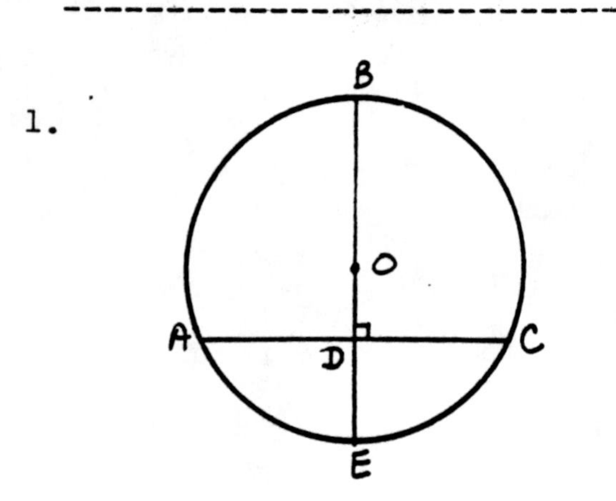

Given: $\overline{BE}$ is a diameter.
$\overline{BE} \perp \overline{AC}$
$\overline{DC}$ = 4 m.
$\overset{\frown}{AE} = 70^{\circ}$

Find: $\overline{AC}$, $\overset{\frown}{EC}$, and $\overset{\frown}{AC}$.

2.

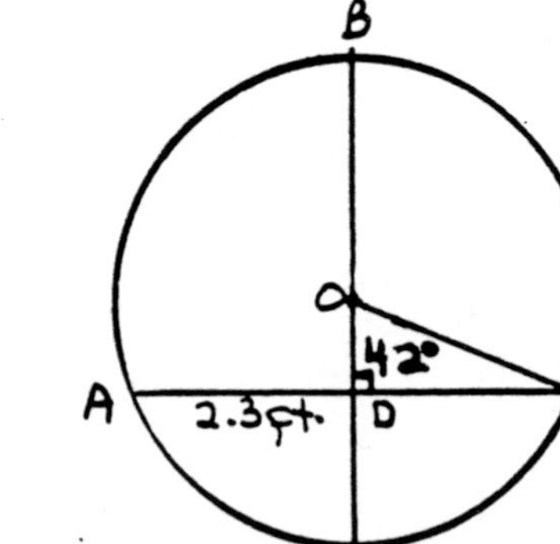

Given: $\overline{OE} \perp \overline{AC}$
$\overline{AD}$ = 2.3 ft.
$\angle DOC = 42^{\circ}$

Find: $\overline{AC}$, $\overset{\frown}{EC}$, $\overset{\frown}{AE}$, $\overset{\frown}{AC}$, and $\angle OCD$.

3.

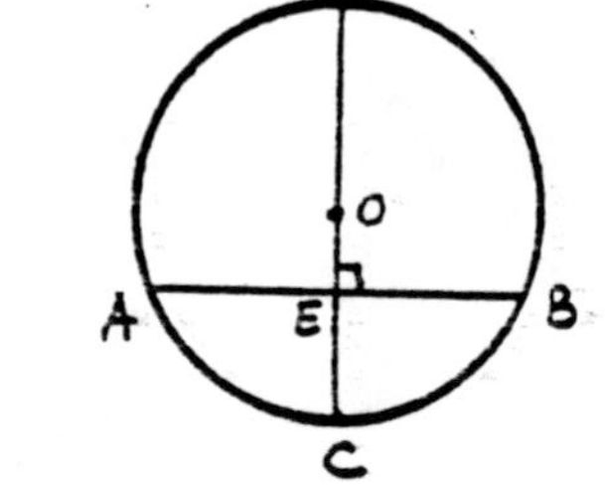

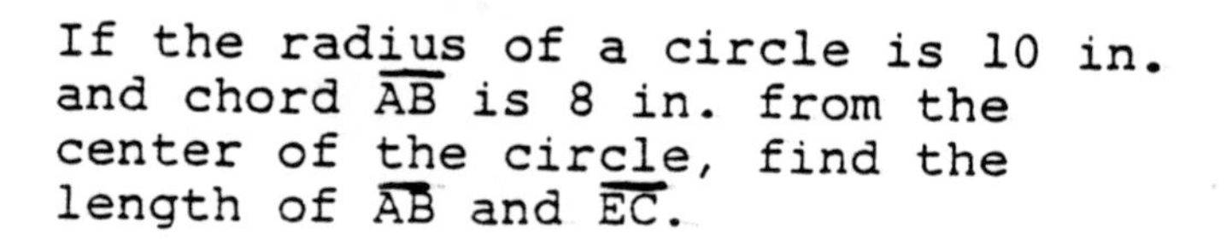

If the radius of a circle is 10 in. and chord $\overline{AB}$ is 8 in. from the center of the circle, find the length of $\overline{AB}$ and $\overline{EC}$.

4.

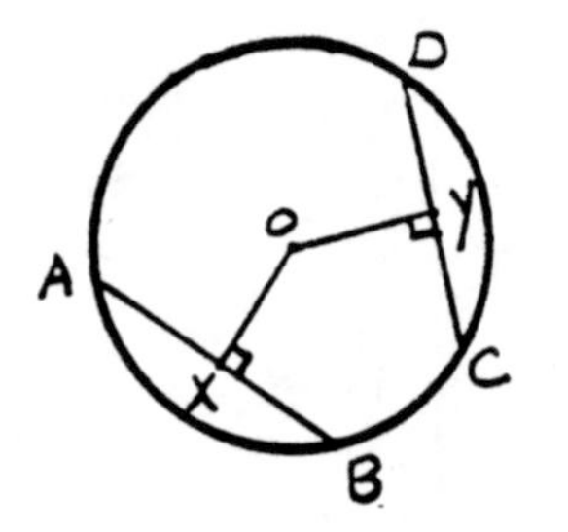

If $\overline{AB} = \overline{DC}$ and $\overline{OX}$ = 4 ft., find the length of $\overline{OY}$.

5.

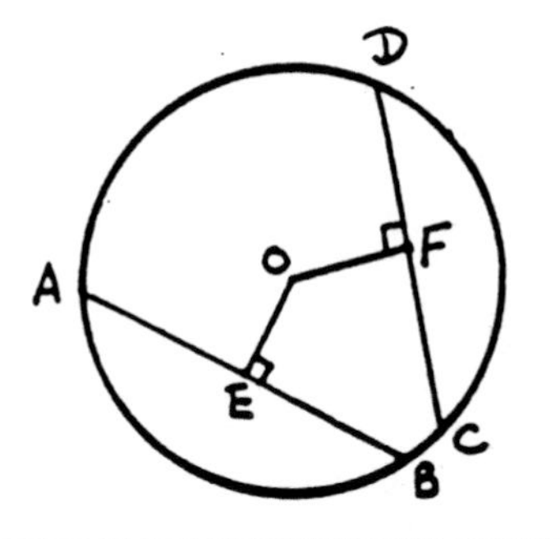

If $\overline{OE} = \overline{OF}$ and $\overline{AB}$ = 2.5 mm., find $\overline{CD}$.

To Inscribe and Circumscribe Circles

Theorem 8: The perpendicular bisectors of the sides of a triangle meet at a point which is equidistant from the verticies of the triangle.

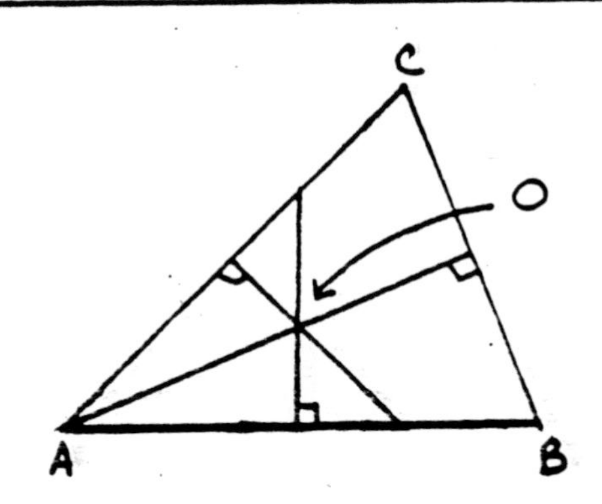

Figure 12.6

In ΔABC, point O is the intersection of three perpendicular bisectors. This point is exactly the same distance from each vertex of the triangle.

Definition:	The point of intersection of the perpendicular bisectors of the sides of a triangle is called the <u>circumcenter</u> of the triangle.

Point O in Figure 12.6 is called the circumcenter of the triangle.

Using this theorem it is possible to construct a point equidistant from the verticies and to circumscribe a circle about the triangle.

Construction # 1 To Circumscribe a Circle About a Triangle

Step 1: Construct a triangle ABC and label the verticies (or copy triangle ABC below.)

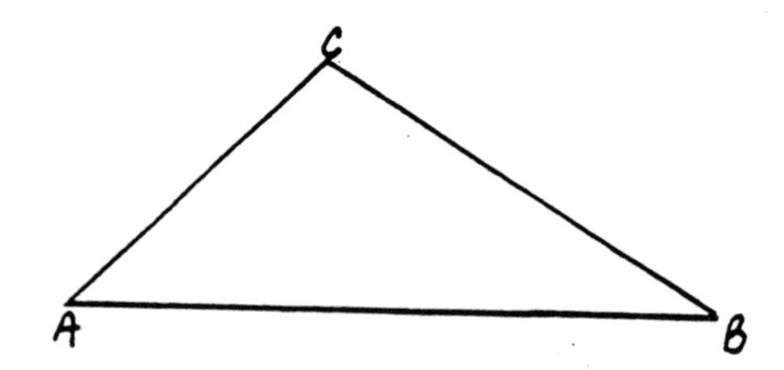

Step 2: Now construct the perpendicular bisector of each side as demonstrated in Unit 1. Actually, it is sufficient to bisect only two sides of the triangle in locating the circumcenter. By constructing the third, a check on accuracy is obtained. If the third bisector intersects the first two in a different point, it is advisable to start over.

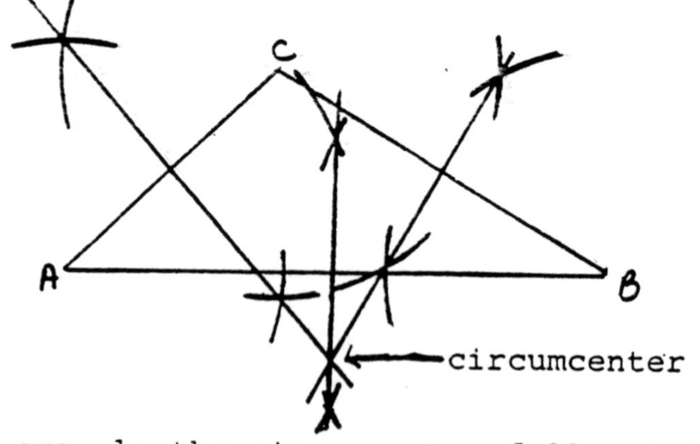

In this example the circumcenter fell outside the triangle. However, this does not always occur. In obtuse triangles the circumcenter may fall in the exterior of the triangle; in acute triangles it may fall in the interior of the triangle; and, in right triangles, it may lie on the hypotenuse.

Step 3: Using the circumcenter as center of your compass and the distance from the circumcenter to either A, B, or C as the radius, draw a circle completely about the triangle.

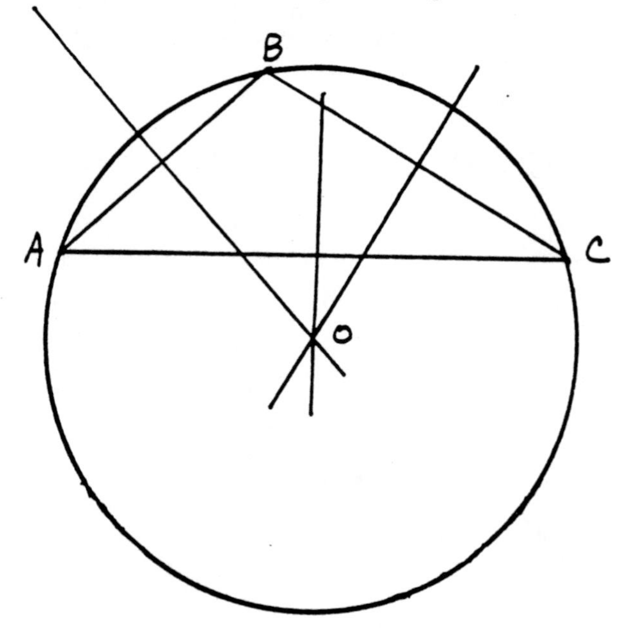

20. Using the three given points below, follow the three steps in Construction # 1 and locate the circumcenter. Then, construct a circle about the figure passing through these points.

.D

E. . F

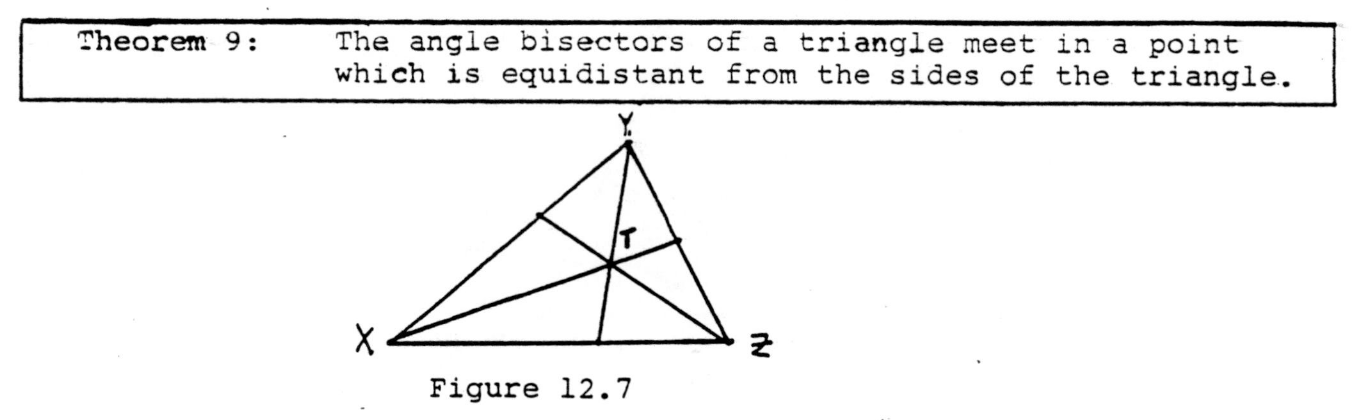

Theorem 9: The angle bisectors of a triangle meet in a point which is equidistant from the sides of the triangle.

Figure 12.7

In $\triangle XYZ$, point T is the intersection of the angle bisectors. This point is exactly the same distance from each of the sides of the triangle.

Definition:	The point of intersection of the angle bisectors of a triangle is called an <u>incenter</u>.

Point T in Figure 12.9 is called the incenter of Δ XTZ.

This theorem is used to inscribe a circle within a triangle.

Construction # 2: To Inscribe a Circle Within a Triangle

Step 1: Construct a triangle ABC and bisect each angle.

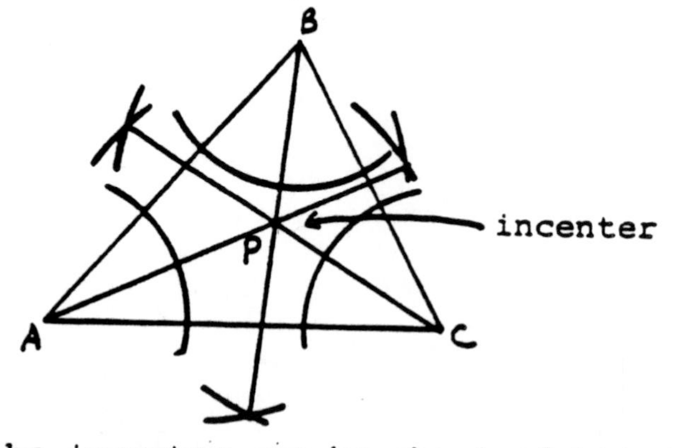

Again the incenter can be obtained by constructing only two angle bisectors. However, a third is an excellent check for accuracy. The incenter will always be inside the triangle.

Step 2: Construct a perpendicular from point P to any one of the sides.

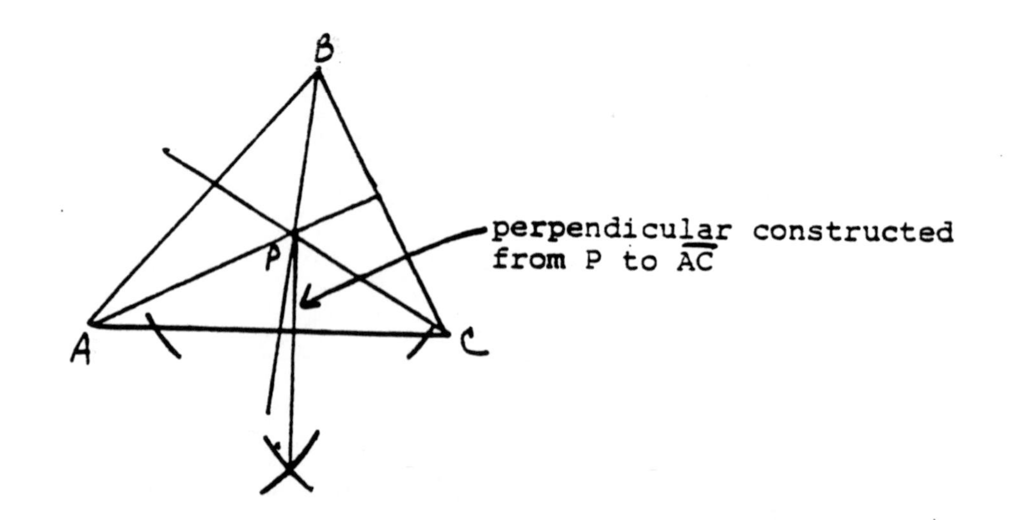

Step 3: Using P as center and the length of $\overline{PD}$ as radius, construct a circle within triangle ABC.

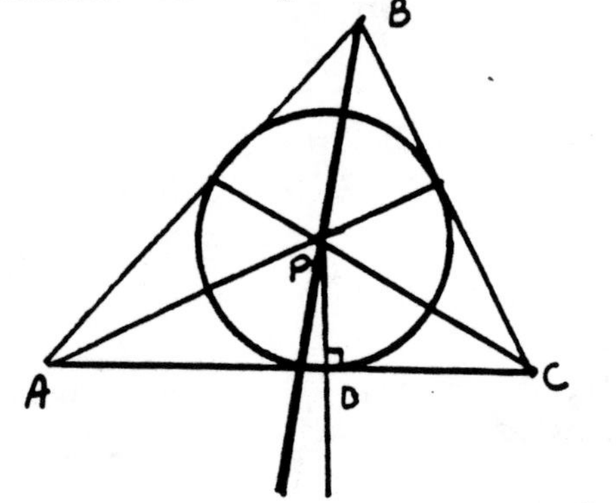

21. Using the steps in Construction # 2, find the incenter of the given triangle and inscribe a circle within.

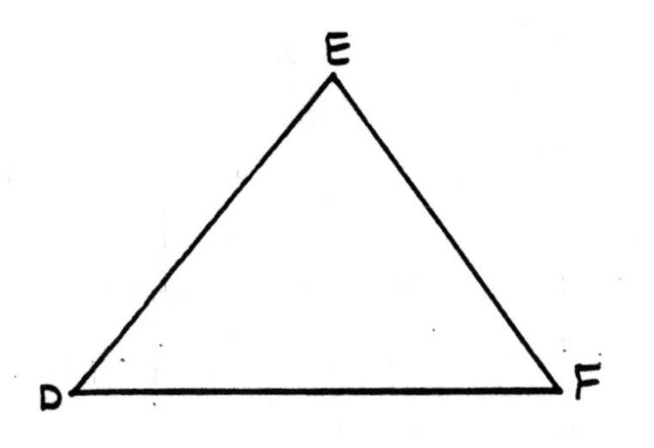

<u>Unit 12 Review</u>

I. Fill in the blanks with an appropriate word or phrase as discussed in this unit.

1. A ____________ is the set of points a given distance from a fixed point. The given distance is a ________ of the circle and the fixed point is the ________________ .

2. If two radii are put together to form a straight line segment, this line segment is called a __________ of the circle.

3. A line segment whose endpoints are both on the circle is called a ______________.

4. A line which intersects a circle in only one place is called a ____________. The point where the line and circle intersect is called the _________________.

5. A ____________ is a line which intersects a circle in two distinct points.

6. A ____________ of a circle is an angle with vertex at the center of the circle and with sides the radii of the circle.

7. An __________ is an angle whose vertex is on the circle and whose sides are chords of the circle.

8. A ______________ of a circle is the set of all possible points bounded by two radii and their intercepted arc.

9. The point of intersection of the perpendicular bisectors of the sides of a triangle is called the ______________________.

10. The point of intersection of the angle bisectors of a triangle is called an ______________________.

II. Find the solution to each of the following:

1.

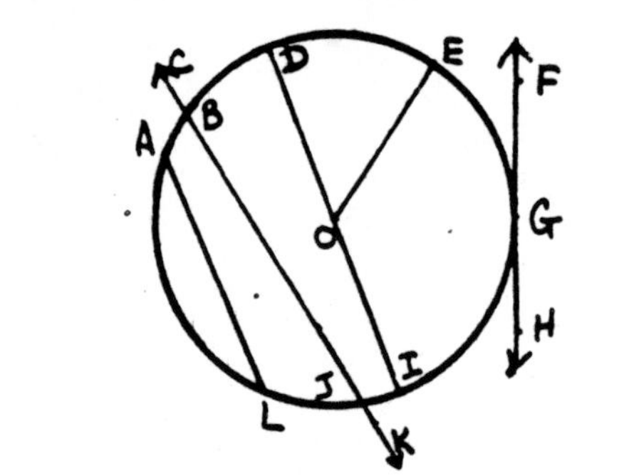

a. Name the center.
b. Name 3 (three) radii.
c. Name a diameter.
d. Name 3 chords.
e. Name a secant.
f. Name a tangent.
g. Name a point of tangency.

2.

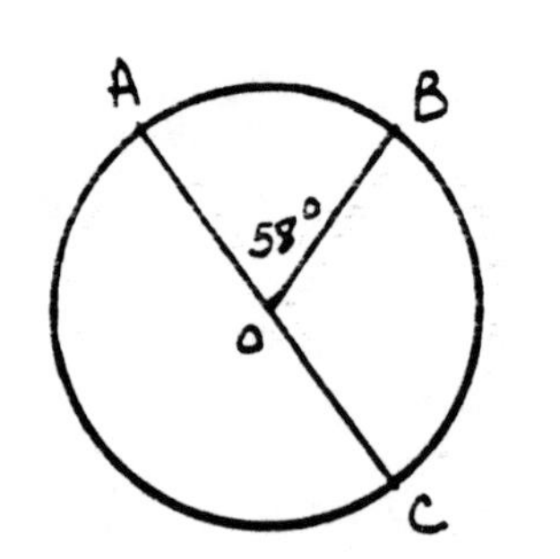

If $\angle AOB$ is a central angle and $\overline{AC}$ is a diameter, find $\overset{\frown}{AB}$, $\overset{\frown}{BC}$ and $\overset{\frown}{AC}$. Name a major arc, a semicircle and a minor arc.

3.

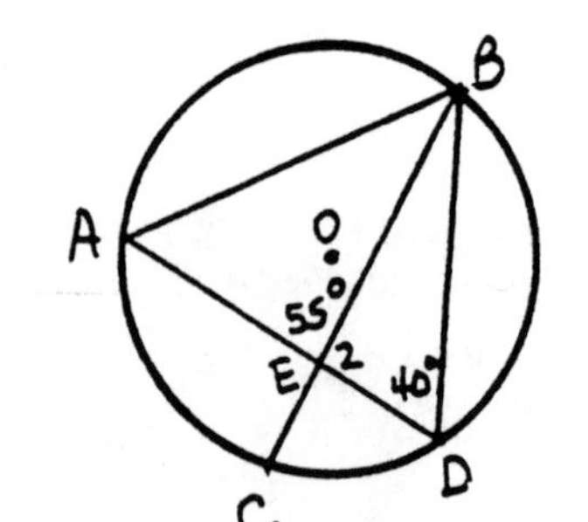

Given: $\angle AEB = 55^\circ$, $\angle D = 40^\circ$, $\overset{\frown}{BD} = 90^\circ$

Find: $\overset{\frown}{CD}$, $\overset{\frown}{AB}$, $\angle BAD$, $\angle 2$, $\angle CBD$, $\overset{\frown}{AC}$, $\angle ABD$, and $\angle ABC$

(Hint: Find the measures of $\angle BAD$, $\angle ABC$, $\angle 2$, and $\angle CBD$ first.)

4. Find the length of the arc and the area of the sector in each of the following:

   a. $r = 6$ ft., central angle $= 45^\circ$

   b. $d = 10\sqrt{2}$ in., central angle $= 80^\circ$

   c. $r = 9$ ft., central angle $12\frac{1}{2}^\circ$

5.

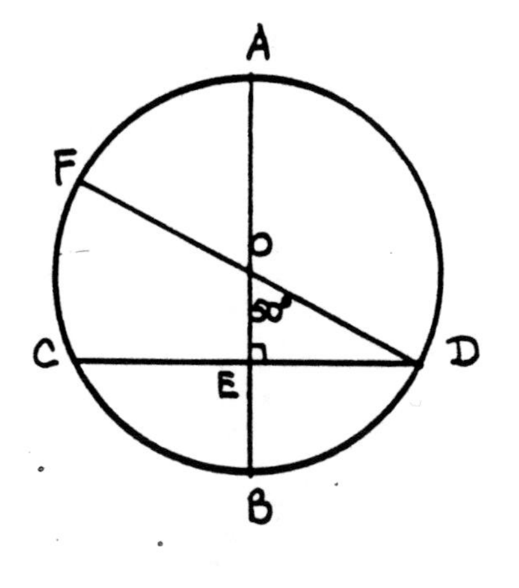

Given: $\overline{AB} \perp \overline{CD}$ and O is the center of the circle. $\angle EOD = 50^\circ$, $\overline{OE} = 3$ in., $\overline{OD} = 5$ in.
Find: $\overline{DE}$, $\overline{CD}$, $\overset{\frown}{BD}$, $\overset{\frown}{CD}$, $\overset{\frown}{CF}$, $\overset{\frown}{AD}$ and $\overset{\frown}{AF}$.

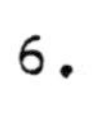
6.

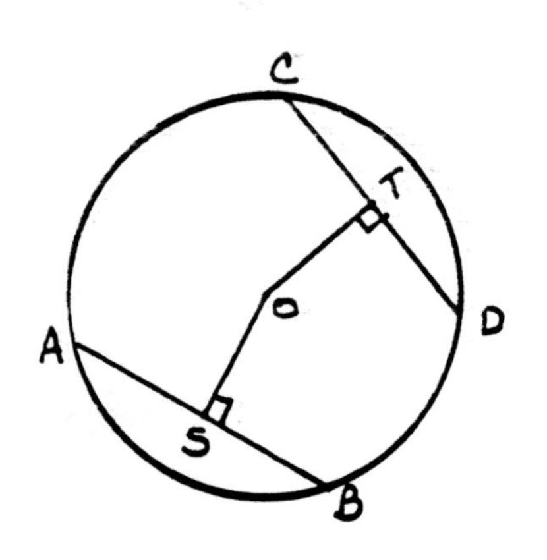

If $\overline{AB} = \overline{CD}$ and O is the center and $\overline{OS} = 9$ yd., find the length of $\overline{OT}$.

7.

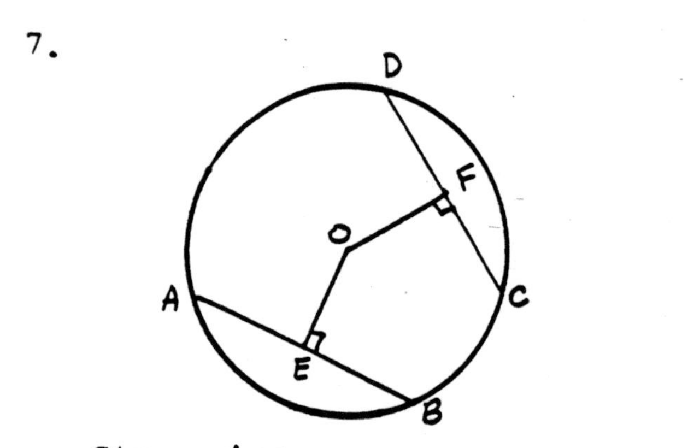

If $\overline{OE} \cong \overline{OF}$ and $\overline{AB} = 2\sqrt{3}$ mm., find $\overline{CD}$.

8. Given ΔABC, circumscribe a circle about it.

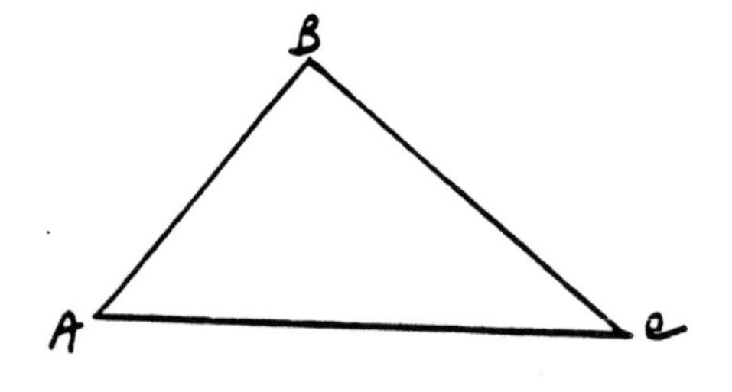

9. Given ΔDEF, inscribe a circle within it.

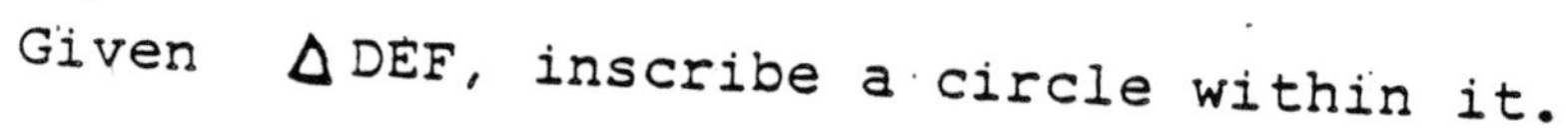

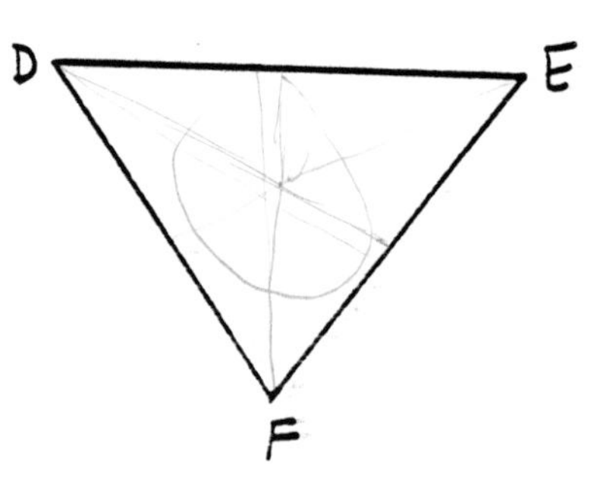

Unit 13

Coordinate Geometry

Learning Objectives:

1. The student will demonstrate his mastery of the following definitions, formulas, and implications by writing them and by applying them to solutions of selected problems.

Definitions: origin, ordinate, abscissa, quadrant, x-axis, y-axis, linear equation, slope, x-intercept, and y-intercept.

Formulas:

$$\text{slope} = \frac{\text{rise}}{\text{run}}$$

$$\text{slope} = \frac{y_2 - y_1}{x_2 - x_1}$$

$$\text{midpoint} = \left(\frac{x_1 + x_2}{2}, \frac{y_1 + y_2}{2}\right)$$

$y = mx + b$ (slope, y-intercept form of straight line)

$y - y_1 = m(x - x_1)$ (point, slope form of straight line)

$Ax + By = C$ (standard form of the straight line)

$$\text{distance} = \sqrt{(x_2 - x_1)^2 + (y_2 - y_1)^2}$$

Theorems:

1. Two lines are parallel if and only if they have the same slope.

2. Two lines are perpendicular if and only if their slopes are negative reciprocals of each other.

3. If (x_1, y_1) and (x_2, y_2) are the endpoints of a line segment, the midpoint of the line segment is determined by the formula:

$$\text{midpoint} = \left(\frac{x_1 + x_2}{2}, \frac{y_1 + y_2}{2}\right)$$

4. If (x_1,y_1) and (x_2,y_2) are the endpoints of a line segment, the undirected distance from (x_1,y_1) to (x_2,y_2) is given by the formula:

$$d = \sqrt{(x_2 - x_1)^2 + (y_2 - y_1)^2}$$

2. The student will graph the equation of a straight line by three methods.
   a. selection of points
   b. slope, y-intercept
   c. x-intercept, y-intercept

3. The student will write the equation of a straight line

   a. given two points
   b. given a point and the slope
   c. given a point and a line to which it is parallel
   d. given a point and a line to which it is perpendicular
   e. given the x-intercept and y-intercept
   f. given the slope and y-intercept
   g. given the slope and x-intercept

4. The student will plot points and describe their location in the Cartesian Coordinate System.

5. The student will identify the slope of a line as positive, negative, zero, or no slope by inspecting the graph of the equation of the line.

COORDINATE GEOMETRY

Coordinate geometry ties together some concepts from algebra and geometry. Rene' Descartes, a Frenchman, is credited with having done a lot of work in this area.

In much the same way a roadmap helps a person along his journey, a Cartesian coordinate system helps to locate points. This system has a horizontal axis, called the x-axis, and a vertical axis, called the y-axis. These axes meet at a point called the origin. The plane is divided into four parts called quadrants. Quadrants are numbered counterclockwise. The positive side of the x-axis is to the right of the y-axis. The positive side of the y-axis is above the x-axis and the negative side of the y-axis is below the x-axis. Points are plotted from ordered pairs. In an ordered pair, the x-coordinate comes first and the y-coordinate comes second. This is why it is called an ordered pair; the order in which the coordinates are placed must be the same always. In an ordered pair, (x,y), the x-coordinate is called the abscissa and the y-coordinate is called the ordinate.

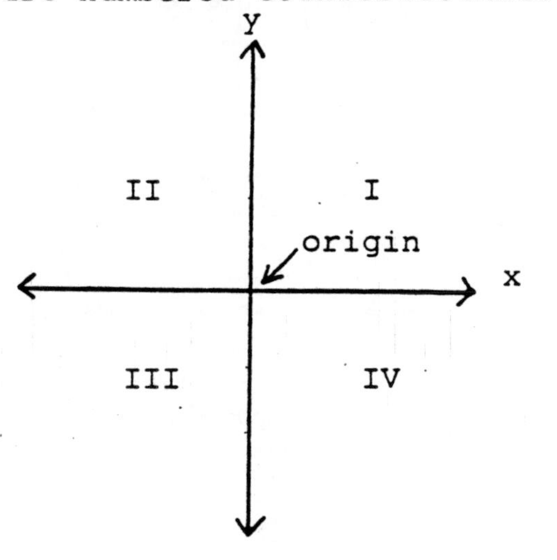

In quadrant I both coordinates are positive. In quadrant II the abscissa is negative and the ordinate is positive. In quadrant III both coordinates are negative. In quadrant IV, the abscissa is positive and the ordinate is negative. Refer to Figure 13.1.

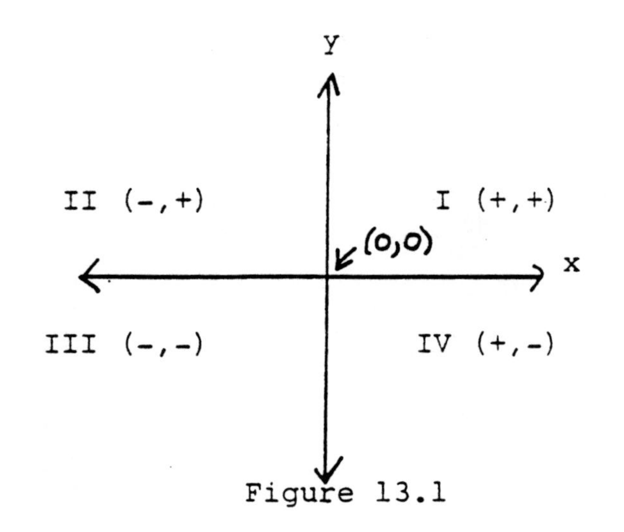

Figure 13.1

Plotting Points

Any point must fall in one of the four quadrants or on the axes. There are eight possibilities for locations of points: positive x-axis, first quadrant, positive y-axis, second quadrant, negative x-axis, third quadrant, negative y-axis, and fourth quadrant.

Taking an example of each case, let's begin with the positive x-axis. The point (4,0) falls on the positive x-axis a distance of four units to the right of the origin.

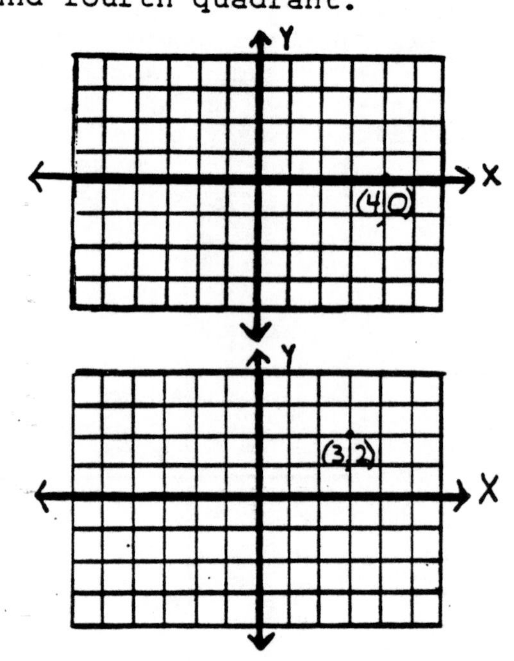

The point (3,2) falls in the first quadrant a distance of three units to the right of the origin and two units above the positive x-axis.

The point (0,3) falls on the positive y-axis a distance of three units above the origin.

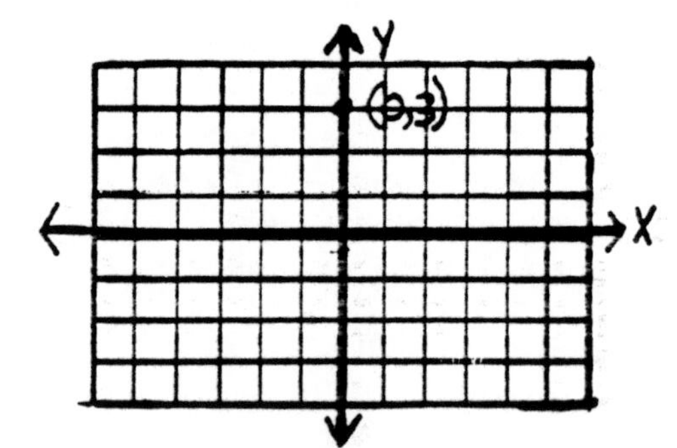

In the second quadrant the abscissa is negative. Thus, (-2,4) would be two units to the left of the origin and four units above the negative x-axis.

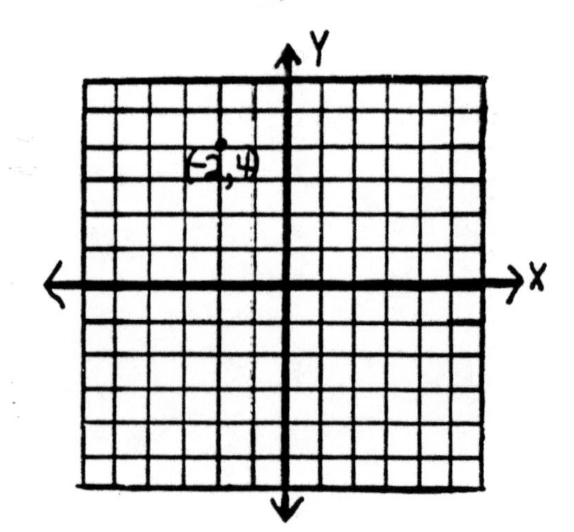

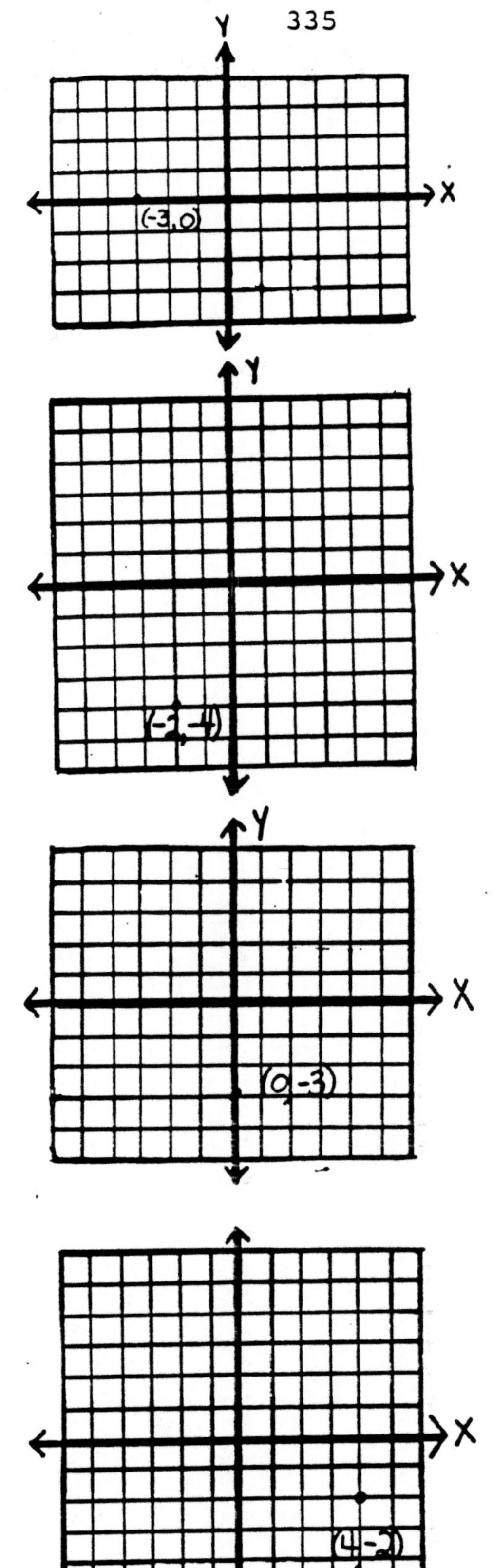

The point (-3,0) falls on the negative x-axis a distance of three units from the origin.

In the third quadrant both abscissa and ordinate are negative. To plot a point here move to the left of the origin and move below the negative x-axis. For example, (-2,-4) is two units to the left of the y-axis and four units below the negative x-axis.

The point (0,-3) falls three units below the origin on the negative y-axis.

The point (4,-2) falls in the fourth quadrant since the abscissa is positive and the ordinate is negative. In plotting go to the right four spaces and go down two spaces from here.

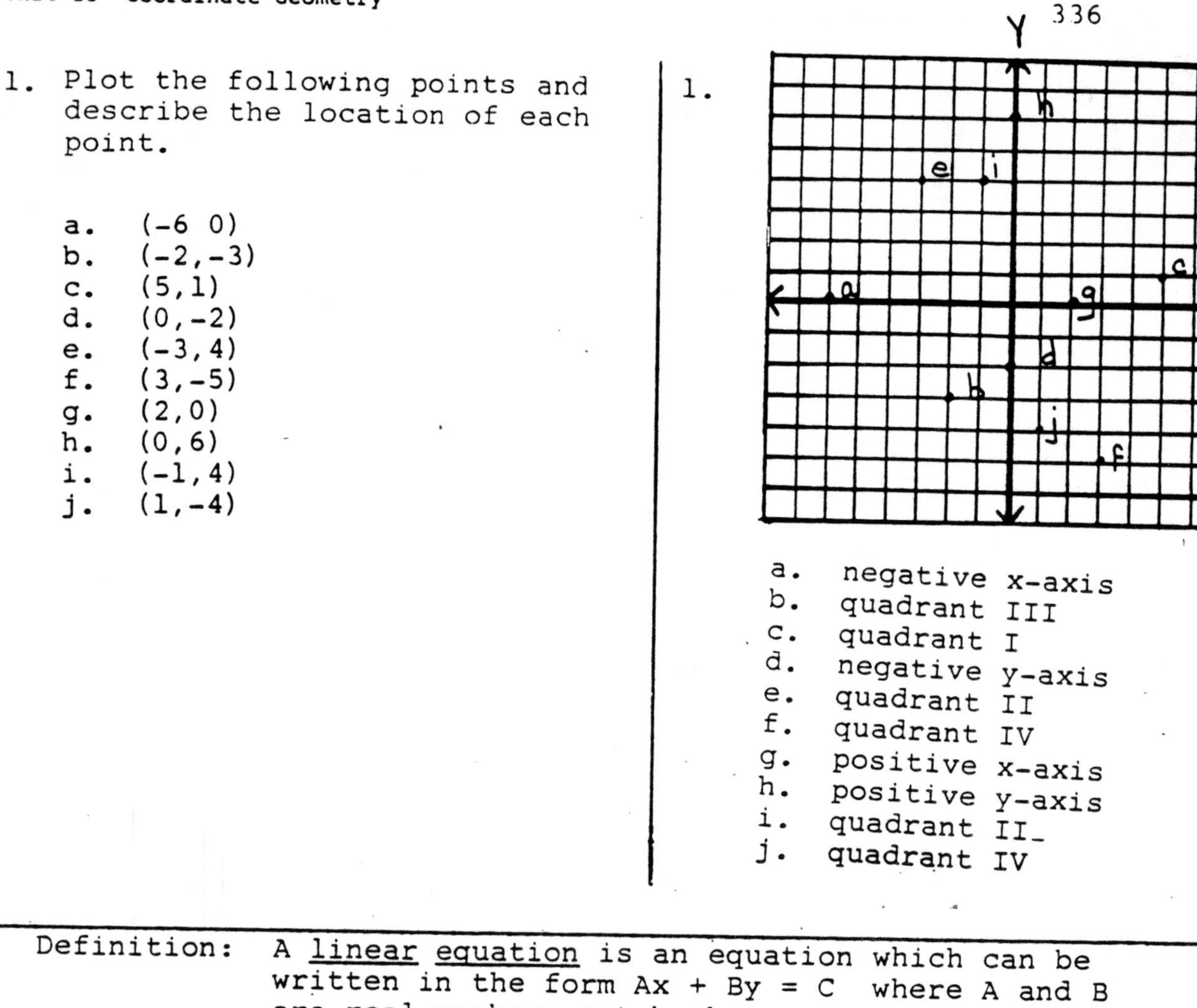

1. Plot the following points and describe the location of each point.

   a. (-6 0)
   b. (-2,-3)
   c. (5,1)
   d. (0,-2)
   e. (-3,4)
   f. (3,-5)
   g. (2,0)
   h. (0,6)
   i. (-1,4)
   j. (1,-4)

1.

   a. negative x-axis
   b. quadrant III
   c. quadrant I
   d. negative y-axis
   e. quadrant II
   f. quadrant IV
   g. positive x-axis
   h. positive y-axis
   i. quadrant II_
   j. quadrant IV

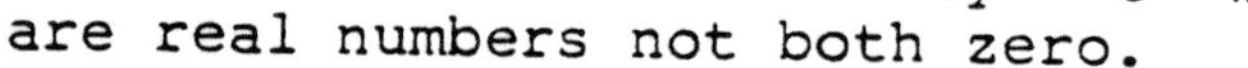

Definition: A <u>linear</u> <u>equation</u> is an equation which can be written in the form $Ax + By = C$ where A and B are real numbers not both zero.

<u>Slope</u>

Geometrically, any two points determine one and only one line in a plane. Hence, (3,1) and (5,3) determine a line as shown below.

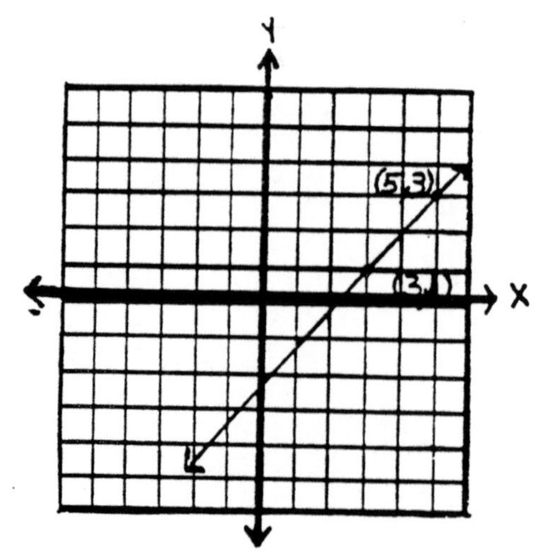

Sometimes we will need to discuss the steepness of a line or whether the line rises or falls. The term used to describe this concept is slope.

> Definition: Slope is defined to be the rise (number of units one goes up or down) divided by the run (number of units one goes left to right or vice versa).
>
> $$\text{Slope} = \frac{\text{rise}}{\text{run}}$$

Example: **Find the slope of the line through (3,1) and (5,3).**

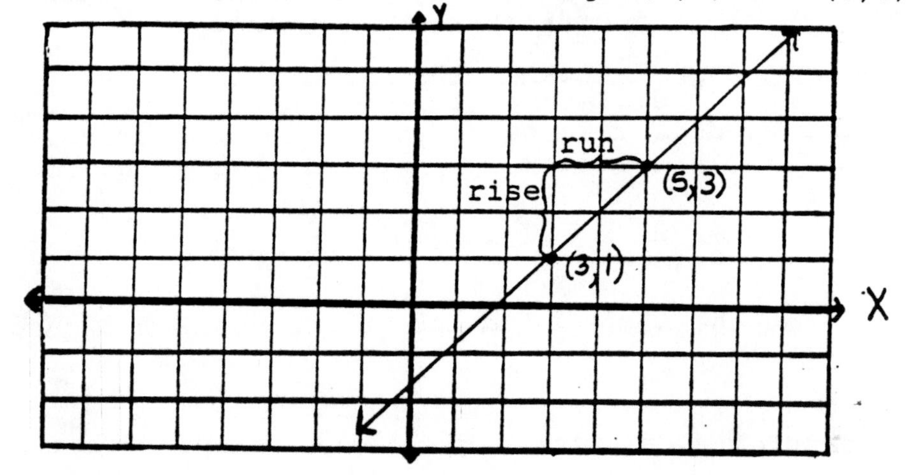

Solution: **To go from the point (3,1) to (5,3) we go up two units and to the right two units. Now up and to the right are both positive directions so**

$$\text{slope} = m = \frac{\text{rise}}{\text{run}} = \frac{2}{2} = 1 \quad .$$

This means the slope is positive so the line rises as you view it from left to right.

2. Find the slope of the line through (-2,3) and (1,5).

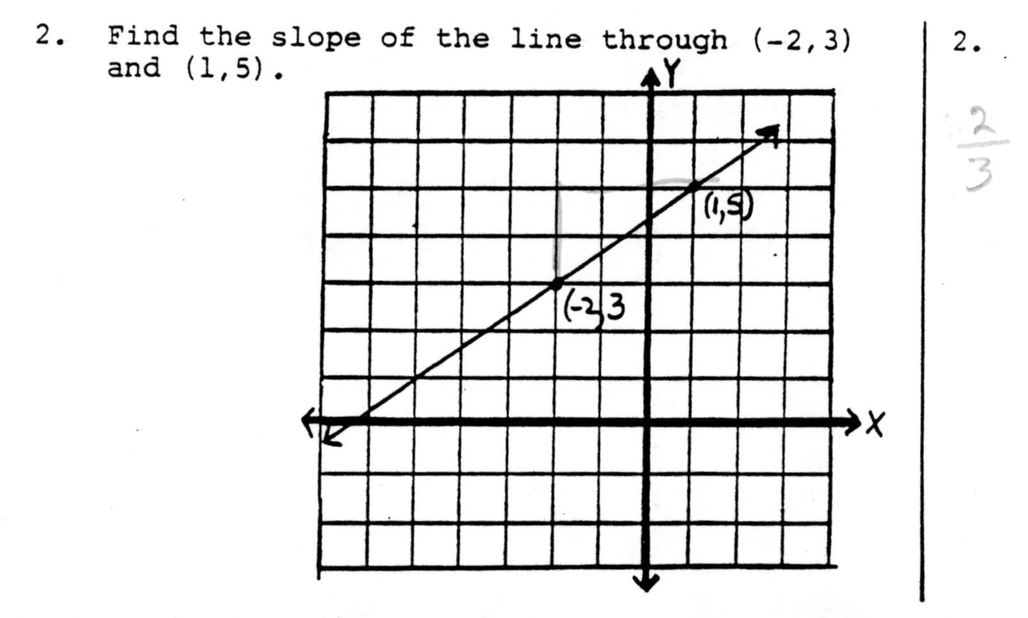

2. $\frac{2}{3}$

From (-2,3) to (1,5) go ______ (up or down) ______ units and go ______ (left or right) ______ units.

Consequently, the slope $= m = \frac{\text{rise}}{\text{run}} =$ ______

up
2, right
3

$\frac{2}{3}$

This is one method for finding slope but depends heavily on being able to easily count units on a graph. Also, notice the slope formula, $m = \frac{\text{rise}}{\text{run}}$, can be obtained by systematically using another formula. The rise, the amount of up and down movement, is actually the difference in the y coordinates. The run, the amount of horizontal movement (left and right movement), is actually the difference in the x coordinates. Consequently,

$$\text{slope} = m = \frac{\text{difference in the y coordinates}}{\text{difference in the x coordinates}}$$

Note the differences must be taken in the same order. In formula notation, if (x_1, y_1) and (x_2, y_2) are any two points, then

$$\text{slope} = m = \frac{y_2 - y_1}{x_2 - x_1}$$

Example: Find the slope of the line through (-2,3) and (1,5).

Solution: Call the first point (x_1, y_1). That is, $(x_1, y_1) = (-2,3)$, so $x_1 = -2$ and $y_1 = 3$. Call the second point (x_2, y_2). That is $x_2 = 1$ and $y_2 = 5$. Substituting into the formula gives:

$$m = \frac{y_2 - y_1}{x_2 - x_1} = \frac{5 - 3}{1 - (-2)} = \frac{2}{3}$$

Does it matter which point is (x_1, y_1) and which is (x_2, y_2)?

No, reversing the original roles, let $(x_2, y_2) = (-2,3)$ and $(x_1, y_1) = (1,5)$. Substituting into the formula,

$$m = \frac{3 - 5}{-2 - 1} = \frac{-2}{-3} = \frac{2}{3}.$$

Notice the same slope was obtained either way.

Example: Determine the slope of the line through (3,-4) and (2,-6) both (a) graphically, $\frac{\text{rise}}{\text{run}}$, and (b) algebraically $m = \frac{y_2 - y_1}{x_2 - x_1}$.

Solution:

(a)

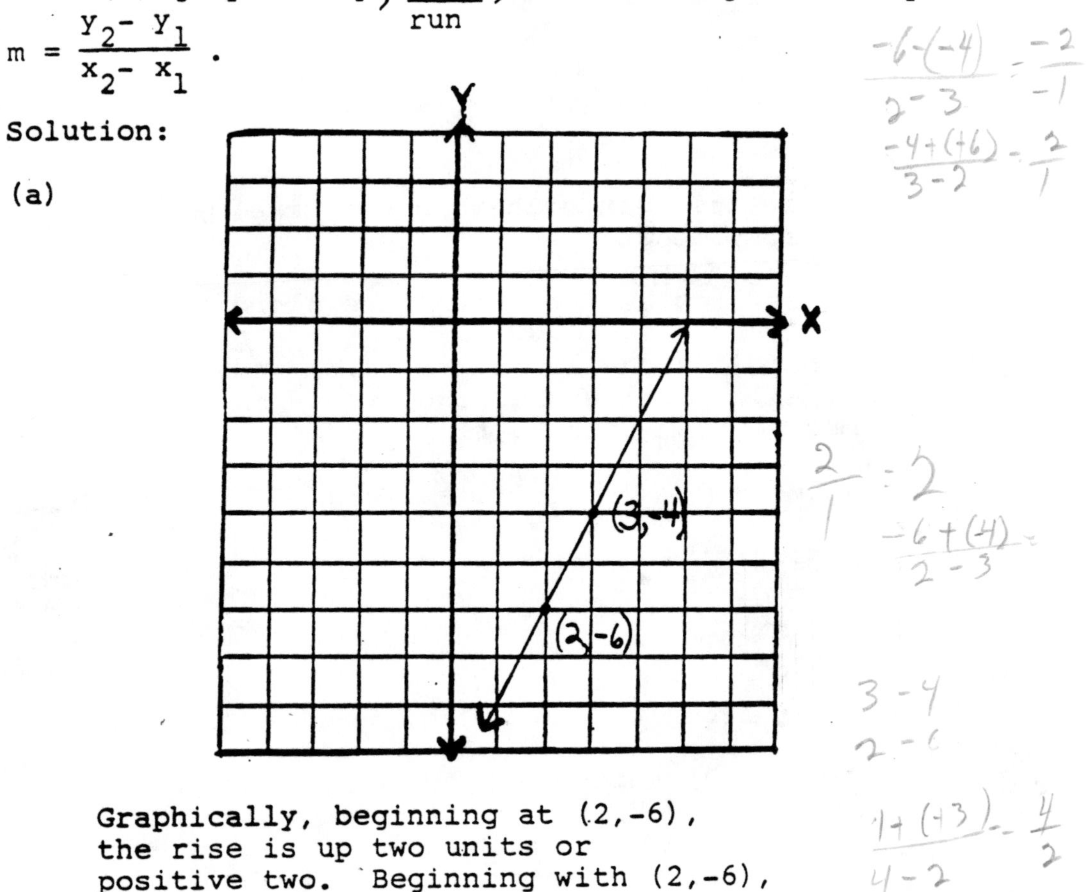

Graphically, beginning at (2,-6), the rise is up two units or positive two. Beginning with (2,-6), the run is over one or positive one. The slope is $m = \frac{\text{rise}}{\text{run}} = \frac{2}{1} = 2$.

(b) Algebraically, let $(x_1,y_1) = (3,-4)$ and $(x_2,y_2) = (2,-6)$. Substituting into the formula,

$$m = \frac{y_2 - y_1}{x_2 - x_1} = \frac{-6 - (-4)}{2 - 3} = \frac{-2}{-1} = 2$$

3. Determine the slope of the line through (4,1) and (2,-3) both (a) graphically and (b) algebraically.

3.

(a)

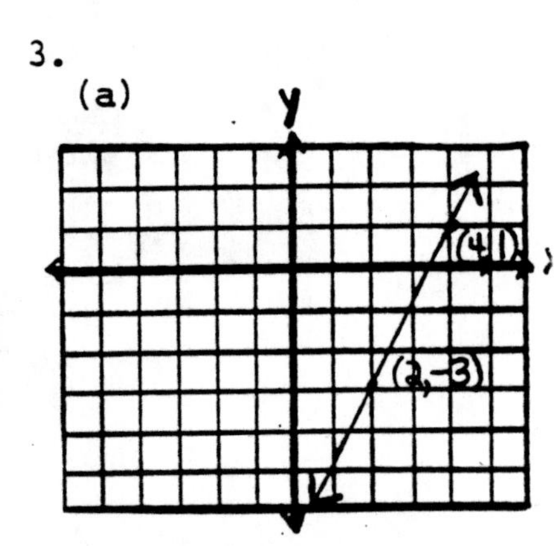

rise = 4 units
run = 2 units

$$m = \frac{rise}{run} = \frac{4}{2} = 2$$

(b)

$$m = \frac{-3 - 1}{2 - 4} = \frac{-4}{-2} = 2$$

The slope of a line must either be positive, negative, zero, or undefined, usually denoted as no slope.

If a line rises from left to right the slope is positive.

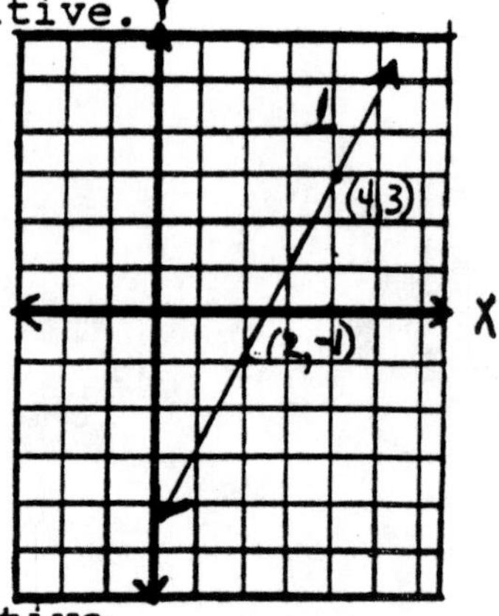

Example: Two points on line l are (4,3) and (2,-1). Notice the line rises moving from left to right. The

$$\text{slope} = m = \frac{3 - (-1)}{4 - 2} = \frac{4}{2} = 2$$

is positive.

If a line falls from left to right the slope is negative.

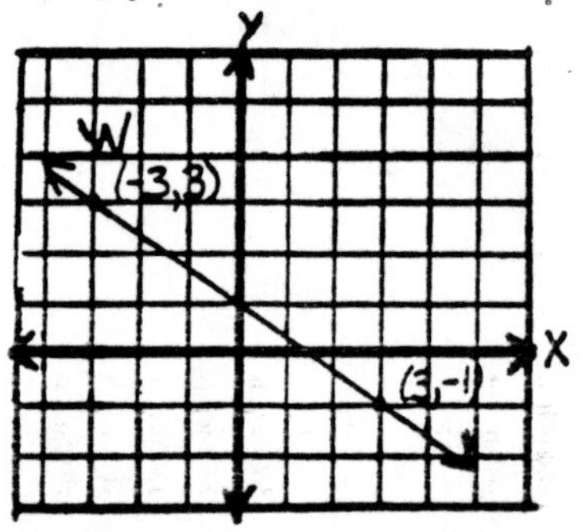

Example: Let (-3,3) and (3,-1) be two points on line w. Using the slope formula,

the slope, $m = \frac{3 - (-1)}{-3 - 3} = \frac{4}{-6} = -\frac{2}{3}$

Observe, the line falls from left to right and the slope is negative.

If a line is vertical, it has no slope.

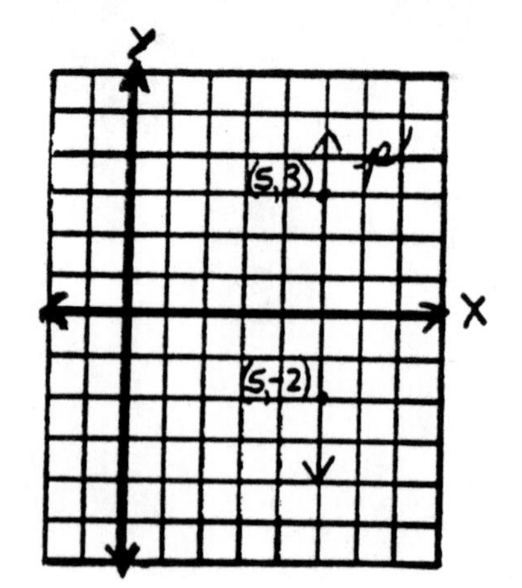

Example: Let (5,3) and (5,-2) be two points on a vertical line, p. Substituting

into the formula $m = \frac{3 - (-2)}{5 - 5} = \frac{5}{0}$

which is undefined. This is called no slope. Notice that on this vertical line the x coordinates are the same. This is true for all vertical lines. Because of this, the denominator of the slope formula becomes zero. Consequently, there is no slope.

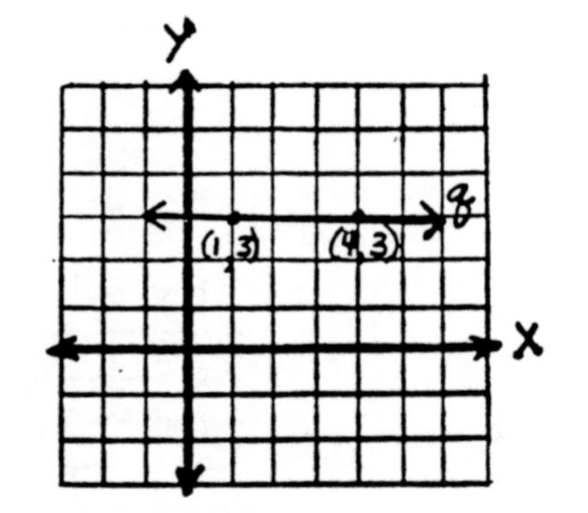

If a line is horizontal, it has zero slope.

Example: Let (1,3) and (4,3) be two points on line q. The slope is

$$m = \frac{3 - 3}{1 - 4} = \frac{0}{-3} = 0.$$

Observe here that the y coordinates are the same which renders the numerator zero. This, in turn, makes the quotient zero.

In summary, lines with positive slopes rise from left to right; lines with negative slopes fall from left to right; lines with no slope are vertical lines; and, lines with zero slope are horizontal lines.

Example: From the following diagrams, determine if the line has a positive slope, zero slope, negative slope, or no slope.

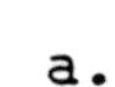

a.

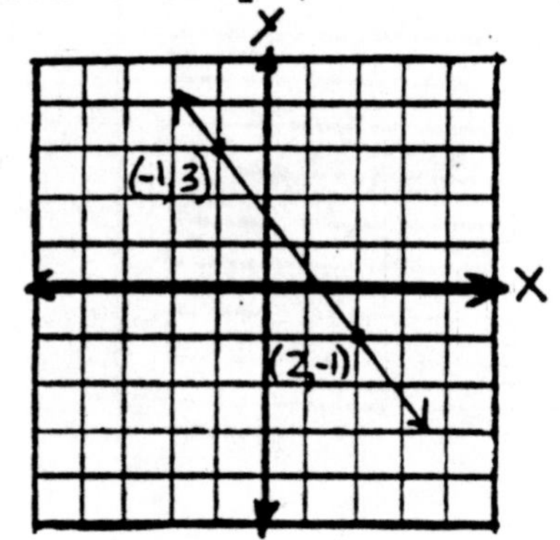

b.

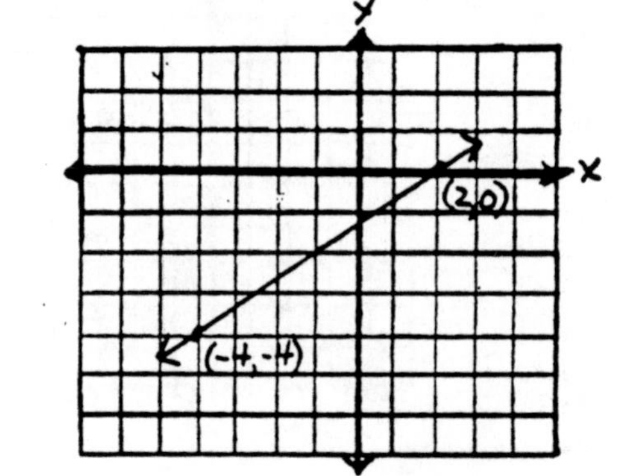

c.

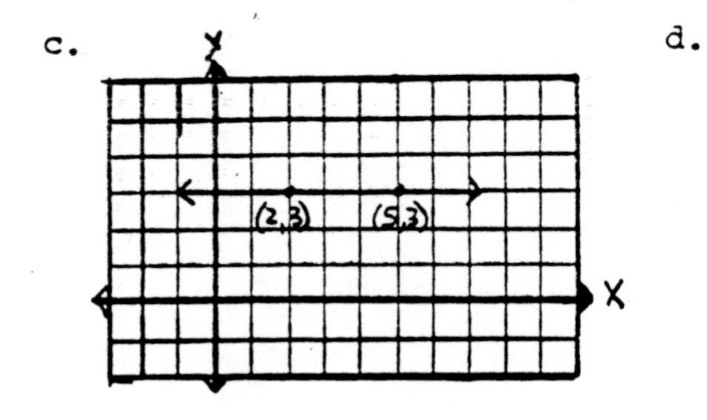

d.

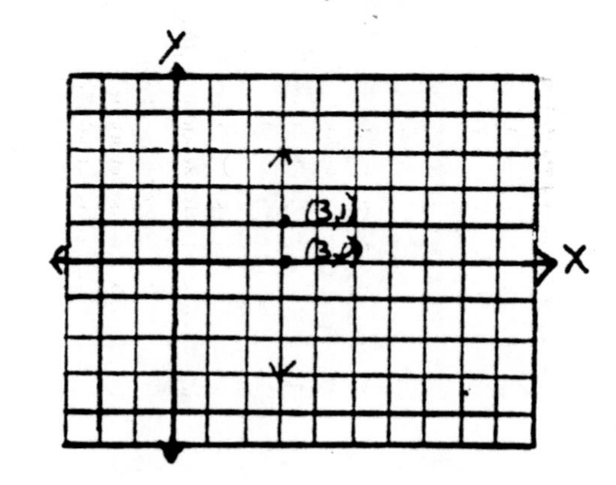

Solution: a. The line falls from left to right so has negative slope.

b. The line rises from left to right so has positive slope.

c. The line is horizontal and has zero slope.
d. The line is vertical and has no slope.

4. From the diagrams below, determine if the lines have positive, negative, zero, or no slope.

a.

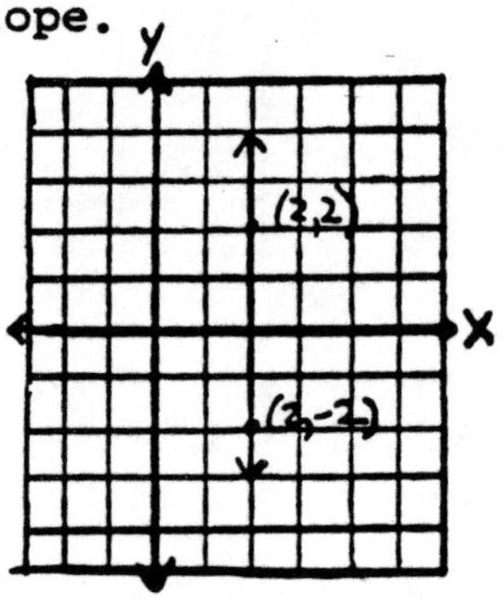

b.

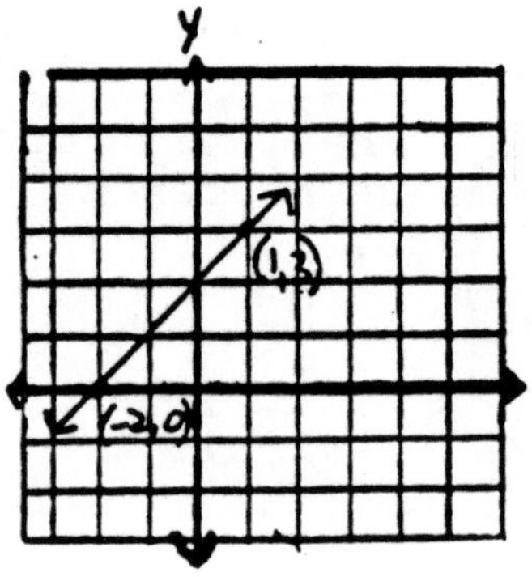

c.

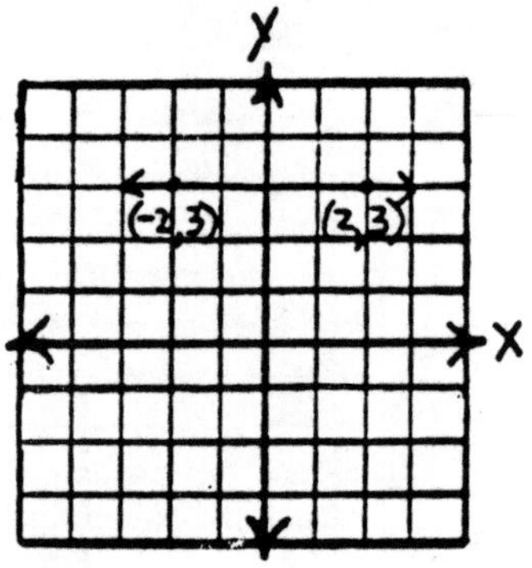

d.

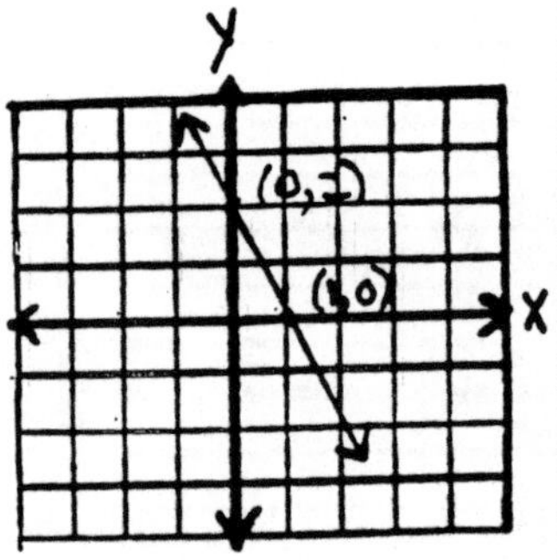

4.

a. no slope

b. positive slope

c. 0 slope

d. negative slope

Exercise 1

1. Plot the following points and describe in words their location.

a. (3,4)
b. (-2,-5)
c. (-3,0)
d. (0,-3)
e. (-2,4)
f. (2,-4)
g. (0,3)
h. (4,0)
i. (3,-1)
j. (6,3)

2. Find the slope of the line determined by the given points.

a. (0,4), (5,6)
b. (4,2), (7,-1)
c. (-3,4), (3,-4)
d. (4,7), (4,2)
e. (2,-3), (-3,-3)

3. Determine from the diagrams below if the lines have positive, negative, zero, or no slope.

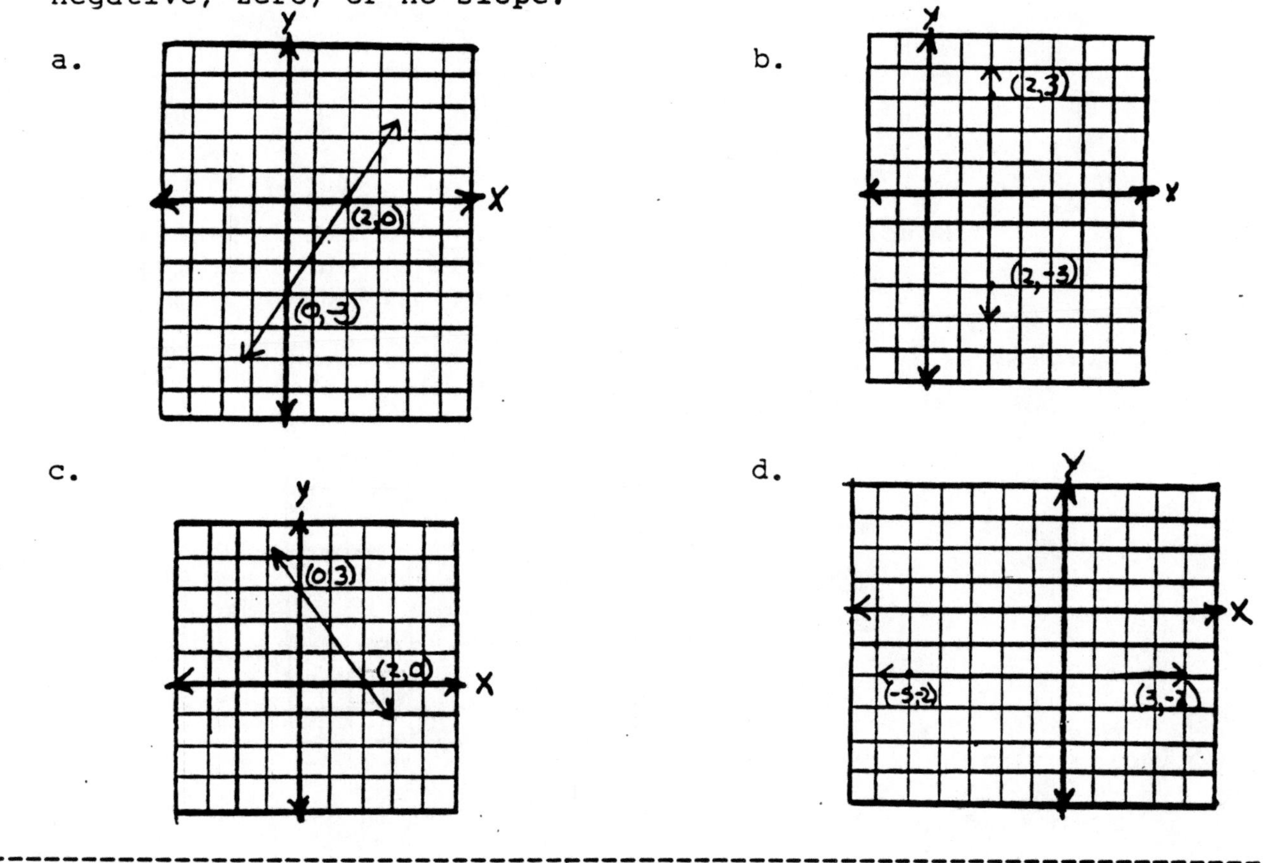

Graphing Lines

The equation Ax + BY = C, where A and B are not both zero, is the equation for a straight line and its graph is a straight line. Three methods for graphing lines will be presented: selection of points; slope y-intercept method; and, x-intercept and y-intercept method.

Method 1: Selection of points method
In this method:
1. Solve the equation for y in terms of x.
2. Select three ordered pairs which satisfy the equation.
3. Plot the points and draw a line through them.

Example: Graph $3x + y = 4$

Step 1: Solve the equation for y in terms of x.

$$\begin{array}{rcl} 3x + y & = & 4 \\ -3x \quad & & -3x \\ \hline y & = & -3x + 4 \end{array}$$

Step 2: Select three ordered pairs which satisfy the equation. Select "x's" and solve for y.

x	$y = -3x + 4$	y	ordered pair
0	$y = -3 \cdot 0 + 4$	4	(0,4)
1	$y = -3 \cdot 1 + 4$	1	(1,1)
-1	$y = -3 \cdot -1 + 4$	7	(-1,7)

Note: Care should be taken that the points are chosen near the origin so the relative position of the line to the origin can be seen.

Step 3: Plot the points.

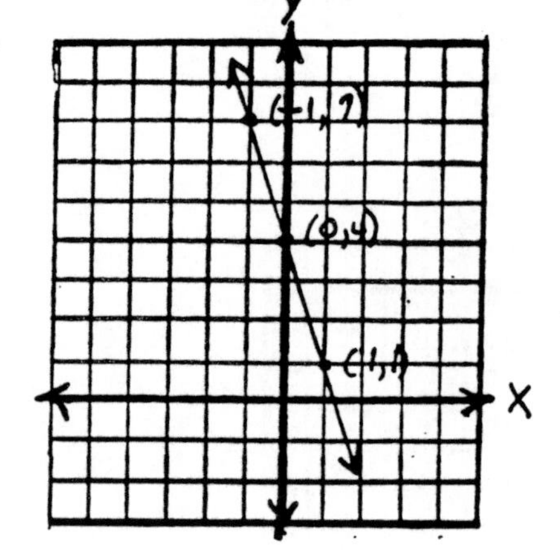

Example: Graph $3x + 2y = 6$.

Step 1:

$$\begin{aligned} 3x + 2y &= 6 \\ -3x \quad &\quad -3x \\ 2y &= -3x + 6 \\ \frac{1}{2}(2y) &= \frac{1}{2}(-3x + 6) \\ y &= -\frac{3}{2}x + 3 \end{aligned}$$

Step 2: Choose x values that are divisible by 2 so that the y values will be integers. This is not an essential requirement but does yield points which are easier to plot.

x	$y = -\frac{3}{2}x + 3$	y	(x,y)
0	$y = -\frac{3}{2} \cdot 0 + 3$	3	(0,3)
2	$y = -\frac{3}{2} \cdot 2 + 3$	0	(2,0)
-2	$y = -\frac{3}{2} \cdot -2 + 3$	6	(-2,6)

Step 3: Plot the points.

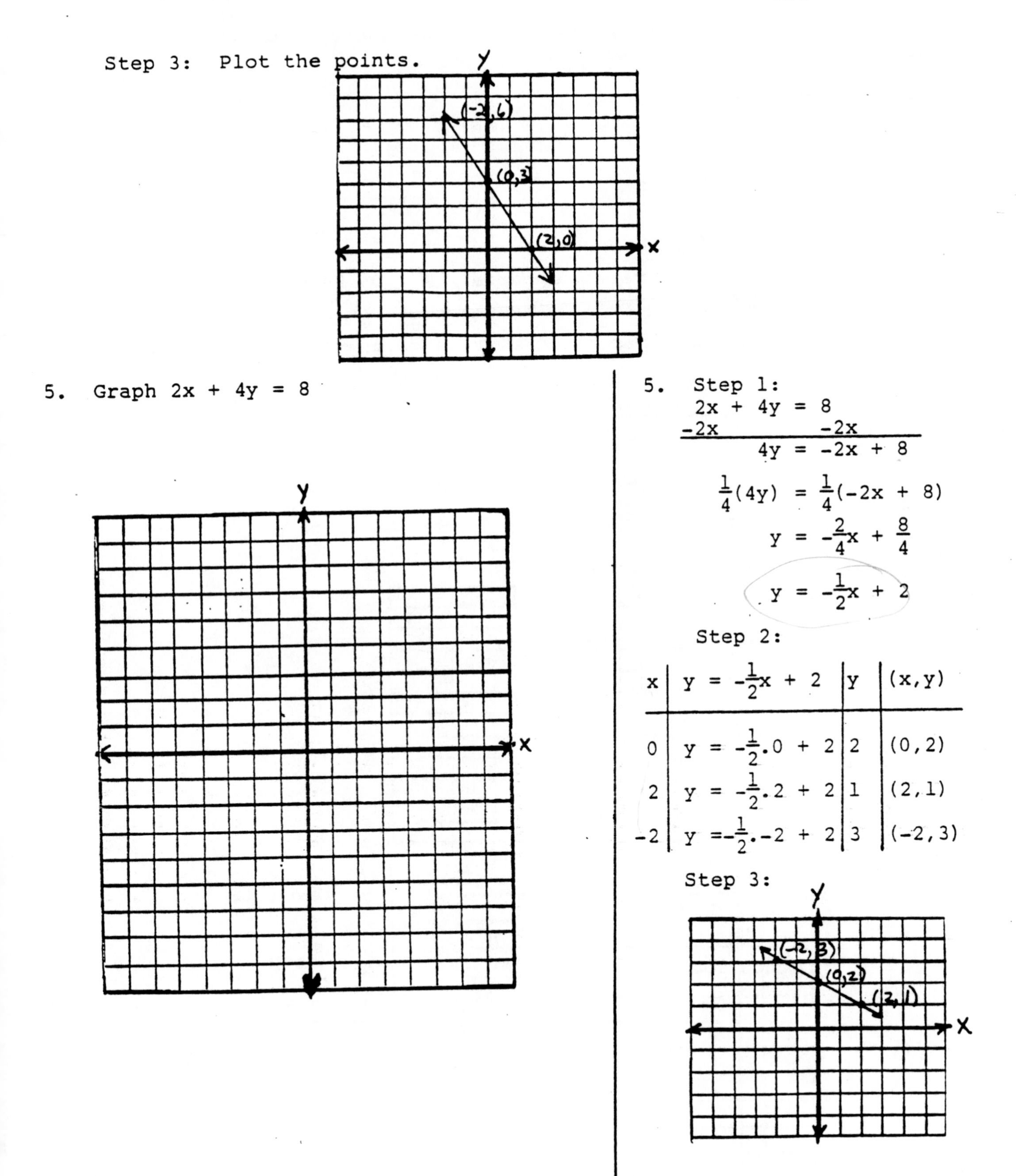

5. Graph $2x + 4y = 8$

5. Step 1:

$$2x + 4y = 8$$
$$\underline{-2x \qquad\quad -2x}$$
$$4y = -2x + 8$$
$$\frac{1}{4}(4y) = \frac{1}{4}(-2x + 8)$$
$$y = -\frac{2}{4}x + \frac{8}{4}$$
$$y = -\frac{1}{2}x + 2$$

Step 2:

x	$y = -\frac{1}{2}x + 2$	y	(x,y)
0	$y = -\frac{1}{2}\cdot 0 + 2$	2	(0,2)
2	$y = -\frac{1}{2}\cdot 2 + 2$	1	(2,1)
-2	$y = -\frac{1}{2}\cdot -2 + 2$	3	(-2,3)

Step 3:

When A = 0, the equation becomes BY = C. The graphs of lines in this form are always horizontal.

Example: Graph 2y = 5.

Step 1: $2y = 5$

$$\frac{1}{2}\cdot 2y = \frac{1}{2}\cdot 5$$

$$y = \frac{5}{2}$$

Step 2: Notice no x term appears. A = 0. Hence, the equation is $y = 0\cdot x + \frac{5}{2}$. Any value can be chosen for x. Since 0 times any number is zero, y will always be $\frac{5}{2}$.

x	$y = 0\cdot x + \frac{5}{2}$	y	(x,y)
0	$y = 0\cdot 0 + \frac{5}{2}$	$\frac{5}{2}$	$(0,\frac{5}{2})$
2	$y = 0\cdot 2 + \frac{5}{2}$	$\frac{5}{2}$	$(2,\frac{5}{2})$
-2	$y = 0\cdot -2 + \frac{5}{2}$	$\frac{5}{2}$	$(-2,\frac{5}{2})$

Step 3: Plot the points.

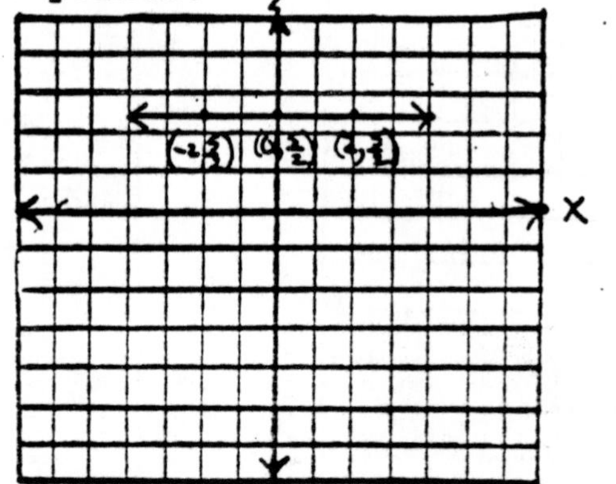

6. Graph 3y = -6.

6. Step 1:
$3y = -6$
$y = -2$

Step 2:

x	$y = 0\cdot x - 2$	y	(x,y)
0	$y = 0\cdot 0 - 2$	-2	(0,-2)
3	$y = 0\cdot 3 - 2$	-2	(3,-2)
-3	$y = 0\cdot -3 - 2$	-2	(-3,-2)

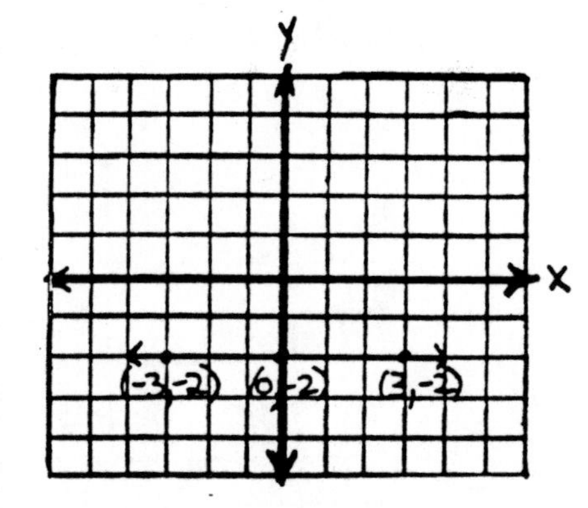

When B = 0, Ax + By = C becomes Ax = C. The graph of lines of this type is a vertical line. In this case y can be any number and x is always the same.

Example: Graph 2x = 6.

Step 1: This presents a problem as we cannot solve for y. There is no y term. Hence, solve for x:

$$2x = 6$$
$$\frac{1}{2}(2x) = \frac{1}{2}(6)$$
$$x = 3$$

Step 2: In choosing values notice that any value may be chosen for y. X is always the same. X's are not chosen and y's obtained, as in the other examples, because the equation had to be solved for x instead of y.

y	$0 \cdot y + x = 3$	x	(x,y)
0	$0 \cdot 0 + x = 3$	3	(3,0)
2	$0 \cdot 2 + x = 3$	3	(3,2)
-2	$0 \cdot -2 + x = 3$	3	(3,-2)

Step 3: Plot the points and draw the line.

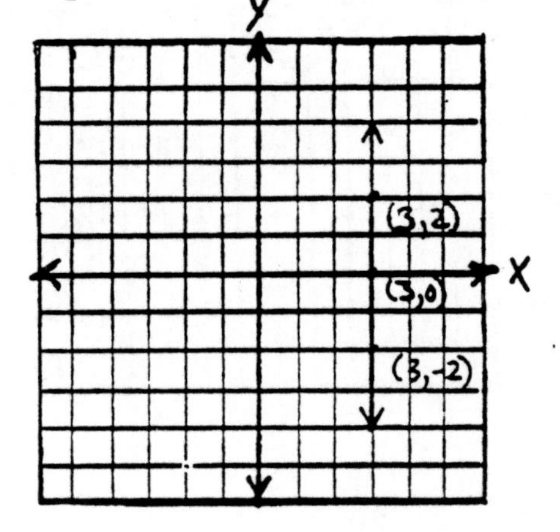

7. Graph $3x + 9 = 0$

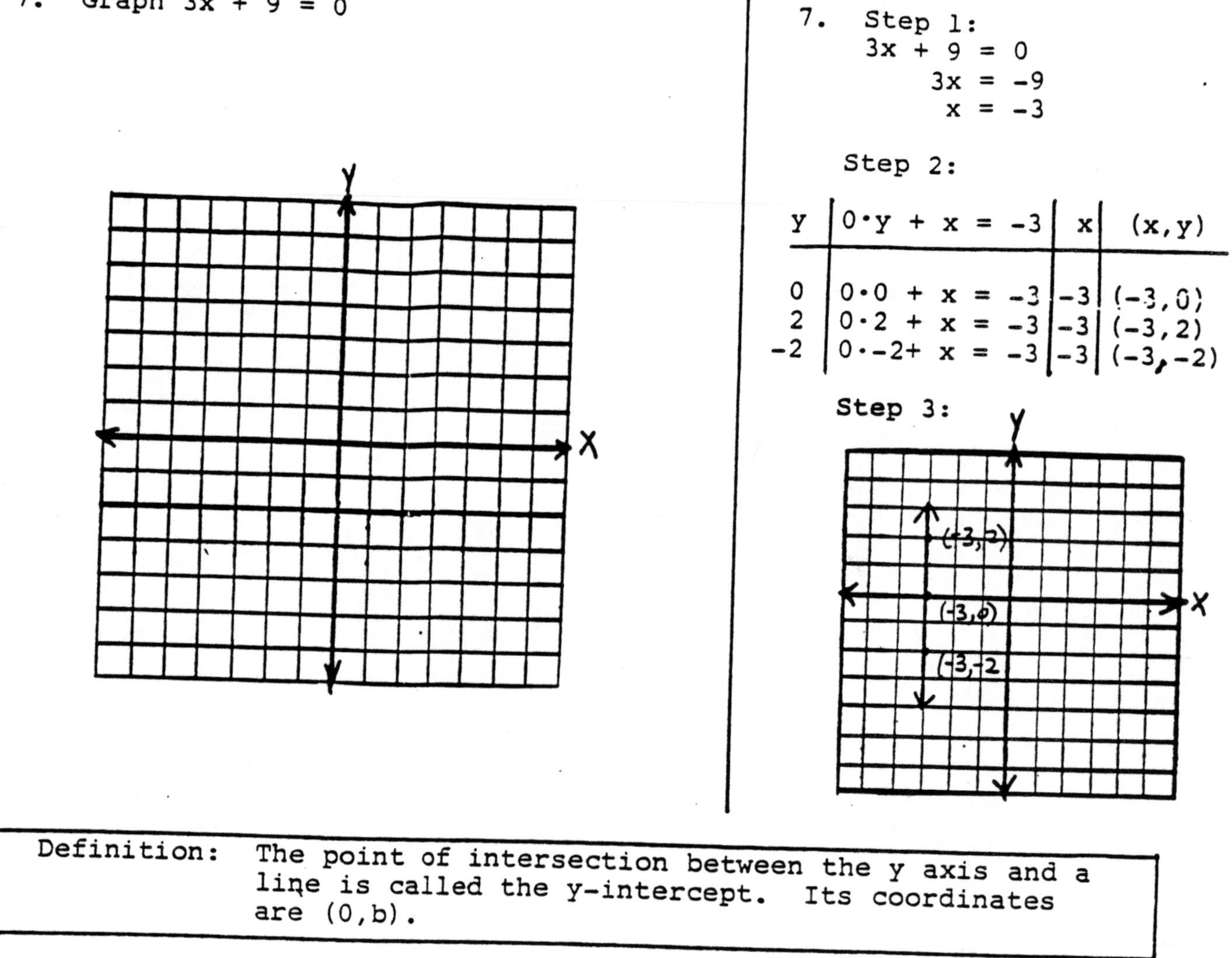

7. Step 1:

$$3x + 9 = 0$$
$$3x = -9$$
$$x = -3$$

Step 2:

y	$0 \cdot y + x = -3$	x	(x,y)
0	$0 \cdot 0 + x = -3$	-3	(-3,0)
2	$0 \cdot 2 + x = -3$	-3	(-3,2)
-2	$0 \cdot -2 + x = -3$	-3	(-3,-2)

Step 3:

Definition:	The point of intersection between the y axis and a line is called the y-intercept. Its coordinates are (0,b).

Definition:	The x-intercept is the point of intersection between the x axis and a line. Its coordinates are (a,0).

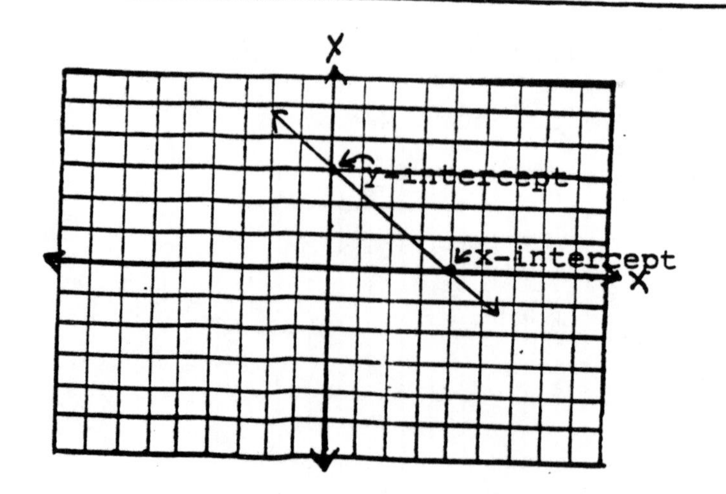

Method 2: Slope, y-intercept method

In this method:

1. Solve for y in terms of x. This transforms the equation into $y = mx + b$ form where m is the slope and b is the y-intercept. The y-intercept is a point whose coordinates are (0,b).
2. Plot the y-intercept.
3. From the y-intercept, use the slope as rise over run to obtain other points. Connect the points.

Example: Graph $3x + 2y = 8$.

Step 1: Solve for y in terms of x.

$$3x + 2y = 8$$
$$2y = -3x + 8$$
$$y = -\frac{3}{2}x + 4$$

Compare $y = -\frac{3}{2}x + 4$ to
$y = mx + b$. Observe the slope

$m = -\frac{3}{2} = \frac{\text{rise}}{\text{run}}$; b, the y-intercept is 4.

Step 2. Plot the y-intercept. The y-intercept is the point (0,4).

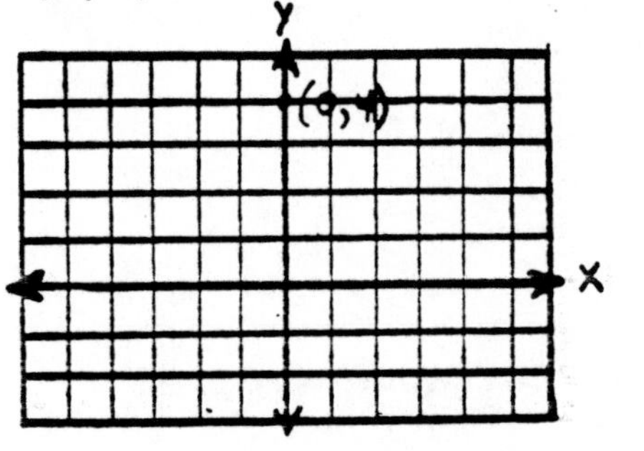

Step 3: The slope is $\frac{-3}{2}$ which means from (0,4) come down 3 units (-3) and go over 2 units (+2). Note that $\frac{-3}{2} = \frac{3}{-2}$; so, equivalently, one could go up 3 units (+3) and back 2 units (-2).

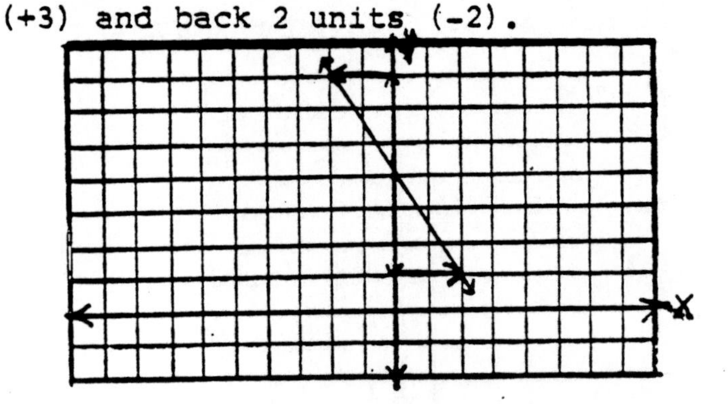

8. Using method 2, graph $2x + 3y = -3$.

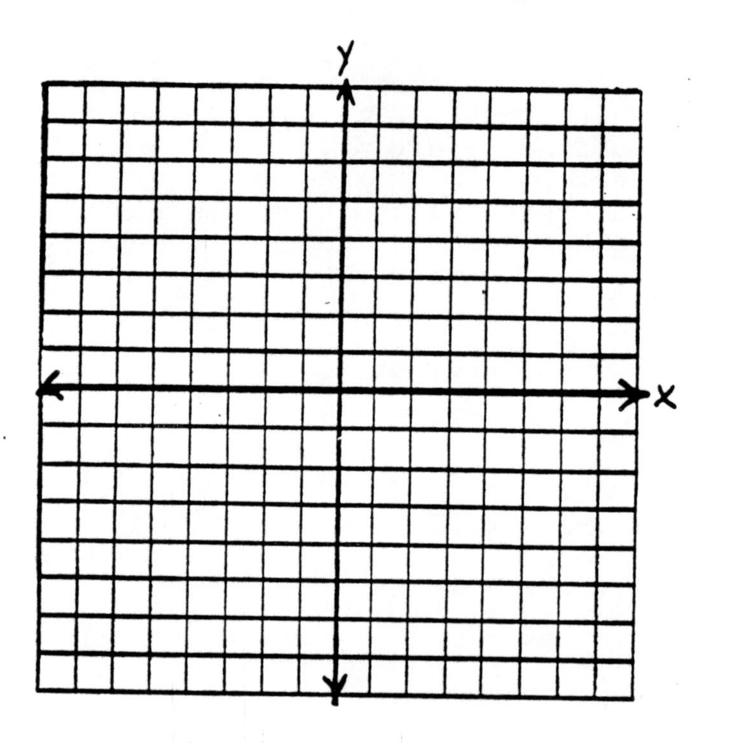

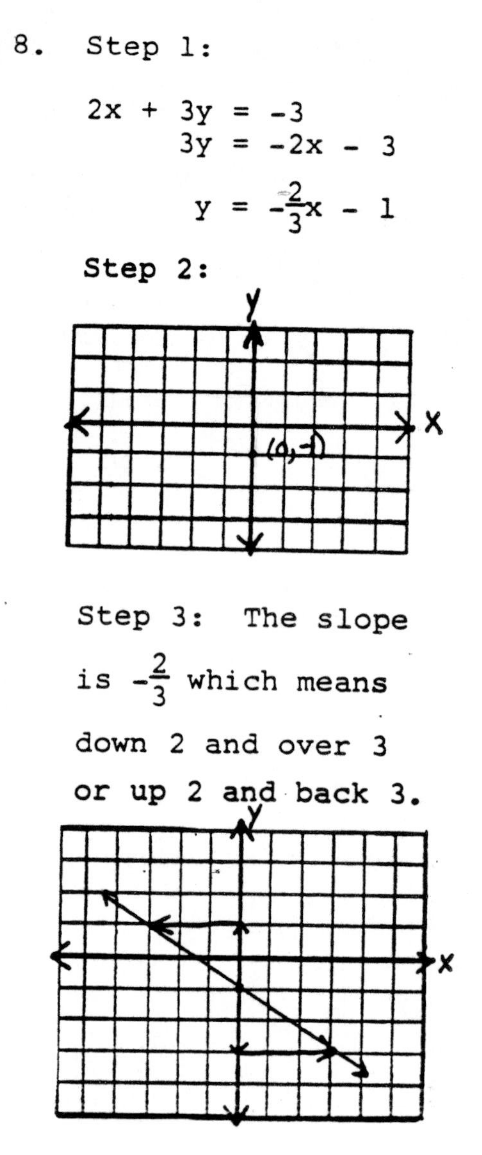

8. Step 1:

$$2x + 3y = -3$$
$$3y = -2x - 3$$
$$y = -\frac{2}{3}x - 1$$

Step 2:

Step 3: The slope is $-\frac{2}{3}$ which means down 2 and over 3 or up 2 and back 3.

This technique can likewise be applied to the case when $A = 0$ and, also, to the case when $B = 0$.

Example: Graph $y = -2$.

In this equation $A = 0$. This equation can be written as $y = 0 \cdot x - 2$ where 0 is the slope and -2 is the y-intercept. A zero slope means there is no rise but run is not zero. Consequently, the graph is a horizontal line.

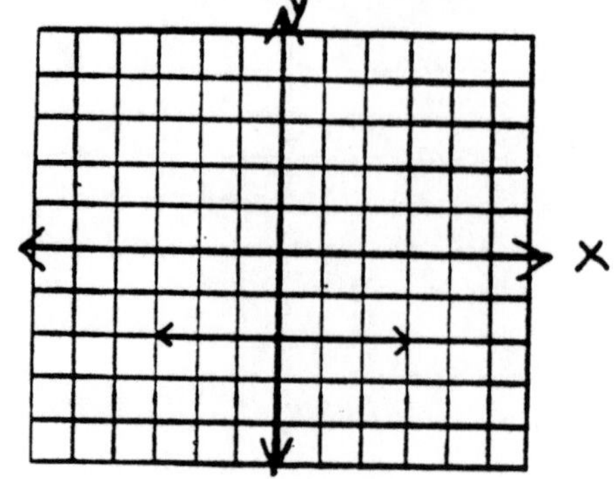

no rise, possible run

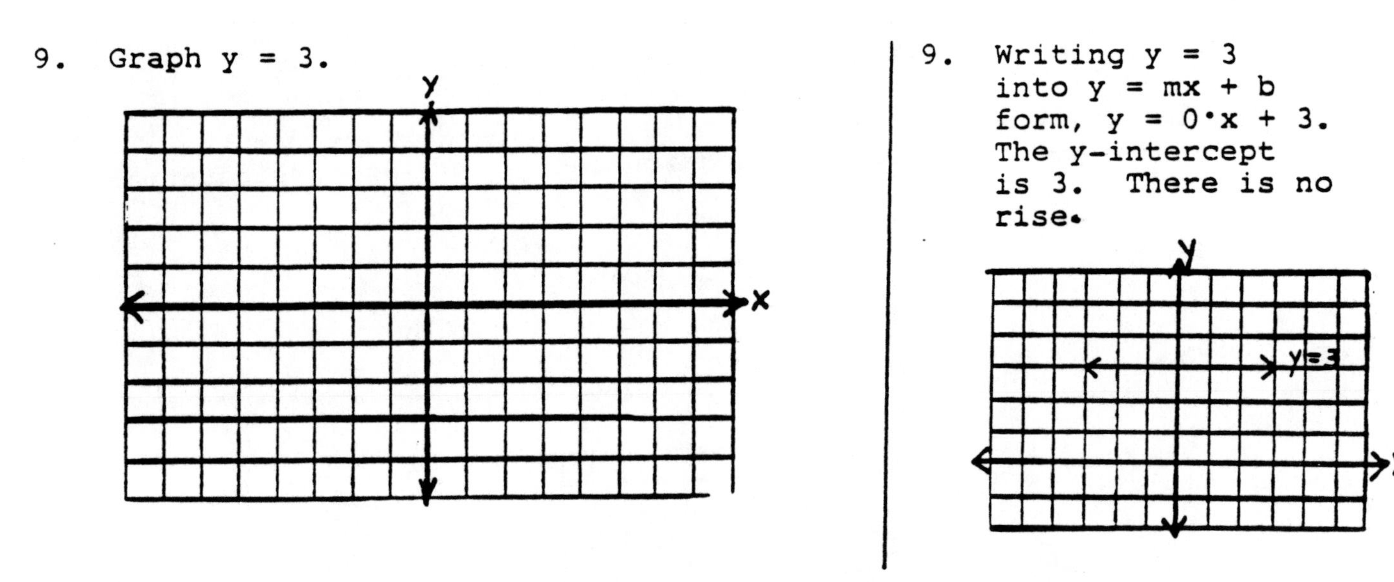

9. Graph y = 3.

9. Writing y = 3 into y = mx + b form, y = 0·x + 3. The y-intercept is 3. There is no rise.

When B = 0, the equation cannot be written in the y = mx + b format. The equation is of the form Ax = C. This is a vertical line which has no slope. Recall the slope formula. To have no slope the denominator was zero which means there is no run. The rise is unequal to zero. There, likewise, is no y-intercept. These lines do not intersect the y-axis. In fact, they are parallel to it.

10. Graph x = -3.

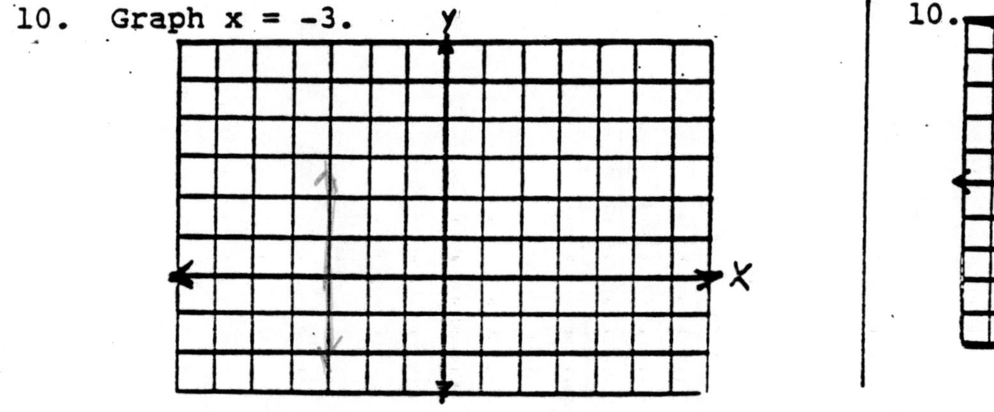

10.

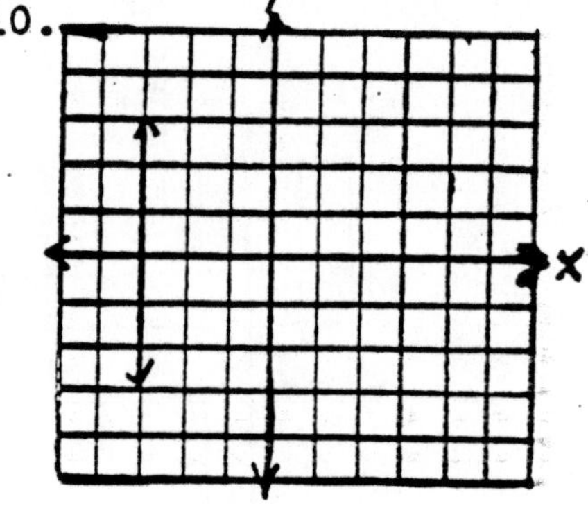

Method 3: x-intercept and y-intercept method

In this method:
1. the y-intercept, the point (0,b), is selected.
2. the x-intercept, the point (a,0), is selected,
3. Plot and connect the points.

Example: Graph 2x + 3y = 6.

Step 1: Select the y-intercept. This point always has a x-coordinate of 0. Substituting x = 0 into the equation,

$$2\cdot 0 + 3y = 6$$
$$3y = 6$$
$$y = 2$$

So, the y-intercept is (0,2).

Step 2: Select the x-intercept. This point always has a y-coordinate of 0. Substituting $y = 0$ into the equation,

$$2x + 3 \cdot 0 = 6$$

$$2x = 6$$

$$x = 3$$

So the x-intercept is the point (3,0).

Step 3:

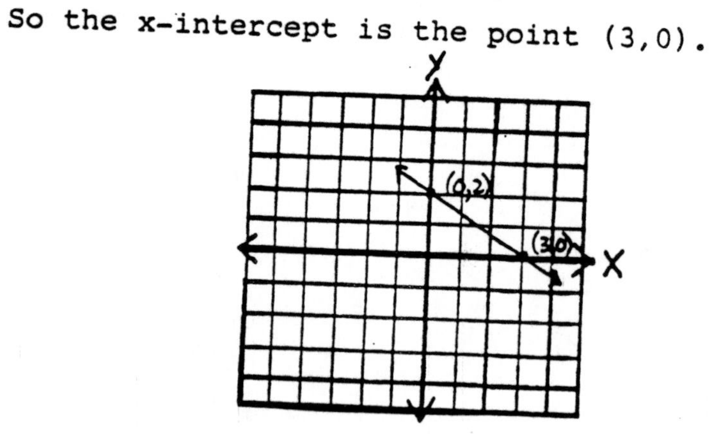

11. Using method 3, graph $4x - 2y = 8$.

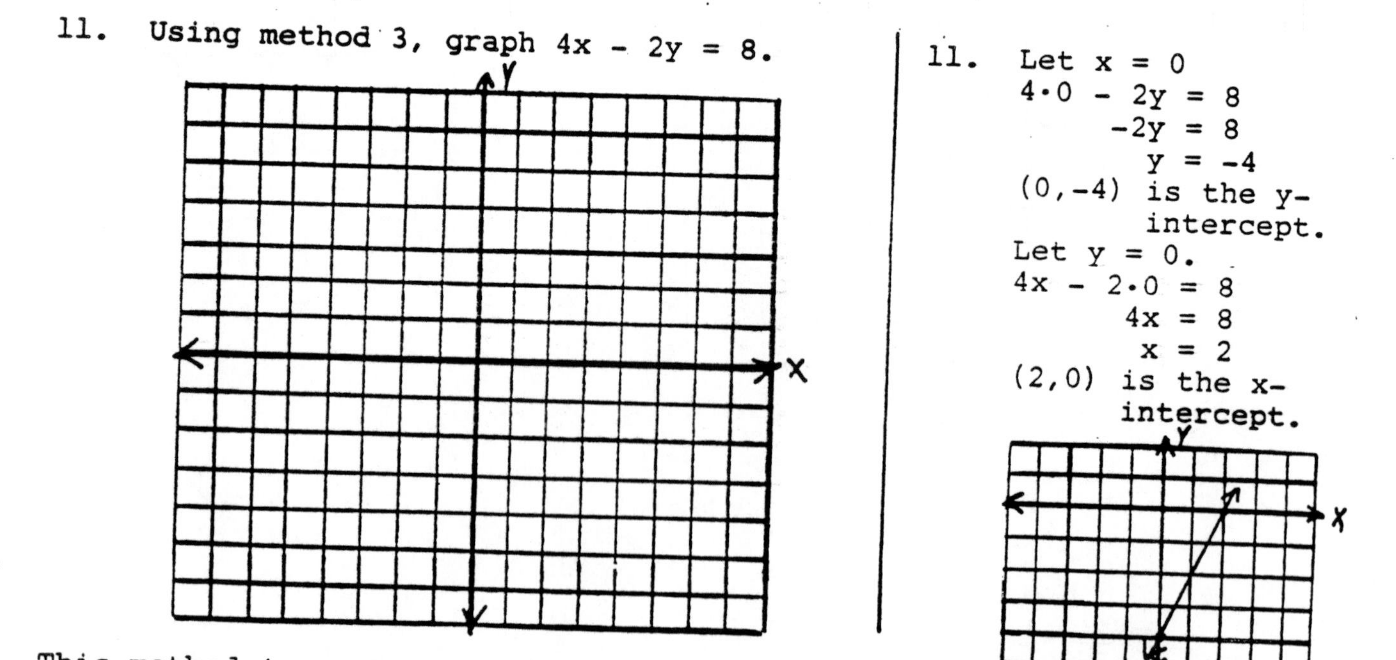

11. Let $x = 0$

$$4 \cdot 0 - 2y = 8$$

$$-2y = 8$$

$$y = -4$$

(0,-4) is the y-intercept.

Let $y = 0$.

$$4x - 2 \cdot 0 = 8$$

$$4x = 8$$

$$x = 2$$

(2,0) is the x-intercept.

This method is particularly useful in that it precisly demonstrates where the line crosses both axes. Frequently, however, the x and y intercepts are fractions and plotting has to be approximate. This method is not especially useful for plotting lines which are parallel to the x-axis, lines where $A = 0$, that is, lines of the form $By = C$. These lines have a y-intercept but no x-intercept. Refer to the graph in the example on page 350. Likewise, lines which are parallel to the y-axis, lines where $B = 0$ which are of the form $Ax = C$ do not have y-intercepts. Refer to the graph in the example on page 351. Despite the non-usefulness of this method for $By = C$ and $Ax = C$, the method is useful in determining where a line crosses the axes, which are essential points in some studies of straight lines.

Exercise 2

1. Graph using method 1 (selection of points).

   a. $2x + 3y = 9$
   b. $5x - y = 2$
   c. $x - 2y = 6$
   d. $4x + 3y = 12$
   e. $3x - 9 = 0$
   f. $2y + 6 = 0$

2. Graph using method 2 (slope y-intercept).

   a. $5x + 2y = 10$
   b. $2x - 3y = 9$
   c. $2x - 8 = 0$
   d. $5y + 10 = 20$
   e. $3x - 5y = 20$

3. Graph using method 3 (x-intercept, y-intercept method).

   a. $2x - 7y = 14$
   b. $3x + 6y = 12$
   c. $4x - 5y = 20$
   d. $-3x + 7y = 21$

Writing the Equation of a Straight Line

In writing the equation of a straight line there are two forms which can be used. The first form we will consider is the point slope form:

$$(y - y_1) = m(x - x_1)$$

where (x_1, y_1) is a point on the line and m is the slope.

The second form of the straight line we will consider is the slope y-intercept form:

$$y = mx + b$$

where m is the slope and b is the y-intercept. (Recall the y-intercept is where the line crosses the y-axis and its coordinates are (0,b).)

How do you decide on which form to use? Frequently, it does not matter. In most cases, both forms can be just as easily used. No matter which form is used in determining the equation, the final answer should be left in <u>standard form</u>, $Ax + By = C$, where A, B, and C are integers and A is positive.

In the following examples, both methods are demonstrated. When you are working the problems, you need only select one method.

Example: Write the equation of the line which passes through the points (4,2) and (5,7).

Solution # 1: Using the point slope form, $y - y_1 = m(x - x_1)$, determine the slope first. The slope,

$$m = \frac{y_2 - y_1}{x_2 - x_1} = \frac{7 - 2}{5 - 4} = \frac{5}{1} = 5.$$

Either point may be used as (x_1, y_1). Using (4,2), the form becomes

$$\begin{aligned} y - y_1 &= m(x - x_1) \\ y - 2 &= 5(x - 4) \\ y - 2 &= 5x - 20 \\ y &= 5x - 18 \\ -5x + y &= -18 \\ -1(-5x + y) &= (-1)(-18) \\ 5x - y &= 18 \end{aligned}$$

is the desired equation.

Note: Suppose we had selected the point (5,7) instead of (4,2) to substitute Would that have made any difference? No, for example, $y - y_1 = m(x - x_1)$ would become

$$\begin{aligned} y - 7 &= 5(x - 5) \\ y - 7 &= 5x - 25 \\ y &= 5x - 18 \\ -5x + y &= -18 \\ 5x - y &= 18 \end{aligned}$$

is the same equation as obtained above.

Solution # 2: Using the slope y-intercept form of the straight line, $y = mx + b$, determine the slope first.

$$m = \frac{y_2 - y_1}{x_2 - x_1} = \frac{7 - 2}{5 - 4} = \frac{5}{1} = 5$$

Substitute this into the form to obtain $y = 5x + b$.

Use either point to obtain b. Using (4,2), $y = 2$, $x = 4$, substitute these into the form to obtain b.

$$2 = 5 \cdot 4 + b$$
$$2 = 20 + b$$
$$-18 = b$$

Now, values for m and b have been obtained so substitute them into the original form.

$$y = mx + b$$
$$y = 5x - 18$$
$$-5x + y = -18$$
$$-1(-5x + y) = -1(-18)$$
$5x - y = 18$ is the desired equation.

12. Write the equation for the line which passes through the points (3,-2) and (2,4).

Select either method you wish. You are encouraged to try both ways in order to decide which method is the easiest for you. Both methods are provided for your convenience.

12. Solution # 1: Using the point slope formula:

$$y - y_1 = m(x - x_1)$$

$$m = \frac{4 - (-2)}{2 - 3} = \frac{6}{-1} = -6$$

Using (2,4),

$$y - 4 = -6(x - 2)$$
$$y - 4 = -6x + 12$$

$6x + y = 16$ is the desired equation.

Solution # 2: Using the slope y-intercept form:

$$y = mx + b$$

$$m = \frac{4 - (-2)}{2 - 3} = \frac{6}{-1} = -6$$

Using (2,4) substitute into $y = mx + b$

$$4 = -6 \cdot 2 + b$$
$$4 = -12 + b$$
$$16 = b$$

Now, replace m and b with their values.

$$y = -6x + 16$$

$6x + y = 16$ is the desired form.

Example: Write the equation of the line which passes through (3,-2) and has slope of $\frac{2}{5}$.

Solution # 1: Using the point slope form

$$y - y_1 = m(x - x_1) \quad \text{substitute directly}$$
$$y - (-2) = \frac{2}{5}(x - 3)$$
$$y + 2 = \frac{2}{5}x - \frac{6}{5}$$

Multiplying both members by 5 to clear the equation of fractions, the equation becomes:

$$5y + 10 = 2x - 6$$
$$5y = 2x - 16$$
$$-2x + 5y = -16$$
$$-1(-2x + 5y) = -1(-16)$$

$2x - 5y = 16$ is the desired equation.

Solution # 2: Using the slope y-intercept form, $y = mx + b$, substitute the slope in first to obtain

$$y = \frac{2}{5}x + b$$

Now use the point (3,-2) to obtain b.

$$-2 = \frac{2}{5} \cdot 3 + b$$
$$-2 = \frac{6}{5} + b$$
$$-10 = 6 + 5b$$
$$-16 = 5b$$
$$\frac{-16}{5} = b$$

Substitute $m = \frac{2}{5}$ and $b = \frac{-16}{5}$ into

$$y = mx + b$$

$$y = \frac{2}{5}x - \frac{16}{5}$$

Simplifying:

$$5y = 2x - 16$$

$$-2x + 5y = -16$$

$$-1(-2x + 5y) = (-1)(-16)$$

$$2x - 5y = 16$$

13. Write the equation of the line which passes through $(-4,-2)$ and has slope $\frac{-3}{8}$.

Select either method you wish. Both solutions are provided for your convenience.

13. Solution # 1:

$$y - y_1 = m(x - x_1)$$

$$y-(-2) = \frac{-3}{8}(x - (-4))$$

$$y + 2 = \frac{-3}{8}(x + 4)$$

$$y + 2 = \frac{-3}{8}x - \frac{12}{8}$$

$$8y + 16 = -3x - 12$$

$$8y = -3x - 28$$

$$3x + 8y = -28$$

Solution # 2:

$$y = mx + b$$

$$y = -\frac{3}{8}x + b$$

$$-2 = -\frac{3}{8}\cdot -4 + b$$

$$-2 = \frac{3}{2} + b$$

$$-4 = 3 + 2b$$

$$-7 = 2b$$

$$-\frac{7}{2} = b$$

$$y = -\frac{3}{8}x - \frac{7}{2}$$

$$8y = -3x - 28$$

$$3x + 8y = -28$$

Example: Write the equation of the line with slope $-\frac{2}{3}$ and y-intercept 6.

Solution # 1: The y-intercept is where the line crosses the y-axis so its coordinates are (0,6). Using $-\frac{2}{3}$ and (0,6), substitute into

$$y - y_1 = m(x - x_1)$$
$$y - 6 = -\frac{2}{3}(x - 0)$$
$$y - 6 = -\frac{2}{3}x$$
$$\frac{2}{3}x + y = 6$$
$$3\left(\frac{2}{3}x + y\right) = 3(6)$$
$$2x + 3y = 18$$

Solution # 2: $y = mx + b$

Substitute directly:

$$y = -\frac{2}{3}x + 6$$
$$\frac{2}{3}x + y = 6$$
$$3\left(\frac{2}{3}x + y\right) = 3(6)$$
$$2x + 3y = 18$$

14. Write the equation of the line with slope 3 and y-intercept 2.

Both solutions are provided for your convenience.

14. Solution #1: The y-intercept is the point whose coordinates are (0,2).

$y - 2 = 3(x - 0)$
$y - 2 = 3x$
$-3x + y = 2$
$3x - y = -2$

Solution # 2:

$y = mx + b$
$y = 3x + 2$
$-3x + y = 2$
$3x - y = -2$

Example: Write the equation of the line with x-intercept 4 and y-intercept, -4.

Solution # 1: The x-intercept is the place where the line crosses the x-axis. The coordinates of this point are (4,0). The y-intercept's coordinates are (0,-4). The slope

$$m = \frac{-4 - 0}{0 - 4} = \frac{-4}{-4} = 1$$

$$y - 0 = 1(x - 4)$$
$$y = x - 4$$
$$-x + y = -4$$
$$x - y = 4$$

Solution # 2: The slope would be obtained as above in Solution # 1. Substituting m = 1 and y-intercept, -4, into the y = mx + b form gives:

$$y = x - 4$$
$$-x + y = -4$$
$$x - y = 4$$

15. Write the equation of the line which has x-intercept, 2, and y-intercept, 3.

You may work the problem by either method. Only one solution is provided.

15. x-intercept = (2,0)

y-intercept = (0,3)

$$m = \frac{3 - 0}{0 - 2} = -\frac{3}{2}$$

$$y = -\frac{3}{2}x + 3$$

$$\frac{3}{2}x + y = 3$$

$$3x + 2y = 6$$

Example: Write the equation of the line which passes through (4,3) and (4,-2).

Solution:

$$m = \frac{3 - (-2)}{4 - 4} = \frac{5}{0} = \text{no slope}$$

Since the slope is "no slope", neither formula can be used. Observe both x-coordinates are 4. Also, vertical lines are of the form, $Ax = C$ or $x = \frac{C}{A}$, where $\frac{C}{A}$ is the first coordinate of all points on the line. Therefore, the desired equation is x = 4.

16. Write the equation of the line which passes through (2,2) and (2,-3).

16. $$m = \frac{2 - (-3)}{2 - 2} = \frac{5}{0}$$

m = no slope

x = 2 is the desired equation.

Example: Write the equation of the line which passes through the points (0,4) and (5,4).

Solution:

$$m = \frac{4 - 4}{0 - 5} = \frac{0}{-5} = 0$$

Use either form to find the equation. Using

$$y - y_1 = m(x - x_1) \text{ and } (0,4)$$
$$y - 4 = 0(x - 0)$$
$$y - 4 = 0$$
$$y = 4$$

Note: This is a horizontal line. The y coordinates are the same for both points. The equation is $y = 4$, the common y-coordinate.

17. Write the equation of the line which passes through (2,2) and (-3,2).

17.

$$m = \frac{2 - 2}{2-(-3)} = \frac{0}{5} = 0$$
$$y - 2 = 0(x - 2)$$
$$y - 2 = 0$$
$$y = 2$$

Theorem 1:	Two lines are parallel if and only if they have the same slope.

Example: Determine if $-2x + 3y = 7$ and $4x = -6y + 3$ are parallel.

Solution: In order to determine if lines are parallel, their slope must be determined first. One way to determine slope is to write the equation in slope y-intercept form. Beginning with the first equation,

$$2x + 3y = 7$$
$$3y = -2x + 7$$
$$y = -\frac{2}{3}x + \frac{7}{3}$$
$$y = m\,x + b$$

The slope is $-\frac{2}{3}$.

Using the same procedure for

$$4x = -6y + 3$$
$$6y + 4x = 3$$
$$6y = -4x + 3$$
$$y = -\frac{4}{6}x + \frac{3}{6}$$
$$y = -\frac{2}{3}x + \frac{1}{2}$$

The slope is $-\frac{2}{3}$. By Theorem 1, the lines are parallel.

18. Determine if the lines $3x - 6y = 2$ and $2y = x + 3$ are parallel.

18. $3x - 6y = 2$

$$-6y = -3x + 2$$
$$y = \frac{-3}{-6}x + \frac{2}{-6}$$
$$y = \frac{1}{2}x - \frac{1}{3}$$

$m = \frac{1}{2}$

$2y = x + 3$

$$y = \frac{1}{2}x + \frac{3}{2}$$

$m = \frac{1}{2}$

By Theorem 1, the lines are parallel.

Example: Find the equation of the line through (1,5) and parallel to the line $x + 6y = 2$.

Solution: First, determine the slope of the given line by writing it in $y = mx + b$ form.

$$x + 6y = 2$$
$$6y = -x + 2$$
$$y = -\frac{1}{6}x + \frac{2}{6}$$
$$y = -\frac{1}{6}x + \frac{1}{3}$$

The slope is $-\frac{1}{6}$.

Substitute the slope and coordinates of the point into the point slope formula.

$$y - y_1 = m(x - x_1)$$
$$y - 5 = -\frac{1}{6}(x - 1)$$
$$y - 5 = -\frac{1}{6}x + \frac{1}{6}$$
$$6y - 30 = -x + 1$$
$$x + 6y = 31$$

19. Determine the equation of the line through (-1,-3) and parallel to the line $3x + 2y = 6$.

19.
$$3x + 2y = 6$$
$$2y = -3x + 6$$
$$y = -\frac{3}{2}x + 3$$
$$m = -\frac{3}{2}$$
$$y-(-3) = -\frac{3}{2}(x-(-1))$$
$$y + 3 = -\frac{3}{2}(x + 1)$$
$$y + 3 = -\frac{3}{2}x - \frac{3}{2}$$
$$2y + 6 = -3x - 3$$
$$2y = -3x - 9$$
$$3x + 2y = -9$$

Theorem 2: Two lines are perpendicular if and only if their slopes are negative reciprocals of each other.

That is, if the slope of one line is m, the slope of the line perpendicular to it is $-\frac{1}{m}$. Observe $m \cdot -\frac{1}{m} = \frac{\cancel{m}}{1} \cdot -\frac{1}{\cancel{m}} = -\frac{1}{1} = -1$.

This gives a way to check if the slopes are negative reciprocals of each other. Find their product. If the product is -1, the lines are perpendicular.

Example: Is the line $x - 4y = 2$ perpendicular to the line $4x + y = 3$?

Solution: Find the slope:
$$x - 4y = 2$$
$$-4y = -x + 2$$
$$y = -\frac{x}{-4} + \frac{2}{-4}$$
$$y = \frac{1}{4}x - \frac{1}{2}$$

$$m = \frac{1}{4}$$

Determine the slope of the second line:

$$4x + y = 3$$
$$y = -4x + 3$$

$m = -4$

Find their product: $\frac{1}{4} \cdot -4 = \frac{1}{\not{4}_1} \cdot -\not{4}^{-1} = \frac{-1}{1} = -1$

Therefore, the given lines are perpendicular.

20. Is the line, $5x + 2y = 10$ perpendicular to the line, $2x - 5y = 10$?

20. $5x + 2y = 10$
$$2y = -5x + 10$$
$$y = -\frac{5}{2}x + 5$$
$m = -\frac{5}{2}$

$2x - 5y = 10$
$$-5y = -2x + 10$$
$$y = \frac{2}{5}x - 2$$
$m = \frac{2}{5}$

Find their product

$$-\frac{5}{2} \cdot \frac{2}{5} = -\frac{\not{5}^1}{\not{2}_1} \cdot \frac{\not{2}^1}{\not{5}_1}$$
$$= -\frac{1}{1} = -1$$

Example: Determine the equation of the line that passes through (4,-2) and is perpendicular to $x - 5y = 4$.

Solution: First determine the slope of $x - 5y = 4$.

$$x - 5y = 4$$
$$-5y = -x + 4$$
$$y = -\frac{x}{-5} + \frac{4}{-5}$$
$$y = \frac{1}{5}x - \frac{4}{5}$$

The slope is $\frac{1}{5}$.

To determine the slope of the desired line, solve the following equation:

$$\frac{1}{5} \cdot m = -1$$

M is the desired slope; $\frac{1}{5}$ is the slope of the given line; and, -1 is the product of any number and its reciprocal. Since we are looking for the reciprocal, let us solve this equation.

$$\frac{1}{5} m = -1$$

$$5\left(\frac{1}{5}m\right) = 5(-1)$$

$$m = -5$$

Now, use the point slope form.

$$y - y_1 = m(x - x_1)$$

$$y - (-2) = -5(x - 4)$$

$$y + 2 = -5x + 20$$

$$5x + y = 18$$

21. Determine the equation of the line perpendicular to the line, $3x + y = 7$ and passing through (-4,2).

21. $3x + y = 7$

$$y = -3x + 7$$

Slope = -3.

$$-3 \cdot m = -1$$

$$\frac{-3m}{-3} = \frac{-1}{-3}$$

$$m = \frac{1}{3}$$

$$y - 2 = \frac{1}{3}(x - (-4))$$

$$y - 2 = \frac{1}{3}(x + 4)$$

$$3y - 6 = x + 4$$

$$3y = x + 10$$

$$-x + 3y = 10$$

$$x - 3y = -10$$

Exercise 3

1. Determine the equation of the following lines which meet the given conditions. Leave all answers in standard form.

   a. Passing through (3,4) and (-2,5).
   b. Passing through (4,5) and (4,-2).
   c. Passing through (3,3) and (-2,3).
   d. Passing through (4,2) and having slope 5.
   e. Passing through (-5,-4) and having slope $-\frac{2}{3}$.

f. Passing through (4,5) and parallel to $2x - 7y = 6$.
g. Passing through (-1,-4) and perpendicular to $x + 4y = 5$.
h. Having x-intercept 3 and y-intercept -5.
i. Having slope -3 and y-intercept -5.
j. Having slope $-\frac{2}{3}$ and x-intercept -4.

Midpoint

Recall the definition of a midpoint. A midpoint is a point which divides a line segment into two equal parts.

> Theorem 3: If (x_1,y_1) and (x_2,y_2) are the endpoints of a line segment, the midpoint of the line segment is determined by the formula,
>
> $$\text{midpoint} = \left(\frac{x_1 + x_2}{2}, \frac{y_1 + y_2}{2}\right)$$

Example: Find the midpoint of the line segment whose endpoints are (3,4) and (-5,3).

Solution:

$$\begin{aligned}\text{Midpoint} &= \left(\frac{x_1 + x_2}{2}, \frac{y_1 + y_2}{2}\right)\\ &= \left(\frac{3 + (-5)}{2}, \frac{4 + 3}{2}\right)\\ &= \left(\frac{-2}{2}, \frac{7}{2}\right)\\ &= \left(-1, \frac{7}{2}\right)\end{aligned}$$

22. Find the midpoint of the line segment whose endpoints are (5,6) and (8,4).

22. $$\begin{aligned}&\left(\frac{x_1+x_2}{2}, \frac{y_1+y_2}{2}\right)\\ &= \left(\frac{5 + 8}{2}, \frac{6 + 4}{2}\right)\\ &= \left(\frac{13}{2}, \frac{10}{2}\right)\\ &= \left(\frac{13}{2}, 5\right)\end{aligned}$$

Example: Find the equation of the line with slope, $\frac{2}{3}$, which passes through the midpoint of (-3,4) and (7,-6).

Solution: First determine the midpoint.

$$\text{Midpoint} = \left(\frac{-3+7}{2}, \frac{4+(-6)}{2}\right) = \left(\frac{4}{2}, \frac{-2}{2}\right) = (2,-1)$$

Using the point slope form:

$$y - (-1) = \frac{2}{3}(x - 2)$$

$$y + 1 = \frac{2}{3}x - \frac{4}{3}$$

$$3y + 3 = 2x - 4$$

$$3y = 2x - 7$$

$$-2x + 3y = -7$$

$$2x - 3y = 7$$

Distance Formula

Theorem 4: If (x_1,y_1) and (x_2,y_2) are the endpoints of a line segment, the undirected distance from (x_1,y_1) to (x_2,y_2) is given by the formula:

$$d = \sqrt{(x_2 - x_1)^2 + (y_2 - y_1)^2}$$

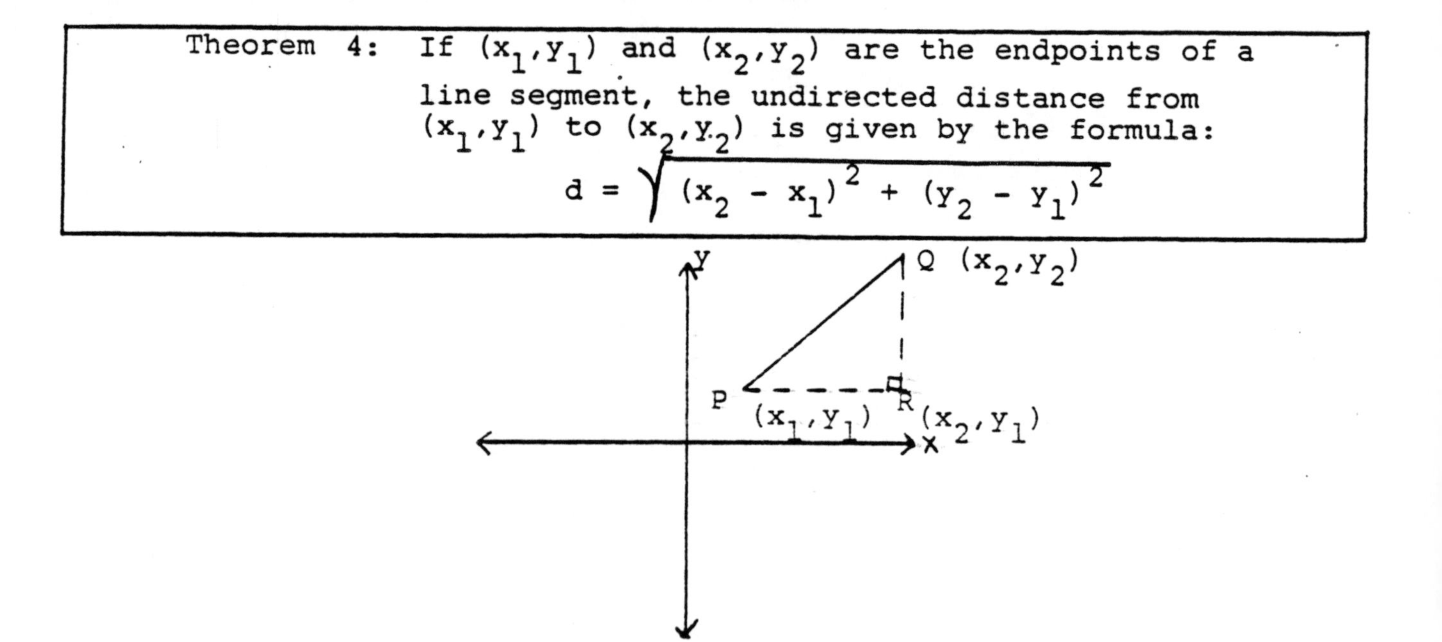

The formula can be derived as follows: Let P = (x_1,y_1) and Q = (x_2,y_2). Draw $\overline{PR}$ parallel to the x-axis and $\overline{QR}$ parallel to the y-axis. $\overline{QR}$ is perpendicular to $\overline{PR}$ so $\triangle PQR$ is a right triangle. The coordinates of R are (x_2, y_1) since all points on a vertical line have the same x-coordinates and all points on a horizontal line have the same y coordinates. The length of $\overline{PR}$ will be the difference in the x coordinates, $x_2 - x_1$. The length of $\overline{QR}$ will be the difference in the y coordinates, $y_2 - y_1$. Since Δ PQR is a right triangle, the Pythagorean Theorem applies.

$$d(\overline{PQ})^2 = d(\overline{PR})^2 + d(\overline{QR})^2$$

where $d(\overline{PQ})$, $d(\overline{PR})$, and $d(\overline{QR})$ is used to mean the distance from P to Q, the distance from P to R, and the distance from Q to R respectively.

$$d(\overline{PQ})^2 = (x_2 - x_1)^2 + (y_2 - y_1)^2$$

$$d(\overline{PQ}) = \sqrt{(x_2 - x_1)^2 + (y_2 - y_1)^2}$$

Example: Find the undirected distance from (3,-2) to (-5,7).

Solution: Let $(x_1,y_1) = (3,-2)$ and $(x_2,y_2) = (-5,7)$.

$$d = \sqrt{(-5 - 3)^2 + (7 - (-2))^2}$$

$$d = \sqrt{(-8)^2 + 9^2}$$

$$d = \sqrt{64 + 81}$$

$$d = \sqrt{145}$$

Note: The selection of (x_1,y_1) and (x_2, y_2) is arbitrary and may be reversed without changing the distance.

23. Find the length of the line segment connecting (-4,2) and (8,7).

23.

$$d = \sqrt{(8-(-4))^2 + (7 - 2)}$$

$$d = \sqrt{12^2 + 5^2}$$

$$d = \sqrt{144 + 25}$$

$$d = \sqrt{169}$$

$$d = 13$$

Example: Find the perimeter of the triangle whose verticies are (10, -1), (7,3), and (-3,2).

Solution:

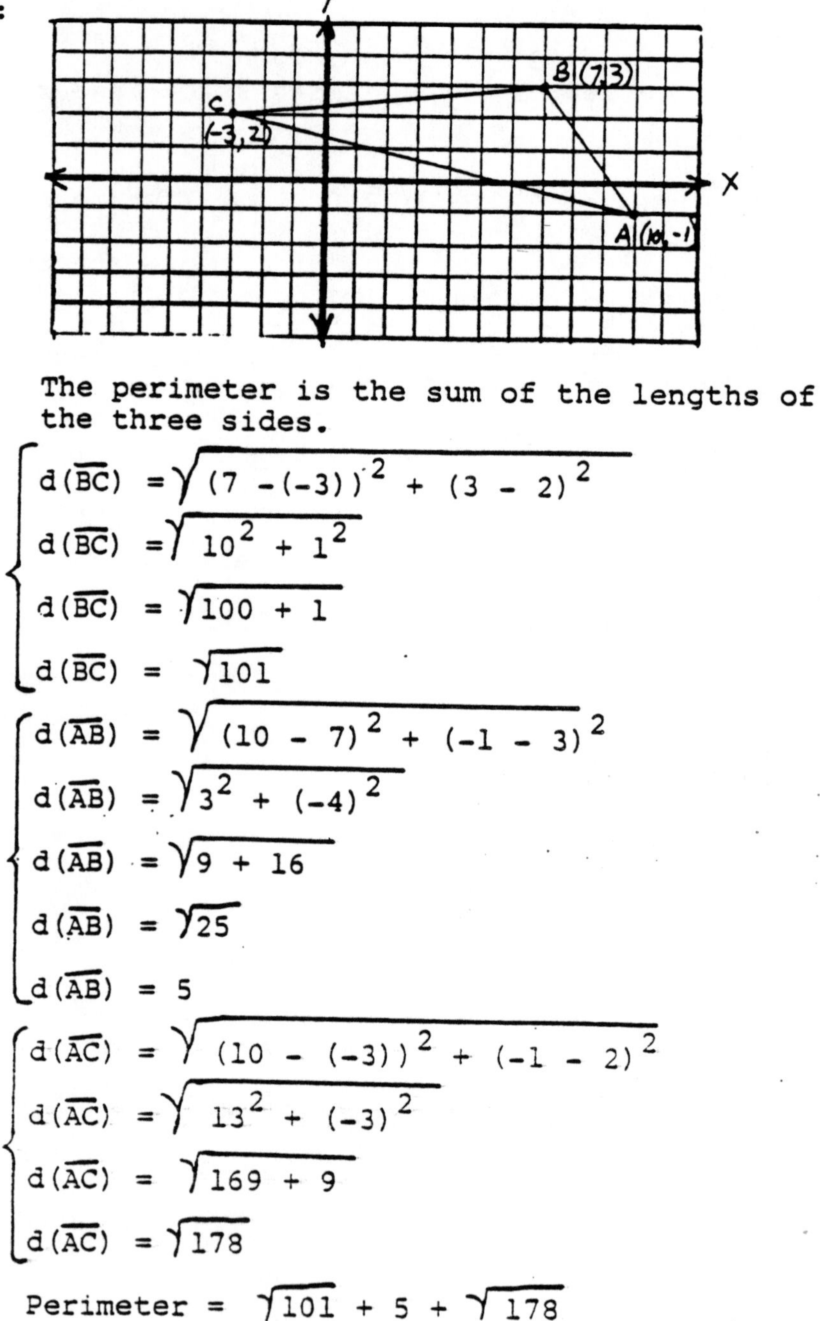

The perimeter is the sum of the lengths of the three sides.

$$d(\overline{BC}) = \sqrt{(7-(-3))^2 + (3-2)^2}$$
$$d(\overline{BC}) = \sqrt{10^2 + 1^2}$$
$$d(\overline{BC}) = \sqrt{100 + 1}$$
$$d(\overline{BC}) = \sqrt{101}$$

$$d(\overline{AB}) = \sqrt{(10-7)^2 + (-1-3)^2}$$
$$d(\overline{AB}) = \sqrt{3^2 + (-4)^2}$$
$$d(\overline{AB}) = \sqrt{9 + 16}$$
$$d(\overline{AB}) = \sqrt{25}$$
$$d(\overline{AB}) = 5$$

$$d(\overline{AC}) = \sqrt{(10-(-3))^2 + (-1-2)^2}$$
$$d(\overline{AC}) = \sqrt{13^2 + (-3)^2}$$
$$d(\overline{AC}) = \sqrt{169 + 9}$$
$$d(\overline{AC}) = \sqrt{178}$$

$$\text{Perimeter} = \sqrt{101} + 5 + \sqrt{178}$$

24. Find the perimeter of the triangle with verticies A = (8,6), B = (-3,3) and C = (1,-1).

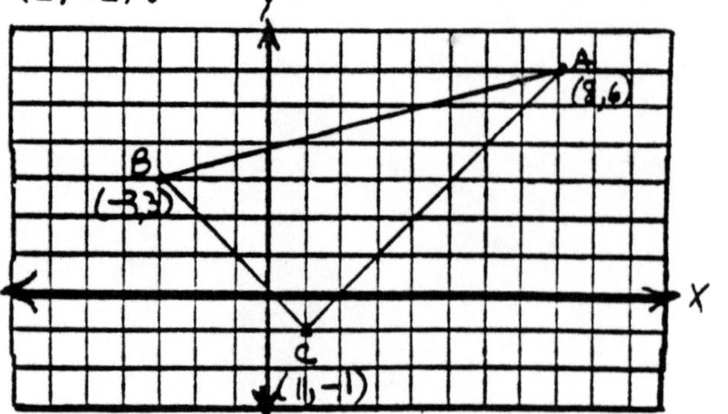

24.

$$d(\overline{AB}) = \sqrt{(-3-8)^2 + (3-6)^2}$$
$$= \sqrt{(-11)^2 + (-3)^2}$$
$$= \sqrt{121 + 9}$$
$$= \sqrt{130}$$

$d(\overline{BC}) = \sqrt{(-3-1)^2+(3-(-1))^2}$

$d(\overline{BC}) = \sqrt{(-4)^2 + 4^2}$

$= \sqrt{16 + 16}$

$= \sqrt{32}$

$= 4\sqrt{2}$

$d(\overline{AC}) = \sqrt{(8-1)^2+(6-(-1))^2}$

$= \sqrt{7^2 + 7^2}$

$= \sqrt{49 + 49}$

$= \sqrt{98}$

$= 7\sqrt{2}$

Perimeter =

$\sqrt{130} + 4\sqrt{2} + 7\sqrt{2} =$

$11\sqrt{2} + \sqrt{130}$

Exercise 4

1. Find the midpoint of the following line segments.

   a. $(4,2)$, $(-7,6)$
   b. $(4\sqrt{2}, 3\sqrt{7})$, $(-2\sqrt{2}, 5\sqrt{7})$
   c. $\left(\frac{1}{2}, -3\right)$, $\left(\frac{2}{3}, 5\right)$

2. Find the length of the following line segments.

   a. $(3,4)$, $(5,-2)$
   b. $(3\sqrt{2}, 4)$, $(6\sqrt{2}, -4)$
   c. $(4,-2)$, $(6,-3)$

3. Find the perimeter of the triangle whose verticies are $(4,7)$, $(4,10)$, and $(6,7)$.

Unit 13 Review

1. Describe each of the following terms:
   a. x-axis
   b. y-axis
   c. origin
   d. quadrant
   e. abscissa
   f. ordinate
   g. linear equation
   h. slope
   i. x-intercept
   j. y- intercept

2. Plot the following points and describe their location.
   a. (3,0)
   b. (4,6)
   c. (0,-3)
   d. (-3,2)
   e. (-3,0)
   f. (-2,-2)
   g. (0,4)
   h. (4,-2)

3. Find the slope of the line through (4,6) and (-5,2).

4. Determine if the slope of the following lines is positive, negative, zero, or no slope.

a.

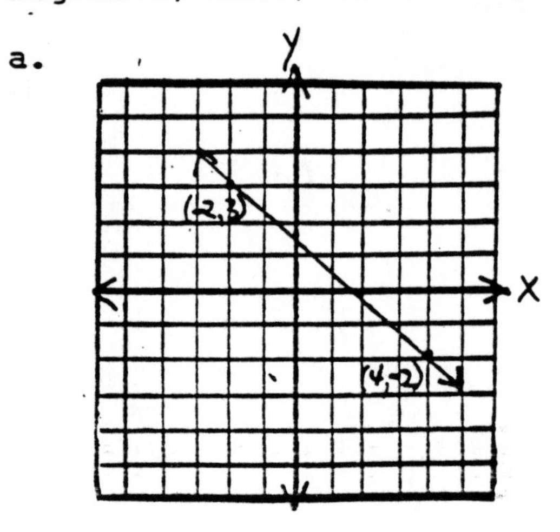

b.

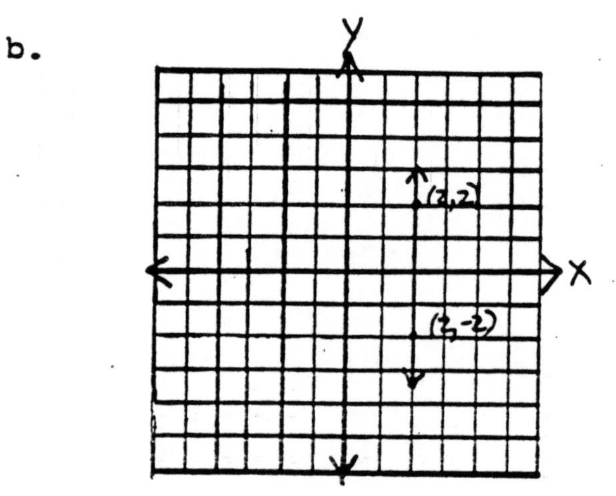

c.

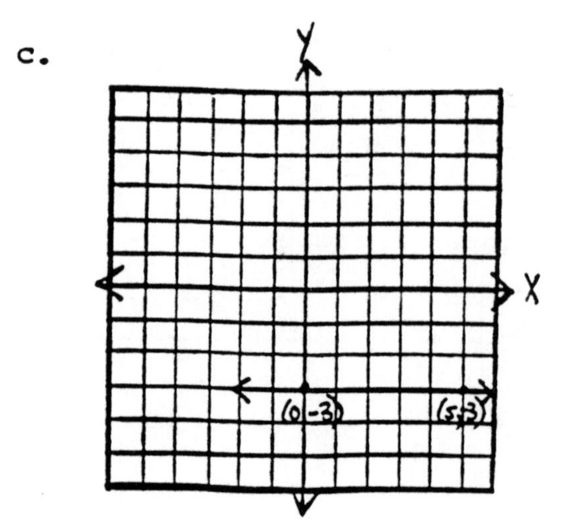

d.

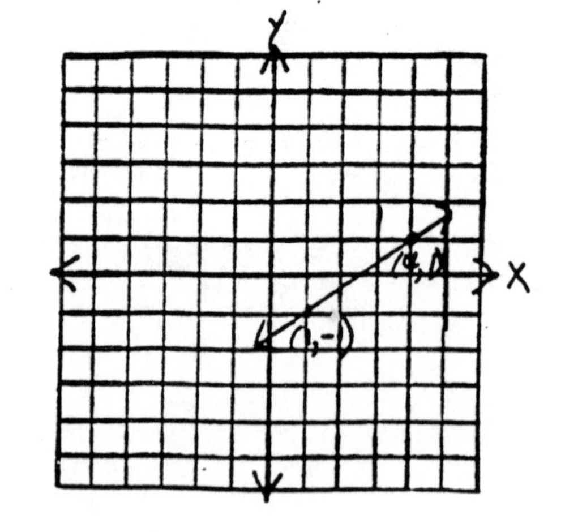

5. Graph $3x - 2y = 12$ using the selection of points method.

6. Graph $4x - y = 8$ using the slope y-intercept method.

7. Graph $2x + 3y = 12$ using the x-intercept, y-intercept method.

8. Determine the equation of the lines which meet the following conditions:
   a. Passes through (4,2) and (-4,3).
   b. Passes through (5,-1) with slope 2.
   c. Has slope $\frac{1}{2}$ and y-intercept 6.
   d. Has x-intercept 3 and y-intercept -2.
   e. Parallel to $2x - 4y = 2$ and passes through (1,1)
   f. Perpendicular to $2x + 3y = 6$ and passes through (-4,2).
   g. Passes through the midpoint of (-2,4) and (4,5) with slope 7.

9. Determine the length of the line segment joining (2,5) and (-4,-3).

Appendix A

Review of the Operations on Fractions

I. Addition and Subtraction

a. Find the least common denominator (LCD).

b. Change each fraction to an equivalent fraction having the LCD as the denominator.

c. Add or subtract the numerators.

d. Reduce your answer to lowest terms whenever possible.

Example 1: $\boxed{\frac{5}{6} - \frac{2}{9}}$ Subtraction

a. $6 = 2 \cdot 3$
$9 = 3 \cdot 3$

LCD $= 2 \cdot 3 \cdot 3 = 18$

b. $\frac{5}{6}(18) \longrightarrow \frac{5}{6} \times \frac{3}{3} = \frac{15}{18}$

$\frac{2}{9}(18) \longrightarrow \frac{2}{9} \times \frac{2}{2} = \frac{4}{18}$

c. $\frac{15 - 4}{18} = \frac{11}{18}$

Example 2: $\boxed{4\frac{2}{5} + 2\frac{2}{3} + 8}$

We will demonstrate two different methods with this example.

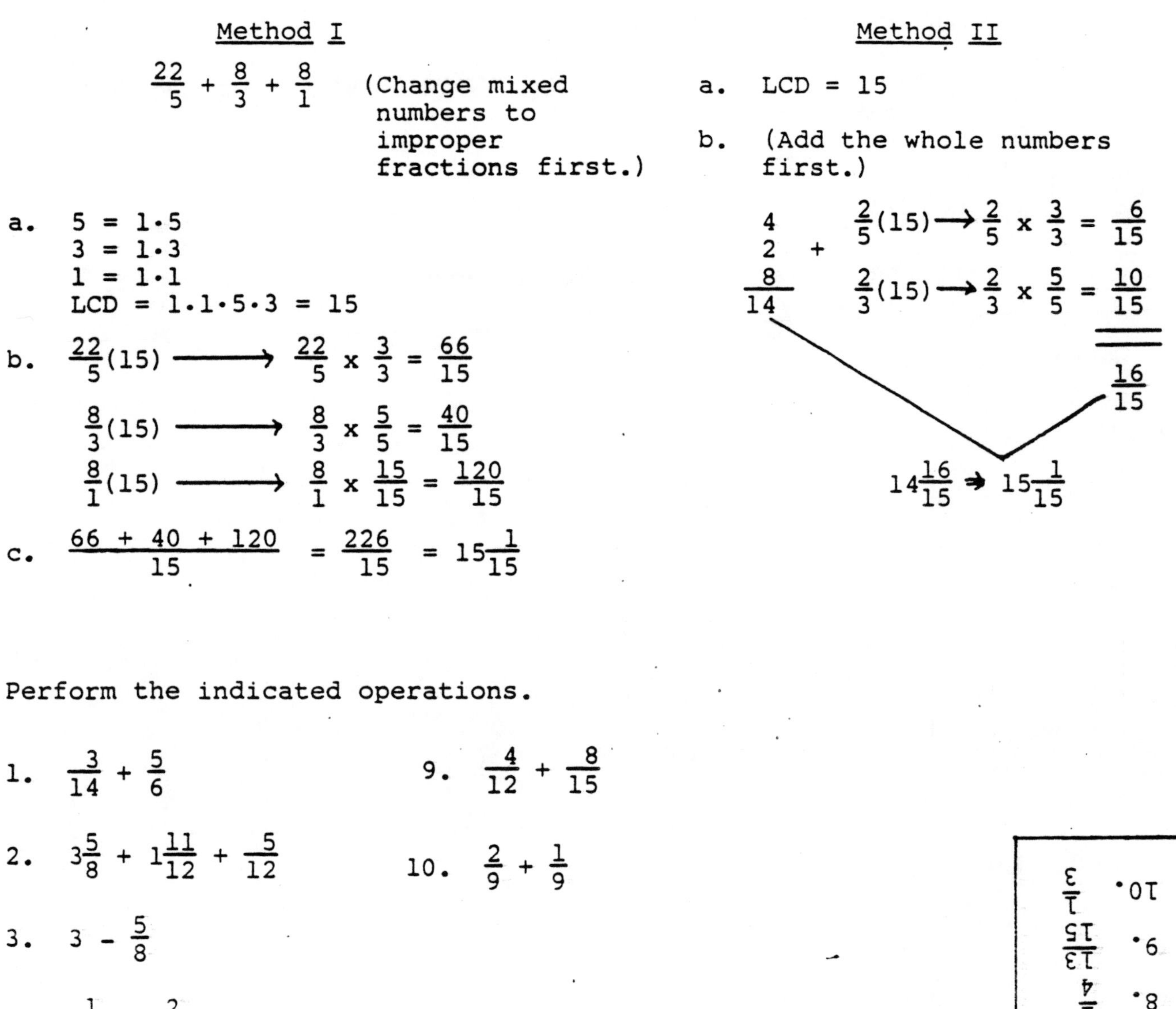

Method I

$\frac{22}{5} + \frac{8}{3} + \frac{8}{1}$ (Change mixed numbers to improper fractions first.)

a. $5 = 1 \cdot 5$
$3 = 1 \cdot 3$
$1 = 1 \cdot 1$
$\text{LCD} = 1 \cdot 1 \cdot 5 \cdot 3 = 15$

b. $\frac{22}{5}(15) \longrightarrow \frac{22}{5} \times \frac{3}{3} = \frac{66}{15}$

$\frac{8}{3}(15) \longrightarrow \frac{8}{3} \times \frac{5}{5} = \frac{40}{15}$

$\frac{8}{1}(15) \longrightarrow \frac{8}{1} \times \frac{15}{15} = \frac{120}{15}$

c. $\frac{66 + 40 + 120}{15} = \frac{226}{15} = 15\frac{1}{15}$

Method II

a. LCD = 15

b. (Add the whole numbers first.)

$4 + 2 + 8 = 14$

$\frac{2}{5}(15) \longrightarrow \frac{2}{5} \times \frac{3}{3} = \frac{6}{15}$

$\frac{2}{3}(15) \longrightarrow \frac{2}{3} \times \frac{5}{5} = \frac{10}{15}$

$\frac{6}{15} + \frac{10}{15} = \frac{16}{15}$

$14\frac{16}{15} \rightarrow 15\frac{1}{15}$

Perform the indicated operations.

1. $\frac{3}{14} + \frac{5}{6}$
2. $3\frac{5}{8} + 1\frac{11}{12} + \frac{5}{12}$
3. $3 - \frac{5}{8}$
4. $6\frac{1}{3} - 1\frac{2}{3}$
5. $15 + 2\frac{2}{5}$
6. $3\frac{4}{5} - 2\frac{1}{2}$
7. $\frac{7}{8} - \frac{3}{4}$
8. $\frac{5}{8} - \frac{3}{8}$
9. $\frac{4}{12} + \frac{8}{15}$
10. $\frac{2}{9} + \frac{1}{9}$

Answers

1. $1\frac{1}{21}$
2. $5\frac{23}{24}$
3. $2\frac{3}{8}$
4. $4\frac{2}{3}$
5. $17\frac{2}{5}$
6. $1\frac{3}{10}$
7. $\frac{1}{8}$
8. $\frac{1}{4}$
9. $\frac{13}{15}$
10. $\frac{1}{3}$

II. Multiplication of Fractions

a. Change each whole number or mixed number to an improper fraction.

b. Cancel wherever possible.

c. Multiply numerators together and denominators together.

d. If the answer is an improper fraction, then change to a mixed number.

Example 1:

$$1\frac{2}{3} \times 1\frac{4}{5} \times \frac{2}{7}$$

a. $\frac{5}{3} \times \frac{9}{5} \times \frac{2}{7}$

b. $\frac{\overset{1}{\cancel{5}}}{\underset{1}{\cancel{3}}} \times \frac{\overset{3}{\cancel{9}}}{\underset{1}{\cancel{5}}} \times \frac{2}{7}$

c. $\frac{1 \times 3 \times 2}{1 \times 1 \times 7} = \frac{6}{7}$

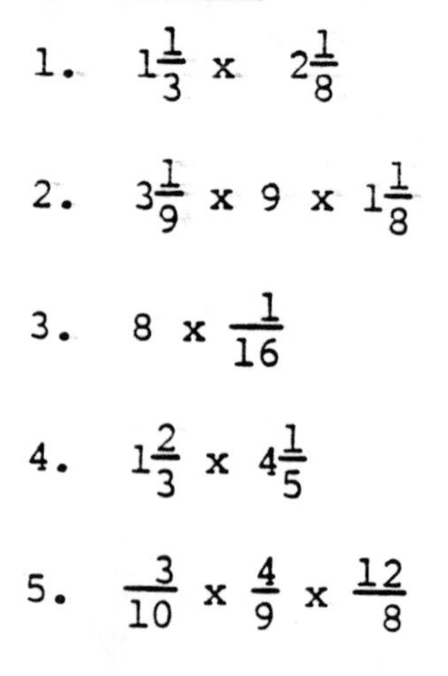

Exercises

1. $1\frac{1}{3} \times 2\frac{1}{8}$

2. $3\frac{1}{9} \times 9 \times 1\frac{1}{8}$

3. $8 \times \frac{1}{16}$

4. $1\frac{2}{3} \times 4\frac{1}{5}$

5. $\frac{3}{10} \times \frac{4}{9} \times \frac{12}{8}$

Answers

1. $2\frac{5}{6}$
2. $31\frac{1}{2}$
3. $\frac{1}{2}$
4. 7
5. $\frac{1}{5}$

III. Division of Fractions

The steps for division are similar to those for multiplication (with the exception of step (b) in part II.). For division, it is imperative that you <u>invert your divisor in the second step</u> and then cancel wherever possible.

Example: $10\frac{1}{2} \div 7$

a. $\frac{21}{2} \div \frac{7}{1}$

*b. $\frac{21}{2} \times \frac{1}{7}$ (invert the divisor)

$\frac{\overset{3}{\cancel{21}}}{2} \times \frac{1}{\underset{1}{\cancel{7}}}$ (cancel)

c. $\frac{3 \times 1}{2 \times 1} = \frac{3}{2}$

d. $\frac{3}{2} = 1\frac{1}{2}$ (change improper fraction to a mixed number)

<u>Exercises</u>

1. $\frac{2}{3} \div \frac{7}{8}$
2. $5\frac{2}{5} \div 6\frac{2}{5}$
3. $\frac{3\frac{1}{3}}{2\frac{1}{2}}$
4. $4 \div \frac{1}{4}$
5. $2\frac{1}{2} \div \frac{5}{2}$

Answers

1. $\frac{16}{21}$
2. $\frac{27}{32}$
3. $1\frac{1}{3}$
4. 16
5. 1

Appendix B

Solving Linear Equations Using the Set of Rational Numbers

The solutions to the examples are shown below:

Steps	Reasons
Example 1:	
$7x - \frac{5}{3} = 8$	
$7x - \frac{5}{3} + \frac{5}{3} = 8 + \frac{5}{3}$	Addition property of equality
$7x = \frac{8}{1} + \frac{5}{3}$	Substitution of a number fact. $-\frac{5}{3}+\frac{5}{3}=0$.
$7x = \frac{24}{3} + \frac{5}{3}$	Find the LCM for adding fractions.
$7x = \frac{29}{3}$	Simplification
$\frac{1}{7}\cdot 7x = \frac{1}{7}\cdot \frac{29}{3}$	Multiplication property of equality
$x = \frac{29}{21}$ or $1\frac{8}{21}$	
Example 2:	
$-3x + 7 = 6x - 4$	
$-3x - 6x + 7 = 6x - 6x - 4$	Addition property of equality
$-9x + 7 = -4$	Substitution of a number fact. $-3x - 6x = -9x$
$-9x + 7 - 7 = -4 - 7$	Addition property of equality
$-9x = -11$	Substitution of a number fact. $-4 - 7 = -11$
$-\frac{1}{9}\cdot -9x = -\frac{1}{9}\cdot -11$	Multiplication property of equality
$x = \frac{11}{9}$ or $1\frac{2}{9}$	

Example 3:

$$\frac{7}{2}x + 5 = \frac{13}{3} - \frac{3}{4}x$$

The fractions may be eliminated by multiplying each term of the equation by the least common denominator.

$\left(\frac{7}{2}x\right) + (5) = \left(\frac{13}{3}\right) - \left(\frac{3}{4}x\right)$	
$12\left(\frac{7}{2}x\right) + 12(5) = 12\left(\frac{13}{3}\right) - 12\left(\frac{3}{4}x\right)$	Multiplication property of equality.
$6(7x) + 12(5) = 4(13) - 3(3x)$	Simplification
$42x + 60 = 52 - 9x$	Substitution of a number fact.
$42x = -8 - 9x$	Addition property of equality.
$51x = -8$	Addition property of equality.
$\frac{1}{51} \cdot 51x = \frac{1}{51} \cdot -8$	Multiplication property of equality.
$x = \frac{-8}{51}$	

Example 4:

$$5(2x - 3) - 4 = 3x + 7$$

$10x - 15 - 4 = 3x + 7$	Distributive law
$10x - 19 = 3x + 7$	Substitution of a number fact.
$10x = 3x + 26$	Addition property of equality
$7x = 26$	Addition property of equality.
$\frac{1}{7} \cdot 7x = \frac{1}{7} \cdot 26$	Multiplication property of equality.
$x = \frac{26}{7} = 3\frac{5}{7}$	

Exercises

1. $-3x - 5 = 2x + 6$

2. $2x - (x + 6) = 10$

3. $6 + 2x = 4(x - 2)$

4. $9 + 7x = 9x - 7$

5. $5x + \frac{1}{6} = 8$

6. $3x - \frac{1}{7} = 15$

7. $\frac{2}{9}y + 2 = \frac{y}{18} + 4$

8. $\frac{4x}{3} + 7 = \frac{5x}{2} - 1$

9. $\frac{1}{2}x - 10 = 12$

10. $\frac{9x}{5} + \frac{2}{3}x = \frac{2}{4}$

Answers

1. $-2\frac{1}{5}$
2. 16
3. 7
4. 8
5. $1\frac{17}{30}$
6. $5\frac{1}{21}$
7. 12
8. $6\frac{6}{7}$
9. 44
10. $\frac{15}{74}$

Table 1 SQUARE ROOTS (0 to 199)

n	$\sqrt{n}$	n	$\sqrt{n}$	n	$\sqrt{n}$	n	$\sqrt{n}$
0	0.000	50	7.071	100	10.000	150	12.247
1	1.000	51	7.141	101	10.050	151	12.288
2	1.414	52	7.211	102	10.100	152	12.329
3	1.732	53	7.280	103	10.149	153	12.369
4	2.000	54	7.348	104	10.198	154	12.410
5	2.236	55	7.416	105	10.247	155	12.450
6	2.449	56	7.483	106	10.296	156	12.490
7	2.646	57	7.550	107	10.344	157	12.530
8	2.828	58	7.616	108	10.392	158	12.570
9	3.000	59	7.681	109	10.440	159	12.610
10	3.162	60	7.746	110	10.488	160	12.649
11	3.317	61	7.810	111	10.536	161	12.689
12	3.464	62	7.874	112	10.583	162	12.728
13	3.606	63	7.937	113	10.630	163	12.767
14	3.742	64	8.000	114	10.677	164	12.806
15	3.873	65	8.062	115	10.724	165	12.845
16	4.000	66	8.124	116	10.770	166	12.884
17	4.123	67	8.185	117	10.817	167	12.923
18	4.243	68	8.246	118	10.863	168	12.961
19	4.359	69	8.307	119	10.909	169	13.000
20	4.472	70	8.367	120	10.954	170	13.038
21	4.583	71	8.426	121	11.000	171	13.077
22	4.690	72	8.485	122	11.045	172	13.115
23	4.796	73	8.544	123	11.091	173	13.153
24	4.899	74	8.602	124	11.136	174	13.191
25	5.000	75	8.660	125	11.180	175	13.229
26	5.099	76	8.718	126	11.225	176	13.266
27	5.196	77	8.775	127	11.269	177	13.304
28	5.292	78	8.832	128	11.314	178	13.342
29	5.385	79	8.888	129	11.358	179	13.379
30	5.477	80	8.944	130	11.402	180	13.416
31	5.568	81	9.000	131	11.446	181	13.454
32	5.657	82	9.055	132	11.489	182	13.491
33	5.745	83	9.110	133	11.533	183	13.528
34	5.831	84	9.165	134	11.576	184	13.565
35	5.916	85	9.220	135	11.619	185	13.601
36	6.000	86	9.274	136	11.662	186	13.638
37	6.083	87	9.327	137	11.705	187	13.675
38	6.164	88	9.381	138	11.747	188	13.711
39	6.245	89	9.434	139	11.790	189	13.748
40	6.325	90	9.487	140	11.832	190	13.784
41	6.403	91	9.539	141	11.874	191	13.820
42	6.481	92	9.592	142	11.916	192	13.856
43	6.557	93	9.644	143	11.958	193	13.892
44	6.633	94	9.659	144	12.000	194	13.928
45	6.708	95	9.747	145	12.042	195	13.964
46	6.782	96	9.798	146	12.083	196	14.000
47	6.856	97	9.849	147	12.124	197	14.036
48	6.928	98	9.899	148	12.166	198	14.071
49	7.000	99	9.950	149	12.207	199	14.107

Table 2 TRIGONOMETRIC RATIOS

degrees	sin	cos	tan	degrees	sin	cos	tan
1	.017	1.000	.017	46	.719	.695	1.035
2	.035	.999	.035	47	.731	.682	1.072
3	.052	.999	.052	48	.743	.669	1.111
4	.070	.998	.070	49	.755	.656	1.150
5	.087	.996	.087	50	.766	.643	1.192
6	.105	.995	.105	51	.777	.629	1.235
7	.122	.993	.123	52	.788	.616	1.280
8	.139	.990	.141	53	.799	.602	1.327
9	.156	.988	.158	54	.809	.588	1.376
10	.174	.985	.176	55	.819	.574	1.428
11	.191	.982	.194	56	.829	.559	1.483
12	.208	.978	.213	57	.839	.545	1.540
13	.225	.974	.231	58	.848	.530	1.600
14	.242	.970	.249	59	.857	.515	1.664
15	.259	.966	.268	60	.866	.500	1.732
16	.276	.961	.287	61	.875	.485	1.804
17	.292	.956	.306	62	.883	.469	1.881
18	.309	.951	.325	63	.891	.454	1.963
19	.326	.946	.344	64	.899	.438	2.050
20	.342	.940	.364	65	.906	.423	2.145
21	.358	.934	.384	66	.914	.407	2.246
22	.375	.927	.404	67	.921	.391	2.356
23	.391	.921	.424	68	.927	.375	2.475
24	.407	.914	.445	69	.934	.358	2.605
25	.423	.906	.466	70	.940	.342	2.747
26	.438	.899	.488	71	.946	.326	2.904
27	.454	.891	.510	72	.951	.309	3.078
28	.469	.883	.532	73	.956	.292	3.271
29	.485	.875	.554	74	.961	.276	3.487
30	.500	.866	.577	75	.966	.259	3.732
31	.515	.857	.601	76	.970	.242	4.011
32	.530	.848	.625	77	.974	.225	4.331
33	.545	.839	.649	78	.978	.208	4.705
34	.559	.829	.675	79	.982	.191	5.145
35	.574	.819	.700	80	.985	.174	5.671
36	.588	.809	.727	81	.988	.156	6.314
37	.602	.799	.754	82	.990	.139	7.115
38	.616	.788	.781	83	.993	.122	8.144
39	.629	.777	.810	84	.995	.105	9.514
40	.643	.766	.839	85	.996	.087	11.430
41	.656	.755	.869	86	.998	.070	14.301
42	.669	.743	.900	87	.999	.052	19.081
43	.682	.731	.933	88	.999	.035	28.636
44	.695	.719	.966	89	1.000	.017	57.290
45	.707	.707	1.000				

APPENDIX E

USING A PROTRACTOR TO MEASURE AN ANGLE

Step 1:

To measure an angle, place the cross mark of the protractor on the vertex of the angle and the straightedge of the protractor on one side of the angle.

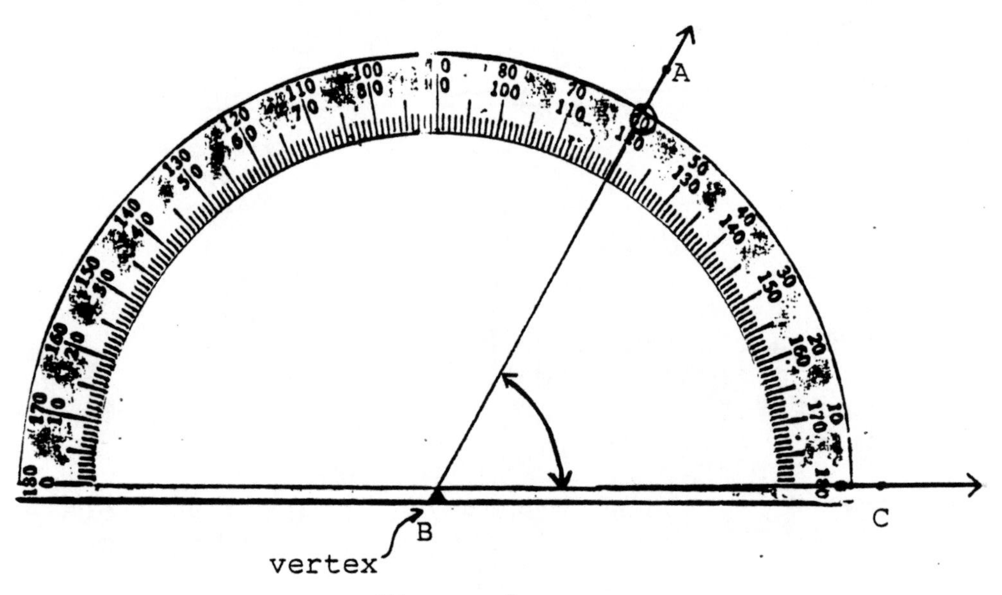

Figure 1

Step 2:

As in Figure 1, the side of the angle along the straightedge of the protractor is to the right of the cross mark and the upper scale on the protractor is used to measure the angle. ∠ABC measures 60°.

Note: In the case that your protractor's scales are reversed, then reverse the instructions and use the lower scale on your protractor. ∠ABC will still be 60°.

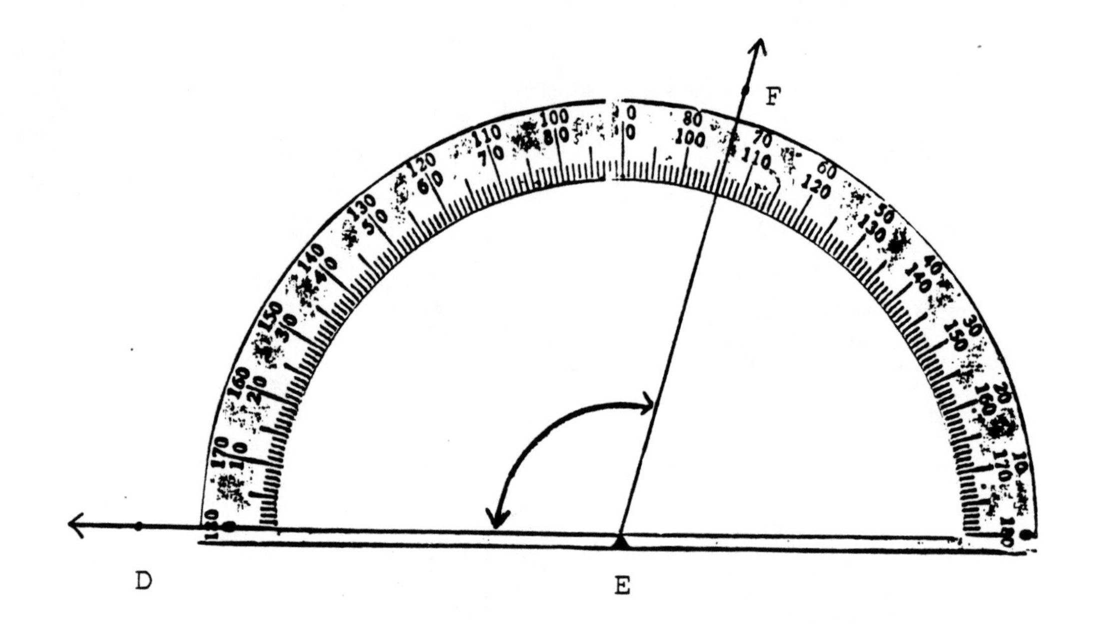

Figure 2

Step 3:

As in Figure 2, the side of the angle along the straightedge of the protractor is to the left of the cross mark. Use the lower scale on the protractor to measure the angle. $\angle$ DEF = 106°.

Remember: In the case that your protractor's scales are reversed, then use your upper scale. $\angle$DEF should still equal 106°.

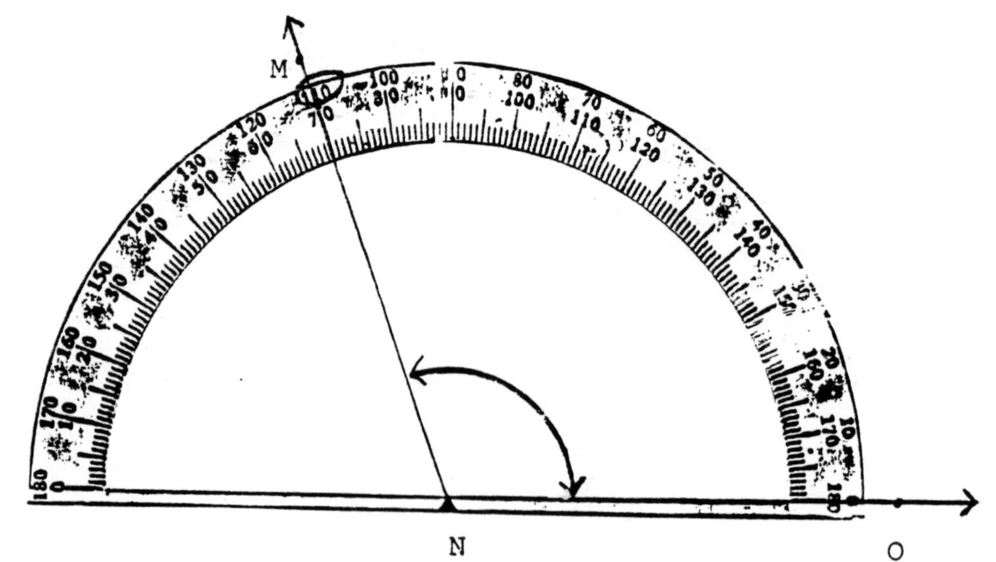

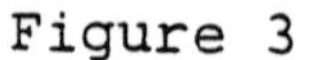

Figure 3

Figure 3, as in Figure 1, the side of the angle along the straightedge of the protractor is to the right of the cross mark. Using the upper scale, ∠MNO measures 110°.

Exercises:

1. Measure each of the following angles.

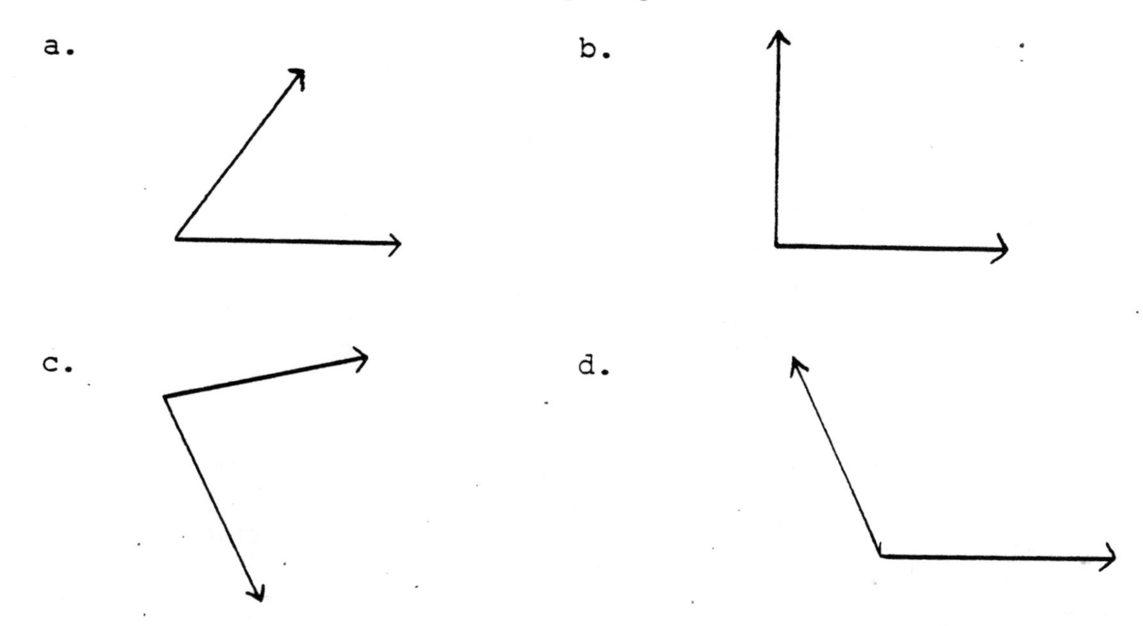

2. Draw each angle as described below:

   a. ∠RML = 40° b. ∠MNQ = 125°

   c. ∠C = 172° d. ∠1 = 89°

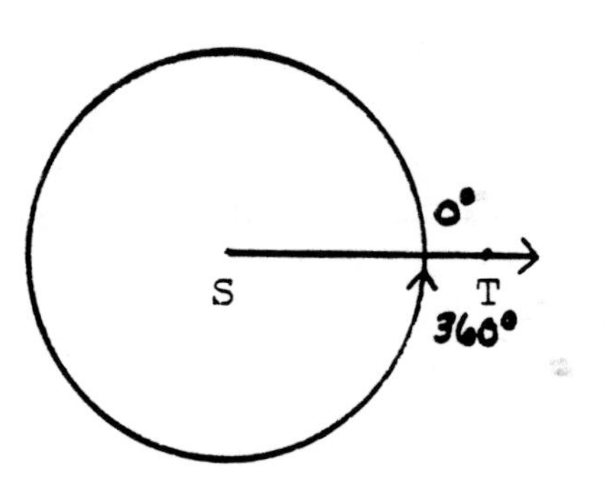

A ray that is rotated about its its endpoint in a flat surface will eventually get back to its starting point. The measure of the ray's starting position is 0°. The measure of one complete rotation is 360°. The measure of ½ rotation is exactly ½ of 360° or 180°.

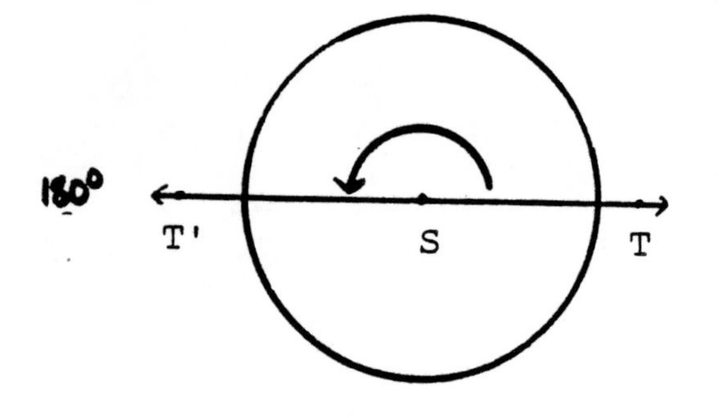

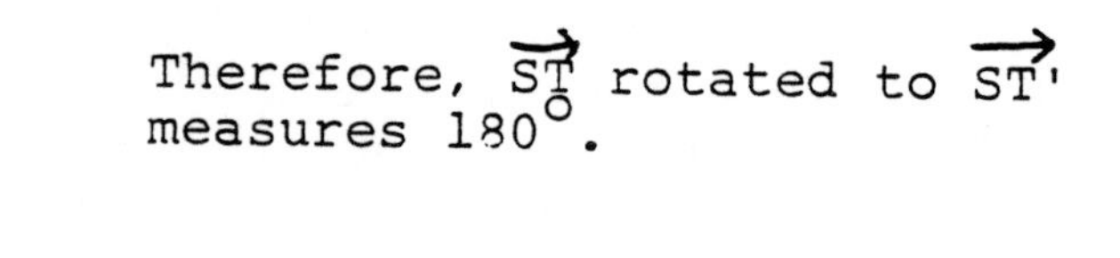

Therefore, $\overrightarrow{ST}$ rotated to $\overrightarrow{ST'}$ measures 180°.

Using a protractor, ∠T'ST would still measure 180°.

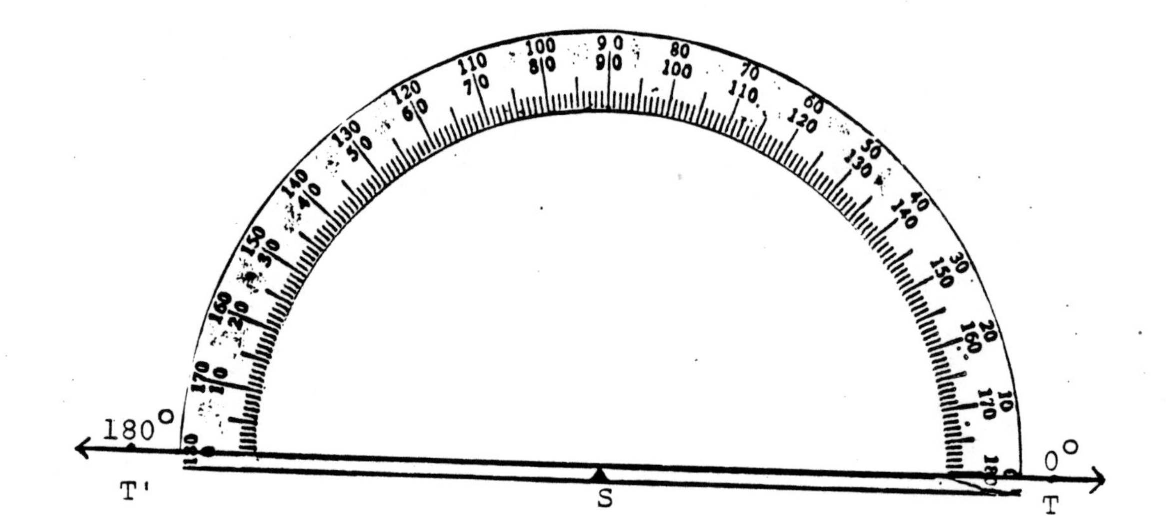

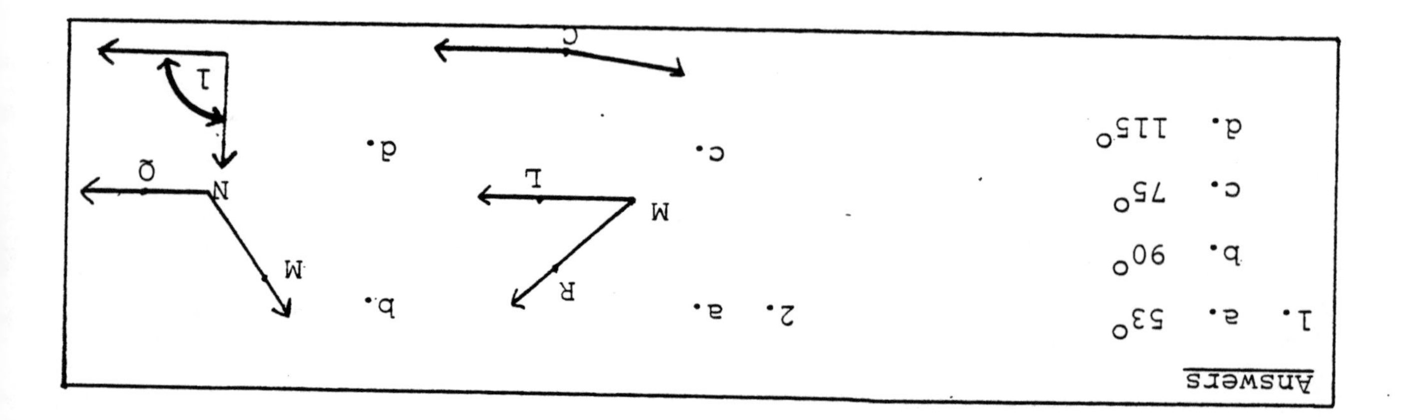

APPENDIX F

Unit 1 Answers

Exercise 1

1. a. A and B

   b. yes

   c. 4 cm.; yes

2. $2c + (c + 3) + (3c - 3) = 12$

$$6c = 12$$

$$c = 2$$

$\overline{XY} = 4$ cm.; $\overline{YZ} = 5$ cm.; $\overline{ZW} = 3$ cm.

Exercise 2

1. $\angle 1$, $\angle ABC$, $\angle CBA$, $\angle B$

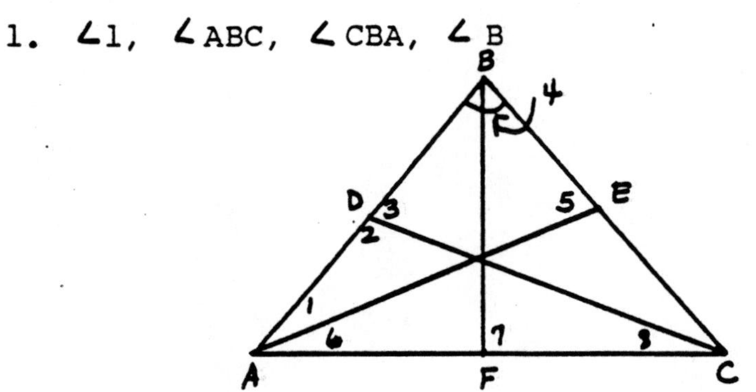

3. a. 72°, acute

   b. 180°, straight

   c. 60°, acute

   d. 120°, obtuse

4.

	Supplement	Complement
a.	150°	60°
b.	60°	none
c.	$(180 - x)^\circ$	$(90 - x)^\circ$
d.	135°	45°

5. $(5N + 7) + (2N) + (N + 29) = 180$

$$8N + 36 = 180$$

$$8N = 144$$

$$N = 18^\circ$$

$\angle XYT = 36^\circ$, $\angle TYS = 97^\circ$, $\angle SYZ = 47^\circ$

6. $x + (x - 30) = 90$

$$x = 60^\circ \text{ and } (x - 30) = 30^\circ$$

7. $x + (5x + 12) = 90$

$$x = 13^\circ \text{ and } (5x + 12) = 77^\circ$$

8. $x + x = 180$

$$2x = 180$$

$$x = 90^\circ \quad \text{both angles are } 90^\circ$$

9. $\left(\frac{2}{3} x + 10\right) + x = 180$

$$x = 102^\circ \text{ and } \frac{2}{3} x + 10 = 78^\circ$$

10. $(3x - 20) + (x) = 90$

$$4x - 20 = 90$$

$$4x = 110$$

$$x = 27\frac{1}{2}^\circ \text{ and } (3x - 20) = 62\frac{1}{2}^\circ$$

Exercise 3

1. Refer to Construction # 3
2. Refer to Construction # 1
3. Refer to Construction # 2

Unit 1 Review

1. Check definitions throughout unit.

2. a. angle, vertex, ∠

   b. complementary

   c. supplementary

3. defined terms, undefined terms, theorems and postulates

4. a. $x + \frac{2}{3}x = 90$

   $x = 54^\circ$ and $\frac{2}{3}x = 36^\circ$

   b. $x + 5x = 180$

   $x = 30^\circ$ and $5x = 150^\circ$

   c. $(4x + 20) + x = 90$

   $x = 14^\circ$ and $(4x + 20) = 76^\circ$

   d. $x + 2x = 180$

   $x = 60^\circ$ and $2x = 120^\circ$

5. a. Follow the steps in Construction # 1

   b. Follow the steps in Construction # 2

   c. Follow the steps in Construction # 3

6. i. 90°, right

   ii. 50°, acute

   iii. 115°, obtuse

   iv. 180°, straight

7. i. points D and X

   ii. yes

   iii. midpoint

8. endpoint; no, the rays not only have a different direction but also have different endpoints.

9. a. $(30x-60)+(2x+20)+(x)=180^\circ$

$\angle DBE = 6\frac{2}{3}^\circ$

$\angle CBE = 33\frac{1}{3}^\circ$

$\angle ABC = 140^\circ$

b. $(2x-9)+(x+3)=90$

$x=32^\circ$

$\angle ABD = 55^\circ$, $\angle DBC = 35^\circ$

$\angle ABC = 90^\circ$

10. a. $\angle 2 = \angle 3$ because supplements of equal angles are equal.

$\angle 1 = 110^\circ$

$\angle 2 = 70^\circ$

$\angle 3 = 70^\circ$

b. $\angle 4 = 95^\circ$, supplementary to $\angle 5$.

$\angle 6 = 95^\circ$, vertical to $\angle 4$.

$\angle 7 = 85^\circ$, vertical to $\angle 5$.

Unit 2 Answers

Exercise 1

1. a. X, Z, N

   b. $\triangle$XZN, $\triangle$YZM, $\triangle$YMO, $\triangle$MNO, $\triangle$NOX, $\triangle$YOX, $\triangle$XYM, $\triangle$YMN, $\triangle$YXN, $\triangle$MXN, $\triangle$ZMX, $\triangle$ZYN

2. 18 cm.

3. No; this figure is not a closed three sided figure.

Exercise 2

1. See Construction # 2

2. See Construction # 1

3. 45 inches

4. $2x + 8 = 32$

   $x = 12$ inches

5. $\frac{1}{2}x + 2x = 41\frac{1}{2}$

   $x = 16\frac{3}{5}$ and the base = $8\frac{3}{10}$

6. a. No, it wouldn't be a closed figure.

   b. No, this figure is not closed.

   c. Yes , Δ acute triangle

Exercise 3

1. $\angle M = \angle P$, $\angle MNO = \angle PNO$, $\angle PON = \angle MON$, $\overline{OP} = \overline{OM}$, $\overline{MN} = \overline{PN}$, $\overline{ON} = \overline{ON}$

2. a. $\overline{PR} = \overline{AB}$, $\overline{RQ} = \overline{BC}$, $\overline{PQ} = \overline{AC}$, $\angle P = \angle A$, $\angle R = \angle B$, $\angle Q = \angle C$

   b. $\overline{LM} = \overline{CU}$, $\overline{MN} = \overline{UT}$, $\overline{LN} = \overline{CT}$, $\angle L = \angle C$, $\angle M = \angle U$, $\angle N = \angle T$

3. $\Delta DCF \cong \Delta ABE$

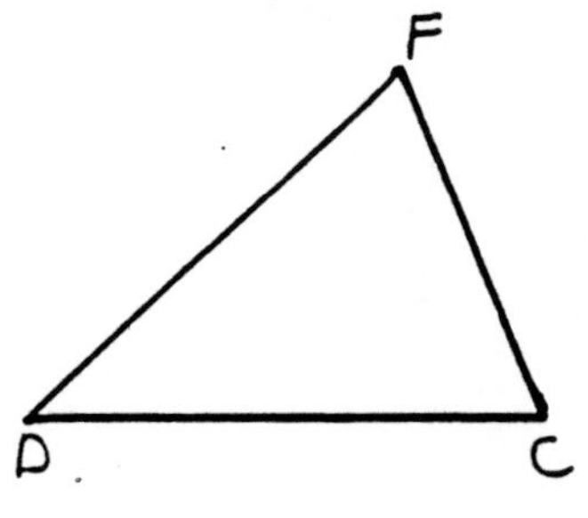

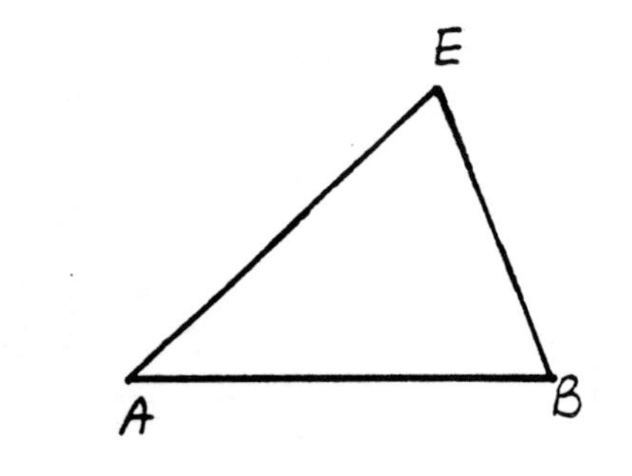

Exercise 4

1.

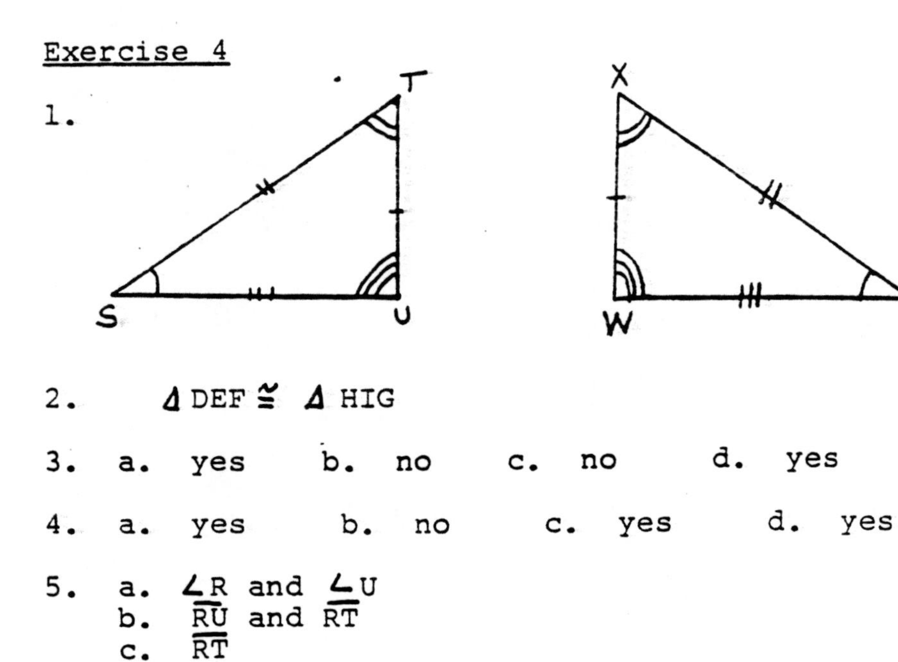

2. $\Delta DEF \cong \Delta HIG$

3. a. yes b. no c. no d. yes

4. a. yes b. no c. yes d. yes

5. a. $\angle R$ and $\angle U$
   b. $\overline{RU}$ and $\overline{RT}$
   c. $\overline{RT}$
   d. $\angle T$

Exercise 5

1. yes; SSS = SSS
2. yes; SSS = SSS
3. yes; SAS = SAS
4. yes; ASA = ASA
5. no
6. no

Unit 2 Review

1. The definitions may be found throughout the unit.

2. a. no
   b. no
   c. yes; SAS = SAS
   d. yes, SSS = SSS
   e. yes, ASA = ASA
   f. yes; SAS = SAS

3. a. equilateral triangle
   b. isosceles triangle
   c. equiangular triangle
   d. right triangle
   e. obtuse triangle
   f. acute triangle

4.

	equal sides	equal angles
a.	$\overline{DC} = \overline{DC}$	$\angle A = \angle B$
	$\overline{BD} = \overline{AC}$	$\angle ADC = \angle BCD$
	$\overline{DA} = \overline{CB}$	$\angle ACD = \angle BDC$
b.	$\overline{PT} = \overline{RS}$	$\angle P = \angle R$
	$\overline{TQ} = \overline{SQ}$	$\angle PQT = \angle RQS$
	$\overline{PQ} = \overline{RQ}$	$\angle QTP = \angle QSR$

	Equal sides	Equal angles
c.	$\overline{PS} = \overline{RT}$	$\angle P = \angle R$
	$\overline{SQ} = \overline{TQ}$	$\angle PQS = \angle RQT$
	$\overline{PQ} = \overline{RQ}$	$\angle PSQ = \angle RTQ$

5. $\Delta AFC \cong \Delta DEB$ same as $\Delta FCA \cong \Delta EBD$ or $\Delta CAF \cong \Delta BDE$

6. $2x + 5 = 23$
   $x = 9$ inches

7. $(x + 5) + (x + 7) + (x + 2) = 59$
   $x = 15$
   20 ft., 22 ft., 17 ft.

8. See Construction #1 and Construction # 2

9. a. ΔABC overlaps ΔDCB
   b. ΔAEC overlaps ΔDEB

Unit 3 Answers

Exercise 1

1. Converse: If you live in the United States, then you live in Virginia.
   Contrapositive: If you don't live in the United States, then you don't live in Virginia.
   Inverse: If you don't live in Virginia, then you don't live in the United States.

2. Converse: If you can fly a plane, then you can operate a car.
   Contrapositive: If you can't fly a plane, then you can not operate a car.
   Inverse: If you can't operate a car, then you can not fly a plane.

3. Converse: If it is equilateral, then the triangle is not scalene.
   Contrapositive: If it isn't equilateral, then the triangle is scalene.
   Inverse: If the triangle is scalene, then it isn't equilateral.

4. Converse: If you are not intelligent, then you are a blond.
   Contrapositive: If you are intelligent, then you are not a blond.
   Inverse: If you aren't blond, then you are intelligent.

Exercise 2

1. Part A #3 - $\angle a$ and $\angle c$ are equal because they are complements of the same angle ($\angle b$).

2. Part B #4 - Reflexive Law: Any quantity is equal to itself.

3. Part A #5 - $\angle 1$ and $\angle 2$ are vertical angles and vertical angles are equal.

4. Part A #2 - $\angle 1$ and $\angle 2$ and $\angle 3$ and $\angle 4$ are pairs of supplementary angles. If $\angle 2 = \angle 3$ and $\angle 1 = \angle 4$ since supplements of equal angles are themselves equal.

5. a. Part B #3 - They are equal sides of an isosceles triangle.
   b. Part B #6 - Halves of equal line segments are equal.

Exercise 3

1.

Statements	Reasons
1. $\angle S = \angle T$, $\overline{SR} = \overline{TU}$ $\overline{SR} \perp \overline{RU}$ and $\overline{TU} \perp \overline{RU}$	1. Given
2. $\angle R$ and $\angle U$ are right angles.	2. Definition of perpendicular lines.
3. $\angle R = \angle U$	3. All right angles are equal.
4. $\triangle SRP \cong \triangle TUP$	4. ASA = ASA

2.

Statements	Reasons
1. $\overline{MO}$ and $\overline{NP}$ bisect one another.	1. Given
2. $\overline{NQ} = \overline{QP}$ and $\overline{MQ} = \overline{QO}$	2. Definition of bisector
3. $\angle NQM$ and $\angle OQP$ are vertical angles.	3. Definition of vertical angles.
4. $\angle NQM = \angle OQP$	4. Vertical angles are equal.
5. $\triangle NQM \cong \triangle OQP$	5. SAS = SAS

3.

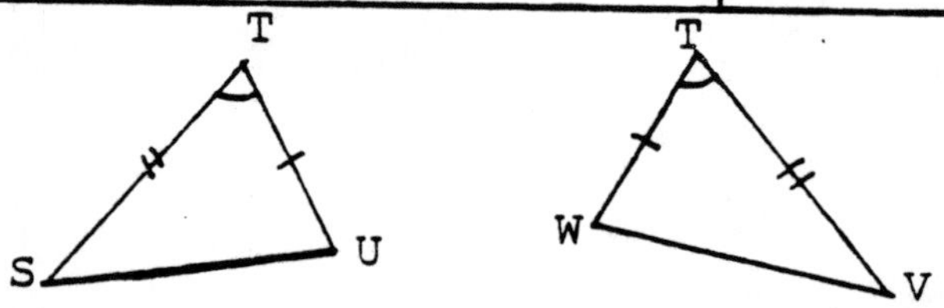

Statements	Reasons
1. $\overline{WT} = \overline{UT}$, $\overline{ST} = \overline{TV}$	1. Given
2. $\angle T = \angle T$	2. Reflexive Law
3. $\triangle WTV \cong \triangle UTS$	3. SAS = SAS

4.

Statements	Reasons
1. $\angle 2 = \angle 3$, $\overline{AB} = \overline{CD}$ $\angle A = \angle D$	1. Given
2. $\angle 1$ and $\angle 2$ are supplementary. $\angle 3$ and $\angle 4$ are supplementary.	2. Definition of supplementary.
3. $\angle 1 = \angle 4$	3. Supplements of equal angles are themselves equal.
4. $\triangle ABE \cong \triangle DCE$	4. ASA = ASA

5.

Statements	Reasons
1. $\triangle$ XYZ is isosceles. $\overline{XO} = \overline{OZ}$	1. Given
2. $\overline{XY} = \overline{ZY}$	2. Def. of isosceles.
3. $\overline{YO} = \overline{YO}$	3. Reflexive Law
4. $\triangle$ XOY ≅ $\triangle$ ZOY	4. SSS = SSS

Exercise 4

1. 1. Given
   2. Definition of vertical angles
   3. Vertical angles are equal.
   4. ASA = ASA
   5. CPCTE

2. 1. Given
   2. If two angles in a triangle are equal, then the sides opposite these angles are equal.
   3. ASA = ASA
   4. CPCTE

3. 1. Given
   2. Reflexive Law
   3. SSS = SSS
   4. CPCTE

4. 1. Given
   2. If two angles are equal, then the sides opposite those angles are equal.
   3. Reflexive Law.
   4. Equal quantities subtracted from equal quantities are equal. (See page 80, Item # 5.)
   5. Definition of supplementary angles.
   6. Supplements of equal angles are equal.
   7. SAS = SAS
   8. CPCTE

5. 1. Given
   2. Definition of midpoint
   3. Definition of midpoint
   4. Definition of vertical angles
   5. Vertical angles are equal.
   6. SAS = SAS
   7. CPCTE

6. 1. Given
   2. Definition of bisector
   3. Reflexive Law
   4. ASA = ASA
   5. CPCTE

7. 1. Given
   2. Definition of perpendicular lines (or) perpendicular lines form right angles.
   3. All right angles are equal.
   4. Definition of perpendicular bisector.
   5. Reflexive Law
   6. SAS = SAS
   7. CPCTE

<u>Unit 3 Review</u>

1. hypothesis; conclusion

2. Converse: If I will pass geometry, then I will study hard.

   Inverse: If I won't study hard, then I won't pass geometry.

   Contrapositive: If I won't pass geometry, then I won't study hard.

3. Yes-----If a theorem is proven valid, its converse is not necessarily true. Its converse is not a logical equivalent of the theorem; and, therefore, must be proven itself valid.

4. SSS = SSS, SAS = SAS, ASA = ASA

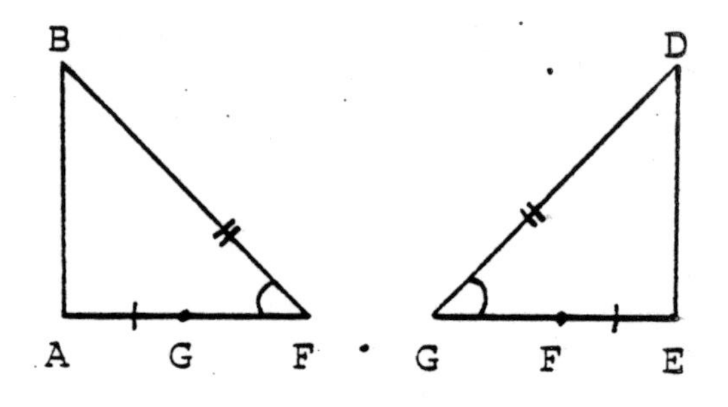

5. 1. Given
   2. Reflexive Law
   3. The sums of equal quantities are equal. ($\overline{AG} + \overline{GF} = \overline{GF} + \overline{FE}$)
   4. SAS = SAS

6. 1. Given
   2. Reflexive
   3. ASA = ASA

7. 1. Given
   2. Definition of midpoint
   3. SAS = SAS
   4. CPCTE
   5. If two sides of a triangle are equal, then the angles opposite these sides are equal.

8. 1. Reflexive Law
   2. Given
   3. Definition of bisector
   4. Given
   5. Definition of perpendicular lines
   6. All right angles are equal.
   7. SAS = SAS
   8. CPCTE
   9. Definition of isosceles triangle.

9. 1. Given
   2. Definition of bisector
   3. Definition of supplementary angles
   4. Supplements of equal angles are equal.
   5. Reflexive law
   6. ASA = ASA
   7. CPCTE

Unit 4 Answers

Exercise 1

1. $\angle 1 = 123^\circ$ Supp. to $\angle 2$. $\angle 5 = 123^\circ$ Alt. int. to $\angle 4$.
   $\angle 3 = 57^\circ$ Vertical to $\angle 2$. $\angle 6 = 57^\circ$ Corresponding to $\angle 2$
   $\angle 4 = 123^\circ$ Vertical to $\angle 1$. $\angle 7 = 57^\circ$ Vertical to $\angle 6$.
   $\angle 8 = 123^\circ$ Vertical to $\angle 5$.
   (Note: Other reasons may be acceptable.)

2. $\angle 4$ and $\angle 7$ are supplementary angles.

   $2x + (x + 4) = 180^\circ$

   $x = 58\frac{2}{3}^\circ$ $2x = 117\frac{1}{3}^\circ$ $x + 4 = 62\frac{2}{3}^\circ$

3. $\angle 2$ and $\angle 7$ are alternate interior angles.

   $3x - 2 = x + 4$
   $x = 3$ $3x - 2 = 7^\circ$ $x + 4 = 7^\circ$

4. $\angle 3$ and $\angle 7$ are corresponding angles.

   $2x - 10 = x + 4$
   $x = 14$ $x + 4 = 18^\circ$ $2x - 10 = 18^\circ$

5. skew lines

6. alternate interior angles, alternate exterior angles, corresponding angles.

7. 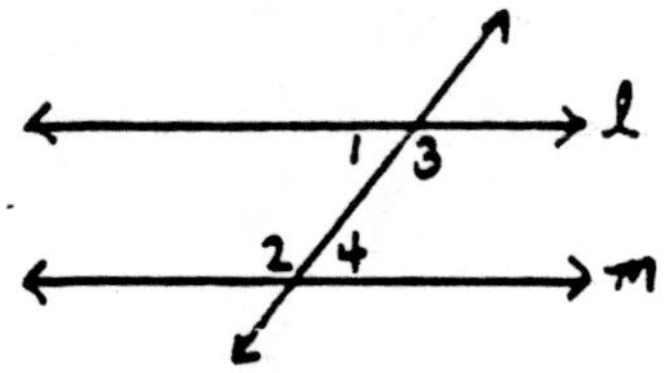 $l \parallel m$ where $\angle 1$ and $\angle 2$ are supplementary angles. $\angle 3$ + $\angle 4$ are supplementary angles.

8. $\overline{AD} = \overline{CB}$ is given.
   $\angle 3 = \angle 2$ If two lines are parallel, then the alternate interior angles are equal.
   $\overline{AC} = \overline{AC}$ by the reflexive property.
   $\triangle ADC \cong \triangle CBA$ by SAS = SAS.

9. $\angle A = \angle C$ If two lines are parallel, then the alternate interior angles are equal.

   $\angle DBA = \angle CBE$ Vertical angles are equal.
   $\overline{AB} = \overline{BC}$ is given.
   $\triangle ADB \cong \triangle CBE$ by ASA = ASA
   $\overline{AD} = \overline{CE}$ by CPCTE.

Exercise 2

1. $\angle BCA = 50^\circ$; $\angle BCD = 130^\circ$

2. 27°, 35°, 118° since $x = 37$

3. $21\frac{1}{9}^\circ$, $42\frac{2}{9}^\circ$, $116\frac{2}{3}^\circ$

4. 58°, 32° since $x = 19$

5. $\angle DAC = 45^\circ$; $\angle DAC$ and $\angle ACB$ are alternate interior angles.
   $\angle CAB = 30^\circ$; $\angle CAB$ and $\angle ACD$ are alternate interior angles.
   $\angle D = 105^\circ$; theorem 5.
   $\angle B = 105^\circ$; theorem 5.

6. $\angle C = 43^\circ$; $\angle B = 94^\circ$ Triangle $\triangle ABC$ is isosceles.

7. $\angle 1 = 60^\circ$
   $\angle 2 = 120^\circ$
   $\angle 3 = 71^\circ$
   $\angle 4 = 109^\circ$
   $\angle 5 = 49^\circ$
   $\angle 6 = 131^\circ$
   $\angle 7 = 49^\circ$
   The exterior angles of $\triangle CGE$ are $\angle 4$, $\angle DEC$, $\angle 2$, $\angle ACG$, $\angle 6$, and $\angle EGH$.

8. Reasons
   a. Given
   b. The interior angles of a triangle sum to 180°
   c. Substitution
   d. Subtraction property
   e. ASA = ASA*

   *See explanation at the bottom of page 120.

Exercise 3

1. $\overline{AC} = 5.4$ cm.; $\overline{BC} = 6$ cm.

2. 40 cm.

3. $\angle T = 47^\circ$, $\overline{LH} = \frac{9}{2}\sqrt{6}$ ft., $\angle ULH = 58^\circ$, $\angle U = 75^\circ$

Exercise 4

1. Only one line through point S is parallel to the given line. (See steps in construction # 1 for drawing.)

2. Postulate 2-If two lines are perpendicular to the same line, then these lines are themselves parallel.

3. See steps in construction #1 for drawing.

Unit 4 Review

1. Equal alternate interior angles, equal alternate exterior angles, equal corresponding angles, and interior angles on the same side of the transversal are supplementary.

2. $\angle 2 = 30^\circ$, $\angle 4 = 60^\circ$, $\angle 5 = 90^\circ$, $\angle 6 = 90^\circ$

3. $\angle 1 = 60^\circ$, $\angle 2 = 30^\circ$, $\angle 3 = 60^\circ$, $\angle 4 = 30^\circ$

4. Refer to body of the unit.

5. Refer to construction # 2.

6. $\angle 1 = 155^\circ$, $\angle 2 = 25^\circ$, $\angle 3 = 25^\circ$, $\angle 4 = 155^\circ$, $\angle 5 = 136^\circ$, $\angle 7 = 44^\circ$, $\angle 8 = 136^\circ$, $\angle 9 = 136^\circ$, $\angle 10 = 44^\circ$, $\angle 11 = 136^\circ$, $\angle 12 = 44^\circ$, $\angle 13 = 155^\circ$, $\angle 15 = 25^\circ$, $\angle 16 = 155^\circ$, $\angle 17 = 111^\circ$, $\angle 18 = 111^\circ$, $\angle 19 = 69^\circ$

7.
   2. Definition of perpendicular.
   3. All right angles are equal.
   4. Lines perpendicular to the same line are parallel.
   5. Definition of a transversal.
   6. Definition of alternate interior angles.
   7. If two lines are parallel, alternate interior angles are equal.
   8. ASA = ASA

8. $2x + 6 = 4x$
   $x = 3$; the angle measures are both 12°.

9. The perimeter of $\Delta STU = 15$ in.

10. Refer to construction # 1.

11. $28\frac{1}{3}^\circ$, 96°, $55\frac{2}{3}^\circ$

12. $(2x - 7) + (3x + 2) = 90^\circ$
$x = 19$; $\angle B = 31^\circ$, $\angle C = 59^\circ$

13. $\angle 9 = 130^\circ$ $\angle 4 = 40^\circ$
$\angle 8 = 50^\circ$ $\angle 3 = 130^\circ$
$\angle 7 = 50^\circ$ $\angle 2 = 90^\circ$
$\angle 6 = 40^\circ$ $\angle 1 = 50^\circ$
$\angle 5 = 50^\circ$

Unit 5 Answers

Exercise 1

1. $\angle B = 86^\circ$, $\angle A = 94^\circ$, $\angle C = 94^\circ$
2. 15 in.
3. 43°
4. side = $18\frac{3}{4}$ m., angle = 16°
5. $\angle A = 1^\circ$, $\angle C = 1^\circ$, $\angle D = 179^\circ$, $\angle B = 179^\circ$
6. a. Given
   b. Def. of parallelogram.
   c. Definition of alternate interior angles.
   d. If two lines are parallel, alternate interior angles are equal.
   e. Reflexive Property
   f. ASA = ASA
   g. CPCTE
7. By the definition of parallelogram, $\overline{AB} = \overline{DC}$.
$(x - 1) = (2x - 9)$
$x = 8$
and $\overline{AB} = \overline{DC} = 7$.
If parallelogram ABCD is a rhombus, then $\overline{BC}$ must equal $\overline{DC}$.
$\overline{BC} = 3x - 17 = 24 - 17 = 7$
Therefore, $\overline{BC} = \overline{DC} = 7$.
By the definition of parallelogram $\overline{BC} = \overline{AD} = 7$. From above, since $\overline{AB} = \overline{DC} = \overline{BC} = \overline{AD} = 7$, the figure is a rhombus.

Exercise 2

1. a. $\overline{EF} = 14$ cm. b. $\overline{BC} = 18$ ft. c. $\overline{WX} = 17$ dm.
2. a. 11 b. 2 c. 25

Unit 5 Review

I. 1. quadrilateral
2. A, B, C, and D
$\overline{DC}$, $\overline{CB}$, $\overline{AB}$, and $\overline{AD}$
opposite; consecutive
3. parallelogram
4. rectangle
5. square
6. rhombus
7. trapezoid; bases
8. isosceles trapezoid
9. median

II. a. 37
b. $x = 4$, $\angle B = 105^\circ$, $\angle A = \angle DAC + \angle CAB = 75^\circ$
c. $\overline{AB} = \overline{DC} = 22$ dm., $\overline{BC} = \overline{AD} = 8$ dm.
d. $x = 61\frac{1}{3}^\circ$, $(x + 3) = 64\frac{1}{3}^\circ$, $(2x - 7) = 115\frac{2}{3}^\circ$
e. $x = 47^\circ$
f. $\overline{AB} = 13$ cm., $\angle 2 = 60^\circ$ since $\overline{AB} \parallel \overline{PQ}$ and $\angle 1$ and $\angle 2$ are corresponding angles.
g. $\overline{AB} = 30$ ft., $\overline{AE} = \overline{BD} = 30$ ft.
h. $x = 2$, $\overline{WR} = 21$, $\overline{YZ} = 7$

Unit 6 Answers

Exercise 1

1.a. Transitive Law
 b. Addition Property-Part 1
 c. Subtraction Property-Part 1
 d. Multiplication Property-Part 1
 e. Multiplication Property-Part 2
 f. Multiplication Property-Part 2
 g. Trichotomy Law
 h. Addition Property-Part 1
 i. Transitive Law
 j. Subtraction Property-Part 2
 k. Addition Property-Part 2
 l. Multiplication Property-Part 2
 m. Trichotomy Law
 n. Subtraction Property-Part 1
 o. Transitive Law
 p. Multiplication Property-Part 1

2.a. $-6 < 5$
 b. $x < y$, or $x = y$, or $x > y$
 c. $-12 > -16$
 d. $(3x + 1) - 1 > 0 - 1$ or $3x > -1$

3.a. $2x + 2 > 0$
 b. $y \geq 0$
 c. $w \leq 0$
 d. $0 < z \leq 78$
 e. $2 < x < 10$
 f. $x^2 > 4$
 g. $-3 < x \leq 7$

Exercise 2

a. $x > -2$
b. $x > -6$
c. $-2 \geq x$
d. $x \geq -4$
e. $x \geq \frac{7}{12}$
f. $x < \frac{-3}{8}$
g. $x \leq -8\frac{2}{5}$
h. $\frac{7}{12} \geq y$
i. $x \geq 1\frac{1}{5}$
j. $x > 0$
k. $x > 0$

Exercise 3

1. $\overline{DC}$ longest; $\overline{BC}$ shortest by Theorem 3
2. $\overline{AB}$ longest; $\overline{BC}$ shortest
3. $\overline{PQ}$
4. $\angle ZNY$ largest; $\angle Y$ is the smallest by Theorem 2
5. Yes, by Theorem 1 and the theorem that gives CBD = $\angle A + \angle C$.

Unit 6 Review

1.a. Transitive Law
b. Addition Property-Part 1
c. Subtraction Property-Part 1
d. Multiplication Property-Part 1
e. Multiplication Property-Part 2
f. Subtraction Property-Part 2

2.a. $-4 < 8$
b. $c < d$, or $c = d$ or $c > d$
c. $12 < 32$

3.a. $3x - 5 > 0$
b. $y \leq 0$
c. $-21 \leq z < 0$
d. $3 < y < 11$

4.a. $x > \frac{2}{5}$
b. $y \leq 15\frac{1}{3}$
c. $x \geq -7$
d. $y < -6$
e. $\frac{-1}{2} < x$
f. $x > 1$
g. $x \leq 3\frac{1}{2}$
h. $3 \leq c$
i. $-6\frac{2}{3} > x$

5.a. $\overline{DC}$ shortest side which is opposite the smallest angle.
$\overline{AC}$ longest side which is opposite the largest angle.
Theorem 3
b. $\overline{AB}$
c. Largest angle is $\angle F$ which is opposite the longest side.
Smallest angle is $\angle E$ which is opposite the shortest side.(Thm.2)
d. By the definition of exterior angle, $\angle 9 = \angle 5 + \angle XYZ$ and $\angle XYZ = 54^\circ$. Since $\angle 5 = 54^\circ$, $\angle YXZ = 72^\circ$, and $\angle XYZ = 54^\circ$, $\angle YXZ$ is the largest angle. By Theorem 3, $\overline{YZ}$ must be longer than $\overline{XZ}$ or $\overline{XY}$.

Unit 7 Answers

Exercises-Radicals Review

I.
1. $4\sqrt{2}$
2. $6\sqrt{2}$
3. $\frac{\sqrt{3}}{4}$
4. $8\sqrt{2}$
5. $\frac{\sqrt{3}}{3}$
6. $\sqrt{7}$
7. $20\sqrt{6}$
8. 18
9. $3\sqrt{5}$
10. $\sqrt{3}$
11. $15\sqrt{3}$
12. $40\sqrt{6}$
13. $\sqrt{5}$
14. $9\sqrt{3}$
15. 360
16. $30\sqrt{6} + 48$
17. $\frac{11}{4}\sqrt{3} + 3$
18. $140\sqrt{3}$
19. $256\sqrt{3}$
20. $\sqrt{64}\sqrt{7} = 8\sqrt{7}$

Exercises-Quadratic Equations Review

I.
1. $\{3, 5\}$
2. $\left\{\frac{1}{3}, \frac{5}{4}\right\}$
3. $\left\{0, -\frac{2}{5}\right\}$
4. $\left\{\frac{3}{2}, -\frac{10}{3}\right\}$
5. $\left\{-\frac{3}{2}, \frac{3}{2}\right\}$
6. $\left\{2, -\frac{3}{2}\right\}$
7. $\left\{\frac{3}{2}, \frac{3}{2}\right\}$
8. $\left\{-\frac{5}{2}, \frac{2}{5}\right\}$

II.
1. $\{-1, -5\}$
2. $\left\{\frac{\sqrt{14}}{2}, -\frac{\sqrt{14}}{2}\right\}$
3. $\left\{\frac{3}{2}, 1\right\}$
4. $\{-1 + 2\sqrt{3}, -1 - 2\sqrt{3}\}$
5. $\left\{\frac{\sqrt{21}}{3}, -\frac{\sqrt{21}}{3}\right\}$
6. $\left\{\frac{1 + \sqrt{19}}{3}, \frac{1 - \sqrt{19}}{3}\right\}$
7. $\left\{\frac{-2 + \sqrt{2}}{2}, \frac{-2 - \sqrt{2}}{2}\right\}$
8. $\left\{\frac{1}{2}, \frac{1}{2}\right\}$

Exercise 1

1. a. 120 sq. in.
   b. 120 sq. ft.
   c. 7 sq. yd.
   d. 53.9 sq. m.

2. a. 2 in.
   b. $\frac{\sqrt{14}}{5}$ cm.
   c. $1\frac{1}{3}$ mm.

Exercise 1 Continued

3. a. 12 sq. in.

   b. 13 sq. in.

   c. $9\frac{3}{8}$ sq. in.

4. 12 ft., 15 ft.

5. 288 ft.

6. 25 sq. ft.

7. $2\frac{1}{4}$ sq. in.

Exercise 2

1. a. 24 sq. in.

   b. $22\frac{2}{3}$ sq. ft.

   c. $126\sqrt{10}$ sq. yd.

   d. $25\frac{2}{3}$ sq. cm.

   e. 205 sq. dm.

2. a. 8 m.

   b. 17 cm.

   c. $3\frac{1}{3}$ mi.

   d. 570 m.

Exercise 3

1. a. 30 sq. in.

   b. $7\frac{1}{2}$ sq. ft.

   c. $\frac{\sqrt{35}}{2}$ sq. cm.

   d. 2 sq. m.

   e. $\frac{1}{6}$ sq. mi.

2. a. 8 ft.

   b. $\frac{28\sqrt{3}}{3}$ in.

   c. 9 yd.

3. 18.92 sq. m.

4. $90 = \frac{1}{2}x(x + 3)$
   base = 15 in.

5. 16 cm.

6. 24 tiles

Exercise 4

1. a. 91 sq. m.

   b. 350 sq. cm.

   c. 120 sq. mm.

   d. $2\frac{17}{24}$ sq. in.

2. a. 5 mi.

   b. 16 m.

3. 6 ft., 4 ft.

4. 42 ft.

5. $40 = \frac{1}{2}x\left[(2x - 1) + (x)\right]$
   Altitude = $5\frac{1}{3}$ cm.

Exercise 5

1\.

	Area	Circumference
a.	225π m.^{2}	30π m.
b.	12π cm.^{2}	$4\sqrt{3}\pi$ cm.
c.	$\frac{9}{4}\pi$ ft.^{2}	3π ft.

2\. a. 12 m.

b. $22\frac{1}{2}$ cm.

c. 5 m.

d. $4\sqrt{3}$ in.

3\. $A_{track} = 16\pi$ sq. ft. + 160 sq. ft.

Unit 7 Review

1\. a. 9 sq. in.

b. 16 sq. yd.

c. 38 sq. m.

d. 96 sq. m.

e. 16 sq. m.

2\. a. $2\sqrt{3}$ in.

b. $8\sqrt{3}$ in.

c. 21 m.

3\. a. 6 cm.

b. 5 ft.

c. $40\frac{1}{2}$ sq. in.

4\. a. $7 + \sqrt{69}$

b. 42 sq. ft.

c. $12x^2 + 35 = (4x + 1)(3x + 2)$
base = 13
altitude = 11

5\. a. 140 sq. in.

b. 2 in., 4 in.

c. $1\frac{2}{9}$ ft.

6\.

	Circumference	Area
a.	$4\frac{2}{3}\pi$ cm.	$\frac{49}{9}\pi$ cm.^{2}
b.	$10\sqrt{2}\pi$ ft.	50π ft.^{2}
c.	1.6π dm.	$.64\pi$ dm.^{2}
d.	12π in.	36π in.^{2}

7\. a. 3 m.

b. 5 ft.

c. 4.7 cm.

d. $8\sqrt{2}$ in.

8\. Area of the field =
77 ft.^{2} + $\frac{49}{8}\pi$ ft.^{2}

9\. a. circle, radius, center
b. diameter
c. circumference

Unit 8 Answers

Exercise 1

1. a. True

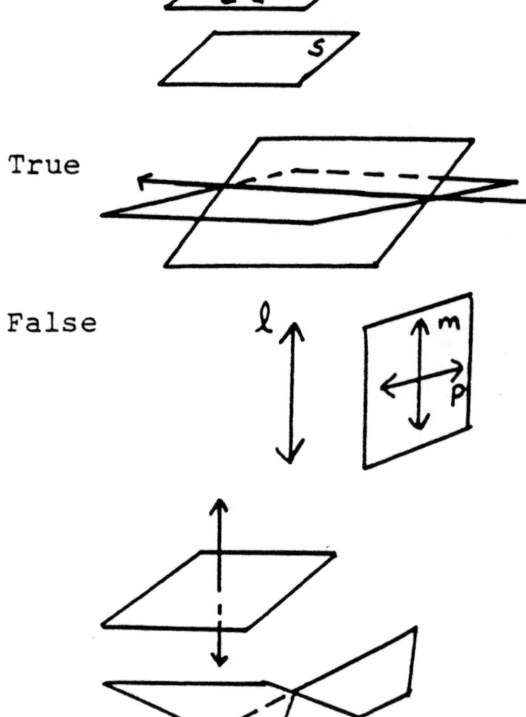

Both lines are parallel to the same plane S.

b. True

c. False

Line l is parallel to line M but line l is not parallel to line p.

2. a.

b.

c.

d.

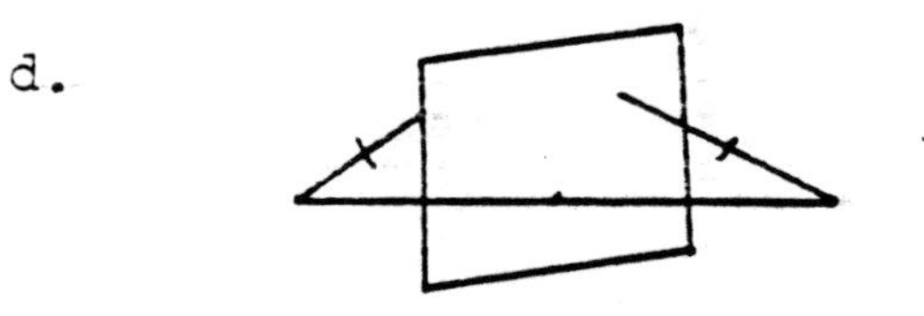

3. a. non collinear

b. a point not on a line

4. a. a straight line

b. a point

Exercise 2

1. yes

2. a. yes

   b. ice cubes, toy blocks, boxes, etc.

3. no

4. circle

Exercise 3

1. a. 140 cu. ft.

   b. 60 cu. m.

   c. $19\frac{1}{8}$ cu. cm.

2. 4 dm.

3. 2 ft.

4. 6 in.

5. 1,875 cu. ft.

Exercise 4

1. a. 8 cu. in.

   b. $28\sqrt{7}$ cu. m.

2. a. $6\frac{2}{3}$ in.

   b. $3\frac{5}{21}$ cm.

3. $12\frac{6}{7}$ in.

Exercise 5

1. a. 300π cu. in.
   b. 360π cu. yd.

2. a. 10 in.
   b. 8 cm.
   c. 12 yd.
   d. 4 ft.
   e. $\sqrt{3}$ m.

3. 12 ft.

Exercise 6

1. a. 108π cu.m.

   b. $81\frac{2}{3}\pi$ cu.in.

2. a. 5 in.

   b. 6 ft.

   c. 12 yd.

   d. 5 m.

   e. 3 cm.

3. 12π cubic inches

Unit 8 Review

1. a. 245π cu.in.

   b. 27π cu. cm.

   c. 972π cu.dm.

2. a. $10\frac{1}{8}$ yd.

   b. $\sqrt{2}$ m.

   c. 14 ft.

3. Refer to body of text.

4. a.

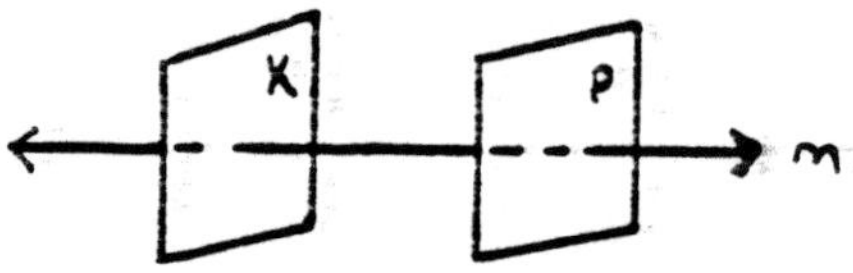

   K P m

   b.

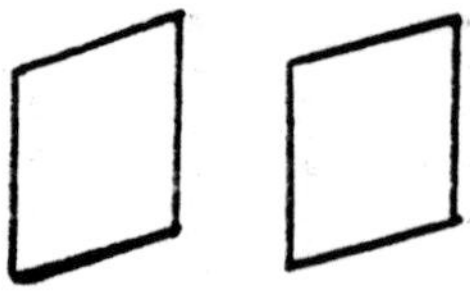

   c.

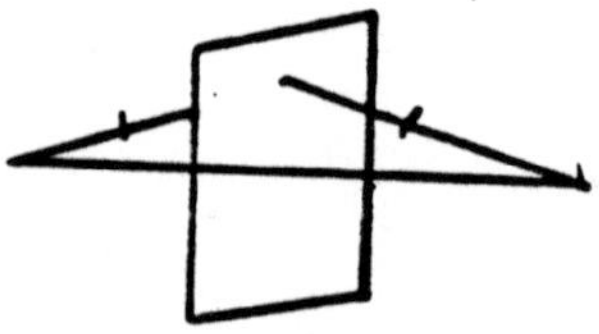

5. a. straight line
   b. a circle
   c. a point

Unit 8 Review Continued

6. $r = 5$ ft., $V = 375\pi$ cu. ft.

7. $V = 3\frac{3}{4}\pi$ cu. in.

8. a. $V = 262\frac{1}{2}\pi + 630$ cu. ft.

   b. 98 cu. in.

9. Volume of sphere $= 10\frac{2}{3}\pi$ cu. in.

   Volume of glass $= 37\frac{1}{2}\pi$ cu. in.

   Volume of water needed $= 37\frac{1}{2}\pi - 10\frac{2}{3}\pi = 26\frac{5}{6}\pi$ cu. in.

10. 5400 cu. in.

Unit 9 Answers

Exercise 1

1. a. $3^2 + 4^2 = 5^2$

   $9 + 16 = 25$

   Right triangle

   b. $1^2 + \left(1\frac{1}{3}\right)^2 = \left(1\frac{2}{3}\right)^2$

   $1^2 + \left(\frac{4}{3}\right)^2 = \left(\frac{5}{3}\right)^2$

   $1 + \frac{16}{9} = \frac{25}{9}$

   $\frac{9}{9} + \frac{16}{9} = \frac{25}{9}$

   Right triangle

   c. $(2\sqrt{2})^2 + (3\sqrt{2})^2 = (4\sqrt{2})^2$

   $8 + 18 = 32$

   $26 = 32$

   Not a right triangle

   d. $8^2 + 15^2 = 17^2$

   $64 + 225 = 289$

   $289 = 289$

   Right triangle.

2. a. $2\sqrt{5}$ cm. $= c$ b. $b = 3\sqrt{3}$ m. c. $a = \sqrt{19}$ in.

3. side $= 5\sqrt{2}$ ft. area $= 25$ sq. ft.

4. $4\sqrt{7}$ ft.

Exercise 2

1. 2 inches, $2\sqrt{3}$ inches

2. $\frac{7\sqrt{3}}{3}$ inches, $\frac{14\sqrt{3}}{3}$ inches

3. $\frac{49\sqrt{3}}{6}$ sq. in.
4. altitude = $\frac{3\sqrt{3}}{2}$ meters, area = $\frac{9}{4}\sqrt{3}$ sq. m.
5. x = 5 ft., area = 45 sq. ft.

Exercise 3

1. $13\sqrt{2}$ cm.
2. $4\sqrt{2}$ ft.
3. x = $3\sqrt{2}$ m. = one side, 9 sq. m. = area
4. x = 41 inches
5. x = 7 ft.
6. area = $\frac{529}{2}$ sq. dm.

Exercise 4

1. altitude = $6\sqrt{3}$ meters; area = $36\sqrt{3}$ sq. m.
2. s = $\frac{16\sqrt{3}}{3}$ in.
3. altitude = $\frac{5\sqrt{3}}{2}$ ft.

Unit 9 Review

I. 1. $a^2 + b^2$
2. x, $x\sqrt{3}$; where the hypotenuse = 2x
3. $x\sqrt{2}$
4. $\frac{1}{2}x\sqrt{3}$

II. 1. no
2. yes.
3. yes
4. yes
5. yes
6. yes

III. 1. $\sqrt{89}$
2. 2
3. $\frac{\sqrt{73}}{12}$
4. $2\sqrt{3}$
5. .5
6. 8

IV. 1. 5 inches
2. $\frac{7}{2}\sqrt{3}$ ft.
3. $10\sqrt{2}$ cm.
4. 16 meters
5. $5\sqrt{2}$ ft.
6. $169\sqrt{3}$ sq. in.
7. $\frac{9}{2}\sqrt{2}$ cm., $\frac{9}{2}\sqrt{6}$ cm.

V. 1. x = 30
10 ft. = height
A = 580 sq. ft.
2. 6 yd.
Area = 240 sq. yd.
3. $\sqrt{589}$ ft. = x

Unit 10 Answers

Exercise 1

1. a. ratio
   b. proportion
   c. constant of proportionality

2. a. $\frac{4}{3}$

   b. $\frac{3}{7}$

   c. $\frac{1}{80}$

   d. $\frac{10}{1}$

3. a. No
   b. Yes, both rations reduce to 2.
   c. Yes, both ratios reduct to $\frac{3y}{x}$.
   d. No

4. a. $\frac{1}{3}$

   b. none

   c. $\frac{3}{2}$

Exercise 2

1. a. property 4
   b. property 3
   c. property 1
   d. property 2

2. a. 20

   b. $4\frac{1}{2}$

   c. $\frac{3}{16}$

3. a. 4
   b. $2\sqrt{21}$

4. a. $19\frac{4}{5}$
   b. -20
   c. -23
   d. 6, -7
   e. $-4 \pm \sqrt{21}$

Exercise 3

1. Refer to definitions in text.

2. CD = 12.5 in.
   ED = 7.5 in.

3. $\frac{3}{12} = \frac{5}{20} = \frac{7}{28} = k$

   Yes, $k = \frac{1}{4}$.

4. $\frac{8}{3} = \frac{x}{48}$

   x = 128 ft.

5. $\frac{5}{6} = \frac{x}{30}$

   x = 25 ft.

6. $\frac{11}{y} = \frac{15}{30}$

   y = 22 in.

   and

   $\frac{13}{z} = \frac{15}{30}$

   z = 26 in.

7. $\frac{\overline{SR}}{\overline{QS}} = \frac{\overline{TR}}{\overline{PT}}$

   $\frac{18}{10} = \frac{126}{x}$

   x = 70 ft.

Exercise 4

1. No, the proportion should be $\frac{\overline{WT}}{8} = \frac{5}{11}$.

2. $\frac{10}{14} = \frac{\overline{BE}}{15}$ $\qquad$ $\overline{EC} = 15 - 10\frac{5}{7} = 4\frac{2}{7}$ in.

   $10\frac{5}{7}$ in. $= \overline{BE}$ $\qquad$ $\angle A = 25^\circ$

3. $\overline{DE}$ and $\overline{AC}$ are parallel since the corresponding angles formed by transversal $\overline{AB}$ are equal. Therefore, $\angle BED = 80^\circ$ because it is corresponding to $\angle C$. $\triangle BDE$ and $\triangle BAC$ can be proved to be similar by Theorem 3, page 277.

$$\frac{4}{9} = \frac{x}{7}$$

$$3\frac{1}{9} = x$$

4. a. $c = 21$ m.

   b. $4\frac{2}{3}$ m.

   c. $a = 13\frac{11}{13}$

Unit 10 Review

1. a. ratio
   b. proportion
   c. extremes, means, fourth proportional
   d. similar, $\sim$
   e. mean proportion

2. Yes, the sides are proportional. $\frac{3}{12} = \frac{5}{20} = \frac{7}{28} = k,\ k = \frac{1}{4}$.

3. $\overline{AC} = 4$ cm. $\qquad$ $\overline{DE} = 7\frac{1}{2}$ cm.

4. $\frac{3}{9} = \frac{17}{d}$
   $d = 51$

5. $\frac{11}{x} = \frac{x}{21}$
   $x = \sqrt{231}$

6. a. $3\frac{2}{3}$ $\qquad$ d. 2

   b. $-6 \pm \sqrt{7}$ $\qquad$ e. 1, -2

   c. $3\sqrt{30}$ $\qquad$ f. -30

7. a. No
   b. Yes

8. $\frac{7}{120}$

9. 49 ft.

10. $x = 9.6$ ft or $9\frac{3}{5}$ ft.

    $y = 14.4$ ft or $14\frac{2}{5}$ ft.

11. $\overline{DC} = 15$ ft., $\overline{AC} = 16$ ft., $\overline{CE} = 24$ ft., $\angle CBD = 40^\circ$, $\angle E = 60^\circ$

    $\angle C = 80^\circ$

12. $\overline{MY} = 14$ in., $\overline{YO} = 7$ in., $\overline{MN} = 24$ in.

UNIT 11 ANSWERS

Exercise 1

1. $\cos \angle H = \frac{8}{17}$
2. $\sin \angle O = \frac{3}{5}$
3. $\tan \angle H = \frac{15}{8}$
4. $\sin \angle T = \frac{8}{17}$
5. $\tan \angle O = \frac{3}{4}$
6. $\cos \angle W = \frac{12}{15} = \frac{4}{5}$
7. $\tan \angle W = \frac{9}{12} = \frac{3}{4}$
8. $\cos \angle M = \frac{3}{5}$
9. $\sin \angle V = \frac{12}{15} = \frac{4}{5}$
10. $\cos \angle V = \frac{9}{15} = \frac{3}{5}$
11. $\tan \angle T = \frac{8}{15}$
12. $\sin \angle M = \frac{4}{5}$

Exercise 2

1. a) $\cos \angle Z = \frac{2}{\sqrt{10}} = \frac{\sqrt{10}}{5}$

   b) $\tan \angle X = \frac{2}{\sqrt{6}} = \frac{\sqrt{6}}{3}$

   c) $\sin \angle Z = \frac{\sqrt{6}}{\sqrt{10}} = \frac{\sqrt{15}}{5}$

2. $\tan 40^\circ = .839$; $\sin 72^\circ = .951$; $\cos 34^\circ = .829$

3. $\angle M = 21^\circ$; $\angle N = 87^\circ$; $\angle P = 1^\circ$

4. $\cos \angle Z = \frac{\sqrt{10}}{5} = .632$ $\tan \angle X = \frac{\sqrt{6}}{3} = .816$

   $\angle Z = 51^\circ$ $\angle X = 39^\circ$

Exercise 3

1. a. $\sin 23^\circ = \frac{h}{\overline{MN}}$

   5.474 in.= h

   $A = \frac{1}{2}(39)(5.474) = 106.743$ sq. in.

   b. $\angle A + \angle B + \angle C = 180$

   $x + 70 + x = 180$

   $2x = 110$

   $x = 55^\circ$

   $\sin 55^\circ = \frac{h}{\overline{AB}}$

   9.009 ft. = h

   $A = \frac{1}{2}(9)(9.009) = 40.541$ sq.ft.

   c. $\sin 17^\circ = \frac{h}{\overline{RU}}$

   1.752 cm. = h

   $A = (1.752)(5) = 8.76$ sq. cm.

2. $\cos 48^\circ = \frac{175}{\overline{XZ}}$

   $\overline{XZ} = 261.58$ ft.

Unit 11 Review

1. a) a d) b

   b) a e) b

   c) c

2. $\sin 29^\circ = \frac{a}{c}$ $\cos 29^\circ = \frac{b}{c}$ $\tan 29^\circ = \frac{a}{b}$

   $\sin 61^\circ = \frac{b}{c}$ $\cos 61^\circ = \frac{a}{c}$ $\tan 61^\circ = \frac{b}{a}$

3. $\sin 29^\circ = .485$ $\cos 29^\circ = .875$ $\tan 29^\circ = .554$

   $\sin 61^\circ = .875$ $\cos 61^\circ = .485$ $\tan 61^\circ = 1.804$

4. $\cos \angle M = \frac{8}{17} = .471$ $\sin \angle N = \frac{8}{17} = .471$

$\tan \angle N = \frac{8}{15} = .533$ $\angle N = 28^\circ$ (since $\tan \angle N = .533$)

$\angle M = 62^\circ$ (since $\cos \angle M = .471$)

5. a. $\sin 15^\circ = \frac{h}{10}$ $A = (18)(2.59) = 46.62$ sq. ft.

2.59 ft. = h

b. $\sin 37^\circ = \frac{h}{3}$ $A = \frac{1}{2}(5)(1.806) = 4.515$ sq. cm.

1.806 cm. = h

c. $\sin 43^\circ = \frac{h}{6}$ $A = \frac{1}{2}(7 + 15)(4.092) = 45.012$ sq.ft.

4.092 ft. = h

6. $\cos \angle B = \frac{\overline{BC}}{\overline{BA}}$

$\cos 25^\circ = \frac{\overline{BC}}{200}$

181.2 ft. = $\overline{BC}$

Unit 12 Answers

Exercise 1

1. A diameter has endpoints on the circle but its segment also passes through the center of the circle whereas other chords do not pass through the center.

2. A diameter is formed by two radii placed end to end. The measure of a diameter therefore equals the measure of two radii.

3. A chord is contained within a secant line. A secant is a line that passes through a circle intersecting it in two points. However, a chord is a line segment that has its endpoints at the intersection of the secant and the circle.

4. A secant line intersects a circle in two points and a tangent line has only one point of intersection.

5. a. center-point O
   b. radii-$\overline{OH}$, $\overline{OI}$, $\overline{OD}$
   c. diameter-$\overline{ID}$
   d. chords-$\overline{AB}$, $\overline{JC}$, $\overline{ID}$
   e. secant-$\overleftrightarrow{KL}$
   f. tangent-$\overleftrightarrow{GE}$
   g. point of tangency-point F

Exercise 2

1. $\angle AOB = 78^\circ$, $\angle ACB = 39^\circ$ minor arcs: $\widehat{AB}$, $\widehat{BC}$, $\widehat{AC}$; major arcs: $\widehat{ABC}$, $\widehat{ACB}$, $\widehat{CAB}$.
2. $\widehat{DC} = 68^\circ$, $\angle DOE = 180^\circ$, $\widehat{CE} = (180 - 68^\circ) = 112^\circ$, $\angle COE = 112^\circ$
3. $\widehat{AC} = 102^\circ$
4. $\widehat{AC} = \widehat{AB} = \widehat{BC} = 120^\circ$
5.

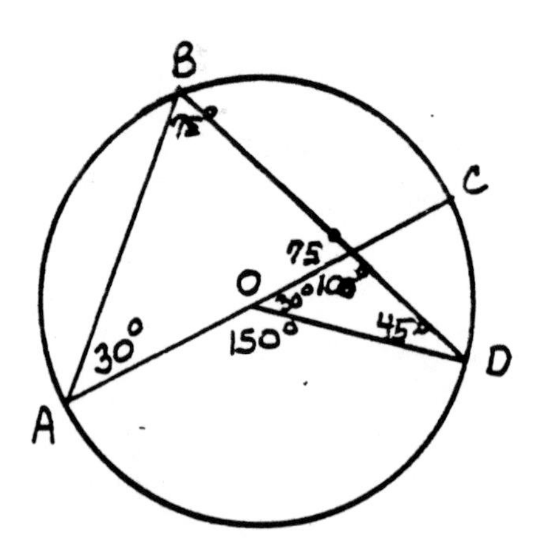

$\angle ABD = 75^\circ$
$\angle BDO = 45^\circ$
$\widehat{BC} = 60^\circ$
$\widehat{CD} = 30^\circ$
$\widehat{AD} = 150^\circ$
$\widehat{AB} = 360^\circ - (\widehat{BC} + \widehat{CD} + \widehat{DA}) = 120^\circ$

Exercise 3

1.

	Arc Length	Area of the Sector
a.	5π ft.	25π ft.^{2}
b.	$\frac{10\sqrt{2}}{9}\pi$ cm.	$\frac{40}{9}\pi$ cm.^{2}
c.	$\frac{3}{10}\pi$ m.	$\frac{27}{40}\pi$ m.^{2}

2. central angle = 60°

3. central angle = 120°

4. length of arc, $\frac{5}{8}\pi$ in.; area of sector, $\frac{45}{64}\pi$ sq. in.

5. $A_{patio} = A_{rectangle} - A_{sector}$

   A_{patio} = 154 sq. ft. $- 1\pi$ sq. ft.

Exercise 4

1. $\overline{AC}$ = 8 m.
   $\overset{\frown}{EC} = 70^\circ$
   $\overset{\frown}{AC} = 140^\circ$

2. $\overline{AC}$ = 4.6 ft.
   $\overset{\frown}{EC} = 42^\circ$
   $\overset{\frown}{AE} = 42^\circ$
   $\overset{\frown}{AC} = 84^\circ$
   $\angle OCD = 48^\circ$

3. $(\overline{EB})^2 = (\overline{OB})^2 - (\overline{OE})^2$

   $(\overline{EB})^2 = 36$

   $\overline{EB}$ = 6 in.

   $\overline{AB}$ = 12 in.

   $\overline{EC} = \overline{OC} - \overline{OE}$

   $\overline{EC}$ = 2 in.

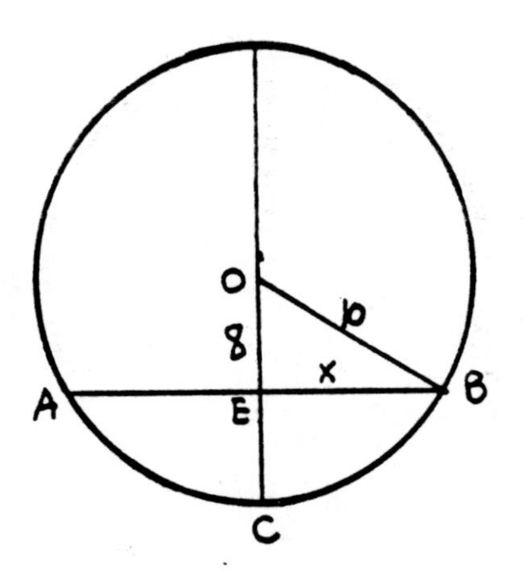

4. $\overline{OY}$ = 4 ft.

5. $\overline{CD}$ = 2.5 mm.

Unit 12 Review

I. 1. circle, radius, center
2. diameter
3. chord
4. tangent, point of tangency
5. secant
6. central angle
7. inscribed angle
8. sector
9. circumcenter
10. incenter

II. 1. a. point O
b. $\overline{OD}$, $\overline{OE}$, $\overline{OI}$
c. $\overline{DI}$
d. $\overline{AL}$, $\overline{DI}$, $\overline{JB}$
e. $\overline{KC}$
f. $\overline{FH}$
g. point G

2. $\overset{\frown}{AB} = 58^\circ$
$\overset{\frown}{AC} = 180^\circ$
$\overset{\frown}{BC} = 122^\circ$

major arc = $\overset{\frown}{ACB}$
minor arc = $\overset{\frown}{AB}$ or $\overset{\frown}{BC}$
semicircle = $\overset{\frown}{ABC}$ or $\overset{\frown}{AC}$

3. $\overset{\frown}{CD} = 30^\circ$, $\overset{\frown}{AB} = 80^\circ$, $\angle BAD = 45^\circ$, $\angle 2 = 125^\circ$, $\angle CBD = 15^\circ$, $\overset{\frown}{AC} = 160^\circ$, $\angle ABD = 95^\circ$, $\angle ABC = 80^\circ$

4.

		A_{sector}
a.	$\frac{3}{2}\pi$ ft.	$\frac{9}{2}\pi$ ft.^{2}
b.	$\frac{20\sqrt{2}}{9}\pi$ in.	$\frac{100}{9}\pi$ in.^{2}
c.	$\frac{5}{8}\pi$ ft.	$\frac{45}{16}\pi$ ft.^{2}

5. $\overline{DE}$ = 4 in., $\overline{CD}$ = 8 in., $\overset{\frown}{BD} = 50^\circ$, $\overset{\frown}{CD} = 100^\circ$, $\overset{\frown}{CF} = 80^\circ$, $\overset{\frown}{AD} = 130^\circ$
$\overset{\frown}{AF} = 50^\circ$

6. $\overline{OT}$ = 9 yd.

7. $\overline{CD} = 2\sqrt{3}$ mm.

8. See Construction # 1

9. See Construction # 2

Unit 13 Answers

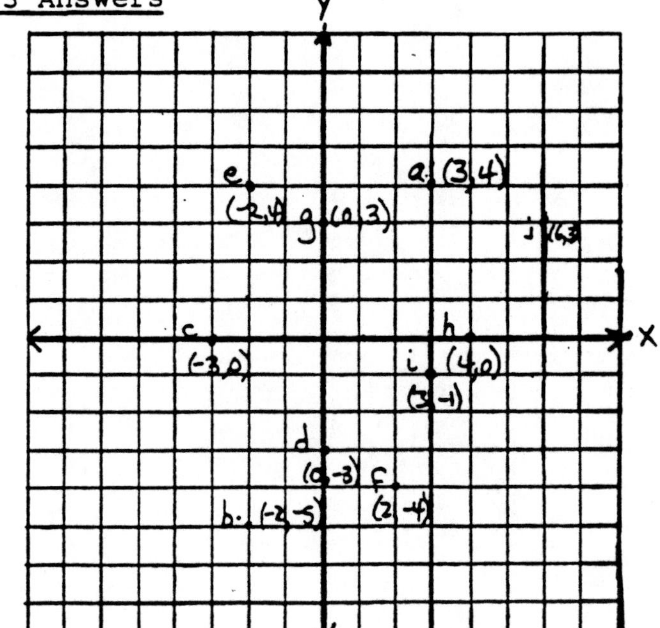

Exercise 1

1. a. Quadrant I
   b. Quadrant III
   c. Negative x-axis
   d. Negative y-axis
   e. Quadrant II
   f. Quadrant IV
   g. Positive y-axis
   h. Positive x-axis
   i. Quadrant IV
   j. Quadrant I

2. a. $\frac{2}{5}$
   b. -1
   c. $\frac{-4}{3}$
   d. no slope
   e. 0

3. a. positive
   b. no slope
   c. negative
   d. zero

Exercise 2

1. a.

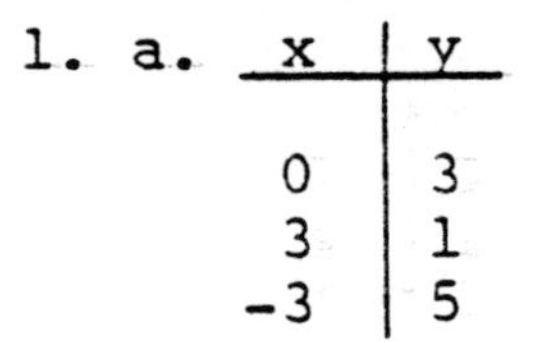

x	y
0	3
3	1
-3	5

b.

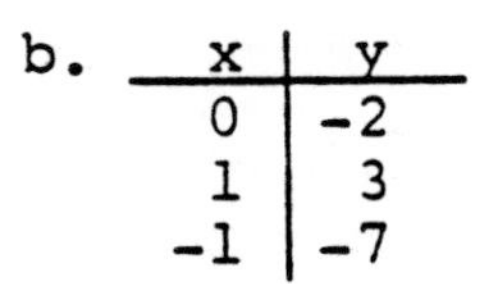

x	y
0	-2
1	3
-1	-7

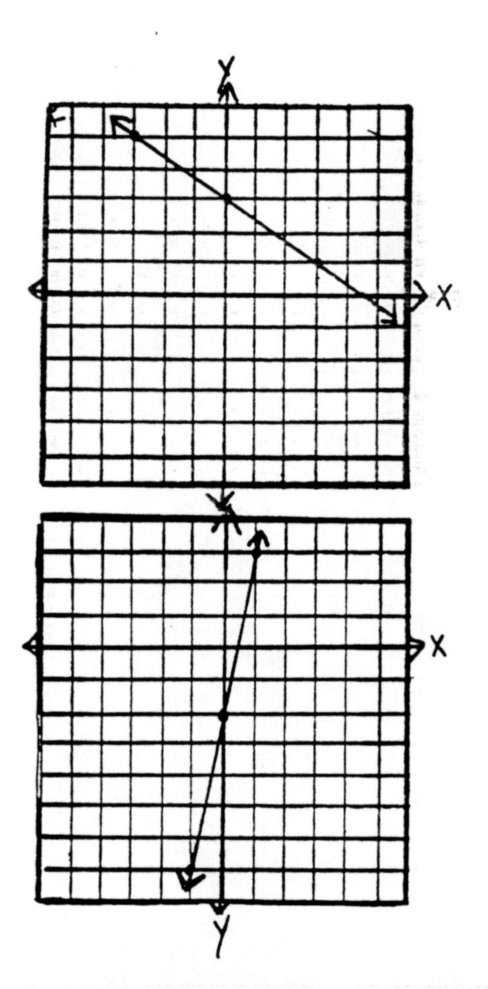

c.

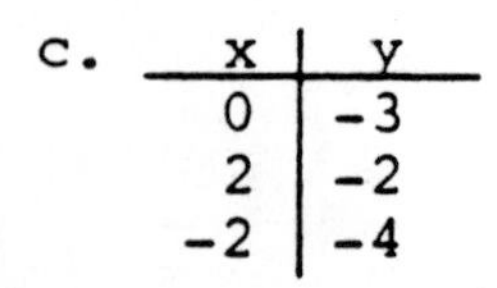

x	y
0	-3
2	-2
-2	-4

d.

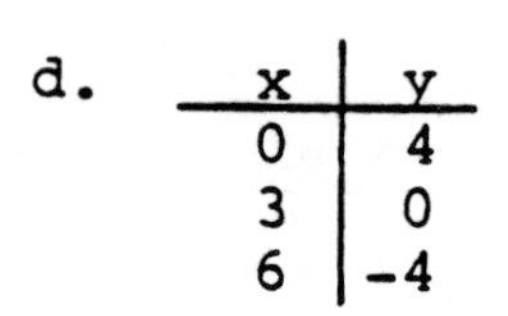

x	y
0	4
3	0
6	-4

e.

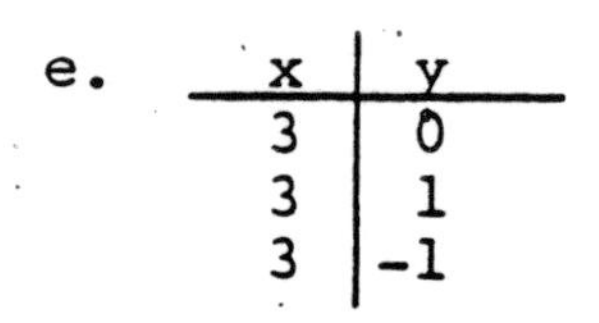

x	y
3	0
3	1
3	-1

f.

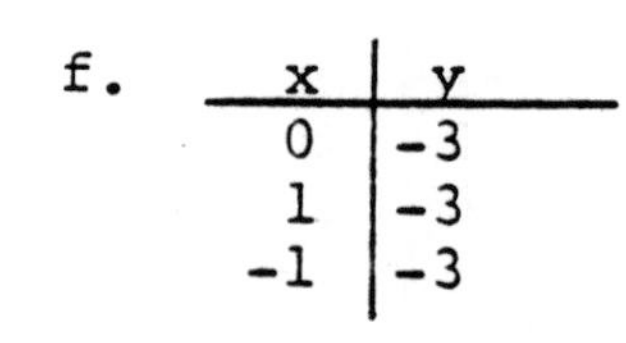

x	y
0	-3
1	-3
-1	-3

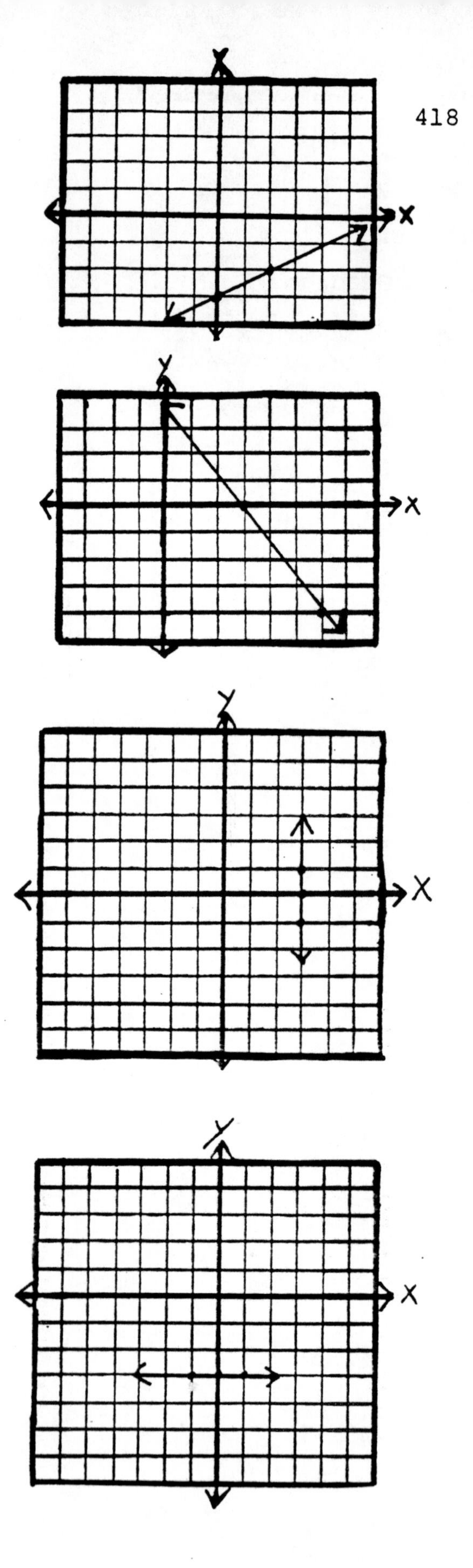

2. a. slope = $\frac{-5}{2}$, y-intercept, 5

b. slope = $\frac{2}{3}$, y-intercept, -3

c. no slope, no y-intercept, x-intercept is 4.

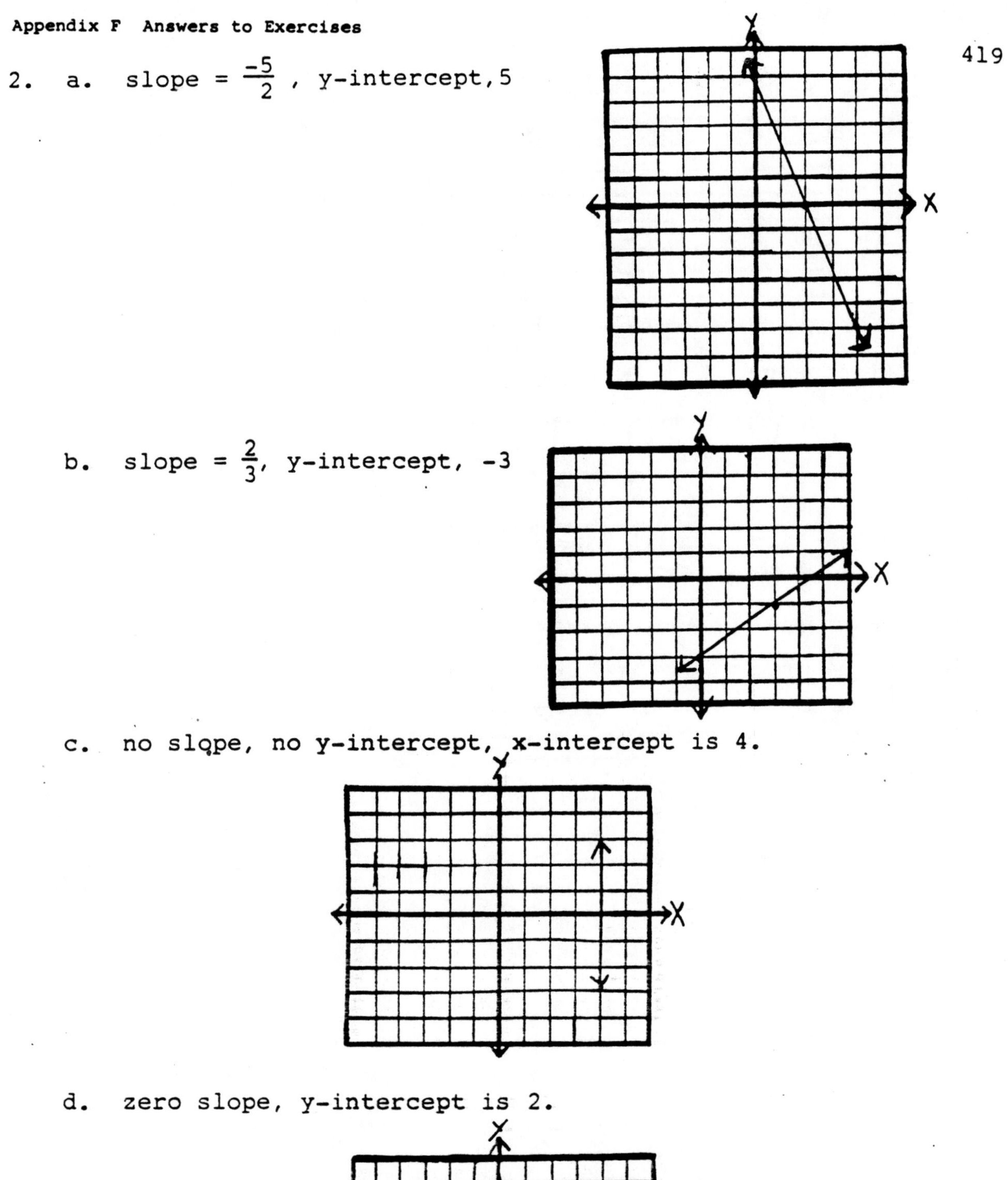

d. zero slope, y-intercept is 2.

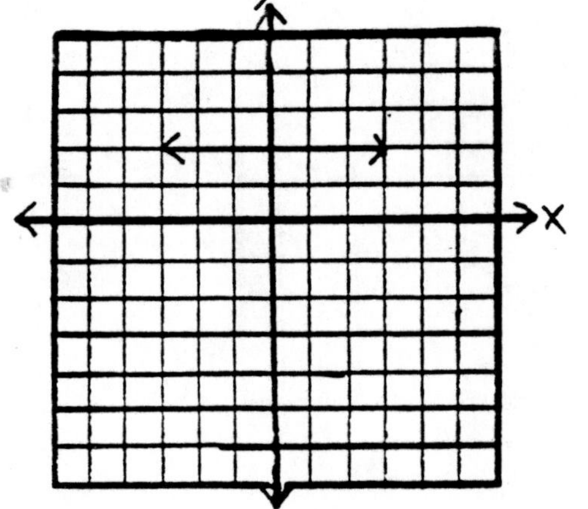

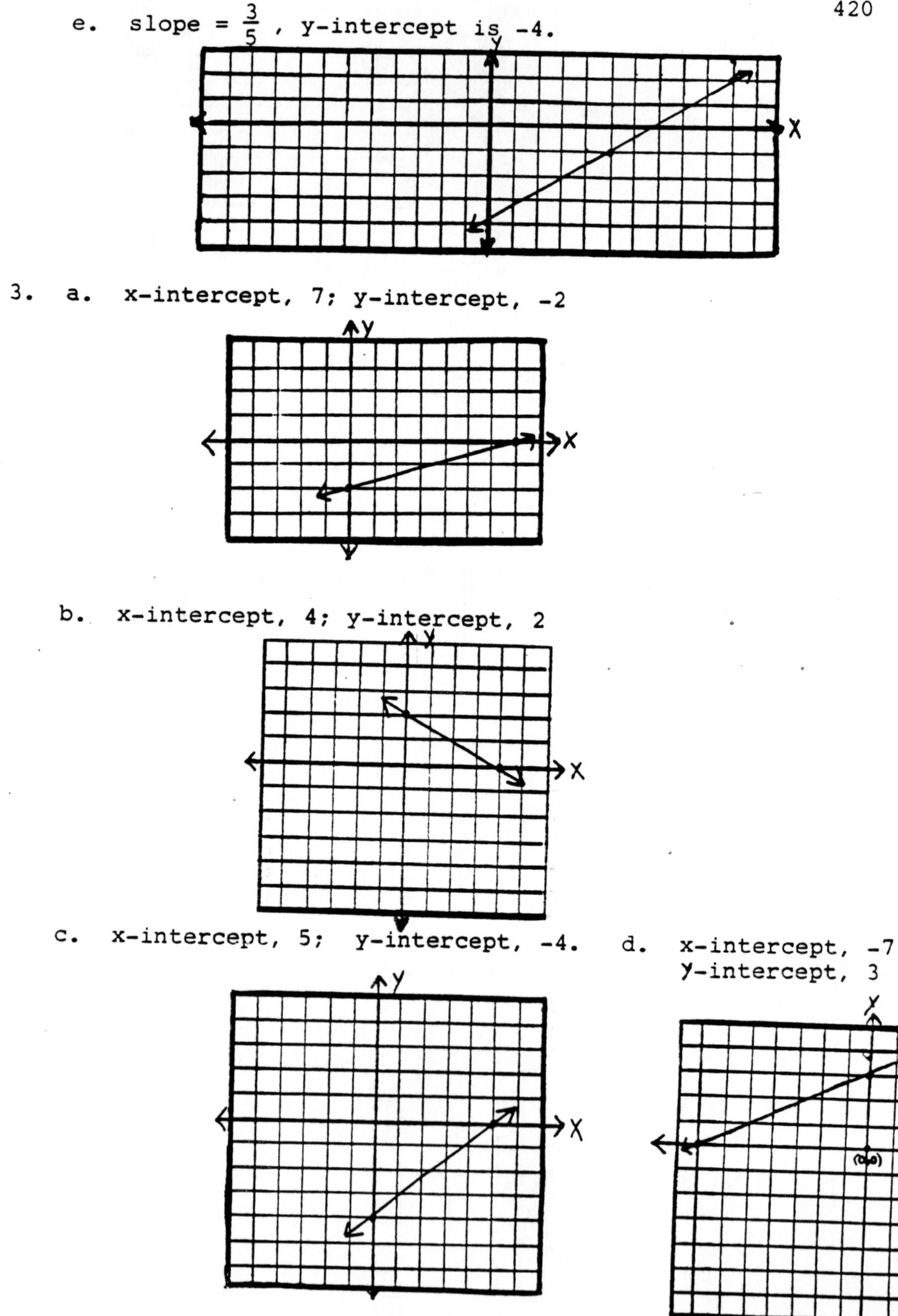

e. slope = $\frac{3}{5}$, y-intercept is -4.

3. a. x-intercept, 7; y-intercept, -2

b. x-intercept, 4; y-intercept, 2

c. x-intercept, 5; y-intercept, -4.

d. x-intercept, -7
y-intercept, 3

Exercise 3

1. a. $x + 5y = 23$
   b. $x = 4$
   c. $y = 3$
   d. $5x - y = 18$
   e. $2x + 3y = -22$
   f. $2x - 7y = -27$
   g. $4x - y = 0$
   h. $5x - 3y = 15$
   i. $3x + y = -5$
   j. $2x + 3y = -8$

Exercise 4

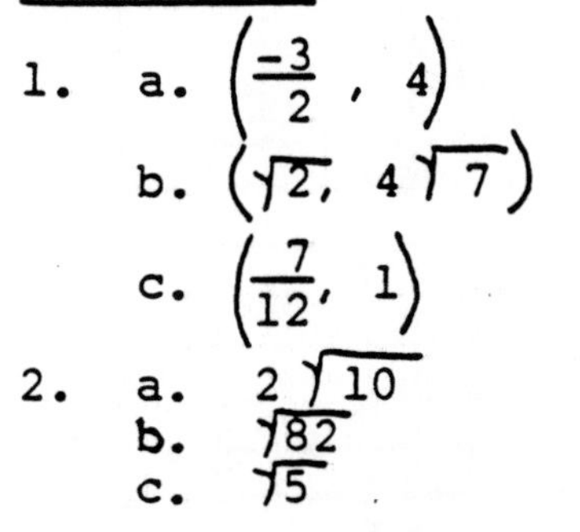

1. a. $\left(\frac{-3}{2}, 4\right)$
   b. $\left(\sqrt{2}, 4\sqrt{7}\right)$
   c. $\left(\frac{7}{12}, 1\right)$
2. a. $2\sqrt{10}$
   b. $\sqrt{82}$
   c. $\sqrt{5}$

3. Perimeter = $5 + \sqrt{13}$

Unit 13 Review

1. Refer to body of the unit for the definitions.

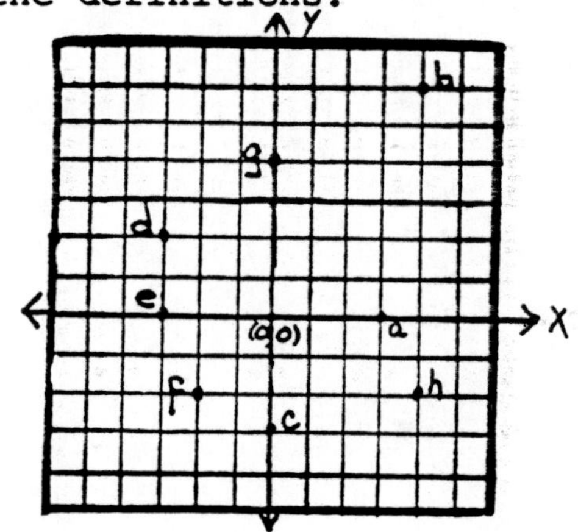

2. a. Positive x-axis
   b. Quadrant I
   c. Negative y-axis
   d. Quadrant II
   e. Negative x-axis
   f. Quadrant III
   g. Positive y-axis
   h. Quadrant IV

3. $\frac{4}{9}$

4. a. negative
   b. no slope
   c. 0 slope
   d. positive

5.

x	y
0	-6
2	-3
4	0

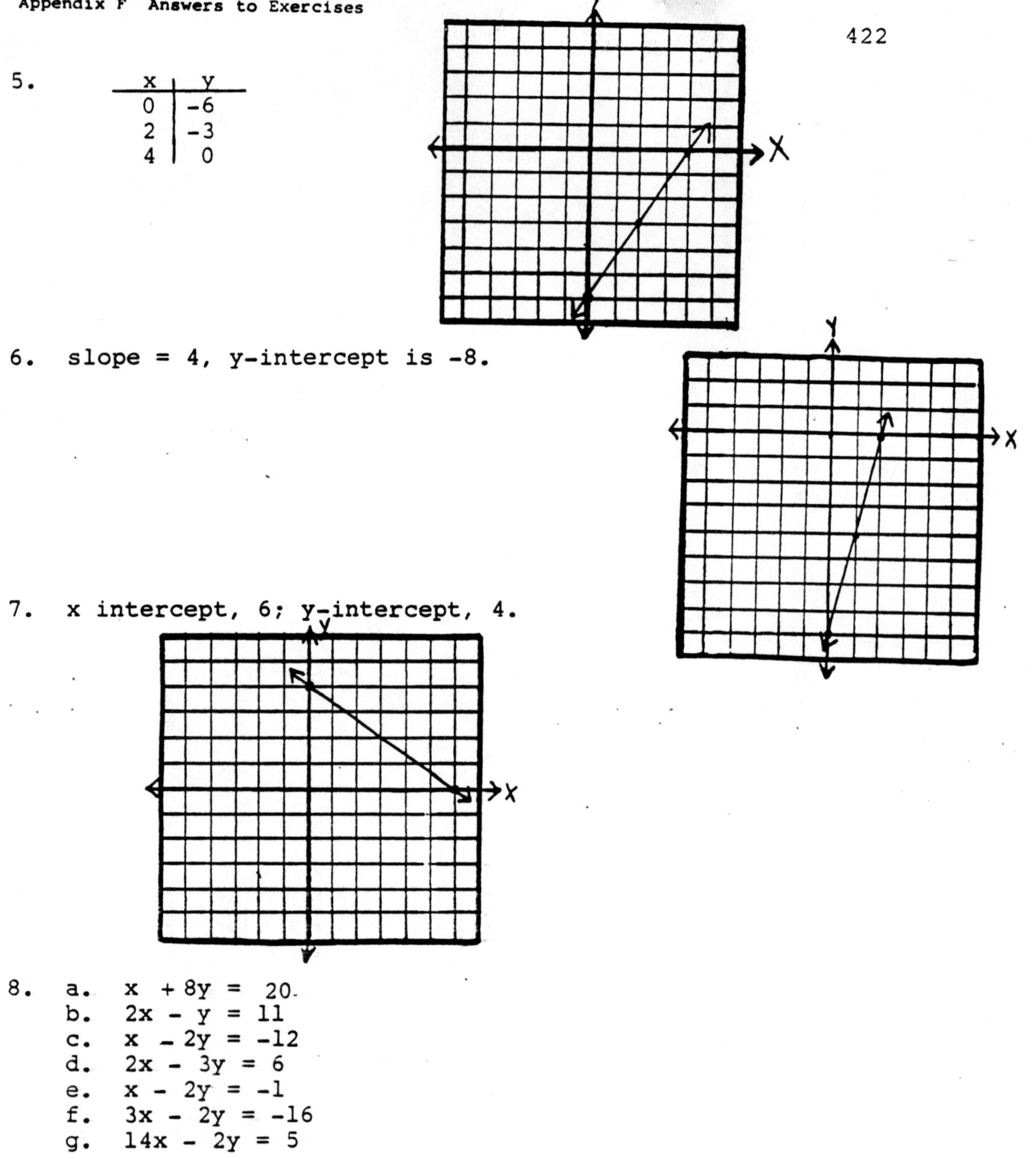

6. slope = 4, y-intercept is -8.

7. x intercept, 6; y-intercept, 4.

8. a. $x + 8y = 20$
   b. $2x - y = 11$
   c. $x - 2y = -12$
   d. $2x - 3y = 6$
   e. $x - 2y = -1$
   f. $3x - 2y = -16$
   g. $14x - 2y = 5$

9. 10

INDEX